Parasitoid Viruses
Symbionts and Pathogens

Parasitoid Viruses
Symbionts and Pathogens

Nancy E. Beckage
Professor Emerita
Departments of Entomology &
Cell Biology and Neuroscience
University of California-Riverside
Riverside, CA, USA

Jean-Michel Drezen
Directeur de Recherche CN
Institut de Recherche sur la Biologie de l'Insecte
CNRS UMR 6035
Université François Rabelais
Tours, France

AMSTERDAM • BOSTON • HEIDELBERG • LONDON • NEW YORK • OXFORD • PARIS
SAN DIEGO • SAN FRANCISCO • SINGAPORE • SYDNEY • TOKYO

Academic Press is an imprint of Elsevier

Academic Press is an imprint of Elsevier
32 Jamestown Road, London NW1 7BY, UK
225 Wyman Street, Waltham, MA 02451, USA
525 B Street, Suite 1800, San Diego, CA 92101-4495, USA

First edition 2012

Notice
No responsibility is assumed by the publisher for any injury and/or damage to persons or property as a matter of products liability, negligence or otherwise, or from any use or operation of any methods, products, instructions or ideas contained in the material herein. Because of rapid advances in the medical sciences, in particular, independent verification of diagnoses and drug dosages should be made.

British Library Cataloguing-in-Publication Data
A catalogue record for this book is available from the British Library

Library of Congress Cataloging-in-Publication Data
A catalog record for this book is available from the Library of Congress

ISBN : 978-0-12-384858-1

For information on all Academic Press publications
visit our website at elsevierdirect.com

Typeset by MPS Limited, a Macmillan Company, Chennai, India
www.macmillansolutions.com

Printed and bound in United States of America

10 11 12 13 14 15   10 9 8 7 6 5 4 3 2 1

Contents

George Salt. Photo: © Godfrey Argent Studio

The Discovery of Polydnaviruses and the Influence of Dr. George Salt

S. Bradleigh Vinson

Department of Entomology, Texas A&M University, Texas, USA

This journey began when I started working in 1963 on my Ph.D. at Mississippi State University under Dr. James Brazzel. Dr. Brazzel had just come from Texas A&M where he had focused on the management of cotton pests, particularly on *Heliothis* sp., and he wanted me to look at DDT resistance in *Heliothis virescens*. Dr. Brazzel also had an MS student, Wallace Joe Lewis, who was looking at and documenting the various natural mortality factors associated with *Heliothis* populations. We both were located in the same office area and we often talked about our research. I became interested in one of the parasitoids, *Cardiochiles nigriceps* (now known as *Toxoneuron nigriceps*) that Joe was collecting, particularly as to how it locates its host. We worked together on this project and published a paper that was one of the first to show that the parasitoid can respond to an extractable host chemical (Vinson and Lewis, 1965). In the course of this research we became aware of Dr. George Salt's research (at Cambridge University) and his papers concerning host recognition and parasitoid development. I became fascinated with the interactions and was excited by the many interrelationships that appeared to exist.

When Joe Lewis completed his MS (Lewis, 1965) he acknowledged my input and included our paper in his thesis. Since I now had my Ph.D. and had been hired as a faculty member in the department, Joe decided to do his Ph.D under my direction looking at the suitability of several *Heliothis* species for several parasitoid species. He completed his Ph.D. (Lewis, 1968) in which a number of papers by Dr. G. Salt played an important role. Following his Ph.D, Dr. Lewis took a job at the USDA (Tifton Ga.)

where he went on to publish another 114+ papers focused on host location, and in 2008 was one of the winners of the prestigious 'Wolf Prize'.

In the late 1950s and early 1960s, Salt (1938, 1956, 1963a, b) had described the defense reactions of insects to metazoan parasites (parasitoids). By the 1970s it was clear that insects mounted an immune response to foreign invaders (Lewis and Vinson, 1968; Salt, 1970; Nappi, 1973a, b; 1975 a, b, c; Nappi and Streams, 1969). Salt (1965, 1966) had provided evidence that parasitized eggs placed in a permissive host had a coating, a mucopolyssachride, that protected them. He also noted that these coated eggs were not encapsulated while other foreign material was (Salt, 1965).

In my laboratory, we (Lewis and Vinson, 1968a, b; 1971) also reported that the eggs of *T. nigriceps* were encapsulated in *Heliothis zea* but not in *Heliothis virescens*. We provided evidence that the eggs did not secrete a repellant or inhibit the hemocytes' ability to respond to foreign materials. Salt (1965) had suggested that the eggs of *Venturia* were altered in some way, as they passed from the ovariole into the ovarian calyx, that conferred protection. Rotheram (1967) reported that the calyx contained numerous 'particles' that were also on the egg surface (Rotheram 1973a, b). Salt (1973) suggested that these 'particles' protected the egg by coating the surface. This concept of particles coating the foreign surface was influenced by King and Richards (1960) and King *et al.* (1969).

I moved to Texas A&M University in 1969, and between 1969 and 1974 I published 32 papers on parasitoids, but I now also had access to an Electron Microscopy Center that would allow my laboratory to look at the ultra-structure of the reproductive system. Salt (1970) had noted that mucopolysacchrides do not elicit a response from insect hemocytes and Osman (1974) provided evidence that eggs of the parasitoid *Pimpla turionellae* are coated with a mucopolysacchride. We (Vinson and Scott, unpublished observations, but noted in Vinson, 1977b) confirmed Osman's results with the eggs of *T. nigriceps*. Earlier, Walters and Williams (1966) noted that hemocytes were unable to encapsulate cation exchange granules such as Sephadex C-50 beads. So I had a student, F.D. Brewer, look at chemicals that might affect encapsulation. He found that inhibitors of melanization and the chelator's of divalent ions (ethylenediaminetetraacetic acid (EDTA)) reduced encapsulation (Brewer and Vinson, 1971). These results suggested that the ionic nature of the surface might play a role, so I looked at the encapsulation of beads with positive, neutral, or negative surfaces injected into *H. virescens* and *H. zea* and found that the negative (anionic) surfaces, represented by C-50 Sephadex containing few or no positive sites, were difficult to encapsulate, confirming the work of Walters and Williams (1966). In contrast, I noted that the positive (cationic) sites (A-50 Sephadex) beads were readily encapsulated (Vinson, 1974). But for me the question was: why are the eggs of *T. nigriceps* not encapsulated in *H. virescens* but are encapsulated in *H. zea*, yet the response of both host insect species to the ionic exchange granules was the same? We (Vinson and Scott, 1974) noted that the fibrous layer coating the eggs of *T. nigriceps* placed in *H. zea* (a non-permissive host) disappeared rapidly in contrast to eggs placed in *H. virescens*.

As discussed (Vinson, 1977b), the particulate calyx fluid must in some way suppress the ability of the hemocytes to encapsulate the parasitoid egg while retaining their ability to respond to other foreign material. This suggested that the particles are in some way altering the egg or, more likely, the hemocytes' encapsulation response. I found no changes in the differential hemocyte count following parasitism by *T. nigriceps* (Vinson, 1971), thus the destruction of a particular type of hemocyte did not appear to play a role.

As noted above, when I moved to Texas A&M, and due to the work of G. Salt, I wanted to look at the reproductive system of the three *Heliothis* parasitoids that I had been rearing and had brought with me to Texas (*T. nigriceps* and *Microplitis croceipes* [Braconidae] and *Campolitis sonorensis* [Ichneumonidae]); the last two were successful in parasitizing both *H. virescens* and *H. zea*. I took on a student, Nick Norton, who was interested in electron microscopy, and I also had the opportunity to work with an EM technician, Mr Randy Scott. Norton began looking at the reproductive system, focusing on the calyx of *C. sonorensis*. As seen in Fig. 1 (Fig. 3 in Norton *et al.*, 1974), particles were seen in the nucleus of calyx cells where they appeared to form. This suggested that the particle's center might contain

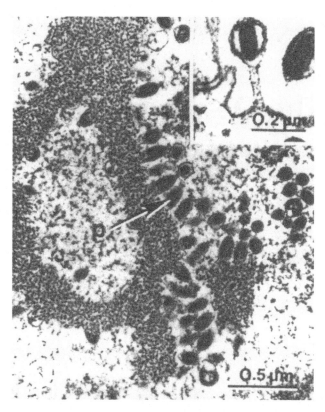

FIGURE 1 Repeat of Fig. 3 from the 1974 Thirty-second Annual EMSA meeting (pp.140–141) showing the calyx nucleus of *Campoletis sonorensis* (Ichneumonidae) by W.N. Norton, S.B. Vinson and E.L. Thurston (Director of the EM Center).

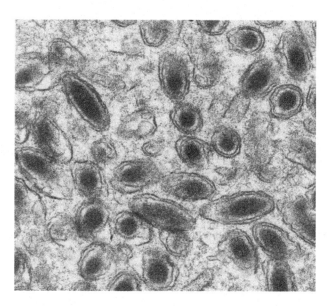

FIGURE 2 EM picture of *Campoletis sonorensis* calyx particles in longitudinal and cross section, but not used in the 1974 meeting paper.

DNA, and they budded through the nuclear membrane and the cell membrane into the calyx lumen. Thus, these particles had two membranes around a central dense core (Fig. 2).

At the same time I had Scott look at the reproductive system of *T. nigriceps*. He came to me within a few weeks with a number of pictures (example shown in Fig. 3) and said there was something wrong. He thought they were either poorly fixed or contaminated with some particulate material. He was referring to the particles with the thick core. When I looked at the 'contamination' it looked like a virus to me and did not appear to be a contamination. I suggested he treat some sections with RNAase and others with DNAase. The RNAase did not have much effect, but the result with the DNAase was that the dense centers of the particles disappeared. This supported my belief that these 'particles' were a virus or at least a virus-like material and this led to several presentations and the first paper to suggest that a virus might be involved (Vinson and Scott, 1975). With the information regarding the large membrane-bound particles with a very electron-dense core that appeared to be assembled in the nucleus of the calyx cells of *C. sonorenesis* (Norton *et al.*, 1974) and the loss the core of the *T. nigriceps* particles treated with *DNAase* suggested to me that DNA was being incorporated or at least comprised into the core of the particles.

I knew that no one would consider me to be a virologist, nor did I have the experience with viruses, so I began to look for someone with experience that could confirm that these were viruses. I had also decided to initially focus on *C. nigriceps,* as it offered more issues to look at. We gave presentations at several conferences regarding

our suggestion that these particles are viruses prior to our paper (submitted to F.S. Sjöstrand, editor *Journal of Ultrastructural Research*, March 22, 1973) being accepted and published. Scott said after his presentation at a meeting that a Dr. Donald Stoltz came up and talked with him and expressed an interest. He said that he was at the University of Houston and gave Scott an address. In July 1973, I wrote to Don to see if he would look at the ovary and calyx area of the reproductive system and the 'particles' (Fig. 4). He replied positively and I traveled to Houston with parasitoids. Don replied after looking at the sections (letters sent on August 22, 1973 and September 28, 1973) and said that he was less than convinced the particles were related to a

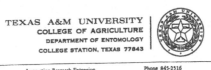

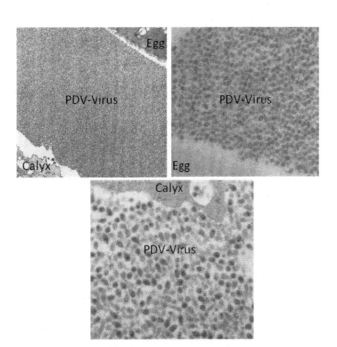

FIGURE 3 Pictures of *Cardiochiles* (now *Toxineuria nigriceps*) calyx and pictures of the 'particles' that were not used in the Vinson and Scott 1974 paper.

FIGURE 4 Letter to Dr. Don Stoltz to see if he was interested in helping me to confirm that these were viruses.

virus. However, in a letter sent on May 6, 1974) he said he was convinced. This led to an important collaborative paper suggesting the particles contain DNA (Norton *et al.,* 1975). It took some time to look at the four species that I had and to get the proof of what I believed was virus-like material or actual virions.

This led to a long collaboration with Don and a number of students, postdocs, and collaborators that rapidly expanded the 'field of Poly-DNA-viruses'. What I am proud of is that I not only strongly suspected that a 'virus-like' particle was involved in the parasitoid–host interaction, but that different 'particles' were present in the four parasitoid species I was working with. The species included three genera of Braconidae (*Toxoneuron* (*Cardiochiles*) *nigriceps* (Viereck); *Microplitis croceipes* (Cresson); and *Chelonis texana* (Cresson)) (Stoltz *et al.,* 1976) and one representative of the Ichneumonidae, *Campolitis sonorenesis* (Cameron). Each of these species displayed different 'particles', thus revealing some of the diversity of these viruses. These four species have been the backbone of the species that have been prominent in the study of the 'polydnaviruses'. In fact, I showed that the 'virus-like particles' of *M. croceipes* could, in fact, protect the eggs of *T. nigriceps* in *H. zea* (Vinson, 1977a). I reflect back to the last paragraph of the Vinson and Scott (1975) paper: 'The evolution and origin of these virus-like particles and the target of these particles in the host are interesting questions under investigation'. As also noted by Fisher (1986) in his review of George Salt's work and the field of experimental insect parasitology: 'The presence of cell produced nuclear DNA secretion in the calyx fluid which after injection into a host appears to regulate the behavior, growth, and internal defense reactions of the host offers new concepts in the parasitoid–host relationship'. This continues today as shown by the chapters in this book.

But I also reflect back to Dr. G. Salt. Dr. Charles Godfray, in an e-mail dated March 14, 2003, informed Joop van Lenteren and me of Dr. G. Salt's passing (published in the *Independent* on March 4, 2003). I wrote a two-page reply beginning, 'I am saddened to learn of the passing of Dr. George Salt' and I then talked about his importance to parasitoid biology. Although we never met, a regret I still harbor, he had a profound impact on my life. I had and have the greatest respect for Dr. George Salt. He was the person that, through his excellent publications, stimulated me to enter the field of parasitoid–host biology. George Salt retired in 1971 (Bateson, 2003) but he continued to publish papers concerning parasitoids (Salt, 1973, 1975, 1976) that resulted in some 34 papers on parasitoids (however, it should also be noted that he published some 32 papers on other subjects). Salt was honored in 1986 with the dedication (Fisher, 1986) of a Symposium Issue of *The Journal of Insect Physiology* [Vol. 32 (4), 1986] put together by N. Beckage and S.B. Vinson. Salt was honored

again in 1990 (along with R.L. Doutt, S.E. Flanders, and H.S. Smith) in the 'Preface' to an issue of the *Archives of Insect Biochemistry and Physiology* 13[1–27] 1990 concerned with parasitoids by Pauline Lawrence and S.B. Vinson.

Although Salt never published anything concerning the polydnaviruses, he knew about them, and sometime between 1976 and the early 1980s he wrote me a letter stating that I was wrong to suggest that the particles in the calyx of parasitoids contain DNA. I wrote him a letter and said that while his particles may not contain DNA, it was also clear that some species do have particles in the calyx and that these do contain DNA and appear to be derived from a virus. (In fairness, the papers that were in support of the view that the particles contain DNA in some species were just beginning to come out. Also, while I personally felt that his particles might have evolved from a PDV-like particle but lacked the DNA or RNA which could instead be incorporated into the host's DNA, I did not put it in the reply.) A few years before his death Salt wrote and said 'We are both right'. It was a short note but a very much appreciated one.

There was no reason to believe that the calyx particles themselves in *Nemeritis* contained DNA. However, it is not clear to me what the source of the *Nemeritis* particle may be. As outlined in the following chapters of this volume, a diversity of viruses and virus-like particles are associated with wasp endoparasitoids, and their physical characteristics show variable features. How all these viruses or virus-like particles evolved is an intriguing question.

REFERENCES

Bateson, P., 2003. George Salt. 12 December 1903–17 February 2003. Biogr. Mems Fell. Royal Soc. 49, 447–459.

Brewer, F.A., Vinson, S.B., 1971. Chemicals affecting the encapsulation of foreign material in an insect. J. Invert. Pathol. 18, 287–289.

Fisher, R.C., 1986. George Salt and the development of experimental insect parasitology. J. Insect Physiol. 32, 249–253.

King, P.E., Ratcliffe, N.A., Copland, M.J.W., 1969. The structure of the egg membranes in *Apanteles glomeratus* (L.) (Hymenopteera: Braconidae). Proc. R. Ent. Lond. (A) 44, 137–142.

King, P.E., Richards, J.G., 1960. Oogenesis in *Nasonia vitripennis* (Walker) (Hymenoptera: Pteromalidae). Proc. R. Ento. Soc. Lond. A 44, 143–157.

Lewis, W.J. 1965. Seasonal occurance and biological studies of certain parasites of the *Heliothis* complex in Mississippi. MS Thesis. Mississippi State University.

Lewis, W.J. 1968. Studies on the suitability of certain *Heliothis* spp. as a host for *Cardiochiles nigriceps* Viereck. Ph.D. Dissertation. Mississippi State University.

Lewis, W.J., Vinson, S.B., 1968a. Egg and larval development of *Cardiochiles nigriceps* Vierick. Ann. Entomol. Soc. Amer. 61, 561–564.

Lewis, W.J., Vinson, S.B., 1968b. Immunological relationships between the parasite *Cardiochiles nigriceps* Vierick and certain *Heliothis* spp. J. Insect Physiol. 14, 613–626.

Lewis, W.J., Vinson, S.B., 1971. Suitability of certain *Heliothis* (Lepidoptera: Noctuidae) as hosts for the parasite *Cardiochiles nigriceps*. Ann. Entomol. Soc. Amer. 64, 970–972.

Nappi, A.J., 1973a. Hemocytic changes associated with the encapsulation and melanization of some insect parasites. Exp. Parasit. 33, 285–302.

Nappi, A.J., 1973b. The role of melanization in the immune reaction of larvae of *Drosophila algonquin* against *Pseudeucoila bochei*. Parasitology 66, 23–32.

Nappi, A.J., 1975c. Cellular immune-reactions of larvae of *Drosophila algonquin*. Parasitology 70, 189–194.

Nappi, A.J., 1975b. Effects of ligation on cellular immune-reactions of *Drosophila algonquin* against hymenopterous parasite *Pseudeucoila bochei*. J. Parasit. 61, 373–376.

Nappi, A.J., 1975a. Parasite encapsulation in insects. In: Maramorosch, K., Shope, R.E. (Eds.), Invertebrate Immunity. Academic Press, New York.

Nappi, A.J., Streams, F.A., 1969. Haemocytic reactions of *Drosophila melanogaster* to the parasitoid *Pseudeucoila mellipes* and *P. bochei*. J. Insect Physiol. 15, 1551–1566.

Norton, W.N., Vinson, S.B., Thurston, E.L., 1974. An ultrastructural study of the female reproductive tract of *Campoletis sonorensis* (Hymenoptera). Proceed. Electron Microscopy Soc. Amer. 1975: 140–141.

Norton, W.N., Vinson, S.B., Stoltz, D.B., 1975. Nuclear secretory particles associated with the calyx cells of the ichneumonid parasitoid *Campoletis sonorensis* (Cameron). Cell Tissue Res. 162, 195–208.

Osman, S.E., 1974. Parasitentoleranz von Schmetterlingspuppen Maskierung der Parasiteneier mit Mucopolysacchariden. Naturwissenschaften 61, 453–454.

Rotheram, S., 1967. Immune surface of eggs of a parasitic insect. Nature, 214, 700.

Rotheram, S., 1973a. The surface of the egg of a parasitic insect. I. The surface of the egg and first-instar larva of *Nemeritis*. Proc. Soc. Lond. B 183, 179–194.

Rotheram, S., 1973b. The surface of the egg of a parasitic insect. II. The ultrastructure of the particulate coat on the egg of *Nemeritis*. Proc. Soc. of Lond. B 183, 195–204.

Salt, G., 1938. Experimental studies in insect parasitism. VI. Host Suitability. Bull. Ent. Res. 29, 223–246.

Salt, G., 1956. Experimental studies in insect parasitism. IX. The reactions of a stick insect to an alien parasite. Proc. R. Soc. Lond. B 146, 93–108.

Salt, G., 1963a. Experimental studies in insect parasitism. XII. The reactions of six exopterygote insects to an alien parasite. J. Insect Physiol. 9, 647–669.

Salt, G., 1963b. The defence reactions of insects to metazoan parasites. Parasitology 53, 527–642.

Salt, G., 1965. Experimental studies in insect parasitism. XIII. The haemocytic reaction of a caterpillar to eggs of its habitual parasite. Proc. R. Soc. Lond. B 162, 303–318.

Salt, G., 1966. Experimental studies in insect parasitism. XIV. The haemocytic reaction of a larva to its habitual parasite. Proc. R. Soc. Lond. B 165, 155–178.

Salt, G., 1970. The Cellular Defense Reactions of Insects. Cambridge University Press, Cambridge.

Salt, G., 1973. Experimental studies in insect parasitism. XVI. The mechanism of the resistance of *Nemeritis* to defense reactions. Proc. R. Soc. Lond. B 183, 337–350.

Salt, G., 1975. The fate of an internal parasitoid, *Nemeritis canescens*, in a varity of insects. Trans. R. Ent. Soc. Lond. 127, 141–161.

Salt, G., 1976. The hosts of *Nemeritis canescens*, a problem in the host specificity of insect parasitoids. Ecol. Ent. 1, 63–67.

Stoltz, D.B., Vinson, S.B., Mackinnon, E.A., 1976. Baculovirus like particles in the reproductive tracts of female parasitoid wasps. Can. J. Microbiol. 22, 1013–1023.

Vinson, S.B., 1971. Defense reaction and hemocytic changes in *Heliothis virescens* in response to its habitual parasitoid *Cardiochiles nigriceps*. J. Invert. Pathol. 18, 94–100.

Vinson, S.B., 1974. The role of the foreign surface and female parasitoid reactions on the immune response of an insect. Parasitology 68, 27–33.

Vinson, S.B., 1977a. Insect host responses against parasitoids and the parasitoids resistance with emphasis on the Lepidoptera Hymenoptera association. In: Bulla, L.A., Cheng, T.C. (Eds.), Comparative Pathobiology, vol. 3 (pp. 103–105). Plenum Press, New York.

Vinson, S.B., 1977b. Microplitis *croceipes*: Inhibition of the *Heliothis zea* defense reaction to *Cardiochiles nigriceps*. Exper. Parasitol. 41, 112–117.

Vinson, S. B. 1983. Parasitoid viruses. In: McGraw Hill Yearbook of Science and Technology. (pp. 193–196).

Vinson, S.B., 1990a. How parasitoids deal with the immune system of their host: an overview. Arch. Insect Biochem. Physiol. 13, 3–27.

Vinson, S.B., 1990b. Immunosuppression. In: Baker, R.R., Dunn, P.E. (Eds.), New Directions in Biological Control: Alternatives for Suppressing Agricultural Pests and Diseases (pp. 517–535). Alan R. Liss, Inc.

Vinson, S. B. 1990c. The interaction of two parasitoid polydnaviruses with the endocrine system of *Heliothis virescens*. In: Vth International Colloquium on Invertebrate Pathology and Microbial Control, Proceedings and Abstracts. Adelaide, Australia. Aug. 20–24, 1990 (pp. 195–199).

Vinson, S.B., Lewis, W.J., 1965. A method of host selection by *Cardiochiles nigriceps*. J. Econ. Entomol. 58, 869–871.

Vinson, S.B., Scott, J.R., 1974. Parasitoid egg shell changes in a suitable and unsuitable host. J. Ultrastruct. Res. 47, 1–15.

Vinson, S.B., Scott, J.R., 1975. Particles containing DNA associated with oocyte of an insect parasitoid. J. Invert. Pathol. 25, 375–378.

Walters, D.R., Williams., C.M., 1966. Re-aggregation of insect cells as studied by a new method of tissue and organ culture. Science 154, 516–517.

While there have been several journal special issues published on parasitoid viruses, this volume is the first book-length publication dedicated entirely to parasitoid viruses and venoms and we hope it will inspire a long series of books on this topic. Among parasitoid viruses, the fascinating models of polydnaviruses (PDVs) were discovered in the 1970s and the new field of polydnavirology was thus opened. This field has been moving very fast since the beginning of the century thanks to the use of genomic approaches and rapid expansion of accessible databases on insect and viral gene sequences. Parasitoid and viral genomic studies have confirmed that PDVs are functionally gene transfer agents used by parasitoid wasps to manipulate the physiology of their parasitized lepidopteran hosts by introducing modified versions of their own genes into host cells. In the case of PDVs from braconid wasps, this kind of gene therapy (detrimental for the patient, which is in this case the lepidopteran host!) originated from the integration of a virus genome in a wasp genome ca. 100 million years ago.

This virus has been modified to incorporate wasp genes instead of its own viral genome in the nucleocapsids inside the viral particles. Such use of viruses as vectors has been selected several times independently during the evolution of parasitoid wasps. The PDVs associated with braconid and ichneumonid wasps (Campopleginae subfamily) are unrelated as judged from the machinery producing the particles, and they represent an example of convergent evolution with different viral origins. A third association event is suspected to have occurred in the Banchinae subfamily of ichneumonid wasps. In essence, the parasitoids have 'captured' viral elements that have evolved a host regulatory role that benefits the parasitoid to facilitate successful parasitism. Other associations with viruses or virus-like particles might have evolved with different organisms but they have not been unraveled yet, and parasitic amoebae that have associations with mammalian viruses are just one example. Many insects have evolved associations with a large number of species of bacteria such as *Wolbachia* and associations with viruses have been less well studied to date compared to bacterial symbionts.

A number of different viruses are found in the genital tract of the parasitoid wasps and conceivably they could be transferred to female wasps by behavioral traits, such as host feeding, initiated following ovipositor puncture of the surface of the integument. Host feeding may thus be an advantageous behavior for viruses which facilitates their spread within insect populations and this intimate association with viruses might have favored interactions leading in some cases to integration of viral sequences into the wasp genome, although most of the wasps described in this book are no longer host feeders. These viruses include RNA viruses of insect parasitoids and most of them appear nonpathogenic. Could these likewise have evolved a symbiotic relationship with their host? Future research may reveal such an intimate relationship with the wasp host carrying them but, currently, we have little information about their functional role as symbionts or pathogens in the virus–wasp–insect host relationship.

While the study of polydnaviruses was initially inspired by the pioneering studies of George Salt and Susan Rotheram at Cambridge University, more recent studies of *Venturia* (formerly *Nemeritis*) *canescen*s particles (the virus-like agent studied by George Salt) by Otto Schmidt and Sassan Asgari documented that these virus-like virions lack both DNA and RNA; the particles are comprised of proteins encoded by parasitoid genes. The multiplicity of different molecular forms seen in these viruses and virus-like particles is truly amazing but, compared to polydnaviruses, we have less information about the biology of virus-like particles and how they function.

Finally, not all parasitoid species are associated with viruses and most in this category have to rely on virulence factors produced by their ovaries and venom glands instead of using the host to produce them like for PDV-encoded gene products. PDV-associated species also produce venom that was shown in some cases to synergize the effect of the virus. New sequencing approaches are more comprehensive and will thus allow comparisons of the arsenal of proteins used in different species, which will enhance our understanding of the dynamics of evolution of parasitoid virulence strategies. Ectoparasitic wasps have not been examined yet for the presence of viral symbionts, and appear to have exploited venoms as a source of host regulatory molecules instead. Comparative studies on paralyzing versus nonparalytic venoms are lacking, and screening ectoparasitic species for viral elements should also be a future research priority.

In addition to enhancing our knowledge of parasitoid strategies and increasing our understanding of the importance of symbiotic relationships in species evolution, parasitoid viruses and venoms may constitute a source of new molecules to control insect pests. This might be a revolutionary outcome of research on PDVs and other parasitoid viruses, since the safety of many chemical pesticides with respect to their detrimental impacts on human health and key species in the environment such as bees and other beneficial insects, is questioned. We anticipate that harvesting biopesticidal molecules from parasitoid venoms will likewise prove fruitful.

Finally, we hope that this book will satisfy the reader by presenting an overview of the most recent findings on all these topics presented by an international assemblage of authors. In addition, we aim to inspire many future researchers to choose polydnavirology or studies of other parasitoid viruses or viral-like elements and venoms as their focus field.

Jean-Michel Drezen
Nancy Beckage

Contributors

Numbers in parentheses indicate the pages on which the author's contributions begin.

Sassan Asgari (217), School of Biological Sciences, The University of Queensland, St Lucia, QLD 4072, Australia

Nancy E. Beckage (163), Departments of Entomology & Cell Biology and Neuroscience, University of California-Riverside, Riverside, CA 92521, USA

Catherine Béliveau (79), Laurentian Forestry Centre, Natural Resources Canada, 1055 rue du P.E.P.S., Québec, G1V 4C7, Canada

Annie Bézier (15), Institut de Recherche sur la Biologie de l'Insecte, CNRS UMR 6035, Université François Rabelais, Faculté des Sciences et Techniques, Parc de Grandmont, 37200 Tours, France

Antoine Branca (127), Institute for Evolution and Biodiversity, Hüfferstraße 1, 48149 Münster, Germany

Anas Cherqui (257), Laboratoire de Biologie des Entomophages, Université de Picardie–Jules Verne, 33 rue Saint Leu, 80039 Amiens cedex, France

Dominique Colinet (181), UMR Interactions Biotiques et Santé Végétale, INRA (1301), CNRS (6243), Université Nice Sophia Antipolis; Sophia Agrobiotech, 400 Route des Chappes, 06903 Sophia Antipolis, France

Michel Cusson (33, 47, 79), Laurentian Forestry Centre, Natural Resources Canada, 1055 rue du P.E.P.S., Québec, G1V 4C7, Canada

Ellen L. Danneels (233), Laboratory of Zoophysiology, Ghent University, B-9000 Ghent, Belgium

Dirk C. de Graaf (233), Laboratory of Zoophysiology, Ghent University, B-9000 Ghent, Belgium

Christopher A. Desjardins (115), The Broad Institute, 320 Charles Street, Cambridge, MA 02141, USA

Géraldine Doury (257), Laboratoire de Biologie des Entomophages, Université de Picardie–Jules Verne, 33 rue Saint Leu, 80039 Amiens cedex, France

Jean-Michel Drezen (15, 33, 79), Institut de Recherche sur la Biologie de l'Insecte, CNRS UMR 6035, Faculté des Sciences et Techniques, Université François Rabelais, Parc Grandmont, 37200 Tours, France

Stéphane Dupas (127), IRD UR072 CNRS, Laboratoire Evolution Genome et Speciation, 91190 Gif-sur-Yvette, France and Université Paris-Sud, 91400 Orsay, France

Catherine Dupuy (47), Institut de Recherche sur la Biologie de l'Insecte, CNRS UMR 6035, Université François-Rabelais, Faculté des Sciences et Techniques, Parc Grandmond, 37200 Tours, France

Patrice Eslin (257), Laboratoire de Biologie des Entomophages, Université de Picardie–Jules Verne, 33 rue Saint Leu, 80039 Amiens cedex, France

Frédéric Fleury (203), Université de Lyon; F-69000, Lyon; Université Lyon 1; CNRS, UMR5558; Laboratoire de Biométrie et Biologie Evolutive; F-69622, Villeurbanne, France

Ellen M. Formesyn (233), Laboratory of Zoophysiology, Ghent University, B-9000 Ghent, Belgium

Sylvain Gandon (203), Centre d'Ecologie Fonctionnelle et Evolutive (CEFE); UMR 5175; 1919 route de Mende; F-34293 Montpellier, France

Jean-luc Gatti (181), UMR Interactions Biotiques et Santé Végétale, INRA (1301), CNRS (6243), Université Nice Sophia Antipolis; Sophia Agrobiotech, 400 Route des Chappes, 06903 Sophia Antipolis, France

Barbara Giordana (271), Dipartimento di Protezione dei Sistemi Agroalimentare e Urbano e Valorizzazione della Biodiversità , Università di Milano, Italy

Catherine Gitau (127), EH Graham Centre for Agricultural Innovation, Charles Sturt University, PO Box 883 Orange, NSW 2800, Australia

Dawn Gundersen-Rindal (47, 99), US Department of Agriculture, Agricultural Research Service, Invasive Insect Biocontrol and Behavior Laboratory, Room 267, 10300 Baltimore Avenue, Bldg 011A BARC-WEST, Beltsville, Maryland, MD 20705, USA

Elisabeth Anne Herniou (15), Institut de Recherche sur la Biologie de l'Insecte, CNRS UMR 6035, Université François Rabelais, Faculté des Sciences et Techniques, Parc de Grandmont, 37200 Tours, France

Elisabeth Huguet (63), Institut de Recherche sur la Biologie de l'Insecte, CNRS UMR 6035, Université François-Rabelais, Faculté des Sciences et Techniques, Parc Grandmont, 37200 Tours, France

Martha Kaeslin (169), Institute of Cell Biology, University of Bern, Baltzerstrasse 4, CH-3012 Bern, Switzerland

Peter J. Krell (5), Department of Molecular and Cellular Biology, University of Guelph, Guelph, ON N1G2W1, Canada

Beatrice Lanzrein (169), Institute of Cell Biology, University of Bern, Baltzerstrasse 4, CH-3012 Bern, Switzerland

Renée Lapointe (79), Sylvar Technologies Inc., P.O. Box 636, Stn. "A", Fredericton, NB, E3B 5A6, Canada

David Lepetit (203), Université de Lyon; F-69000, Lyon; Université Lyon 1; CNRS, UMR5558; Laboratoire de Biométrie et Biologie Evolutive; F-69622, Villeurbanne, France

Roger C. Levesque (79), Institut de Biologie Intégrative et des Systèmes, Université Laval, 1030 Avenue de la Médecine, Québec, QC, G1V 0A6, Canada

Halim Maaroufi (79), Institut de Biologie Intégrative et des Systèmes, Université Laval, 1030 Avenue de la Médecine, Québec, QC, G1V 0A6, Canada

Donato Mancini (249), Dipartimento di Entomologia e Zoologia Agraria 'F Silvestri', Università di Napoli 'Federico II', Napoli, Italy

Julien Martinez (203), Université de Lyon; F-69000, Lyon; Université Lyon 1; CNRS, UMR5558; Laboratoire de Biométrie et Biologie Evolutive; F-69622, Villeurbanne, France

Sébastien J.M. Moreau (63, 257), Institut de Recherche sur la Biologie de l'Insecte, CNRS UMR 6035, Université François-Rabelais, Faculté des Sciences et Techniques, Parc Grandmont, 37200 Tours, France

Audrey Nisole (79), Laurentian Forestry Centre, Natural Resources Canada, 1055 rue du P.E.P.S., Québec, G1V 4C7, Canada

Jaqueline M. O'Connor (89), Department of Entomology, 320 Morrill Hall, 505 S. Goodwin Avenue, University of Illinois, Urbana, IL 62801, USA

Sabine Patot (203), Université de Lyon; F-69000, Lyon; Université Lyon 1; CNRS, UMR5558; Laboratoire de Biométrie et Biologie Evolutive; F-69622, Villeurbanne, France

Francesco Pennacchio (249, 271), Dipartimento di Entomologia e Zoologia Agraria 'F Silvestri', Università di Napoli 'Federico II', Napoli, Italy

Rita Pfister-Wilhelm (169), Institute of Cell Biology, University of Bern, Baltzerstrasse 4, CH-3012 Bern, Switzerland

Marylène Poirié (81), UMR Interactions Biotiques et Santé Végétale, INRA (1301), CNRS (6243), Université Nice Sophia Antipolis; Sophia Agrobiotech, 400 Route des Chappes, 06903 Sophia Antipolis, France

Geneviève Prevost (257), Laboratoire de Biologie des Entomophages, Université de Picardie–Jules Verne, 33 rue Saint Leu, 80039 Amiens cedex, France

Rosa Rao (271), Dipartimento di Scienza del Suolo, della Pianta dell'Ambiente e della Produzione Animale, Università di Napoli 'Federico II', Italy

Sylvaine Renault (193), Université François Rabelais, UMR CNRS 6239 GICC Génétique, Immunothérapie, Chimie et Cancer, UFR des Sciences et Techniques, Parc Grandmont, 37200 Tours, France

Thomas Roth (169), Institute of Cell Biology, University of Bern, Baltzerstrasse 4, CH-3012 Bern, Switzerland

Antonin Schmitz (181), UMR Interactions Biotiques et Santé Végétale, INRA (1301), CNRS (6243), Université Nice Sophia Antipolis; Sophia Agrobiotech, 400 Route des Chappes, 06903 Sophia Antipolis, France

Céline Serbielle (63), GEMI, UMR CNRS IRD 2724, IRD, 911 avenue Agropolis, BP 64501, 34394 Montpellier CEDEX 5, France

Don Stoltz (5, 79), Department of Microbiology and Immunology, Dalhousie University, Halifax, NS B3H 1X5, Canada

Michael R. Strand (139, 149), Department of Entomology, University of Georgia, Athens, Georgia 30602-7415, USA

Julien Varaldi (203), Université de Lyon, F-69000, Lyon; Université Lyon 1; CNRS, UMR5558, Laboratoire de Biométrie et Biologie Evolutive, F-69622, Villeurbanne, France

S. Bradleigh Vinson (xi), Department of Entomology, Texas A&M University, Texas, USA

Anne-Nathalie Volkoff (33, 79), UMR 1333 INRA — Université Montpellier 2, Diversité, Génomes et Interactions Microorganismes–Insectes, Place Eugène Bataillon, cc101, 34095 Montpellier Cedex 5, France

Bruce A. Webb (33), Department of Entomology, S-225 Agricultural Science Center N, University of Kentucky, Lexington, KY 40546-0091, USA

Gabriela Wespi (169), Institute of Cell Biology, University of Bern, Baltzerstrasse 4, CH-3012 Bern, Switzerland

James B. Whitfield (89), Department of Entomology, 320 Morrill Hall, 505 S. Goodwin Avenue, University of Illinois, Urbana, IL 62801, USA

Parasitoid Polydnaviruses: Evolution, Genomics, and Systematics

Part I

Insights into Polydnavirus Evolution and Genomics

Chapter 1

The Origins and Early History of Polydnavirus Research

Don Stoltz[*] and Peter Krell[†]

[*]*Department of Microbiology and Immunology, Dalhousie University, Halifax, NS B3H 1X5, Canada*
[†]*Department of Molecular and Cellular Biology, University of Guelph, Guelph, ON N1G 2W1, Canada*

SUMMARY

In this brief review, we trace the origins and early history of polydnavirus (PDV) research, linking that to recent progress in the field. In so doing, we have tried to identify some of the important milestones in PDV research (see Table 1), nonetheless recognizing the inherently subjective nature of any such enterprise. We have made no attempt to provide a comprehensive coverage of events (that, arguably, is for the book as a whole to accomplish), and have not dealt with literature describing the effects of PDVs (and/or wasp venoms) on host physiology, since this information can be found in other chapters. We conclude with some thoughts on the question of where PDVs fit into the world of 'classical' virology.

ABBREVIATIONS

BV	bracovirus
CF	calyx fluid
IV	ichnovirus
PDV	polydnavirus

PERSONAL REMINISCENCES

Don's Perspective

While polydnavirus (PDV) research, as we know it today, can be said to have started in my laboratory, it clearly had its roots elsewhere, primarily in the Salt (1968) and Vinson (see below and Vinson, the Foreword for this volume) labs. What I brought, for the first time, was the perspective of a virologist and, as will be seen, the unanticipated freedom to go in a new direction.

As is often the case, it was purely by chance that I got into this field! What follows may be of interest to the PDV community in the particular sense that the narrative provides an interesting example of the arguably huge role of serendipity in science. In early 1970, with funding from the Canadian National Research Council, I joined Max Summers' lab (then with the Cell Research Institute at the University of Texas, in Austin) as a post-doctoral fellow. Nor was that the original plan, which would instead have had me working with K.M. Smith, then a visiting scholar in the same building. Many readers might recall that Dr. Smith wrote the first book on our subject (Smith, 1967); what you may not know, but was of interest to me at the time, is that he was actually the first to show that some virus particles had the form of an icosahedron. Before I was to join him in Texas, however, he had returned to England, I think for reasons of health. Max offered space and very good EM facilities to me, and so I became his first post-doc; not then famous, he kindly encouraged me to do my own thing, which had little to do with baculoviruses and everything to do with iridovirus structure. At some point in 1972, I moved to Houston, having been invited to set up an electron microscope facility in the Department of Biology at Texas Southern University.

Houston, as expected, was oppressively hot and humid, which had the effect of encouraging regular visits to the more tolerable venue of Galveston. At the time, I had a friend living there (one Norman Granholm, also an electron microscopist), and we would take turns visiting each other. One day, I think in early 1973, he called to inform me that there was to be a joint meeting of the Texas and Louisiana electron microscopy societies, in Galveston, and did I wish to attend? Having never heard of either society, I was not particularly interested in the venture until, that is, he brought up the subject of beer, which he anticipated would flow in copious amounts; I confess that it was *this* prospect, not the meeting, which initially held my attention! At some point I learned, contrary to all expectations, that there would actually be a presentation that I might have some interest in attending...something to do with virus-like particles in a parasitic wasp. The wasp was *Cardiochiles* (now *Toxoneuron*) *nigriceps*.

The researchers, S. Bradleigh Vinson and J.R. Scott, out of Texas A and M University (TAMU), were completely unknown to me. As it happened, Vinson was not at the meeting, so the talk was given by Scott who, I later discovered, was employed in the Electron Microscopy Center (TAMU Biology Department). I remember feeling an immediate buzz of excitement: the particles on display bore a faint resemblance to herpesvirions! Herpesviruses, I knew, had never before been seen in an insect, and to my knowledge still haven't. Wanting to take a look at this material first-hand, I quickly got in touch with Brad Vinson, who was only too happy to oblige; in fact, he hand-delivered some live parasitoids to my lab at TSU in short order, and things took off from there. From this initial encounter, I was immediately able to draw two conclusions: (1) these things were certainly not herpesviruses, and I remember being somewhat disappointed with that outcome; and (2) if anything, they looked a lot like an unusual form of non-occluded baculovirus. The only significant difference I could discern was that the *Cardiochiles* particles had nucleocapsids of variable length; in addition, having learned from Brad that calyx fluid (CF) was present in *all Cardiochiles* females, it was apparent that these things were in some manner being transmitted vertically. In keeping with this assumption, work then being pursued in Brad's lab was confirming the notion, initially developed by Salt (Salt, 1968), that CF played a role in successful parasitism (e.g., Vinson, 1974).

In 1974, I was hired to teach microbial ultrastructure, among other things, in the Department of Microbiology at Dalhousie University, where I have remained to this day. By this time, I had begun to collaborate with Brad, and was taking an interest in looking for CF particles in additional parasitoid species. I moved into an old house just outside of Halifax. The property comprised five acres of land bordered by a stream to the south and secondary forest on the north; the place was crawling with parasitoids of every description, and I always wondered how any lepidopteran species could manage to survive in this environment! So, I would often spend the weekend collecting wasps, later processing their ovaries in the lab for thin-sectioning, and always saving the cadavers for identification. In those days, the Biosystematics Research Institute (Agriculture Canada, based in Ottawa) employed a superb cadre of insect taxonomists, among whom was the then reigning expert on braconid taxonomy, namely W.R.M. (Bill) Mason. Mason is perhaps best remembered for his monograph on the *Microgastrinae* (Mason, 1981), in which the genus *Apanteles* was revised extensively. For one thing, many former *Apanteles* spp. were assigned to the new genus, *Cotesia* (a genus now well known to PDVers!).

While Mason was working on this project, I was coincidentally looking at CF particles from a variety of *Apanteles* spp., and had noticed that in some the nucleocapsids were individually enveloped, whereas in others there were several nucleocapsids per virion (Stoltz and Vinson, 1977). I recall discussing this at some point with Mason, who was delighted to find congruence with some aspects of his soon-to-be-published monograph. Interestingly, the current authority on microgastrine parasitoids, Jim Whitfield, was actually in Mason's office in Ottawa when copies of the newly published monograph were being unpacked! I first met Jim at a scientific meeting held sometime between 1988 and 1989.

All the above having been said, the calyx fluid project was a sideline. My primary focus was on iridovirus structure, concerning which I had by then amassed a reasonable publication record. So, it made good sense to base my first grant application (to the Medical Research Council of Canada) on said record; if memory serves, I believe I recruited my first graduate student (Peter Krell), who joined me in September 1975, specifically to work on iridoviruses, with which he too had already had some experience (Krell and Lee, 1974; this was Peter's first scientific paper).[1] Unfortunately, depending on how one wants to look at this, my initial attempt to write a grant application was not successful, and so I decided to try something else. Having published my first CF paper as a first author by this time (Stoltz *et al.*, 1976), I quickly put something together for an alternative granting agency, namely the Canada National Research Council. This was a success, and it's fair to say that I never looked back. Peter assured me that he was up to a new and different challenge.

Our first task was to establish some suitable parasitoid colonies, and for this we chose the braconid, *Cotesia melanoscela* (at that time known as *Apanteles melanoscelus*), a gypsy moth parasitoid, and the ichneumonid, *Hyposoter exiguae*, which parasitizes the cabbage looper; off and on, we also maintained a colony of *H. fugitivus*, using the tent caterpillar as host. These in hand, I used electron microscopy to examine the fate of CF particles in parasitized host larvae; this approach yielded a number of interesting observations (Stoltz and Vinson, 1977, 1979a), and among other things made it quite clear that these particles entered but did not replicate in host cells. Concurrently, Peter and I began to apply our complementary skills to the characterization of purified CF particles, publishing our observations on braconid and ichneumonid particles in 1979 and 1980, respectively (Krell and Stoltz, 1979, 1980).

---

[1] I first met Peter in 1964, when he was still a young, impressionable high school student; we were at the time both working at the Forest Pest Management Institute, in Sault Ste. Marie.

It was during this period that, in addition to PDVs, filamentous viruses were observed in the ovaries of some species (Stoltz and Vinson, 1979b); of course, we now know that all kinds of viruses are to be found in PDV-bearing wasps (reviewed in Stoltz and Whitfield, 1992).

It was also during this period that I noticed a striking resemblance between the tail-like appendage associated with the nucleocapsids of braconid CF particles and a similar structure attached to one end of the *Oryctes* virus capsid. At the time, this led me to predict a probable evolutionary relationship between the 'braconid baculoviruses' and 'subgroup C' of the *Baculoviridae* (Stoltz and Vinson, 1979b); thirty years later, this prediction is no longer viewed as speculative (Bézier *et al.*, 2009).

In 1979, Peter graduated with what would later be seen as the first ever Ph.D. degree in polydnavirology; in fact, his work can truthfully be said to have launched that field of study. In early 1980, he joined Max Summers' lab, which had by then been moved to Texas A and M University, as a post-doctoral fellow; that summer, I went there too, on sabbatical leave, taking my technician, Doug Cook, with me. Fortuitously, Brad Vinson's lab was located in the same building, and so there was huge potential for synergy! One of the first things that had to happen sooner or later was a demonstration to the effect that *purified* CF particles (in this case from *Campoletis sonorensis*) were immunosuppressive; this work, a collaboration between the Summers and Vinson labs, was published in 1981 (Edson *et al.*, 1981; see Table 1 for a compilation of selected milestones in PDV research). It was during my year-long sojourn at TAMU that the name 'polydnavirus' came to me. Somewhat later, recognizing that the International Committee on Taxonomy of Viruses (ICTV) was suffering from an unaccountable aversion to the use of viral names derived from sigla (such as pico-RNA-virus, for picornavirus), I decided that an end-run strategy might render the issue moot.[2] Accordingly, in 1984 we made a formal taxonomic proposal, not to the ICTV but rather to the journal, Intervirology (Stoltz *et al.*, 1984). It was also during this period that the stage was set for a new direction in PDV research: the application of modern molecular approaches, including DNA cloning and sequencing, to the further characterization of these very unusual viruses. In order to get things rolling, Max brought in two new

graduate students, Gary Blissard and David Theilmann, and two post-doctoral fellows, Jo-Ann Fleming and, of course, from the outset, Peter, who remained there until being recruited by the University of Guelph in November of 1981; all of these individuals were to make significant contributions going forward. In 1980, Mike Strand had started graduate study with Brad, working on the biology of an egg parasitoid. I distinctly remember meeting Mike at that time, and being impressed with his potential; regrettably, we would have to wait until the early 1990s to see that potential being applied to PDV biology.

Peter's Perspective

My interest in insect viruses began in the summer of 1968 at the Insect Pathology Research Institute (IPRI) in Sault Ste. Marie, where I worked with Fred T. Bird on insect cytoplasmic polyhedrosis viruses affecting the spruce budworm, a major Canadian forest pest. IPRI, since replaced, was the first biocontainment laboratory in the world dedicated to research on insect pathogens; it was a good incubator in terms of instilling a lifelong interest in virology. Later, during my MSc program at Carleton University, I worked with Peter E. Lee on *Tipula* iridescent virus, a large dsDNA iridovirus (Krell and Lee, 1974). After completing my MSc, and in search of adventure along Canada's east coast, I contacted Don, then a newly minted faculty member at Dalhousie University. I wanted to carry out Ph.D. research in his laboratory, continuing to work on iridoviruses. That I had a publication to my name by then must have influenced him to accept me as a student (his first!), and so I started in September 1975. No sooner had I arrived in Halifax, than Don began showing me EM images of his virus-like particles in the calyx tissue of the braconid parasitoid, *Apanteles melanoscelus*. Having worked on insect viruses in the 'Soo', I immediately recognized their similarity to the nuclear polyhedrosis viruses (now known as baculoviruses). Thus, I readily accepted the opportunity to work on the biochemical characterization of a new 'baculovirus'. At the time, I thought this would be a simple virus characterization study leading to some publications and a relatively stress-free Ph.D.

The first thing I had to learn was how to raise sufficient parasitoids so as to serve as a source of virus. That also necessitated learning how to rear the parasitoid host. What little I knew of entomology at the time had been picked up mostly by osmosis, and so it was critical that Don was there to educate me on the insect side of polydnaviruses. I set out initially to analyze the proteins and nucleic acid of the *A. melanoscelus* virus (AmV, now referred to as CmeBV). The polypeptide profile generated by polyacrylamide gel electrophoresis of purified CmeBV comprised at least 20 bands, more than the 10 typical of a

---

[2] Peter was invited to give a presentation on the polydnaviruses at the ICTV triennial meeting in Strasbourg, France, in 1981. The ICTV recognized that these were in fact quite different 'baculoviruses' and approved classification of the braconid baculoviruses as a subgroup of the Baculoviridae family. The ICTV did, however, acknowledge a need for the polydnaviruses to be given their own family designation at some point. It was only after the ICTV meeting in Berlin that our proposal to establish a new virus family, Polydnaviridae, was finally accepted and included in the Fifth Report of the International Committee on Taxonomy of Viruses, published in 1991.

baculovirus. About five of these, including two major ones, were found in nucleocapsid preparations (Krell, 1979).

The CmeBV genome was expected to consist of DNA, in keeping with the assumption that it was simply another baculovirus. Vinson and Scott (1975) had also noted that the core of similar particles in the braconid *Cardiochiles nigriceps* was sensitive to DNase, but not RNase, digestion suggestive of a DNA genome. When purified CmeBV DNA was subjected to isopycnic centrifugation in CsCl gradients containing ethidium bromide, two bands were detected, as was typical for circular dsDNAs. From this, I expected that the lower, higher-density band would consist of circular supercoiled DNA and the upper, lower-density band of relaxed (nicked) circular DNA. This was a reasonable assumption, since Max Summers had just detected two bands in CsCl EtBr gradients for DNAs isolated from several nuclear polyhedrosis viruses (Summers and Anderson, 1973). CmeBV DNA isolated from each band when run in CsCl alone were identical in density (at 1.694 gm/ml), and this was determined to be equivalent to a G + C content of 35%. These results demonstrated that CmeBV DNA was double stranded, circular and supercoiled, observations consistent with CmeBV being a baculovirus.

Subsequent DNA analysis, however, led to some unexpected results, forcing us to reassess any idea of considering CmeBV to be a baculovirus. This was not turning out to be a simple run-of-the-mill virus characterization study! Agarose gel electrophoresis as a way of separating DNA, especially following restriction endonuclease digestion, was just coming into vogue. My earliest studies made use of agarose tube gels (flat bed gels had yet to be described). At first, having convinced myself that these parasitoid viruses were simply baculoviruses, I had expected to observe a single viral DNA band. Instead, several bands, of varying intensities, were resolved. As no other dsDNA virus had been shown to have a segmented genome, these results were interpreted as being due to an artifact caused by mechanical shearing, partial digestion from contaminating DNases, or even digestion with a novel restriction enzyme. All efforts to 'improve' DNA isolation proved ineffective in reducing the number of bands. Subsequent gels were poured in a flat horizontal gel electrophoresis tray with multiple wells (before commercial ones had become available). Ethidium bromide stained gels were viewed on a handmade, plywood-based UV transilluminator to visualize the DNA bands. This allowed for more reliable sample-to-sample comparisons of DNAs run in different lanes and was sufficient for publication (Krell and Stoltz, 1979).

In subsequent work, I found that a similar polydispersity of DNA bands was also characteristic of nucleic acids extracted from calyx fluid particles of the ichneumonid parasitoid, *Hyposoter exiguae* (HeIV), except that most DNA bands displayed faster rates of migration (Krell and Stoltz, 1980). The diversity of bands seen using AGE, combined with the double bands detected by isopycnic centrifugation, strongly suggested that the viral DNA was not only circular, double stranded, and mostly superhelical, but was also segmented. In retrospect, I should have expected multiple bands, at least in the case of CmeBV, since the nucleocapsids were of varying lengths. In fact, the relative sizes of the CmeBV DNAs closely correlated to the relative sizes of the nucleocapsids. Nevertheless, there was still skepticism, at least from my Ph.D. advisory committee, that what I was looking at was really a segmented dsDNA genome.

The final proof for the conformation and polydisperse nature of the DNA was provided by electron microscopy, using the Kleinschmidt technique then being honed by Don. The EM data confirmed that the higher-density CsCl EtBr band contained supercoiled DNA and the lower-density CsCl EtBr band contained relaxed circular DNAs; in both cases the DNAs were of varying lengths as initially suggested by the AGE analyses. The different lengths observed by DNA shadowing of CmeBV DNA correlated nicely with both the relative migration rates of bands separated by AGE and the observed variation in nucleocapsid length. Electron microscopy also confirmed the non-equimolar ratio of different viral DNAs initially observed by AGE; the presence of non-equimolar packaged genome segments is now considered to be the norm for polydnaviruses. The existence of a multipartite DNA genome raised an early question about the relationship between the different circles: were the smaller circles simply derived from a larger master circle or were they all unrelated?

The question of DNA relatedness among the circular DNAs was initially addressed by Southern blot hybridization using as probes individual DNA bands isolated directly from gels and labeled with [32]P. For three CmeBV DNA probes, major hybridization was to the supercoiled and relaxed circular DNAs equivalent in molecular mass to that of the probe DNA itself. However, hybridization was also observed to other (but not all) bands (Krell, 1979). This cross-hybridization was acknowledged and initially interpreted as being non-specific due to using too low a stringency during the hybridization reactions. It is of course now known that cross-hybridization among different bands is due to the presence of gene and/or genome segment families sharing some degree of homology. In a comparable study of ichneumonid calyx fluid particles, there was the expected and strong hybridization with DNA bands cognate to the probe, but also minor hybridization to two to three additional bands, depending on the probe used (Krell and Stoltz, 1980).

At this point, with a Ph.D. in virology under my belt, and some respectable publications from my work in Don's laboratory, I was able to convince both Brad Vinson and Max Summers to accept me as a postdoctoral fellow at TAMU, where I joined them in February 1980. Since it was in culture in Brad's laboratory at the time, I concentrated on *Campoletis sonorensis* and its calyx fluid particles (now well known as the icheumonid polydnavirus, CsIV). This arguably laid the ground work for the many studies that were to follow. At TAMU, I benefited greatly from the strength in parasitoid biology that characterized Brad's laboratory and the more molecular and biochemically oriented laboratory of Max Summers and his team, all providing a great synergy of complementary expertise.

While characterizing CsIV, I noted some minor cross-hybridization among DNA circles, as had been observed earlier for other viruses of this type (Krell *et al.*, 1982). At that time, I had already acknowledged that the observed cross-hybridization suggested that some of circles might share common sequences. In November of 1981, I returned to Canada as an assistant professor (University of Guelph) intent on continuing my work on what were now being referred to (as yet unofficially) as PDVs. When David Theilmann (a fellow Canadian) and Gary Blissard initiated their Ph.D. degree programs at TAMU shortly after I left, they demonstrated some of the CsIV cross-hybridized (Blissard *et al.*, 1987) and that this was due to the presence of common tandem 540 bp repeat elements present on several circles (Theilmann and Summers, 1987). The initial sequencing of CsIV cDNAs which mapped to circle W DNA but also shared sequences with circle V DNA confirmed that indeed some of the polydnavirus DNA sequences were shared among different DNA circles (Dib-Haji *et al.*, 1993).

It is now known that genome segments are usually one of two types: unique, sharing little or no sequence homology to any other DNA circle, or 'nested' with smaller genome segments derived from larger ones (Xu and Stoltz, 1993; Cui and Webb, 1997). Beginning at TAMU, I kept thinking of the origin of the genome segments, and whether this was from pre-existing DNA circles or derived from chromosomal DNA (i.e., from an integrated version, like retroviral DNA). Jo-Ann Fleming, who replaced me at TAMU, immediately got to work analyzing wasp chromosomal DNA for the presence of CsIV sequences and demonstrated that DNA from CsIV bands B and Q hybridized to male chromosomal 'off-size' DNA fragments (Fleming and Summers, 1986, 1991, and reviewed in Fleming and Krell, 1993). In my own study, though somewhat obscured because it was published in a review paper (Krell, 1991), I subsequently reported that CsIV segment M DNA hybridized to different sized bands of male *C. sonorensis* DNA digested with different enzymes, suggestive of integrated forms. Later, the regions flanking the integrated DNA of segment B were sequenced and shown to be similar to the 540 bp repeat reported by David (Theilmann and Summers, 1987).

Still in the 'early' days of polydnavirology, both Jo-Ann and Gary, working with Max Summers at TAMU, were the first to demonstrate the expression of polydnavirus mRNA in host insects (Fleming *et al.*, 1983; Blissard *et al.*, 1986). Going beyond the genome and mRNA expression, Bruce Webb, also in Max's laboratory, concentrated on the polydnavirus proteins and their possible role in immunomodulation of the host response to parasitization (Webb and Summers, 1990).

At Guelph, I continued to work on polydnaviruses, first on DtIV, the ichnovirus from *Diadegma terebrans* and tissue organization of virus expressing cells in the host ovary (Krell, 1987a, b). However, my initial attempts to study polydnaviruses from the braconid, *Cardiochiles nigriceps* fell to naught as my host colony of *Heliothis virescens* kept succumbing to baculovirus infections. Instead of fighting this, I decided to launch a career in baculovirology, while still continuing to collaborate (e.g., Stoltz *et al.*, 1988 and Cusson *et al.*, 1998) and engaged in following further advances in polydnavirology.

A QUESTION IS RAISED

From the earliest, it was clear that PDVs were incapable of replicating in the parasitized host (e.g., Theilmann and Summers, 1986), and of course this made sense: a replicative virus might also be a pathogenic virus, with potentially dire consequence for the developing endoparasitoid. While early work from the Summers lab described virus-specific transcriptional activity in various host tissues (Fleming *et al.*, 1983), it was clear that this was occurring in the apparent absence of viral DNA synthesis; accordingly, the assumption made was that viral replication was necessarily abortive, for the reason already given. It was of course known at the time that immunosuppression of the host was a critically important function mediated by CF particles (Edson *et al.*, 1981). Since some of the larger DNA viruses were known to encode immunosuppressive gene products, it seemed reasonable to suspect that one or more PDV genes behaved similarly, and had somehow been captured by an ancestral parasitoid for just this purpose. Some began to wonder why parasitoids needed a complex viral structure to deliver such genes—why not just deliver the gene products? As sequencing projects got underway, it soon became clear that packaged PDV genomes comprised more non coding sequences than was typical for viruses (e.g., Strand *et al.*, 1997; Cui and Webb, 1997; Webb *et al.*, 2006). Later, as complete PDV genome sequences became available, it was evident that *wasp*, rather than viral, genes were being packaged (Espagne *et al.*, 2004; Webb *et al.*, 2006; Tanaka *et al.*, 2007; Desjardins *et al.*, 2008; Cusson *et al.*, Chapter 6 of this volume). This of course explained why viral morphogenesis had never been observed in parasitized host animals, but immediately raised a number of questions, the most important of which, for us, was this: did parasitoids 'invent' virus-like particles *de novo* for use as gene delivery vehicles (one way to do this would have been to gradually acquire replicative and structural genes from a variety of viruses over time, although it is difficult to imagine the sort of selective pressure that would be required to maintain such a strategy), or did they simply hijack an entire conventional viral genome for this purpose? No one really knew; that said, the latter hypothesis was generally preferred. At least initially, an arguably less-important question had to do with nomenclature: even if viral in origin, can these things legitimately still be referred to as bona fide viruses? Put another way, the question became: just what *are* these things?! This issue is discussed below.

QUESTION LARGELY ANSWERED

Again, for us, a central question was: where did these particles come from? Were the PDVs entirely a wasp invention, or did they have one or several viral ancestors? Using complementary genomic and proteomic approaches, Bézier et al. (2009) have now convincingly shown that the bracoviruses originated from an ancestral nudivirus. More recently, Volkoff et al. (2010) have shown that ichnoviruses too have a viral origin; in this case, however, the putative ancestor remains unidentified. The ichnovirus progenitor may belong to some lineage as yet undiscovered, or undescribed, or which is now extinct. It is of interest in this regard that the parasitoid ovary may well represent a 'privileged' site for the replication of viral commensals (Stoltz and Whitfield, 1992); ovarian tissue might therefore be a good place to look for an ichnovirus progenitor. It will be of interest to determine whether there exists an independent origin for the more recently characterized banchine PDVs (Lapointe et al., 2007).

WHERE NEXT?

For those of us with an interest in PDV research, plenty of questions remain; many of these are fairly obvious, and will undoubtedly be raised by others contributing to this volume, and so there's little point in reprising them here. We have chosen instead to take a look at the continuing and as yet unresolved issue of nomenclature/taxonomy; while not (at least in our view) an issue of paramount importance, it should nonetheless be addressed, if only to stimulate a lively discussion. First raised in 2003 (Federici and Bigot, 2003; Whitfield and Asgari, 2003), the question is whether it is now still appropriate to continue to think of PDVs as being, in fact, viruses; should we not instead be viewing them as wasp organelles? (Parenthetically, ichnoviruses had been portrayed in one early paper as 'nuclear secretory particles' (Norton et al., 1975)). In large part, this is a question of semantics, at least until such time as the word 'virus' is clearly defined. That said, there are consequences that follow the use of standard viral terminology; for example, as Federici and

Bigot rightly note, if PDVs are given specific scientific names, then so could mitochondria and chloroplasts, which no one sees as being necessary, and so they aren't listed in Bergey's Manual. Similarly, there's no compelling need to list the PDVs in the ICTV compendium. A related item that we need to exercise some care with is the use of phraseology such as '…the PDV genome encodes genes which…'; if one is referring to *packaged* genes, then they are likely to be *wasp*, not viral, genes! In any case, by way of providing some clarification, we would like to suggest that the ICTV now consider establishing *categories* of viruses and virus-like entities (and in so doing thereby define, once and for all, what the genome of a conventional virus must be able to accomplish. As has been previously argued (Stoltz and Whitfield, 2009), whether PDVs should be considered to be viruses is not the most important question; whether or not they were of viral *origin* is arguably far more important. Whether they should, in a formal sense, be referred to as viruses is entirely a matter for the ICTV to resolve—for which we have herein provided a reasonable solution (see Appendix). Finally, whether or not one would prefer to retain current nomenclature, '*Polydnaviridae*' has to go. The family name was always a bit unusual (albeit convenient!) in the sense that the practice of assigning viruses of quite dissimilar morphology to the same family was unlikely to survive close scrutiny over the long term; whether to replace it with 'Bracoviridae', etc. is for the

Appendix

Suggested categories of viruses and virus-like entities:

1. Classical viruses (parasites and commensals)—infection of a susceptible host results in the production of progeny virions. Put another way, the genomes of such viruses encode whatever is needed to make more virions.

2. Gene transfer agents—the original viral genome has been permanently transferred to the host genome, but remains essentially intact, as virus-like particles are made. These particles cannot, however, replicate if transferred to another host because *packaged* genes are for the most part non viral. Viruses/particles of this type, such as the polydnaviruses found in certain parasitoid lineages and the gene transfer agents of bacteria (reviewed in Stanton, 2007), appear to have evolved to become gene delivery vehicles.

3. Endogenous viruses—the original viral genome has been permanently transferred to the host genome, but has not been maintained intact; accordingly, physical virions can no longer be made (e.g., Villareal, 1997). Genes encoding a specific viral polypeptide, however, may be maintained and expressed, sometimes with quite interesting consequences (e.g., Mi *et al.*, 2000).

TABLE 1 Selected Milestones in Polydnavirus Research[3]

Observation	Reference(s)
The ovarian calyx is recognized as the source of substance(s) which protect parasitoid eggs from encapsulation in host larvae	Salt, 1965
Particulate material derived from the calyx is observed in wasp oviducts	Rotheram, 1967
DNA is associated with braconid calyx fluid (CF) particles, now described as 'virus-like'	Vinson and Scott, 1975
Calyx fluid particles (described as nuclear secretions, or microorganisms) are reported in the ovaries of two ichneumonid parasitoids	Norton et al., 1975; Hess et al., 1975
Invasion of host cells/tissues by putative viral nucleocapsids is documented	Stoltz and Vinson, 1977 and 1979a
Braconid CF particle dsDNA is circular, supercoiled and polydisperse	Krell and Stoltz, 1979
Ichneumonid CF particle dsDNA is also circular, supercoiled and polydisperse	Krell and Stoltz, 1980
Suppression of host cellular immunity by purified CF particles is described	Edson et al., 1981
PDV genes are expressed in the parasitized host	Fleming et al., 1983
Establishment of the virus family, Polydnaviridae, is proposed	Stoltz et al., 1984
PDV DNA persists but does not replicate in host tissues	Theilmann and Summers, 1986
Some PDV genes are spliced	Blissard et al., 1987
Direct evidence provided for chromosomal location and transmission of PDV DNA; indirect evidence for this was provided earlier by Fleming and Summers (1986)	Stoltz, 1990; Fleming and Summers, 1991
Polydnaviridae formally accepted by the ICTV as the family name for polydnaviruses; two genera, Bracovirus (formerly designated as group C of the Baculoviridae) and Ichnovirus, are established	Fifth Report of the International Committee on Taxonomy of Viruses, 1991
Evidence for 'nesting' (generation of one PDV genome segment from another) is provided	Xu and Stoltz, 1993(see also Cui and Webb, 1997)
A model for the excision of bracovirus genome segments by homologous recombination is proposed	Krell, 1991; Fleming and Krell 1993; Gruber et al., 1996
Each bracovirus nucleocapsid packages a single molecule of DNA	Albrecht et al., 1994 (see also Beck et al., 2007)
A bracovirus is used to establish a role for cellular immunity in the defense response versus other viruses	Washburn et al., 2000 (see also Rivkin et al., 2006)
Amplification of bracovirus DNA occurs prior to excision and packaging of genome segments	Pasquier-Barre et al., 2002 (see also Annaheim and Lanzrein, 2007)
The origin of the wasp/bracovirus relationship is dated to ~73.7 myr ago	Whitfield, 2002 (see also Whitfield, 1997)
A bracovirus genome segment is integrative in vitro	Gundersen-Rindal and Lynn, 2003 (see also Doucet et al., 2007)
RNA interference is used to explore the function of a PDV genome segment	Beck and Strand, 2003
First complete sequence of a (packaged) PDV genome is published; sequences described appear to be largely if not exclusively cellular, rather than viral, in nature	Espagne et al., 2004
Sequencing of PDV DNA reveals that genes required to make virions are not packaged, and that braco- and ichnovirus genomes share very few genes	Espagne et al., 2004; Webb et al., 2006; Tanaka et al., 2007
Genome sequencing and proteomics identify a nudiviral ancestor for the bracoviruses	Bézier et al., 2009
In vivo experiments using individual PDV genome segments are described	Kwon et al., 2010; Park and Kim, 2010
Genome sequencing and proteomics suggest a viral origin for the ichnoviruses	Volkoff et al., 2010
Polydnavirus genome segments are integretive in vivo	Beck et al., submitted to J., Virol

[3]We have focused here largely on PDV genomes, since this is central to the question of where PDVs came from. Information relating to the function of genes delivered to parasitized hosts is covered elsewhere in this volume. We omit most of the papers comprising the legacy of George Salt as this has been very nicely dealt with elsewhere (see Fisher, 1986). In addition, with the exception of the first two references, we have not included work on the Venturia (Nemeritis) canescens particle, primarily for the reason that it does not package any DNA.

Polydnavirus Study Group of the ICTV, and then ICTV itself, to decide. The above notwithstanding, there is of course no particular reason why common, or vernacular, names (such as polydnavirus, ichnovirus, etc.), along with their various acronyms, cannot be used exactly as done now; ICTV itself has no policy on the use of vernacular names or acronyms, nor does it wish to.

ACKNOWLEDGMENTS

DBS wishes to acknowledge the many invaluable contributions made over the years by his students and technicians, many of whom worked above and beyond the call of duty; among these is included the incomparable Aricy Pan, whose efforts have been received with both gratitude and affection. In addition, he thanks the PDV community in general for its unwavering enthusiam, support, and collegiality. Research in his lab has been supported by the then Medical and National Research Councils and, more recently, by the Natural Sciences and Engineering Research Council of Canada. PJK is also pleased to acknowledge support from the latter agency, as well as support from Max Summers and Brad Vinson during his postdoctoral days at TAMU.

REFERENCES

Albrecht, U., Wyler, T., Pfister-Wilhelm, R., Gruber, A., Stettler, P., Heiniger, P., et al. 1994. Polydnavirus of the parasitic wasp *Chelonus inanitus* (Braconidae): characterization, genome organization and time point of replication. J. Gen. Virol. 75, 3353–3363.

Annaheim, M., Lanzrein, B., 2007. Genome organization of the *Chelonus inanitus* polydnavirus: excision sites, spacers and abundance of proviral and excised segments. J. Gen. Virol. 88, 450–457.

Beck, M., Strand, M.R., 2003. RNA interference silences *Microplitis demolitor* bracovirus genes and implicates *glc1.8* in disruption of adhesion in infected host cells. Virology 314, 521–535.

Beck, M.H., Inman, R.B., Strand, M.R., 2007. *Microplitis demolitor* bracovirus genome segments vary in abundance and are individually packaged in virions. Virology 359, 179–189.

Beck, M.H., Zhang, S., Bitra, K., Burke, G.R., Strand, M.R. The encapsidated genome of microplitis demoliton bracovirus integrates into the host pseudoplusia includens. J. Virol. (submitted).

Bézier, A., Annaheim, M., Herbiniere, J., Wetterwald, C., Gyapay, G., Bernard-Samain, S., et al. 2009. Polydnaviruses of braconid wasps derive from an ancestral nudivirus. Science 323, 926–930.

Blissard, G.W., Smith, O.P., Summers, M.D., 1987. Two related viral genes are located on a single superhelical DNA segment of the multipartite *Campoletis sonorensis* virus genome. Virology 160, 120–134.

Blissard, G.W., Vinson, S.B., Summers, M.D., 1986. Identification, mapping, and in vitro translation of *Campoletis sonorensis* virus mRNAs from parasitized Heliothis virescens larvae. J. Virol. 57, 318–327.

Cui, L., Webb, B.A., 1997. Homologous sequences in the *Campoletis sonorensis* polydnavirus genome are implicated in replication and nesting of the W segment family. J. Virol. 71, 8504–8513.

Cusson, M., Lucarotti, C, Stoltz, D., Krell, P., Doucet, D., 1998. A polydnavirus from the spruce budworm parasitoid, *Tranosema rostrale* (Ichneumodidae). J. Invertebr. Pathol. 72, 50–56.

Desjardins, C.A., Gundersen-Rindal, D.E., Hostetler, J.B., Tallon, L.J., Fadrosh, D.W., Fuester, R.W., et al. 2008. Comparative genomics of mutualistic viruses of *Glyptapanteles* parasitic wasps. Genome Biol. 9, R183.

Dib-Haji, S.D., Webb, B.A., Summers, M.D., 1993. Structure and evolutionary implications of a "cysteine-rich" *Campoletis sonorensis* polydnavirus gene family. Proc. Natl. Acad. Sci. U.S.A. 90, 3765–3769.

Doucet, D., Levasseur, A., Béliveau, C., Lapointe, R., Stoltz, D., Cusson, M., 2007. *In vitro* integration of an ichnovirus genome segment into the genomic DNA of lepidopteran cells. J. Gen. Virol. 88, 105–113.

Edson, K.M., Vinson, S.B., Stoltz, D.B., Summers, M.D., 1981. Virus in a parasitoid wasp: suppression of the cellular immune response in the parasitoid's host. Science 211, 582–583.

Espagne, E., Dupuy, C., Huguet, E., Cattolico, L., Provost, B., Martins, N., et al. 2004. Genome sequence of a polydnavirus: insights into symbiotic virus evolution. Science 306, 286–289.

Federici, B.A., Bigot, Y., 2003. Origin and evolution of polydnaviruses by symbiogenesis of insect DNA viruses in endoparasitic wasps. J. Insect Physiol. 49, 419–432.

Fisher, R.C., 1986. George Salt and the development of experimental insect parasitology. J. Insect Physiol. 32, 249–253.

Fleming, J.A.G.W., Blissard, G.W., Summers, M.D., Vinson, S.B., 1983. Expression of *Campoletis sonorensis* virus in the parasitized host, *Heliothis virescens*. J. Virol. 48, 74–78.

Fleming, J.G., Krell, P.J., 1993. Polydnavirus genome organization. In: Beckage, N.E., Thompson, S.N., Federici., B.A., (Eds.), Parasites and Pathogens of Insects, (vol. 1) (pp. 189–225). Academic Press, San Diego.

Fleming, J.G., Summers, M.D., 1986. *Campoletis sonorensis* endoparasitic wasps contain forms of *C. sonorensis* virus DNA suggestive of integrated and extrachromosomal polydnavirus DNAs. J. Virol. 57, 552–562.

Fleming, J.G., Summers, M.D., 1991. Polydnavirus DNA is integrated in the DNA of its parasitoid wasp host. Proc. Natl. Acad. Sci. U.S.A. 88, 9770–9774.

Gruber, A., Stettler, P., Heiniger, P., Schumperli, D., Lanzrein, B., 1996. Polydnavirus DNA of the braconid wasp *Chelonus inanitus* is integrated in the wasp genome and excised only in later pupal and adult stages of the female. J. Gen. Virol. 77, 2873–2879.

Gundersen-Rindal, D.E., Lynn, D.E., 2003. Polydnavirus integration in lepidopteran host cells in vitro. J. Insect Physiol. 49, 453–462.

Hess, R.T., Benham, G.S., Poinar G.O., Jr., 1975. The ultrastructure of microorganisms in the tissues of *Casinaria infesta* (Hymenoptera: Ichneumonidae). J. Invertebr. Pathol. 26, 181–191.

Krell, P., Lee, P.E., 1974. Polypeptides in *Tipula* iridescent virus (TIV) and in TIV-infected hemocytes of *Galleria mellonella* (L.) larvae. Virology 60, 315–326.

Krell, P.J., 1979. Viruses in parasitoid hymenoptera: a preliminary characterization. Ph.D. thesis, Department of Microbiology, Dalhousie University.

Krell, P.J., 1987a. Polydnavirus replication and tissue organization of infected cells in the parasitic wasp *Diadegma terebrans*. Can. J. Microbiol. 33, 176–183.

Krell, P.J., 1987b. Replication of long virus like particles in the ovary of the ichneumonid wasp *Diadegma terebrans*. J. Gen. Virol. 68, 1477–1483.

Krell, P.J., 1991. The polydnaviruses: multipartite DNA viruses from parasitic Hymenoptera. In: Kurstak, E. (Ed.), Viruses of Invertebrates (pp. 141–177). Marcel Dekker, Inc., New York.

Krell, P.J., Stoltz, D.B., 1979. Unusual baculovirus of the parasitoid wasp *Apanteles melanoscelus*: isolation and preliminary characterization. J. Virol. 29, 1118–1130.

Krell, P.J., Stoltz, D.B., 1980. Virus-like particles in the ovary of an ichneumonid wasp: purification and preliminary characterization. Virology 101, 408–418.

Krell, P.J., Summers, M.D., Vinson, S.B., 1982. Virus with a multipartite superhelical DNA genome from the ichneumonid parasitoid *Campoletis sonorensis*. J. Virol. 43, 859–870.

Kwon, B., Song, S., Choi, J.Y., Je, Y.H., Kim, Y., 2010. Transient expression of specific *Cotesia plutellae* bracoviral segments induces prolonged larval development of the diamondback moth, *Plutella xylostella*. J. Insect Physiol. 56, 650–658.

Lapointe, R., Tanaka, K., Barney, W.E., Whitfield, J.B., Banks, J.C., Béliveau, C., et al. 2007. Genomic and morphological features of a banchine polydnavirus: comparison with bracoviruses and ichnoviruses. J. Virol. 81, 6491–6501.

Mason, W.R.M., 1981. The polyphyletic nature of *Apanteles* Foerster (Hymenoptera: Braconidae): a phylogeny and reclassification of Microgastrinae. Memoirs Entomol. Soc. Can. 115, 1–147.

Mi, S., Lee, X., Li, X., Veldman, G.M., Finnerty, H., Racie, L., et al. 2000. Syncytin is a captive retroviral envelope protein involved in human placental morphogenesis. Nature 403, 785–789.

Norton, W.N., Vinson, S.B., Stoltz, D.B., 1975. Nuclear secretory particles associated with the calyx cells of the ichneumonid parasitoid *Campoletis sonorensis* (Cameron). Cell Tiss. Res. 162, 195–208.

Park, B., Kim, Y., 2010. Transient transcription of a putative RNase containing BEN domain encoded in *Cotesia plutellae* bracovirus induces an immuno suppression of the diamondback moth, *Plutella xylostella*. J. Invertebr. Pathol. 105, 156–163.

Pasquier-Barre, F., Dupuy, C., Huguet, E., Monteiro, F., Moreau, A., Poirie, M., et al. 2002. Polydnavirus replication: the EP1 segment of the parasitoid wasp *Cotesia congregata* is amplified within a larger precursor molecule. J. Gen. Virol. 83, 2035–2045.

Rivkin, H., Kroemer, J.A., Bronshtein, A., Belausov, E., Webb, B.A., Chejanovsky, N., 2006. Response of immunocompetent and immunosuppressed *Spodoptera littoralis* larvae to baculovirus infection. J. Gen. Virol. 87, 2217–2225.

Rotheram, S., 1967. Immune surface of eggs of a parasitic insect. Nature 214, 700.

Salt, G., 1965. Experimental studies in insect parasitism. XIII. The haemocytic reaction of a caterpillar to eggs of its habitual parasite. Proc. R. Soc. B. 162, 303–318.

Salt, G., 1968. The resistance of insect parasitoids to the defense reactions of their hosts. Biol. Rev. 43, 200–232.

Stanton, T.B., 2007. Prophage-like gene transfer agents—novel mechanisms of gene exchange for *Methanococcus*, *Desulfovibrio*, *Brachyspira*, and *Rhodobacter* species. Anaerobe 13, 43–49.

Smith, K.M., 1967. Insect virology. Academic Press, New York.

Stoltz, D.B., 1990. Evidence for chromosomal transmission of polydnavirus DNA. J. Gen. Virol. 71, 1051–1056.

Stoltz, D.B., Krell, P., Cook, D., MacKinnon, E.A., Lucarotti, C.J., 1988. An unusual virus from the parasitic wasp *Cotesia melanoscela*. Virology 162, 311–320.

Stoltz, D.B., Krell, P., Summers, M.D., Vinson, S.B., 1984. Polydnaviridae—a proposed family of insect viruses with segmented, double-stranded, circular DNA genomes. Intervirology 21, 1–4.

Stoltz, D.B., Vinson, S.B., 1977. Baculovirus-like particles in the reproductive tracts of female parasitoid wasps. II. The genus *Apanteles*. Can. J. Microbiol. 23, 28–37.

Stoltz, D.B., Vinson, S.B., 1979a. Penetration into caterpillar cells of virus-like particles injected during oviposition by parasitoid ichneumonid wasps. Can. J. Microbiol. 25, 207–216.

Stoltz, D.B., Vinson, S.B., 1979b. Viruses and parasitism in insects. Adv. Virus Res. 24, 125–171.

Stoltz, D.B., Vinson, S.B., MacKinnon, E.A., 1976. Baculovirus-like particles in the reproductive tracts of female parasitoid wasps. Can. J. Microbiol. 22, 1013–1023.

Stoltz, D.B., Whitfield, J.B., 1992. Viruses and virus-like entities in the parasitic Hymenoptera. J. Hym. Res. 1, 125–139.

Stoltz, D.B., Whitfield, J.B., 2009. Making nice with viruses. Science 323, 884–885.

Strand, M.R., Witherell, S.A., Trudeau, D., 1997. Two related *Microplitis demolitor* polydnavirus mRNAs expressed in hemocytes of *Pseudoplusia includens* contain a common cysteine-rich domain. J. Virol. 71, 2146–2156.

Summers, M.D., Anderson, D.L., 1973. Characterization of nuclear polyhedrosis virus DNAs. J. Virol. 12, 1336–1346.

Tanaka, K., Lapointe, R., Barney, W.E., Makkay, A.M., Stoltz, D.B., Webb, B.A., 2007. Shared and species-specific features among ichnovirus genomes. Virology 363, 26–35.

Theilmann, D.A., Summers, M.D., 1986. Molecular analysis of *Campoletis sonorensis* virus DNA in the lepidopteran host *Heliothis virescens*. J. Gen. Virol. 67, 1961–1969.

Theilmann, D.A, Summers, M.D., 1987. Physical analysis of the *Campoletis sonorensis* virus multipartite genome and identification of a family of tandemly repeated elements. J. Virol. 61, 2589–2598.

Villareal, L.P., 1997. On viruses, sex, and motherhood. J. Virol. 71, 859–865.

Vinson, S.B., 1974. The role of the foreign surface and female parasitoid secretions on the immune response of an insect. Parasitology 68, 27–33.

Vinson, S.B., Scott, J.R., 1975. Particles containing DNA associated with the oocyte of an insect parasitoid. J. Invertebr. Pathol. 25, 375–378.

Volkoff, A.N., Jouan, V., Urbach, S., Samain, S., Bergoin, M., Wincker, P., et al. 2010. Analysis of virion structural components reveals vestiges of the ancestral ichnovirus genome. PLoS Pathog. 6, 1–10.

Washburn, J.O., Haas-Stapleton, E.J., Tan, F.F., Beckage, N.E., Volkman, L.E., 2000. Co infection of *Manduca sexta* larvae with polydnavirus from *Cotesia congregata* increases susceptibility to fatal infection by *Autographa californica* M nucleopolyhedrovirus. J. Insect Physiol. 46, 179–190.

Webb, B.A., Summers, M.D., 1990. Venom and viral expression products of the endoparaitic wasp *Campoletis sonorensis* share epitopes and related sequences. Proc. Natl. Acad. Sci U.S.A. 87, 4961–4965.

Webb, B.A., Strand, M.R., Dickey, S.E., Hilgarth, R.S., Beck, M.H., Walter, B.E., et al. 2006. Polydnavirus genomes reflect their dual roles as mutualists and pathogens. Virology 347, 160–174.

Whitfield, J.B., 1997. Molecular and morphological data suggest a common origin for the polydnaviruses among braconid wasps. Naturwissenschaften 84, 502–507.

Whitfield, J.B., 2002. Estimating the age of the polydnavirus–braconid wasp symbiosis. Proc. Natl. Acad. Sci. U.S.A. 99, 7508–7513.

Whitfiekd, J.B., Asgari, S., 2003. Virus or not? Phylogenetics of polydnaviruses and their wasp carriers. J. Insect Physiol. 49, 397–405.

Xu, D., Stoltz, D.B., 1993. Polydnavirus genome segment families in the ichneumonid parasitoid, *Hyposoter fugitivus*. J. Virol. 67, 1340–1349.

Chapter 2

Evolutionary Progenitors of Bracoviruses

Jean-Michel Drezen, Elisabeth Anne Herniou, and Annie Bézier

Institut de Recherche sur la Biologie de l'Insecte, CNRS UMR 6035, Université François Rabelais, Facuté des Sciences et Techniques, Parc de Grandmont, 37200 Tours, France

SUMMARY

While viruses usually produce particles in the infected cells, polydnaviruses (PDVs) have a very unusual virus life-cycle. Particle production does not occur in infected tissues of parasitized caterpillars, but is restricted to specialized cells of the wasp ovary. The genome enclosed in the particles encodes almost no viral structural protein, but mostly virulence factors used to manipulate the physiology of the parasitized host. This lack of virus genes in the packaged genome has generated a debate on the viral nature of PDVs. The characterization of a hidden bracovirus genome composed of virus genes, residing permanently in the wasp chromosomes and producing the particles, confirmed that bracoviruses originated from a virus. The viral machinery is comprised of genes typical of nudiviruses, a sister group of baculoviruses. The conservation of nudiviral genes in the various lineages of braconid wasps associated with bracoviruses strongly suggests that their common ancestor acquired a nudivirus genome 100 million years ago, which still controls the production of particles during viral replication.

ABBREVIATIONS

BV	bracovirus
EVE	endogenous viral elements
IV	ichnovirus
IVSPER	ichnovirus structural protein encoding region
LEF	late expression factor
ODV	Occlusion-derived virus
PIF	*per os* infectivity factor
PDV	polydnavirus
PTP	protein tyrosine phosphatase
NV	nudivirus
TE	transposable element
TEM	transmission electronic microscopy
VLF	very late factor

The associations between several thousand parasitic wasps and polydnaviruses (PDVs) represent unique examples of virus domestication by a cellular organism, a parasitic wasp, to manipulate the physiology of another organism, the caterpillar host (Webb and Strand, 2005). PDV particles are injected along with parasite eggs into the host body. The wasps use them to transfer and express virulence genes inside the host (Dupuy *et al.*, 2006). The DNA molecules packaged in the particles encode factors allowing the manipulation of host immune defenses and its developmental program. This enables the wasp larvae to survive and develop in an environment that otherwise would be harmful. Particle production occurs exclusively in specialized calyx cells of the wasp ovaries and PDVs are maintained through vertical transmission of their chromosomally integrated form (Stoltz, 1990; Belle *et al.*, 2002; Fleming and Summers, 1991). Two genera of viruses were originally described, *Ichnoviruses* (IV) and *Bracoviruses* (BV), which are associated respectively with wasp species of the families Ichneumonidae and Braconidae. In addition, a third virus group associated with ichneumonid wasps belonging to the Banchinae subfamily has recently been proposed as a third taxonomic genus, based on the study of the polydnavirus from the wasp *Glypta fumiferanae*. This virus has particular distinctive features compared to the ichnoviruses described in Campopleginae: (1) the morphology of the particles is different; (2) the packaged genome contains genes coding for protein tyrosine phosphatases (PTPs, a characteristic shared with bracoviruses) and (3) it is carried by a wasp lineage separated from Campopleginae by lineages not reported to be associated with PDVs (Lapointe *et al.*, 2007). This suggests that either this virus originated from an independent ancestral wasp/virus association event or that PDVs have been lost or remained unidentified in the lineages that separate Campopleginae and Banchinae (Cusson *et al.*, Chapter 6 of this volume). The description of this unusual ichnovirus will surely stimulate the research of new polydnaviruses in these groups.

Since their characterization in the late seventies by transmission electronic microscopy (TEM) of ovary sections (see Vinson, Foreword for this volume), PDVs have been classified as viruses, because of the evident similarity of their particles with viruses. These particles are produced in the calyx, a specialized region located at the basis of the oviducts. They are released in the ovary lumen and constitute the major component of the calyx fluid injected into the host with the eggs (Stoltz and Vinson, 1979; Stoltz and Krell, Chapter 1 of this volume). Despite the obvious viral morphology of the particles, the confirmation of their relationship with pathogenic viruses has taken more than thirty years before the characterization of the genes involved in particle production. The analysis of the mRNAs present in parasitoid wasp ovaries led recently to the identification of a set of genes typical of nudiviruses expressed in bracovirus-associated species (nudiviral genes; Bézier *et al.*, 2009). Nudiviruses were formerly classified as non-occluded baculoviruses (or baculovirus subgroup C); they are the sister group of baculoviruses and share genes and several properties with these well-studied insect viruses (Wang and Jehle, 2009). The involvement of nudiviral genes in bracovirus particle production was demonstrated by mass spectrometry analysis (LC MS-MS) of purified particles, which clearly indicated that a large number of nudiviral genes that are expressed in braconid wasp ovaries do encode bracovirus particle components (Bézier *et al.*, 2009).

It was expected that a set of genes originating from the bracovirus ancestor should be conserved throughout the group of bracovirus-associated wasps. To assess this hypothesis, two nudiviral genes (*HzNVorf9-like1* and *HzNVorf128-like*) well-conserved in *Cotesia congregata* and *Chelonus inanitus* (their products sharing more than 80% similarities on aligned sequences) were used to design primers for PCR amplification. Homologs of these genes were isolated successfully from a series of wasps belonging to most lineages associated with bracoviruses (Bézier *et al.*, 2009), with the exception of the Khoikhoiinae subfamily. The lack of amplification does not necessarily indicate the absence of the genes in the genome of a wasp but rather that they might be too divergent for the PCR amplification. The overall conservation of the nudiviral genes in bracovirus-associated wasps that form a monophyletic group, known as the microgastroid complex, strongly suggests that they have been inherited from the common ancestor of this group (Bézier *et al.*, 2009). This wasp lived approximately 100 million years ago according to a phylogenetic estimation based on wasps that were discovered in fossil amber (Murpy *et al*, 2008; Whitfield and O'Connor, Chapter 7 of this volume).

In addition to gene content, some nudivirus biological properties point to possible mechanisms that could have been instrumental to the domestication of the bracovirus ancestor. The nudivirus HzNV-1 (*Heliothis zea* nudivirus-1), which appears to be relatively close to the bracovirus ancestor based on its gene content, was shown to persist as a latent infection in cultured cells with both circular and integrated DNA present in the cells (Chao *et al.*, 1998). This suggests that HzNV-1 genome integration in the DNA of infected cells might play a role in its life-cycle. During latency, HzNV-1 express the single noncoding *pat1* gene, which is thought to control the latency (Lin *et al.*, 1999), while the recently identified *hh1* gene might be involved in the virus reactivation (Wu *et al.*, 2010). HzNV-2, formerly known as gonad-specific virus (GSV; Raina *et al.*, 2000) and closely related to HzNV-1 (Wang *et al.*, 2007), is sexually transmitted and infects reproductive organs of *H. zea*. A virus with this particular tropism for the gonads could sometimes enter germ line cells and its integration into the genome of its host would assure its vertical transmission. The bracovirus ancestor might have had a gonad tropism and been able to integrate into the genome of infected cells and undergo latency, like HzNVs.

The organization of nudiviral genes in *C. congregata* genome is consistent with a scenario in which a nudivirus integrated its genome into the genome of a wasp ancestor. Half of the nudiviral genes identified are clustered in the genome of the wasp *C. congregata,* an organization that might have been inherited from the original nudivirus integration site (Fig. 1A).

The other nudiviral genes identified are dispersed in *C. congregata* genomic regions containing wasp genes and mobile elements (Bézier *et al.*, 2009). If there is no selective pressure acting for the genes required for particle production to remain together, after 100 million years some genes could be expected to have separated from the integrated virus genome. Another important modification in bracoviruses compared to pathogenic viruses resides in the fact that the nudiviral genes have lost the ability to incorporate viral particles (Bézier *et al.*, 2009). They appear to have been totally replaced in the DNA packaged in the particles by genes thought to originate from the wasp genome (see below).

With a similar tripartite approach as for the bracoviruses, the genes involved in particle production of ichnoviruses have been identified by combining (1) the analysis of the transcripts from the wasp *Hyposoter dydimator*, (2) protein characterization of the particle components, and (3) resolution of the corresponding gene organization in the wasp genome (Volkoff *et al.*, Chapter 3 of this volume). In contrast to braconid wasps, the genes expressed in the ovaries are not related to nudiviruses. They are also not related to ascoviruses, as it had been proposed based on particles morphology and low similarities in a capsid protein (Bigot *et al.*, 2008). The only viral genes detected (*p12* and *p53*

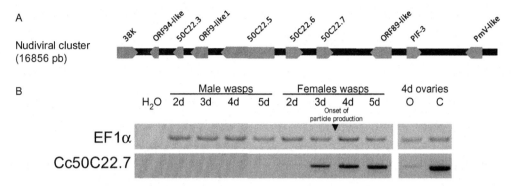

FIGURE 1   Genomic map of the nudiviral cluster and analysis of gene expression. **A:** Nudiviral gene organization in *C. congregata* genome (accession number FM212911). Six nudiviral genes (*38K*, *HzNVorf94-like*, *HzNVorf9-like1*, *HzNVorf89-like*, *pif-3* and *PmV-like*) are organized as a cluster, which is likely to constitute a remnant of the virus integrated in the ancestral wasp genome. **B:** Semi-quantitative RT-PCR analysis of *Cc50C22.7* gene expression during male and female pupae development (d: days) and in ovaries (o: oviduct; c: calyx). EF1α gene expression was used as internal reference to normalize mRNA quantities. The gene Cc50C22.7 is specifically expressed in female wasps in the calyx region of the ovaries where bracovirus particles are produced, but not in males. This gene belonging to the nudiviral cluster has no homolog in sequenced nudivirus genomes. However, its expression correlates temporally with bracovirus particle production. Altogether this suggests that Cc50C22.7 originated from the genome of the ancestral nudivirus captured and belongs to the 'nudivirosphere'.

genes) are homologs of genes coding for two structural proteins of *Campoletis sonorensis* ichnovirus (CsIV) particles that had previously been identified (Deng *et al.*, 2000). The genes coding for particle components do not have introns and are organized in particular gene-rich regions of the wasp chromosomes called the ichnovirus structural protein encoding regions (IVSPERs), and the IVSPERs structure resembles that of the *C. congregata* nudiviral cluster. The three IVSPERs identified share related genes belonging to seven families and could correspond to remnants of a virus genome duplicated in several regions of the wasp genome (Volkoff *et al.*, 2010; Volkoff *et al.*, Chapter 3 of this volume). As nudiviral genes of the microgastroid complex, the IVSPER genes are conserved among ichnovirus-associated wasps. IVSPER gene transcripts have been found in the ovaries of *Tranosema rostrale* (Volkoff *et al.*, 2010) and are present in a cDNA library (whole body) of *C. sonorensis* (available in Genbank; Volkoff *et al.*, Chapter 3 of this volume). Overall the amount of data obtained on the machinery producing ichnovirus particles is comparable to that obtained for bracoviruses. These data suggest that the IVSPER gene set most likely also originated from the genome of a viral ancestor. However, the lack of similarity between these genes and known viral genes did not allow the identification of the ichnovirus ancestor among known pathogenic viruses. This may be accounted for by gaps in our knowledge of viruses infecting insects, resulting in the absence of their sequences from databases. If this interpretation is correct, the situation may change as new generation sequencing approaches could allow broader access to biodiversity. One might predict this will result in the identification of new invertebrate viruses, possibly including members of the ichnovirus ancestor family. However this family might be extinct in the form of pathogenic viruses.

The most striking conclusion from the results already obtained is that the machineries producing ichnovirus and bracovirus particles are completely different. Thus the domestication of viruses has obviously occurred at least twice independently during the evolution of parasitoid wasps. A similar approach could determine whether the particles of *Glypta fumiferanae* virus (GfV), a proposed intermediate between ichnoviruses and bracoviruses, are produced by a machinery resembling that of ichnoviruses or bracoviruses, or a combination of both, or even involve a third virus type and different replication machinery (Cusson *et al.*, Chapter 6 of this volume).

WHY ARE NO GENES FROM THE PACKAGED GENOME RELATED TO NUDIVIRUSES?

Based on current data, we propose a scenario to explain the different transformation events of the ancestral nudivirus originally integrated in the wasp genome leading ultimately to the evolution of contemporary bracoviruses. This is of course highly speculative but useful to stimulate further analyses. Conceivably, the ability of the virus genome to be packaged in the particles could have been lost in relatively few simple molecular events. The translocation of regulatory sequences, allowing for example virus genome excision from the chromosome and/or the DNA entry in the particles during packaging, could explain why no nudiviral sequences are found in bracovirus particles. Since this process could occur independently in different wasp lineages, different stages of genomic transformation could be observed. This is the case in ichnoviruses for a region encoding virion structural genes which is packaged in *C. sonorensis* while the homologous region of *H. dydimator* is amplified during particle production but absent from viral particles (Volkoff *et al.*, 2010) showing evidence for the loss of packaging regulatory sequences in *H. dydimator* sequences. Strikingly, none of sequenced bracovirus packaged genomes (*Microplitis demolitor* bracovirus MdBV, CcBV, *Glyptapanteles indiensis* bracovirus GiBV, *G. flavicoxis* bracovirus GfBV; as described in Desjardins *et al.*, 2008; Espagne *et al.*, 2004; Webb *et al.*, 2006) as well as those partially sequenced (CiBV, CvBV; Choi *et al.*, 2005; Wyder *et al.*, 2002) associated with wasps from different subfamilies of the microgastroid complex contain any nudiviral genes. In comparison, obligatory symbionts of prokaryotic origin still retain in their genome some genes from their free-living ancestor (Moran, 2007). This lack of nudiviral genes suggests the whole ancestral virus genome could have been lost by the particles as a block, following a single event, early in the evolution of bracoviruses. Regulatory sequence translocation might have resulted in the amplification and packaging of a wasp genomic region instead of the virus genome in the particles. This novel chimeric virus could have been maintained if it gave an immediate selective advantage to the wasp during parasitism or had protected the wasp from super-infections by related viruses. For example, low-level expression of wasp genes in the host conferring protection of the wasp larvae developing in the lepidopteran host would have readily been selected for. The fact that the parasite and the host are both insects and use similar transcription factors might have ensured the initial expression of wasp genes in lepidopteran cells, at least at a low level. Mutations of the promoters could later have been selected leading to an increased mRNA expression in the host tissues. The organization of the packaged genome in multiple circles, a feature unique to PDVs among viruses, may have arisen following a series of duplications from the genomic region initially packaged. Several elements can explain the presence of duplicated regions in PDV genomes. (1) Duplications occur spontaneously, more often than usually thought, as shown by comparison of genomes from closely related species (Francino, 2005). They are a common mode of chromosomal gene evolution (Zhang, 2003) and bracovirus proviral sequences are parts of wasp chromosomes. (2) The products of duplications are easily packaged since the bracovirus genome has a modular structure. It has been shown that each bracovirus circular molecule is packaged in a single nucleocapsid for CiBV (Albrecht *et al.*, 1994) and MdBV (Beck *et al.*, 2007). Thus a new segment produced by duplication should be packaged provided that it contains the recognition sequences allowing DNA entry in the particle. This plasticity has been shown for CcBV for which two segments, CcBVS2 and CcBVS31, differ by the integration of several mobile elements. The larger segment is still packaged even though it is twice the size (30kb) of the segment without insertions (Dupuy *et al.*, 2011). In theory, the only limitation for the number of circles that can be packaged

is the quantity of circular DNA and structural proteins that are produced in calyx cells producing the particles. Thus, a newly duplicated region of the provirus could be transferred to the host and possibly contribute to host–parasitoid interaction. (3) Duplications are useful: they play a central role in the model of adaptive evolution for the acquisition of a new gene function (Francino, 2005). According to this theory, duplications could compensate low production or activity of gene products with increased gene copy numbers. Then, through a competitive phase between the different copies, a more efficient product should be selected. Thus, from an early stage in the PDV–wasp association, duplications in the proviral locus may have contributed to the adaptation of PDV molecules originating from the wasp to lepidopteran host targets (Bézier *et al.*, 2008). Later duplications could have been maintained to mediate the adaptive evolution of the packaged genes in response to changes in host molecules either in a host–parasite arms race or with parasitoid host shifts (Bézier *et al.*, 2008).

SOME GENES OF THE BRACOVIRUS PACKAGED GENOME COME FROM THE WASP GENOME

The packaged genes are thought to have a wasp origin, however it should be noted that this hymenopteran origin is rarely documented. PDV packaged genes are highly divergent from insect sequences and they generally do not appear closer to insect than to human sequences. This is probably due to their involvement in a host–parasitoid arms race causing rapid sequence evolution (Bézier *et al.*, 2008).

However, a wasp origin has been shown in the case of the GiBV sugar transporter genes. These genes were specifically acquired by bracoviruses in the genus *Glyptapanteles* (Desjardins *et al.*, 2008). They are not present in bracovirus particles of closely related species from the *Cotesia* genus. They display a clear phylogenetic link with cellular wasp genes. This shows that cellular copies of these genes have been transferred, via an unknown mechanism, to the chromosomal region containing the integrated form of the DNA packaged in the particles (proviral genome, Desjardins *et al.*, 2008). The CiBV packaged genome content also sustains the view that the physiological relationship with the host determines the type of genes maintained in the particles. No homologs of the CiBV gene have been found in any other sequenced bracoviruses (Weber *et al.*, 2007), even though many CiBV circles (nine) have already been sequenced. Conversely *ptp* and *ankyrin* genes are widespread in bracovirus genomes and they are absent from CiBV. This lack of gene content overlap might be explained by the particular life history traits of the Cheloninae. *C. inanitus* females oviposit in the eggs, thus their larvae encounter a different environment to those of wasps ovipositing directly in caterpillar larvae. The only sequences conserved across all bracovirus packaged genomes are the direct repeat junctions (DRJ) regulatory elements present on each circle. These sequences are involved in DNA molecular circularization. Circularization is a late step in the process of virus particle production (Pasquier-Barre *et al.*, 2002) that might be coupled with DNA encapsidation, like in several double-stranded DNA viruses. During the replication, baculoviruses produce genome concatemers (Oppenheimer and Volkman, 1997) as well as highly branched structures, thought to be separated in monomers during encapsidation by very late factor-1 (VLF-1), a tyrosine recombinase located at the extremity of the nucleocapsid (Vanarsdall *et al.*, 2006). Thus it appears of the utmost importance to conserve the regulatory sequences required to incorporate the dsDNA circle in the particles in the packaged genome rather than the genes themselves. Notably the machinery producing the particles (nudiviral gene products), most likely interacting with these regulatory sequences, is also well-conserved.

SOME GENES OF THE BRACOVIRUS PACKAGED GENOME COME FROM MOBILE ELEMENTS OR OTHER VIRUSES

In addition to genes of wasp origin some genes present in the particle dsDNAs originated from mobile elements and viruses (Gundersen-Rindall, Chapter 8 of this volume). Mobile elements have most probably inserted themselves in the chromosomal form of the PDV genome like they do anywhere in the wasp genome. The proviral form of packaged DNAs in the wasp chromosomes constitutes a large target, several hundred kilobases long, for transposable elements (TEs) insertions. Interestingly, when a mobile element is inserted in the proviral form it can be packaged, thus PDVs could be used as vehicles for horizontal transfer of TEs between Hymenoptera and Lepidoptera genomes (Dupuy *et al.*, 2011) provided that the parasitized host could escape parasitism and reproduce. This is not generally the case, but it might occur rarely.

Some TE genes could also have been mobilized to contribute to host–parasite interactions. For example a *Toxoneuron nigriceps* PDV gene encoding a retroviral aspartyl transferase is expressed in parasitized larvae, suggesting it might play a role in host physiology manipulation (Falabella *et al.*, 2003). It is noteworthy that most of the genes described as bracovirus-related in the genome of *G. fumiferanae* polydnavirus (Lapointe *et al.*, 2007) resemble those of mobile elements. This indicates that these elements are present in the genomes of both braconid and ichneumonid wasps, and therefore do not provide clues on the relatedness of this proposed new type of virus (banchine ichnovirus) to bracoviruses.

In addition to retroelements remnants, a few CcBV packaged genes share similarities with virus genes that could have been misleading with respect to the origin of

bracoviruses. In particular, *DNA polymerase B* and *ATPase A32L* genes belonging to the remnants of a *Maverick* transposable element were striking (Drezen *et al.*, 2006). These two genes belong to the nine core genes of the nucleocytoplasmic large DNA viruses (NCLDV) (Iyer *et al.*, 2001) used for viral phylogenies. Upon superficial examination, they could have been misleading clues in considering the bracovirus origin. *Mavericks* are abundant in the genome of *Nasonia vitripennis* and were found in non-PDV sequences of several parasitoid species suggesting that the CcBV truncated *Maverick* is a copy of a mobile element component of *C. congregata* genome that most probably inserted randomly in the chromosomal form of CcBV (Dupuy *et al.*, 2011). The *DNA polymerase B* and *ATPase A32L* genes, therefore do not yield information pertinent to the origin of bracoviruses.

Surprisingly, CcBV also contains genes having similarities with two virus genes not deriving from nudivirus genomes. Two putative products have significant similarity with p94 (48% similarity with p94 of *Autographa californica* multiple nucleopolyhedrovirus, AcMNPV), a non-essential baculovirus protein of unknown function. According to phylogenetic analysis, these p94 protein truncated forms are most closely related to the lepidopteran *Xestia c-nigrum* granulovirus (55% similarity) suggesting that they could have been acquired relatively recently by the PDV from a granulovirus (Drezen *et al.*, 2006). In *Cotesia vestalis, G. indiensis* and *G. flavicoxis* bracoviruses, the *p94* gene is not truncated and the corresponding protein is potentially functional, suggesting it might play a role in host–parasite interactions. Such virus gene acquisition probably occurs rarely following the infection of a parasitized caterpillar by a pathogenic virus. The same scenario could explain the similarity of a third protein (CcBV 32.6) to a hypothetical protein conserved in ascoviruses (39.9% with *Spodoptera frugiperda* ascovirus SfAV1 ORF 14 product) a group of lepidopteran-infecting large dsDNA viruses (Drezen *et al.*, 2006). Lateral gene transfer could also explain the presence of pox-D5 family of nucleoside triphosphatasases shared between ascoviruses and *G. fumiferanae* PDV (Bigot *et al.*, 2008). Moreover, the common origin of ankyrin domain containing proteins of ichnoviruses and bracoviruses (Falabella *et al.*, 2007), designated either as *vankyrins*, *IκB-like* or *cactus-like* gene products, might be explained by the occurrence of lateral gene transfers between PDVs following oviposition in the same host of PDV-associated ichneumonid and braconid wasps. In summary, pathogenic and symbiotic viruses share the same ecological niches defined by their infected host, which could result in lateral gene transfers. Even if they occur infrequently and thus could not be detected experimentally when different PDV associated wasp species were used to parasitize the same host (Stoltz, 1990), such transfers would explain that some genes are shared by ichnoviruses

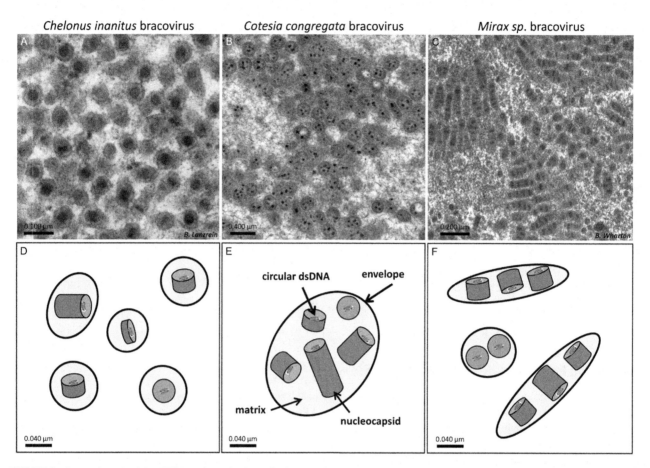

FIGURE 2   Bracovirus particles. TEM sections of calyx cells from ovaries of *C. inanitus* (Cheloninae, A), *C. congregata* (Microgastrinae, B) and *Mirax* sp. (Miracinae, C) pupae. (D to F) Schematic representation of the corresponding bracovirus particles. Virions consist of enveloped cylindrical electron-dense nucleocapsids of uniform diameter (34–40 nm) but of variable length (8–200 nm length) embedded in a protein matrix. They may contain one (D) or several dispersed (E) or organized (F) nucleocapsids.

and bracoviruses or between PDVs and other viruses. The occurrence of horizontal gene transfers emphasizes the necessity to identify a large set of genes from the same origin to draw conclusions about the nature of the ancestral virus from which PDVs originated.

BRACOVIRUSES AND NUDIVIRUSES HAVE SIMILAR NUCLEOCAPSID MORPHOLOGY

Previous attempts based on the sequencing of the genome packaged in the particles had failed to identify the relation between bracoviruses and nudiviruses, simply because no nudiviral genes remain in the particles. This complete dichotomy between the genes involved in virus particle production and those present in virus particles is a characteristic feature of bracoviruses. Indeed at least one gene encoding particles components is packaged in the ichnovirus CsIV (Deng and Webb, 1999) and a gene family (*N-gene* family) code for components of the particles as well as proteins expressed in the host from the packaged genome (Volkoff *et al.*, 2010).

Remarkably, molecular data have at long last confirmed the initial predictions on the nudiviral origin of bracoviruses, even though the sequencing of the DNAs packaged in the particles had previously cast some doubts on the link between PDVs and viruses (Espagne *et al.*, 2004). The fact that PDVs associated with braconid wasps and *Oryctes rhinoceros* nudivirus (OrNV) showed morphological similarities had already been reported thirty years ago, when PDVs were initially characterized and OrNV was described as a non-occluded baculovirus (Stoltz and Vinson, 1979). These results have regained attention since the identification of the nudiviral genes involved in bracovirus particles production was published (Bézier *et al.*, 2009; Stoltz and Whitfield, 2009). In different braconid wasp species, the bracovirus particle morphology differs in the number and organization of nucleocapsids, although they are produced by the same conserved viral machinery. The particles may contain one nucleocapsid (CiBV, MdBV) surrounded by a protein matrix and enclosed by a membrane envelope, or several nucleocapsids in species of the *Cotesia* and *Mirax* genera (Fig. 2). The nucleocapsids are either dispersed (CcBV) or organized in a

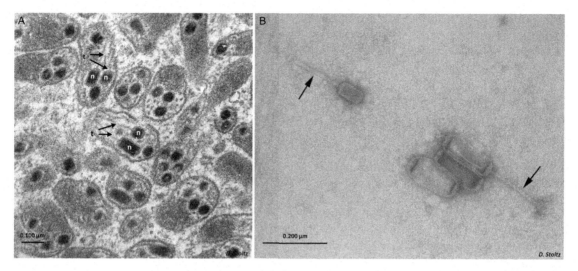

FIGURE 3   Tail-like appendages found in bracovirus and nudivirus particles. A: Bracovirus particles from *Cotesia melanoscela* calyx fluid visualized using TEM. Filamentous structures corresponding to tail-like appendages (t) can be seen attached to nucleocapsids (n). **B:** Electronic micrograph using negative staining that shows the tail-like appendages (arrows) associated with a *C. melanoscela* bracovirus nucleocapsid (upper left) and *Oryctes rhinoceros* nudivirus capsid (lower right, reproduced with permission from *Science*).

structure resembling an ascus (PDV from the *Mirax* species). Each nucleocapsid contains a single circular DNA molecule as shown by TEM analysis of DNA released from particles of *C. inanitus* (Albrecht *et al.*, 1994) and *M. demolitor* (Beck *et al.*, 2007). One striking feature of the nucleocapsids is a tail-like appendage, a particularity that is shared only with nudiviruses among viruses infecting eukaryotes (Stoltz and Whitfield, 2009; Fig. 3). Bracovirus particles enter host cells by penetrating the basement membrane and translocating to nuclear pores where they release their associated DNA. The nucleocapsid 'tail' is supposedly involved in the introduction of bracovirus DNA into the nucleus as suggested by electronic microscopy observations of host cells where tails were inserted in nuclear pores (Stoltz and Vinson, 1979; Webb and Strand, 2005). Except for the larger size of OrNV, the nucleocapsid structures of the bracoviruses from MdBV (Beck *et al.*, 2007) and from *Cotesia melanoscela* bracovirus CmBV (Krell and Stoltz, 1979) are overall very similar to that of the nudivirus (Stoltz and Whitfield, 2009; Fig. 3). Moreover, the protein profiles of OrNV and CmBV showed bands of similar sizes, suggesting the conservation of major capsid components (Krell and Stoltz, 1979).

BRACOVIRUS PARTICLES ARE PRODUCED AND ENVELOPED WITHIN THE NUCLEUS: A CHARACTERISTIC FEATURE SHARED WITH NUDIVIRUSES AND BACULOVIRUSES

Baculoviruses produce two types of viral particles: the budded virus form, involved in the cell-to-cell infection

spreading, and the Occlusion-derived virus form (ODV), which is released in the field for infection of the pest caterpillar upon feeding. ODV production typically induces an extensive elaboration and proliferation of intranuclear membranes that appear as microvesicles and membrane structures within the nucleoplasm (Hong *et al.*, 1994). Trafficking and assembly of intranuclear microvesicles and the ODV envelope have been studied using AcMNPV envelope proteins linked to green fluorescent protein as markers. The results suggest that movements of these proteins are mediated through cytoplasmic membranes and the nuclear envelope (Hong *et al.*, 1997). ODV envelope proteins are incorporated into the endoplasmic reticulum and transported to the outer and inner nuclear membrane then they are detected in microvesicles and finally in ODV envelopes (Hong *et al.*, 1997). Particle production has been reported for OrNV after cultured cells infection; compared to baculovirus the originality resides in the fact that envelope and empty shells appear first in the nucleus before nucleocapsids become visible (Crawford and Sheehan, 1985). This situation has also been described for the nudivirus infecting the hemocytes of the spotted cucumber beetle *Diabrotica undecimpunctata* (Kim and Kitajima, 1984). For bracoviruses, viral envelopes also typically appear in the nucleus before the formation of nucleocapsid-containing virions (De Buron and Beckage, 1992; Pasquier-Barre *et al.*, 2002). As in nudiviruses, no intermediary step of microvesicle formation has been detected (Wyler and Lanzrein, 2003). Many envelope proteins are shared between baculoviruses, nudiviruses, and bracoviruses (see below) suggesting that membrane formation might occur via a common mechanism. Herpesviruses also acquire an envelope in the nucleus but by a different mechanism, involving budding from the inner nuclear membrane (Bibor-Hardy *et al.*, 1982), while baculoviruses, nudiviruses, and bracoviruses acquire their envelope within the nucleoplasm.

NUDIVIRAL mRNAs ARE EXPRESSED IN BRACONID WASP OVARIES DURING PARTICLE PRODUCTION

Molecular analyses recently substantiated the similarities in particle morphology and in the characteristic virion production seen in nudiviruses and bracoviruses. Particles are produced in very large numbers by specialized cells of the calyx, resulting in a fluid filled with particles. Due to this high level of particle production, it was hypothesized that the mRNAs encoding the proteins involved in particle morphogenesis should be highly expressed in calyx cells. Large-scale sequencing of a cDNA library from the ovaries was undertaken to identify these mRNAs. Two braconid wasps belonging to different subfamilies of bracovirus-associated species, namely *Chelonus inanitus* (Ichneumonoidea, Braconidae, Cheloninae) and *Cotesia congregata* (Ichneumonoidea, Braconidae, Microgastrinae) were studied. The two species are distantly related. Within the microgastroid complex (the monophyletic group of bracovirus-associated wasps), the Cheloninae have a basal position in the tree, while the Microgastrinae corresponds to a more derived subfamily (Whitfield, 2002). For each species, 5000 expressed sequence tags (ESTs) from the ovaries were sequenced and the deduced proteins were compared with public database sequences. A large set of mRNAs with similarities with virus sequences was identified (Bézier *et al.*, 2009). The predicted products have similarities with those of 25 nudivirus genes. Fifteen genes belong to the set of 20 core genes conserved in all nudiviruses and baculoviruses, the most important genes for the virus cycle (Table 1). Unfortunately, our knowledge of nudiviral genomes and biology is fairly limited compared to baculoviruses. Three nudivirus genomes have been sequenced recently, *Heliothis zea* nudivirus-1 (HzNV-1), *Gryllus bimaculatus* nudivirus (GbNV) and *Oryctes rhinoceros* nudivirus (OrNV). Little is known about the function of nudivirus proteins except for those conserved in baculoviruses. Hopefully, many nudiviral genes expressed in braconid wasp ovaries have baculovirus homologs, the functions of which have already been experimentally studied in baculoviruses (Table 1); for reviews, see Rohrmann (2008) and Wang and Jehle (2009). It is thus possible to infer that nudiviral genes may have a similar role during bracovirus particle production.

Genes Involved in Transcription

Four gene products are similar to proteins that play an important role in gene transcription in baculoviruses.

Among these, late expression factor-4 (LEF-4; Ac90), LEF-8 (Ac50) and p47 (Ac40) constitute three out of four subunits of the viral RNA polymerase involved in the expression of late and very late genes during baculovirus infection (Guarino *et al.*, 1998). The fourth subunit required to constitute the baculovirus RNA polymerase complex, LEF-9 (Ac62), was not identified in *C. congregata* and *C. inanitus* libraries. However a *lef-9* homolog was recently identified using deep sequencing analysis (454) of cDNAs from *T. nigriceps* ovaries, another bracovirus-associated wasp (Fallabella, unpublished data). Altogether, the data strongly suggest that bracoviruses encode their own viral RNA polymerase that might be used to transcribe bracovirus genes of nudiviral origin in the wasp genome. Bracovirus *p47* gene expression in the wasp *C. congregata* was studied by quantitative real-time PCR. This gene is highly upregulated in females in the calyx region of the ovaries where particles are produced (Bézier *et al.*, 2009). The *p47* maximum gene expression is reached at day four, when the onset of particles morphogenesis is observed using TEM (Pasquier-Barre *et al.*, 2002), but expression is already detected at day three and thus precedes the expression of the other nudiviral genes tested. This different timing is compatible with the hypothesis that the nudiviral RNA polymerase would control the transcription of other nudiviral genes. In baculoviruses, the genes expressed early after infection are transcribed by the RNA polymerase of the infected cells (early genes) and the viral RNA polymerase is involved in the expression of genes transcribed later (late and very late genes). However, the regulatory sequences described involved in baculovirus late and very late gene expression during infection appear too short to confer a specificity in the context of the higher complexity of the whole insect genome and a more precise mechanism appears necessary to restrict the enhanced expression to nudiviral genes in the calyx cells (Bézier *et al.*, 2009).

The protein LEF-5 (Ac99) is an initiation factor of baculovirus late gene transcription (Guarino *et al.*, 2002; Passarelli and Miller, 1993). *Lef-5* is a core gene of baculoviruses and nudiviruses (Wang and Jehle, 2009) that belong to the unique cluster of genes conserved in baculoviruses from Lepidoptera, Diptera and Hymenoptera (Herniou *et al.*, 2003). *C. congregata* LEF-5 shares low but significant similarity with the products of homologs from baculovirus and nudivirus *lef-5* gene. Braconid LEF-5 has a shorter LEF-5 baculovirus domain than baculovirus gene products but the potential zinc ribbon domain is present at the C-terminal end; this domain, thought to interact with nucleic acids, is important for the level of LEF-5 activity (Harwood *et al.*, 1998). It is noteworthy that no nudiviral gene product involved in transcription has been found in purified bracovirus particles, which is consistent with their role during early steps of particle production.

TABLE 1 Nudiviral Genes Identified in Braconid Wasps and their Homologs in Nudiviruses and in the Baculovirus AcMNPV

Protein functions	Gene name	Cotesia congregata				Chelonus inanitus			Orf # or Presence in Baculo- or Nudiviruses				
		cDNA	gene #	integration	part.	cDNA	gene #	part.	AcMNPV	HzNV-1	GbNV	OrNV	PmNV
Nudivirus/Baculovirus Core Genes													
Transcription													
RNA polymerase subunit	p47	+	1	–		nd			40	75	69	20	nd
RNA polymerase subunit	lef-8	+	1	–		+	1		50	90	49	64	nd
RNA polymerase subunit	lef-4	nd				+	1		90	98	96	42	nd
Transcription initiation factor	lef-5	+	1	–		nd			99	101	85	52	+
Packaging and assembly													
Very late factor	vlf-1a	nd				+	1	Δ	77	121	80	30	+
	vlf-1b	nd				+	1			144	57	75	+
Viral capsid protein	vp91/p95	nd				+	1	Δ	83	46	2	106	nd
Viral capsid protein	vp39	+	1	C	▲	+	1	▲	89	89	64	15	nd
Viral phosphatase	38K	+	1	C	▲	+	1	▲	98	10	1	87	+
ODV envelope components													
Per os infectivity factor	pif-0 (p74)	+	1	–		+	1	Δ	138	11	45	126	nd
Per os infectivity factor	pif-1	nd				+	1	Δ	119	55	52	60	nd
Per os infectivity factor	pif-2	nd				+	1	Δ	22	123	66	17	nd
Per os infectivity factor	pif-3	+	1	C		nd			115	88	3	107	nd
Per os infectivity factor	pif-4 (19kDa)	+	1	–		+	1	Δ	96	103	87	33	nd
Occlusion derived virus	pif-5 (odv-e56)	+	2	–		+	1	Δ	148	76	5	115	nd
Unknown	Ac68	nd	1	C		nd			68	74	55	72	nd

(Continued)

TABLE 1 (Continued)

Protein functions	Gene name	Cotesia congregata				Chelonus inanitus			Orf # or Presence in Baculo- or Nudiviruses				
		cDNA	gene #	integration	part.	cDNA	gene #	part.	AcMNPV	HzNV-1	GbNV	OrNV	PmNV
Nudivirus/Lepidopteran Baculovirus Core Genes													
ODV envelope components *Occlusion derived virus*	odv-e66	+	5		▲	+	3	▲△	△ 46	–	–	12	nd
Nudivirus Specific Genes													
Unknown	HzNVorf9-like	+	2	C	▲▲	+	2	▲▲	–	9	–	–	nd
	HzNVorf64-like (p51)	+	1			nd			–	64	–	–	+
	HzNVorf94-like	+	1	C		nd			–	94	–	–	nd
	HzNVorf106-like	+	1		▲	+	1	▲	–	106	–	–	nd
	HzNVorf118-like (PmV-like)	+	1	C	▲	+	1	▲	–	118	–	–	+
	HzNVorf124-like	+	1			+	5		–	124	95	41	nd
	HzNVorf128-like	+	1	I		+	1		–	128	–	–	nd
	HzNVorf140-like	+	1	I		+	1	△	–	140	–	–	nd
Unassigned nudiviral clustered genes	Cc50C22.5	nd	1	C		nd			–	–	–	–	nd
	Cc50C22.6	nd	1	C		nd			–	–	–	–	nd
	Cc50C22.7	+	1	C		+	1	△	–	–	–	–	nd

The predicted ORFs in baculovirus and nudiviruses are indicated by their number (Orf #). The protein function is that determined for the baculovirus gene product. –, gene absent; +, gene present; nd, gene not isolated; gene #, number of paralogs found; I, nudivirual gene isolated in the wasp genome; C, gene of the nudiviral cluster; part., particle component; △, protein identified as bracovirus particle component; ▲, protein identified as particle component in both bracoviruses; ▲▲, proteins of the two paralogs identified as particle component in both bracoviruses; AcMNPV, Autographa californica multiple nucleopolyhedrovirus; HzNV-1, Heliothis zea nudivirus-1; GbNV, Gryllus bimaculatus nudivirus; OrNV, Oryctes rhinoceros nudivirus; PmNV, Penaeus monodon nudivirus.

Genes Involved in Nucleocapsid Packaging and Assembly

Four genes expressed in braconid wasp ovaries are potentially involved in packaging and nucleocapsid assembly. One gene, *HzNVorf10-like*, encodes a protein that shares 51% similarity with its HzNV-1 homologue, 38K (Ac98), a core gene of baculoviruses and nudiviruses (Wang and Jehle, 2009). Deletion of *38K* in knock-out experiments resulted in interruption of nucleocapsid assembly of the deleted baculoviruses (Wu *et al.*, 2006). It is noteworthy that CiBV 38K and HzNVORF89-like products are the major components of CiBV particles based on the intensity of the bands obtained on SDS-Page gels of purified particles (Wetterwald *et al.*, 2010). The latter was proposed to correspond to the homolog of baculovirus *vp39* gene (Wang and Jehle, 2009) but the similarity between the proteins is low. However, the orthologous relationship between these genes is more convincing taking into account the fact that VP39 is the major baculovirus capsid protein. Interestingly, both putative major bracovirus nucleocapsid components, 38K and VP39, are encoded by the nudiviral cluster in the *C. congregata* genome. Another gene encodes a predicted protein that shares 38% similarities with *GbNVorf2* gene product, described as a baculovirus VP91/P95 (Ac83) homolog, a baculovirus core gene that is also conserved in all the fully sequenced nudivirus genomes (Wang and Jehle, 2009). VP91/P95 is a nucleocapsid associated protein (Russell and Rohrmann, 1997; Braunagel *et al.*, 2003; Perera *et al.*, 2007). The relationship between bracovirus, nudivirus, and baculovirus VP91/P95 proteins can be confirmed easily by psi-blast analysis using *C. inanitus* sequence as a query from the first iteration (data not shown).

A fourth gene product shares 43% similarity with the very late factor-1 (VLF-1, Ac77) from OrNV (OrNVORF030). VLF-1 is a putative tyrosine recombinase belonging to the lambda integrase (Int) family, a large group of site-specific DNA recombinases that catalyze DNA rearrangements in a variety of organisms. *Vlf-1* is an essential baculovirus gene (McLachlin and Miller, 1994). It was initially described as a factor involved in baculovirus very late gene transcription and also implicated in genome processing (Yang and Miller, 1998). More recently it was shown to constitute a structural component of nucleocapsids and to play an important role in capsid assembly (Li *et al.*, 2005; Vanarsdall *et al.*, 2006). In a mutant lacking a tyrosine at the recombination catalytic site, the nucleocapsids remain trapped in the nucleus suggesting that VLF-1 could be involved in the final stage of the packaging process. It was proposed that VLF-1 resolves the complex DNA structure (highly branched structures comprising concatemers of genomes) produced during baculovirus genome replication, thus allowing the incorporation of a single molecule

of genome in each nucleocapsid. The bracovirus protein contains a phage integrase domain (e-value = 0.00017 using motif scan program, pfam HMMs at http://myhits. isb-sib.ch/cgi-bin/motif_scan) and has the conserved tyrosine residue at the recombination catalytic site, like nudivirus and baculovirus proteins. It could therefore be involved in bracovirus nucleocapsid assembly and maturation. In particular, this protein may play a role in the production of the DNA circles, which are individually packaged in nucleocapsids following the resolution of larger replication intermediates (Pasquier-Barre *et al.*, 2002).

In nudiviruses, *vlf-1* gene homologs form a multigenic family with various degrees of sequence similarities and/or integrase domain conservation. The braconid protein HzNVORF144-like also displays a phage integrase domain (e-value = 0.00012), with the conserved tyrosine residue while the *HzNVorf140-like* gene products identified in *C. inanitus* and *C. congregata* are similar to that of VLF-1 but they have no detectable integrase domain and no conserved tyrosine. Finally, upon closer phylogenetic examination (Fig. 4), it is not clear which braconid gene (CiBV *vlf-1* and *HzNVorf144-like*) is the genuine baculovirus ortholog. Thus we propose to use the name *vlf-1a* (CiBV *vlf-1*) and *vlf-1b* (*HzNVorf144-like*) to designate the two braconid homologs of the baculovirus *vlf-1* gene coding for a protein with the integrase domain. An interesting possibility is that the product of one of these genes could have the functions of VLF-1 while the other could mediate the integration of DNA circles into host cells (see Gundersen-Rindal, Chapter 8 of this volume).

Genes Expressed in Braconid Wasp Ovaries are Similar to Baculovirus Occlusion Derived Virus Components

Following baculovirus infection and death of infected caterpillars shedding virus, ODVs persist in the field and are ingested by caterpillars feeding on contaminated leaves. Although bracovirus particles do not persist in the field and are not ingested by the caterpillars but injected directly into the host hemolymph or eggs (in the case of egg–larval parasitoids), some of the genes involved in particle production are related to those coding for ODV components in baculoviruses.

Six predicted proteins share similarity with baculovirus ODV envelope components (Braunagel *et al.*, 2003; Braunagel and Summers, 2007) conserved in nudiviruses: ODV-E56 (PIF-5, HzNVORF76, Ac148), p74 (PIF-0, HzNVORF11, Ac138), PIF-1 (HzNVORF55, Ac119), PIF-2 (HzNVORF123, Ac22), PIF-3 (HzNVORF88, Ac115) and 19 kDa (PIF-4, HzNVORF103, Ac96). Thus, bracoviruses produce the six 'per os infectivity factors' (PIFs) required for baculovirus primary infection in the midgut. In baculoviruses, deletion of any of these proteins results in a profound

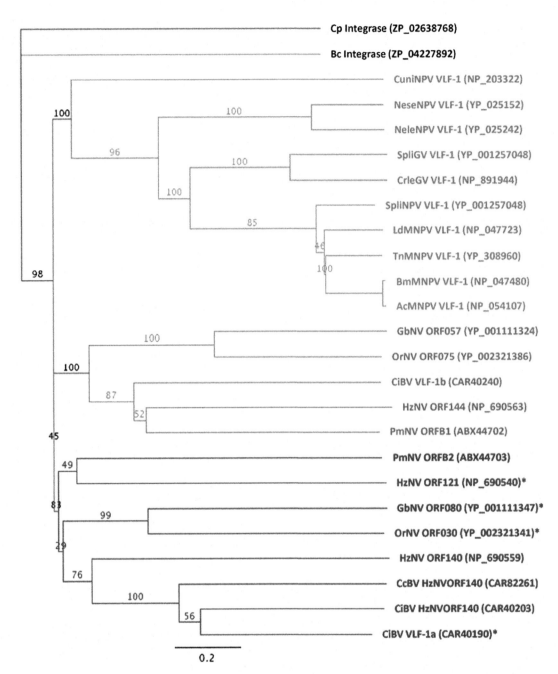

FIGURE 4: **Phylogenic tree of the integrase homolog sequences found in baculoviruses, nudiviruses and bracoviruses.** The maximum likelihood tree was obtained using PhyML software and WAG model (100 bootstraps). Phage integrase sequences from *Clostridium perfringens* (Cp) and *Bacillus cereus* (Bc) were used as outgroup. Accession numbers are indicated in brackets. *Nudivirus and bracovirus ORFs products that have been registered as VLF-1 in databases. PmNV ORFs are numbered according to their order on the sequenced genome fragments (ex: B2: second ORF on BamHI fragment B). CuniNPV, *Culex nigripalpus* nucleopolyhedrovirus; NeseNPV, *Neodiprion sertifer* NPV; NeleNPV, *Neodiprion lecontii* NPV; SpliGV, *Spodoptera litura* granulovirus; CrleGV, *Cryptophlebia leucotreta* GV; SpliNPV, *Spodoptera litura* NPV; LdMNPV, *Lymantria dispar* multiple nucleopolyhedrovirus; TnSNPV, *Trichoplusia ni* single NPV; BmNPV, *Bombyx mori* NPV; AcMNPV, *Autographa californica* MNPV; GbNV, *Gryllus bimaculatus* nudivirus; OrNV, *Oryctes rhinoceros* NV; HzNV, *Heliothis zea* NV; PmNV, *Penaeus monodon* NV; CiBV, *Chelonus inanitus bracovirus*; CcBV, *Cotesia congregata* BV.

decrease in the oral infectivity. It was shown recently that PIF-1, PIF-2, PIF-3 and p74/PIF-0 form a stable protein complex detectable by immunoprecipitation and scattered on the surface of the ODV (Peng *et al.*, 2010). They are also conserved in the recently described salivary gland

hypertrophy viruses (SGHVs) infecting Diptera (Abd-Alla *et al.*, 2008; Garcia-Maruniak *et al.*, 2008). ODV-E56/PIF-5 has been shown recently to be required for baculovirus oral infectivity but its function does not involve binding or fusion (Sparks *et al.*, 2010). The function of PIF-4 (19 kDa) is still unknown. The six PIFs are conserved in nudiviruses. This indicates that their function in virus entry is conserved among different virus families. Remarkably, the PIFs targets appear not to be limited to midgut cells, since the nudivirus HzNV-2 is sexually transmitted (Raina *et al.*, 2000) and since bracoviruses are injected into the hemocoel.

A series of genes coding for different copies of ODV-E66 (Ac46) are expressed in braconid wasp ovaries (five in *C. congregata* and two in *C. inanitus*, data not shown). ODV-E66 proteins are not conserved in all baculoviruses but they are components of ODV envelopes conserved in lepidopteran baculoviruses (Jehle *et al.*, 2006) and a gene is present in the genome of OrNV (*OrNVorf012*; Table 1; Wang *et al.*, 2011). ODV-E66 has a hyaluronan lyase activity thought to be important for the penetration of extracellular barriers to access host cells (Vigdorovich *et al.*, 2007). The presence of ODV-E66 copies is not a specificity of bracovirus genomes since two variants of ODV-E66 are also found in several baculovirus genomes (accession numbers: NC_009011, NC_011616, NC_002169, NC_003529 and NC_004117); however, this trend to diversification is more pronounced in bracoviruses. Several variants of other nudiviral genes are expressed in *C. congregata* and/or *C. inanitus*: two for *HzNVorf9* and *odv-e56* and five for *HzNVorf124*. The diversification of nudiviral genes in families might reflect specific interaction of the particles with different tissues and hosts (a wasp species might parasitize several lepidopteran species in the field). Moreover the chromosomal transmission of bracovirus genes might favor the conservation of duplicated genes because the constraint operating on the size of the viral genome is irrelevant. In contrast to baculoviruses, for which all the genes necessary for their replication need to be packaged in the nucleocapsids, the bracovirus machinery is chromosomally integrated, and therefore the genes are not required in the infectious particles.

BRACOVIRUS PARTICLES ALSO CONTAIN DERIVED CELLULAR PROTEINS AND LINEAGE-SPECIFIC PROTEINS

The study unraveling the contribution of nudiviral genes to particle productions was focused on viral genes but it was not clear whether nudiviral products represented the majority of bracovirus particle proteins or only a portion. A comprehensive analysis of CiBV particles was performed recently indicating that the unique detectable viral contribution to particle components is nudiviral, one third of the proteins having a clear nudiviral origin (Wetterwald *et al.*, 2010). In this study, CiBV particles were purified by sucrose gradient centrifugation, protein components were separated by gel electrophoresis and the content of individual bands was analyzed by mass spectrometry. Peptide sequences were matched to cDNA sequences from the

wasp ovaries to identify the proteins present in purified particles. Since viral particles are known to incorporate protein present in the cells it was important to verify that the proteins identified were true components of virus particles. Thus quantitative RT-PCR (qRT-PCR) was used to compare the expression levels of the corresponding genes at the onset and at the peak of virus particle formation to determine the way they are regulated during virus particle production. Some genes are upregulated during bracovirus production in addition to nudiviral gene products, such as genes coding for a putative metalloproteinase and several Rab-GTPase activating proteins, which are also present in the particles (Wetterwald *et al.*, 2010). It is noteworthy that metalloproteinases are found in the venom or virus-like particles (VLP) of parasitoids (Parkinson *et al.*, 2002; Asgari *et al.*, 2002; Price *et al.*, 2009; Gatti *et al.*, Chapter 15 of this volume).

Rab are small GTP-binding proteins required for membrane trafficking that have been implicated in the lifecycle of various enveloped viruses (Sklan *et al.*, 2007). We hypothesized that these proteins may act as virulence factors and contribute to parasitism success. In particular, they could protect the parasitoid eggs shortly after wasp oviposition, before the expression by parasitized host cells of the genes packaged in the particles. Alternatively, they might mediate particle entry into the cells. In addition to these modified versions of conserved cellular proteins, many proteins are lineage-specific having no homologs in the data banks. These proteins were proposed to be part of the amorphous protein matrix, which surrounds the nucleocapsid in CiBV particles (Wetterwald *et al.*, 2010).

Interestingly, most of the non-nudiviral protein components of CiBV have no homolog expressed in *C. congregata* ovaries except the genes encoding the 17a, 30b and 27b products. The protein 27b homolog is encoded by the nudiviral cluster in the genome of *C. congregata*. It is upregulated during both CiBV and CcBV particle production (Wetterwald *et al.* 2010; Fig. 1B) suggesting it could derive from the nudiviral ancestor, although no homologs have been found in the few sequenced nudivirus genomes. Overall, the gene products of nudiviral origin are mostly conserved among components of bracovirus particles while the others are not, further confirming the importance of these genes and the nudiviral origin of bracoviruses.

DISCUSSION

The identification of the viral face of PDVs revealed a complex history involving at least two different virus acquisitions in the ichneumonid and the braconid wasps. In the coming years a more complete picture will probably emerge with the determination of the viral machineries producing *G. fumiferanae* virus and *Venturia canescens* VLPs. As found for prokaryotic symbionts of Coleoptera (Conord *et al.*, 2008), one might expect that viruses from the same or different families have been captured independently and have been maintained, lost, or replaced in different wasp lineages. One may ask why associations with viruses have been selected several times independently, at least twice from the results already obtained on bracoviruses and ichnoviruses. Virus associations could allow a larger set of factors controlling host physiology to be produced at a lower physiological cost for the wasps, compared to the other strategies generally consisting of the injection of virulence proteins produced in the venom gland or the ovaries (Poirié *et al.*, 2009; Asgari, Chapter 18 of this volume; Formesyn *et al.*, Chapter 19 of this volume). The larger arsenal of virulence factors produced using PDVs could also limit host resistance, since this large panel is likely to target different levels of the host immune response.

Endoparasitoid wasps insert their ovipositors in several individuals of a lepidopteran population either to probe the host quality before oviposition or for host feeding. This behavior may have favored their use as vectors for horizontal transmission of different viruses, as shown in the case of ascoviruses (Govindarajan and Federici, 1990; Tilmann *et al.*, 2004). This could have enhanced the probability of loose associations between wasps and viruses resulting in long-term domestication. It should be noted that in certain cases a virus infection induced during oviposition, if not killing the host by inducing a disease or depriving the parasitoid larvae from their resources, may favor the development of parasitoid larvae for example by inhibiting host defenses (Renault *et al.*, 2002) or controlling apoptosis of hemocytes and other cells. Thus PDVs may conceivably have evolved from an initial step of wasp association with infectious viruses. Some viruses, such as nudiviruses, might have been particularly prone to domestication because of particular features, like infections of the gonads and/or genome integration in host cells as a part of their life-cycle.

Recent reports showed that mammals and insect genomes contain besides endogenous retroviruses a number of endogenous viral elements (EVE) (Katzourakis and Gifford, 2010; Belyi *et al.*, 2010). From the replication mechanisms of the virus families to which they belong, these EVEs were not expected to be found inserted into host DNA. Overall this indicates that genetic material derived from all known viral genome types and replication strategies can enter the animal germ line. In most cases these insertions appear to be nonfunctional but some of EVE sequences code for potential proteins expressed as cDNAs. This suggests a possible physiological role for these proteins (Katzourakis and Gifford, 2010). Moreover, complete viral genomes are found integrated in various genomes and can sometimes be reactivated, such as plant pararetroviruses, which become infectious following

stresses such as wounding or crosses (Gayral *et al.*, 2008). In humans, vertical transmission has been shown for the latent herpes virus, the chromosomal form of which is able to lead to the production of functional particles (Arbuckle *et al.*, 2010). This further indicates that viral genome vertical transmission might be more common than usually thought. Dual strategies of viruses, such as retroviruses and phages, involve particles causing infectious cycle and persistent viruses maintained in host DNA and eventually transmitted vertically. Nudiviruses, such as HzNV-1, are also suspected to use integration in the host genome as a persistence strategy during latent infection (Lin *et al.*, 1999).

Symbiotic viruses can be considered as an extreme case of the vertical transmission strategy, ensuring the success of the host genome to allow its transmission to the next generation. It is striking that except DNA replication, the essential functions and their associated genes common to baculoviruses and nudiviruses are conserved to produce bracovirus particles: viral gene transcription, packaging, nucleocapsid assembly, and entry into host cells. Moreover, the nucleocapsid structure is remarkably similar to that of pathogenic nudiviruses (Stoltz and Whitfield, 2009). This strengthens the view that bracoviruses are not a somewhat degenerated form of nudiviruses, but retain the essential functions of these viruses. Furthermore, the fact that particles are infectious is an important point in the concept of viruses and PDV particles are infectious and cause host disease. In related viruses such as baculoviruses, genes ensuring a better adaptation to host cells have been picked up from various sources including host insects and bacteria (Hughes and Friedman, 2003). The PDV packaged genes may be regarded either as 'wasp genes' or as genes ensuring a better adaptation to lepidopteran host cells, while the nudiviral genes constitute the core genes of bracoviruses. Thus, a PDV, although non-autonomous, remains mostly a virus. Insect obligatory symbionts such as *Buchnera aphidicola* for the pea aphid are still designated as bacteria and thus the term symbiotic virus appears sufficient to account for the fact that PDVs have evolved by symbiogenesis (Federici and Bigot, 2003) while emphasizing the fact that they have retained most of the characteristic features of viruses.

Thus PDVs highlight the fact that viruses should no longer be exclusively considered obligatory parasites, but in certain cases obligatory symbionts. It would be surprising that such kind of symbiotic viruses would have evolved only in parasitoid wasps. Finding other symbiotic associations is a new frontier in virology. A first step to identifing new symbiotic viruses may consist in screening for viruses stably integrated in arthropod genomes correlated with in-depth investigation of viruses pathogenic to arthropods using new sequencing approaches.

ACKNOWLEDGMENTS

We are grateful to Beatrice Lanzrein, Juline Herbinière and Robert Wharton for providing the TEM seen in Fig. 2 and to Donald Stoltz for the Fig. 3 pictures. We thank George Periquet and Nathalie Volkoff for thousands of discussions. Thanks to Donald Stoltz for having stimulated the field to find PDVs' origins and to Brian Federici for his provocative naming of PDVs as 'organelles', providing another stimulation to find the viral genes. We thank the Genoscope team for the sequencing and in particular Gabor Gyapay, who advocated successfully the extension of the project to *Chelonus inanitus*. The work reported was funded in part by the ANR projects 'Evparasitoid' and 'Paratoxose', the Research Federative Institute 136, by the CNRS Research networks on mobile elements (GDR 2157) and symbiotic interactions (GDR 2153) and the European Research Council Grant 205206 'GENOVIR'.

REFERENCES

Abd-Alla, A.M., Cousserans, F., Parker, A.G., Jehle, J.A., Parker, N.J., Vlak, J.M., et al. 2008. Genome analysis of a *Glossina pallidipes* salivary gland hypertrophy virus reveals a novel, large, double-stranded circular DNA virus. J. Virol. 82, 4595–4611.

Albrecht, U., Wyler, T., Pfister-Wilhelm, R., Gruber, A., Stettler, P., Heiniger, P., et al. 1994. Polydnavirus of the parasitic wasp *Chelonus inanitus* (Braconidae): characterization, genome organization and time point of replication. J. Gen. Virol. 75, 3353–3363.

Arbuckle, J.H., Medveczky, M.M., Luka, J., Hadley, S.H., Luegmayr, A., Ablashi, D., et al. 2010. The latent human herpesvirus-6A genome specifically integrates in telomeres of human chromosomes *in vivo* and *in vitro*. Proc. Natl. Acad. Sci. U.S.A. 107, 5563–5568.

Asgari, S., Reineke, A., Beck, M., Schmidt, O., 2002. Isolation and characterization of a neprilysin-like protein from *Venturia canescens* virus-like particles. Insect Mol. Biol. 11, 477–485.

Beck, M.H., Inman, R.B., Strand, M.R., 2007. *Microplitis demolitor* bracovirus genome segments vary in abundance and are individually packaged in virions. Virology 359, 179–189.

Belle, E., Beckage, N.E., Rousselet, J., Poirié, M., Lemeunier, F., Drezen, J.-M., 2002. Visualization of polydnavirus sequences in a parasitoid wasp chromosome. J. Virol. 76, 5793–5796.

Belyi, V.A., Levine, A.J., Skalka, A.M., 2010. Unexpected inheritance: multiple integrations of ancient bornavirus and ebolavirus/marburgvirus sequences in vertebrate genomes. PLoS Pathog. 6, e1001030.

Bézier, A., Annaheim, M., Herbinière, J., Wetterwald, C., Gyapay, G., Bernard-Samain, S., et al. 2009. Polydnaviruses of braconid wasps derive from an ancestral nudivirus. Science 323, 926–930.

Bézier, A., Herbinière, J., Serbielle, C., Lesobre, J., Wincker, P., Huguet, E., et al. 2008. Bracovirus gene products are highly divergent from insect proteins. Arch. Insect Biochem. Physiol. 67, 172–187.

Bibor-Hardy, V., Suh, M., Pouchelet, M., Simard, R., 1982. Modifications of the nuclear envelope of BHK cells after infection with herpes simplex virus type 1. J. Gen. Virol. 63, 81–94.

Bigot, Y., Samain, S., Augé-Gouillou, C., Federici, B.A., 2008. Molecular evidence for the evolution of ichnoviruses from ascoviruses by symbiogenesis. BMC Evol. Biol. 8, 253.

Braunagel, S.C., Russell, W.K., Rosas-Acosta, G., Russell, D.H., Summers, M.D., 2003. Determination of the protein composition of the occlusion-derived virus of *Autographa californica* nucleopolyhedrovirus. Proc. Natl. Acad. Sci. U.S.A. 100, 9797–9802.

Braunagel, S.C., Summers, M.D., 2007. Molecular biology of the baculovirus occlusion-derived virus envelope. Curr. Drug Targets 8, 1084–1095.

Chao, Y.C., Lee, S.T., Chang, M.C., Chen, H.H., Chen, S.S., Wu, T.Y., et al. 1998. A 2.9-kilobase noncoding nuclear RNA functions in the establishment of persistent Hz-1 viral infection. J. Virol. 72, 2233–2245.

Choi, J.Y., Roh, J.Y., Kang, J.N., Shim, H.J., Woo, S.D., Jin, B.R., et al. 2005. Genomic segments cloning and analysis of *Cotesia plutellae* polydnavirus using plasmid capture system. Biochem. Biophys. Res. Commun. 332, 487–493.

Conord, C., Despres, L., Vallier, A., Balmand, S., Miquel, C., Zundel, S., et al. 2008. Long-term evolutionary stability of bacterial endosymbiosis in curculionoidea: additional evidence of symbiont replacement in the dryophthoridae family. Mol. Biol. Evol. 25, 859–868.

Crawford, A.M., Sheehan, C., 1985. Replication of Oryctes baculovirus in cell culture: viral morphogenesis, infectivity and protein synthesis. J. Gen. Virol. 66, 529–539.

De Buron, I., Beckage, N.E., 1992. Characterization of a polydnavirus (PDV) and virus-like filamentous particle (VLFP) in the braconid wasp *Cotesia congregata* (Hymenoptera: Braconidae). J. Invertebr. Pathol. 59, 315–327.

Deng, L., Stoltz, D.B., Webb, B.A., 2000. A gene encoding a polydnavirus structural polypeptide is not encapsidated. Virology 269, 440–450.

Deng, L., Webb, B.A., 1999. Cloning and expression of a gene encoding a *Campoletis sonorensis* polydnavirus structural protein. Arch. Insect Biochem. Physiol. 40, 30–40.

Desjardins, C.A., Gundersen-Rindal, D.E., Hostetler, J.B., Tallon, L.J., Fadrosh, D.W., Fuester, R.W., et al. 2008. Comparative genomics of mutualistic viruses of *Glyptapanteles* parasitic wasps. Genome Biol. 9, R183.

Drezen, J.-M., Bézier, A., Lesobre, J., Huguet, E., Cattolico, L., Periquet, G., et al. 2006. The few virus-like genes of *Cotesia congregata* bracovirus. Arch. Insect Biochem. Physiol. 61, 110–122.

Dupuy, C., Huguet, E., Drezen, J.-M., 2006. Unfolding the evolutionary story of polydnaviruses. Virus Res. 117, 81–89.

Dupuy, C., Periquet, G., Serbielle, C., Bézier, A., Louis, F., Drezen, J.-M., 2011. Transfer of a chromosomal *Maverick* to endogenous bracovirus in a parasitoid wasp. Genetica 139, 489–496.

Espagne, E., Dupuy, C., Huguet, E., Cattolico, L., Provost, B., Martins, N., et al. 2004. Genome sequence of a polydnavirus: insights into symbiotic virus evolution. Science 306, 286–289.

Falabella, P., Varricchio, P., Gigliotti, S., Tranfaglia, A., Pennacchio, F., Malva, C., 2003. *Toxoneuron nigriceps* polydnavirus encodes a putative aspartyl protease highly expressed in parasitized host larvae. Insect Mol. Biol. 12, 9–17.

Falabella, P., Varricchio, P., Provost, B., Espagne, E., Ferrarese, R., Grimaldi, A., et al. 2007. Characterization of the *IkB-like* gene family in polydnaviruses associated with wasps belonging to different Braconid subfamilies. J. Gen. Virol. 88, 92–104.

Federici, B.A., Bigot, Y., 2003. Origin and evolution of polydnaviruses by symbiogenesis of insect DNA viruses in endoparasitic wasps. J. Insect Physiol. 49, 419–432.

Fleming, J.A.G.W., Summers, M.D., 1991. Polydnavirus DNA is integrated in the DNA of its parasitoid wasp host. Proc. Natl. Acad. Sci. U.S.A. 88, 9770–9774.

Francino, M.P., 2005. An adaptive radiation model for the origin of new gene functions. Nat. Genet. 37, 573–577.

Garcia-Maruniak, A., Maruniak, J.E., Farmerie, W., Boucias, D.G., 2008. Sequence analysis of a non-classified, non-occluded DNA virus that causes salivary gland hypertrophy of *Musca domestica*, MdSGHV. Virology 377, 184–196.

Gayral, P., Noa-Carrazana, J.C., Lescot, M., Lheureux, F., Lockhart, B.E., Matsumoto, T., et al. 2008. A single banana streak virus integration event in the banana genome as the origin of infectious endogenous pararetrovirus. J. Virol. 82, 6697–6710.

Govindarajan, R., Federici, B.A., 1990. Ascovirus infectivity and effects of infection on the growth and development of noctuid larvae. J. Invertebr. Pathol. 56, 291–299.

Guarino, L.A., Dong, W., Jin, J., 2002. *In vitro* activity of the baculovirus late expression factor LEF-5. J. Virol. 76, 12663–12675.

Guarino, L.A., Xu, B., Jin, J., Dong, W., 1998. A virus-encoded RNA polymerase purified from baculovirus-infected cells. J. Virol. 72, 7985–7991.

Harwood, S.H., Li, L., Ho, P.S., Preston, A.K., Rohrmann, G.F., 1998. AcMNPV late expression factor-5 interacts with itself and contains a zinc ribbon domain that is required for maximal late transcription activity and is homologous to elongation factor TFIIS. Virology 250, 118–134.

Herniou, E.A., Olszewski, J.A., Cory, J.S., O'Reilly, D.R., 2003. The genome sequence and evolution of baculoviruses. Annu. Rev. Entomol. 48, 211–234.

Hong, T., Braunagel, S.C., Summers, M.D., 1994. Transcription, translation, and cellular localization of ODV-E66: a structural protein of the PDV envelope of *Autographa californica* nuclear polyhedrosis virus. Virology 204, 210–222.

Hong, T., Summers, M.D., Braunagel, S.C., 1997. N-terminal sequences from *Autographa californica* nuclear polyhedrosis virus envelope proteins ODV-E66 and ODV-E25 are sufficient to direct reporter proteins to the nuclear envelope, intranuclear microvesicles and the envelope of occlusion-derived virus. Proc. Natl. Acad. Sci. U.S.A. 94, 4050–4055.

Hughes, A.L., Friedman, R., 2003. Genome-wide survey for genes horizontally transferred from cellular organisms to baculoviruses. Mol. Biol. Evol. 20, 979–987.

Iyer, L.M., Aravind, L., Koonin, E.V., 2001. Common origin of four diverse families of large eukaryotic DNA viruses. J. Virol. 75, 11720–11734.

Jehle, J.A., Blissard, G.W., Bonning, B.C., Cory, J.S., Herniou, E.A., Rohrmann, G.F., et al. 2006. On the classification and nomenclature of baculoviruses: a proposal for revision. Arch. Virol. 151, 1257–1266.

Katzourakis, A., Gifford, R.J., 2010. Endogenous viral elements in animal genomes. PLoS Genet. 6, e1001191.

Kim, K.S., Kitajima, E.W., 1984. Nonoccluded Baculovirus and filamentous virus-like particles in the spotted cucumber beetle, *Diabrotica undecipunctata* (coleoptera: chrysomelid). J. Invertebr. Pathol. 43, 234–241.

Krell, P.J., Stoltz, D.B., 1979. Unusual baculovirus of the parasitic wasp *Apanteles melanoscelus*: isolation and primary characterization. J. Virol. 29, 1118–1130.

Lapointe, R., Tanaka, K., Barney, W.E., Whitfield, J.B., Banks, J.C., Béliveau, C., et al. 2007. Genomic and morphological features of a banchine polydnavirus: comparison with bracoviruses and ichnoviruses. J. Virol. 81, 6491–6501.

Li, Y., Wang, J., Deng, R., Zhang, Q., Yang, K., Wang, X., 2005. *vlf-1* deletion brought AcMNPV to defect in nucleocapsid formation. Virus Genes 31, 275–284.

Lin, C.L., Lee, J.C., Chen, S.S., Wood, H.A., Li, M.L., Li, C.F., et al. 1999. Persistent Hz-1 virus infection in insect cells: evidence for insertion of viral DNA into host chromosomes and viral infection in a latent status. J. Virol. 73, 128–139.

McLachlin, J.R., Miller, L.K., 1994. Identification and characterization of *vlf-1*, a baculovirus gene involved in very late gene expression. J. Virol. 68, 7746–7756.

Moran, N.A., 2007. Symbiosis as an adaptive process and source of phenotypic complexity. Proc. Natl. Acad. Sci. U.S.A. 104, 8627–8633.

Murphy, N., Banks, J.C., Whitfield, J.B., Austin, A.D., 2008. Phylogeny of the parasitic microgastroid subfamilies (Hymenoptera: Braconidae) based on sequence data from seven genes, with an improved time estimate of the origin of the lineage. Mol. Phylogenet. Evol. 47, 378–395.

Oppenheimer, D.I., Volkman, L.E., 1997. Evidence for rolling circle replication of *Autographa californica* M nucleopolyhedrovirus genomic DNA. Arch. Virol. 142, 2107–2113.

Parkinson, N., Conyers, C., Smith, I., 2002. A venom protein from the endoparasitoid wasp *Pimpla hypochondriaca* is similar to snake venom reprolysin-type metalloproteases. J. Invertebr. Pathol. 79, 129–131.

Pasquier-Barre, F., Dupuy, C., Huguet, E., Monteiro, F., Moreau, A., Poirié, M., et al. 2002. Polydnavirus replication: the EP1 segment of the parasitoid wasp *Cotesia congregata* is amplified within a larger precursor molecule. J. Gen. Virol. 83, 2035–2045.

Passarelli, A.L., Miller, L.K., 1993. Identification of genes encoding late expression factors located between 56.0 and 65.4 map units of the *Autographa californica* nuclear polyhedrosis virus genome. Virology 197, 704–714.

Peng, K., van Oers, M.M., Hu, Z., van Lent, J.W., Vlak, J.M., 2010. Baculovirus *per os* infectivity factors form a complex on the surface of occlusion-derived virus. J. Virol. 84, 9497–9504.

Perera, O., Green, T.B., Stevens Jr., S.M., White, S., Becnel, J.J., 2007. Proteins associated with *Culex nigripalpus* nucleopolyhedrovirus occluded virions. J. Virol. 81, 4585–4590.

Poirié, M., Carton, Y., Dubuffet, A., 2009. Virulence strategies in parasitoid Hymenoptera as an example of adaptive diversity. C. R. Biol. 332, 311–320.

Price, D.R., Bell, H.A., Hinchliffe, G., Fitches, E., Weaver, R., Gatehouse, J.A., 2009. A venom metalloproteinase from the parasitic wasp *Eulophus pennicornis* is toxic towards its host, tomato moth (*Lacanobia oleracae*). Insect Mol. Biol. 18, 195–202.

Raina, A.K., Adams, J.R., Lupiani, B., Lynn, D.E., Kim, W., Burand, J.P., et al. 2000. Further characterization of the gonad-specific virus of corn earworm, *Helicoverpa zea*. J. Invertebr. Pathol. 76, 6–12.

Renault, S., Petit, A., Bénédet, F., Bigot, S., Bigot, Y., 2002. Effects of the *Diadromus pulchellus* ascovirus, DpAV-4, on the hemocytic encapsulation response and capsule melanization of the leek-moth pupa, *Acrolepiopsis assectella*. J. Insect Physiol. 48, 297–302.

Rohrmann, G.F., 2008. Baculovirus Molecular Biology. National Library of Medicine, National Center for Biotechnology Information, Bethesda. (http:www.ncbi.nlm.gov/books/NBK1736/)

Russell, R.L., Rohrmann, G.F., 1997. Characterization of P91, a protein associated with virions of an *Orgyia pseudotsugata* baculovirus. Virology 233, 210–223.

Sklan, E.H., Staschke, K., Oakes, T.M., Elazar, M., Winters, M., Aroeti, B., et al. 2007. A Rab-GAP TBC domain protein binds hepatitis C virus NS5A and mediates viral replication. J. Virol. 81, 11096–11105.

Sparks, W.O., Harrison, R.L., Bonning, B.C., 2010. *Autographa californica* multiple nucleopolyhedrovirus ODV-E56 is a *per os* infectivity factor, but is not essential for binding and fusion of occlusion-derived virus to the host midgut. Virology 409, 69–76.

Stoltz, D.B., 1990. Evidence for chromosomal transmission of polydnavirus DNA. J. Gen. Virol. 71, 1051–1056.

Stoltz, D.B., Vinson, S.B., 1979. Viruses and parasitsm in insects. Adv. Virus Res. 183, 125–171.

Stoltz, D.B., Whitfield, J.B., 2009. Making nice with viruses. Science 323, 884–885.

Tilmann, P., Styer, E., Hamm, J., 2004. Transmission of ascovirus from *Heliothis virescens* (Lepidotera: Noctuidae) by three parasitoids and effects of virus on survival of parasitoid *Cardiochiles nigriceps* (Hymenoptera: Braconidae). Environ. Entomol. 33, 633–643.

Vanarsdall, A.L., Okano, K., Rohrmann, G.F., 2006. Characterization of the role of very late expression factor 1 in baculovirus capsid structure and DNA processing. J. Virol. 80, 1724–1733.

Vigdorovich, V., Miller, A.D., Strong, R.K., 2007. Ability of hyaluronidase 2 to degrade extracellular hyaluronan is not required for its function as a receptor for jaagsiekte sheep retrovirus. J. Virol. 81, 3124–3129.

Volkoff, A.N., Jouan, V., Urbach, S., Samain, S., Bergoin, M., Wincker, P., et al. 2010. Analysis of virion structural components reveals vestiges of the ancestral ichnovirus genome. PLoS Pathog. 6, e1000923.

Wang, Y., Bininda-Emonds, O.R., van Oers, M.M., Vlak, J.M., Jehle, J.A., 2011. The genome of *Oryctes rhinoceros* nudivirus provides novel insight into the evolution of nuclear arthropod-specific large circular double-stranded DNA viruses. Virus Genes. 42, 444–456.

Wang, Y., Burand, J.P., Jehle, J.A., 2007. Nudivirus genomics: diversity and classification. Virologica Sinica 22, 128–136.

Wang, Y., Jehle, J.A., 2009. Nudiviruses and other large, double-stranded circular DNA viruses of invertebrates: New insights on an old topic. J. Invert. Pathol. 101, 187–193.

Webb, B., Strand, M.R., 2005. The biology and genomics of polydnaviruses In: Iatrou, K., Gilbert, L., Gill, S. (Eds.), Comprehensive Molecular Insect Science. Vol. 6. (pp. 323–360). Elsevier, Amsterdam.

Webb, B.A., Strand, M.R., Dickey, S.E., Beck, M.H., Hilgarth, R.S., Barney, W.E., et al. 2006. Polydnavirus genomes reflect their dual roles as mutualists and pathogens. Virology 347, 160–174.

Weber, B., Annaheim, M., Lanzrein, B., 2007. Transcriptional analysis of polydnaviral genes in the course of parasitization reveals segment-specific patterns. Arch. Insect. Biochem. Physiol. 66, 9–22.

Wetterwald, C., Roth, T., Kaeslin, M., Annaheim, M., Wespi, G., Heller, M., et al. 2010. Identification of bracovirus particle proteins and analysis of their transcript levels at the stage of virion formation. J. Gen. Virol. 91, 2610–2619.

Whitfield, J.B., 2002. Estimating the age of the polydnavirus/braconid wasp symbiosis. Proc. Natl. Acad. Sci. U.S.A. 21, 7508–7513.

Wu, W., Lin, T., Pan, L., Yu, M., Li, Z., Pang, Y., et al. 2006. *Autographa californica* multiple nucleopolyhedrovirus nucleocapsid assembly is interrupted upon deletion of the *38K* gene. J. Virol. 80, 11475–11485.

Wu, Y.L., Wu, C.P., Lee, S.T., Tang, H., Chang, C.H., Chen, H.H., et al. 2010. The early gene *hhi1* reactivates *Heliothis zea* nudivirus 1 in latently infected cells. J. Virol. 84, 1057–1065.

Wyder, S., Tschannen, A., Hochuli, A., Gruber, A., Saladin, V., Zumbach, S., et al. 2002. Characterization of *Chelonus inanitus* polydnavirus segments: sequences and analysis, excision site and demonstration of clustering. J. Gen. Virol. 83, 247–256.

Wyler, T., Lanzrein, B., 2003. Ovary development and polydnavirus morphogenesis in the parasitic wasp *Chelonus inanitus*. II. Ultrastructural analysis of calyx cell development, virion formation and release. J. Gen. Virol. 84, 1151–1163.

Yang, S., Miller, L.K., 1998. Expression and mutational analysis of the baculovirus very late factor 1 (*vlf-1*) gene. Virology 245, 99–109.

Zhang, J., 2003. Evolution by gene duplication an update. Trends Ecol. Evol. 18, 292–297.

Chapter 3

The Organization of Genes Encoding Ichnovirus Structural Proteins

Anne-Nathalie Volkoff*, Jean-Michel Drezen†, Michel Cusson‡ and Bruce A. Webb§

*UMR 1333 INRA — Université Montpellier 2, Diversité, Génomes et Interactions Microorganismes–Insectes, Place Eugène Bataillon, cc101, 34095 Montpellier Cedex 5, France

†Institut de Recherche sur la Biologie de l'Insecte, CNRS UMR 6035, Faculté des Sciences et Techniques, Université François Rabelais, Parc Grandmont, 37200 Tours, France

‡Laurentian Forestry Centre, Natural Resources Canada, 1055 rue du P.E.P.S., Québec, G1V 4C7, Canada

§Department of Entomology, S-225 Agricultural Science Center N, University of Kentucky, Lexington, KY 40546-0091, USA

SUMMARY

A large number of genes encoding structural components of polydnaviruses associated with ichneumonid wasps of the subfamily Campopleginae have recently been identified (Volkoff et al., 2010). The genes involved in the production of ichnovirus particle proteins were shown to be localized in specific regions embedded in the genome of the wasp *Hyposoter didymator*. These specialized regions were named 'IchnoVirus Structural Proteins Encoding Regions' (IVSPERs). Three IVSPERs, representing over 60kb in length, were identified in the *H. didymator* genome. Although they are not packaged in the virus particles, the IVSPERs can be considered functionally as an integral part of the viral genome and are amplified during virus replication. Their particular genomic organization (high coding sequence density, single-exon predicted genes) suggests that IVSPERs represent fingerprints of an ancestor virus that integrated its own DNA into the genome of an ancestor wasp. The IVSPER genes constitute a set of genes specific to ichnoviruses and conserved among ichnovirus-associated wasps, as shown for the nudivirus-related genes involved in bracovirus particle production. However, their lack of similarity with genes from known pathogenic viruses suggests that ichnoviruses originated from a virus family that has yet to be described. Altogether these recent findings show that ichnoviruses and bracoviruses derive from independent evolutionary events and represent an example of convergent evolution in the use of virus particles to transfer virulence genes into the host caterpillar.

ABBREVIATIONS

BV	bracovirus
EST	expressed sequence tag
IV	ichnovirus
IVSPER	ichnovirus structural proteins encoding region
PDV	polydnavirus

INTRODUCTION

Polydnaviruses (PDVs) are unique mutualistic viruses associated with ichneumonid and braconid wasps developing as endoparasites of caterpillar hosts. They are today classically separated into ichnoviruses (IVs), associated with ichneumonids from the Campopleginae subfamily, and the bracoviruses (BVs), associated with braconids from the Microgastroid complex, a monophyletic group. Additionally, a virus associated with an ichneumonid wasp from the Banchinae subfamily was recently described which may merit classification in a third group as it has characteristics of both braco- and ichneumonid viruses (Lapointe et al., 2007).

PDV genomes are integrated into the wasp genome and hence transmitted vertically from one parasitoid generation to the other (Belle et al., 2002; Fleming and Summers, 1991; Stoltz, 1990). Nonetheless, the virus cycle passes through the production of viral particles. This occurs exclusively in female wasps during pupal and adult development within a specialized ovarian tissue named the calyx located at the apical region of the oviducts (Krell, 1991; Krell and Stoltz, 1980; Stoltz and Vinson, 1977; Volkoff et al., 1995). The viral particles are stored in the oviducts and injected into the lepidopteran hosts along with the egg during oviposition. In this parasitized host, tissue infection is followed by viral gene expression, which is responsible for major physiological modifications leading to the success of parasitism (Beckage, 1998; Beckage and Gelman, 2004; Glatz et al., 2004; Malva et al., 2004; Pennacchio and Strand, 2006; Webb, 1998).

Parasitoid Viruses: Symbionts and Pathogens. DOI: 10.1016/B978-0-12-384858-1.00003-5

Several PDV packaged genomes have been sequenced these last years, revealing that IV and BV particles contain from 15 to more than 100 circular double-stranded DNA molecules (Desjardins *et al.*, 2008; Espagne *et al.*, 2004; Lapointe *et al.*, 2007; Tanaka *et al.*, 2007; Webb *et al.*, 2006). They encode from 60 to 200 genes whose products are believed to ensure successful parasitism by suppressing the host immune response and/or altering host larval development. Many of these genes encoding host regulation factors are typically organized in gene families and resemble eukaryotic genes (Espagne *et al.*, 2004; Lapointe *et al.*, 2007; Tanaka *et al.*, 2007; Webb *et al.*, 2006). These features lead to the hypothesis that they most probably derive from acquisitions from the wasp or the lepidopteran genomes (Desjardins *et al.*, 2008). By contrast, at least by classical similarity searches against generalist sequence databases, PDV packaged genomes contain no typical viral core genes, such as genes involved in DNA replication, transcription, or particle morphogenesis.

Thus, for more than thirty years of PDV studies, the nature and genomic organization of the genes encoding proteins associated with the polydnaviral particles have remained an unresolved question. Until recently, only two such genes had been described, both encoding structural proteins of the IV associated with the Campopleginae wasp *Campoletis sonorensis*. The first, named *p12* (NCBI reference gi|4101554), is reported as encoding a major CsIV virion protein which is synthesized only in ovarian calyx cells during virus replication (Deng and Webb, 1999). The *p12* gene is found on a CsIV viral segment and is unrelated to other known sequences. This *p12* gene remains today the only described PDV structural gene that maps to the packaged viral DNA. The second gene, also described in *C. sonorensis*, is the *p53* gene (gi|4101552), which encodes a protein apparently associated with the CsIV virion inner membrane (Deng *et al.*, 2000). Surprisingly, the *p53* gene was not found in the encapsidated CsIV genome but was shown to reside permanently in the genome of the wasp (Deng *et al.*, 2000). This finding indicated that genes encoding viral structural proteins could reside permanently in the wasp genome. It also raised the hypothesis that since PDVs do not replicate in the lepidopteran host (Stoltz *et al.*, 1986; Theilmann and Summers, 1986), most of the genes required only for viral replication were no longer packaged within the virion (Webb, 1998). Conversely, the genes encoding host regulation factors would have been selected for inclusion in the packaged viral genome to ensure their delivery into lepidopteran host (Webb and Summers, 1990).

Recently, a larger number of genes encoding structural components of PDVs were identified, for both viruses associated with ichneumonid (Volkoff *et al.*, 2010) and braconid wasps (Bézier *et al.*, 2009a; Wetterwald *et al.*, 2010).

These recent works confirmed that virus structural genes were present in the wasp genome but not packaged in the virus particles. Interestingly, deciphering the nature and genomic organization of these genes also shed new light on the origins of PDVs. IVs and BVs differ both in terms of particle morphology and genome content, and BVs are associated with a monophyletic braconid group. These observations had lead to the hypothesis that modern IVs and BVs derive from independent evolutionary events of association between an ancestral wasp and an ancestral virus entity (Murphy *et al.*, 2008; Whitfield and Asgari, 2003). Analysis of the genes encoding structural components of PDVs confirmed this hypothesis since these studies revealed that BVs derive from an ancestral nudivirus (nudiviruses constitute a sister group of baculoviruses) (Bézier *et al.*, 2009a), whereas IVs derive from a different ancestral virus or mobile element not yet identified (Volkoff *et al.*, 2010).

The work on the identification of genes encoding BV structural proteins is reviewed in Chapter 2 whereas the present chapter focuses on the corresponding genes in IVs. IV particles have an ovocylindrical shape and a large electron-dense nucleocapsid measuring ~330 × 85 nm (Stoltz and Vinson, 1979; Webb, 1998) and assembled within calyx cell nuclei from a virogenic stroma (Fig. 1). The viral particle is composed of a nucleocapsid surrounded by two envelopes with an ill-defined matrix region between the two unit membranes. The inner envelope is acquired *de novo* in the nucleus of calyx cells where virions are assembled, whereas the outer envelope is acquired from the plasma membrane of calyx cells, during budding of particles into the oviduct lumen (Volkoff *et al.*, 1995). This complex morphological structure is linked to an equally complex structural polypeptide composition. Indeed, SDS-PAGE profiles indicate that IV particles are composed of over 54 proteins ranging in size from 10 to more than 250 kDa (Deng and Webb, 1999; Krell *et al.*, 1982; Volkoff *et al.*, 2010).

In order to decipher the nature and genomic organization of genes associated with IV particles in Campopleginae parasitic wasps, an extensive analysis was conducted on the ichneumonid wasp, *Hyposoter didymator*. The study was based on the analysis of cDNA libraries from the ovaries (the tissue where the particles are produced), on mass spectrometry analysis of the purified *Hyposoter didymator* ichnovirus (HdIV) particles, and on the sequencing of *H. didymator* genomic clones. This work allowed the identification of a set of genes encoding structural components located in genomic regions specialized in the production of IV particle proteins. These regions were named 'IchnoVirus Structural Proteins Encoding Regions' (IVSPERs). The nature and genomic organization of the IVSPERs strongly suggest that these particular regions represent fingerprints of the IV ancestor.

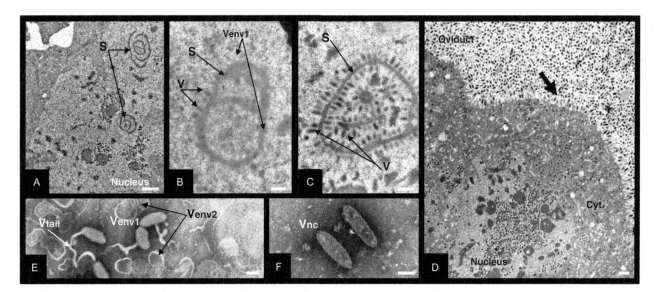

FIGURE 1 Viral particle production in *Hyposoter didymator* calyx cells during pupal development. A: Formation of virogenic stroma (S) in the nuclei of calyx cells in young female pupae. **B:** Detection of the first nucleocapsids surrounded by a viral envelope (V) in the vicinity of the virogenic stroma; some empty viral envelopes (Venv1) can also be observed, suggesting they are formed prior to virus particle assembly. **C:** Later stage of IV particle production in the nucleus. **D:** Shortly before adult emergence, a large number of virus particles are visible in the nucleus; they bud through the nuclear envelope, cross the cytoplasm (Cyt.) and then bud through the calyx cell membrane (arrow), thus acquiring the outer envelope (Venv2) before being stored in the oviduct lumen. **E:** Negative staining of HdIV particles showing nucleocapsids surrounded only by the inner envelope (Venv1) with a characteristic tail appendage (Vtail). External envelopes (Venv2) are located in their vicinity. **F:** Negative staining of HdIV nucleocapsids (Vnc). Scale bars: A and D, 1 μm; B and C, 500 nm; E and F, 100 nm. Micrographs by M. Ravallec.

THE IV STRUCTURAL PROTEINS ENCODING REGIONS (IVSPERS)

The results obtained for *H. didymator* demonstrate that the genes encoding proteins associated with HdIV particles are clustered in specific and atypical regions within the wasp genome, the IVSPERs (Volkoff *et al.*, 2010). These regions are characterized by a higher coding sequence density (62%) compared to the rest of the wasp genome (~21%), on the basis of available sequences. Additionally, all the genes in these regions are predicted to be single-exon while the majority of the wasp genes are predicted to have multiple exons.

Three IVSPERS Identified in the *H. didymator* Genome

Thus far, three IVSPERs have been identified in the ichneumonid wasp *H. didymator* (Fig. 2) and two are located near a sequence corresponding to a packaged viral segment (i.e., detected in the HdIV particle). The sizes of these regions range from 12 to 25 kbp and they encode nine to 16 genes. Interestingly, the gene content of the three IVSPERs is related. Indeed IVSPERs contain a combination of genes belonging to seven gene families, suggesting that they may share a common ancestral genome comprising a member of each of these families. The overall size

of these three IVSPERs is over 60 kb and exceeds previous estimates of the minimum nucleic acid molecular weight required to encode CsIV virion proteins (~38 kb) based on the aggregate molecular weights of the structural proteins (Krell *et al.*, 1982). However, the three IVSPERs are probably not unique in *H. didymator*. Indeed, other sequences related to the IVSPER genes have been identified in the ovarian cDNA library, thus suggesting presence of at least one additional IVSPER in the wasp genome. For *Cotesia congregata* bracovirus, the structural genes that have been identified are either clustered or isolated within the wasp genome (Bézier *et al.*, 2009b). For now, all the genes identified as involved in HdIV particle production reside within IVSPERs, but we cannot completely exclude the possibility that some may be isolated in the wasp genome. Finally, each of the *H. didymator* genomic clones sequenced was 50–60 kbp long and contained a single IVSPER, thus suggesting that the different IVSPERs are not located in close proximity within the wasp genome. However, it is not yet known if IVSPERs are dispersed in the wasp genome or located within the same chromosome.

IVSPER Genes are Conserved in Campopleginae Wasps

Blast searches of expressed sequence tag (EST) databases allowed the identification of IVSPER-related sequences

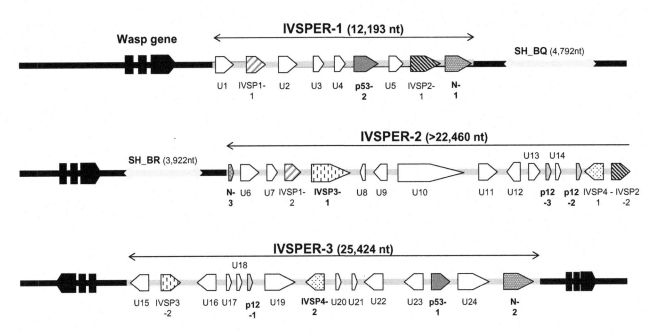

FIGURE 2   **Schematic representation of the three IVSPERs identified in *Hyposoter didymator* genome.** Double-headed arrows indicate the two viral segments (SH_BQ and SH_BR, corresponding to packaged DNA). The IVSPERs are surrounded either by a viral segment and a wasp gene (i.e., a sequence presenting high similarity to hymenopteran sequences) or by two wasp genes. In the IVSPERs, arrowed boxes indicate the predicted coding sequences: U1 to U24, for 'Unknown' genes (i.e., unrelated to other sequences) and the members of the seven identified multigenic families (p12, p53, N, IVSP1, IVSP2, IVSP3 and IVSP4). Adapted from Volkoff *et al.*, 2010.

in other Campopleginae wasps. Homologs of 15 IVSPER genes were found in *Tranosema rostrale* and/or *C. sonorensis* EST databases (Table 1), including homologs of nine of the genes shown by mass spectrometry analysis to encode proteins of HdIV particles. Thus IVSPER genes are conserved amongst *Campopleginae* wasps, which suggests that they were acquired by a common ancestor of the progenitor wasp before radiation of the subfamily.

IVSPERS are Specialized in IV Particle Production

Amongst the 40 genes contained in the three *H. didymator* IVSPERs characterized to date, 26 were shown to encode protein components of purified HdIV particles (Table 2). Among them, 19 were identified by LC-MS/MS (Volkoff *et al.*, 2010) analysis of the major bands observed on SDS-PAGE and an additional seven have been identified more recently by further analyses of HdIV particles (Volkoff, personal data; Fig. 3). The IVSPER genes not yet identified as encoding components of particles were shown to be at least transcribed in a specific manner in the *H. didymator* calyx cells at time of virus replication (Volkoff *et al.*, 2010). In *T. rostrale*, where IVSPER gene homologs have also been identified, the sequences derive from an ovarian cDNA library, thus indicating that IVSPER genes are similarly expressed in the replicative tissue in this campoplegine wasp. The tissue-specific expression pattern

in Campopleginae wasps and the timing of expression correlated with that of virus replication indicates that all IVSPER genes are most likely involved in IV particle production either directly by encoding structural proteins or indirectly by promoting their production.

IVSPERS are not Encapsidated

It was shown that *H. didymator* genes coding for IV structural proteins are not packaged in the viral particles (Volkoff *et al.*, 2010). The regions specialized in IV particle production do actually reside permanently in the wasp genome. This finding confirmed the early hypothesis proposed by Webb (1998) that genes involved in production of IV particles may have been selected during evolution of IV-associated ichneumonids for not being transferred into the parasitized lepidopteran host.

IVSPERS are Amplified During HdIV Particle Production

The IVSPERs are thus large regions specialized in IV particle production and they should be considered functionally as part of the IV genome. A main argument for such a statement was the finding that the IVSPERs have retained the property of being amplified during virus particle production (Volkoff *et al.*, 2010). Amplification of gene copy number has also been reported for the *C. sonorensis p53*

TABLE 1 List of the *Hyposoter didymator* IVSPER Genes for which Homologs Have Been Identified in Two Other *Campopleginae* Wasps, *Tranosema rostrale* and *Campoletis sonorensis* after Blast Similarity Searches Against EST libraries (the ID of the Clones Leading to the Best Match is Given; in Bold, the ID of the p12 and p53 Proteins Rirst Described in CsIV)

Hyposoter didymator IVSPER gene	*Tranosema rostrale*	*Campoletis sonorensis*
U1	TRO1211_I03 (4e-74)	gb|HO082139.1| (1e-52)
U3	TRO122_B21(1e-68)	—
U6	TRO121_E19 (2e-46)	gb|HO081888.1| (1e-12)
U8	—	gb|HO082015.1| (5e-04)
U10	TRO126_I05 (e-126)	—
U13	—	gb|HO082032.1| (2e-23)
U14	—	gb|HO082032.1| (5e-06)
U16	TRO122_H20 (e-119)	—
U19	TRO1210_P09 (e-136)	—
U23	TRO122_K14 (e-119)	—
U24	TRO127_F15 (2e-28)	—
p12 family	TRO1213_H14 (2e-08)	**gb|AAD01200.1| (6e-06)**
p53 family	—	**gb|AAD01199.1| (2e-30)**
IVSP4 family	TRO123_C22 (e-113)	gb|HO082184.1| (1e-117)
N family	TRO1210_H10 (e-119)	gb|HO081888.1| (4e-14)

TABLE 2 List of the *Hyposoter didymator* IVSPER Genes, Including Those for which the Corresponding Protein has been Identified by LC-MS/MS Analysis of Purified HdIV Particles ('Yes' in the 'Peptide' Column)

IVSPER-1 CDS	Peptide	IVSPER-2 CDS	Peptide	IVSPER-3 CDS	Peptide
U1	Yes	N-3	—	U15	Yes
IVSP1-1	Yes	U6	—	IVSP3-2	Yes
U2	Yes	U7	Yes	U16	Yes
U3	Yes	IVSP1-2	Yes	U17	—
U4	Yes	1VSP3-1	Yes	U18	—
p53-2	Yes	U8	Yes	p12-1	Yes
U5	Yes	U9	Yes	U19	—
IVSP2-1	Yes	U10	—	IVSP4-2	Yes
N-1	Yes	U11	—	U20	—
		U12	—	U21	—
		U13	Yes	U22	Yes
		p12-3	—	U23	Yes
		U-14	—	p53-1	Yes
		p12-2	—	U24	—
		IVSP4-1	Yes	N-2	Yes
		IVSP2-2	Yes		

```
>IVSP2-1
MAATVSLAEHQDESAVHDGRLQTLKAVQQFRHNVIGQLEAFEELFGPSNTLMPLSYSLQSKSKYEFVDFGYKLCR
DHEVLIGPIYMYGLCKSTEHDKSIAKLNTDISEGKERLMMLISNCDNISHKPSGSLLSLIFNDMANKLDNVLSNA
EIVKRITSRYPRFDALAFARNKKNILQKYISGKENYSCDPVRLRVDIAPTTDENYTCYEISQGLCRTKGHCYFMP
PMLRRGYLILKMPYIRAVLTWKSTDYLFKLNAKMMMVEKIENTPPPSCFDISTGKAVDDTEAIFMLDGCSDVDWS
DEVENVIVLPSTRELSELTTATSEHSPILPPFCYPRQDLNDNGSKSEPDTFEDDDVDMTADVSVHQGKILSCAIS
DFIKTIRDEGNWSPSIANLSKFVNRFIRVMKRKKASLLYTDGRTVNVNGSVVRELSITDSRTIQTPAAFEDLVEQ
YYDRFRLNCDTDEIMRSCWNETFSHEETTTWHSCQLTNKKKRSATSITHNASKNRQTKR


>IVSP2-2_partial
MATTTNLARYESHSDMNDERLLTLKSLQQIRRNIIGELDTFNKLFGPSNTLIPLSYSFQSKSKYEFVDFGYKLYR
EYEVLIGPIDMYGLCRSMDDYKESTKLEADINEGKERLMMFISNCDNMANKPSGSLLSVILRDLADKL


>U16
MVKSIEDFVFQGLSTKDQPSSHLFPRKCMVTRANTNGFDALDLFAASYKPQIGQYKCFQEKNGCAPEQGYPYRIV
VDLDSSDEHLLQSLLYEIGKLLEKLLCQANTSDFKVMICIMRKQQTGRFHVHLLNVVTNDLTTYKNFLIMLHEKV
QAVDKGAGVNYFMVFGAIKAYKLNVNPTTPAADQCYLPWKLCCAEPNEIDLNNFVDLPGFTGGSCDEYLEDIYNY
FKKHFEPFSCKTLFHALSLHRRYNPDFDVVLPSCQDSLVRCAKRRANEDSDSEGKKKVRTGKSDPINRAKESFFE
NVLFKLPRNYYEEYDSWIGIGKIIAYVKQNYGLHLFHKFSAQSRNKYDAEKVTATYEGLLETIKVNQGEGEDEPA
IRTTSALRTLLLGSNSIIDQIEHKMFYKWKGDTHAIVGAVNCCIQQMSPMLTFPHPLNANYLFAIEFSRSDGAYR
ISHDTFIQSFGGQRATIDAGVERHEHYERILECFIYLAIKYYNDNHRFLRMYTAHGILDSTPFKLKNNHLHAKNNA
EDMRSSQYSFYRRAVVQKLNKLKKRWFLAQANGATTVSRKNNKRLWSKLYAEFEFHRALQSARKCGRYPQSVLVK
AVTRSGPPRTAL


>p53-1
MPTLTIYRPHAPAYAAAWPLRNTISGVNSFLEYNEPREDKKRAIKTSVPEPDSVKKEIDLQVNTQTEITDGRRTE
ADVDATEAIIDKALVTKEFKVNCNKDMQLLKIMKYPNVRHDEPSAILSHVLVKEQDSKPFRMGAIVINDASIKGN
DNITQWNVLSKYPEHSQALEAALKDSTPSTVHAFRATLRYSDDFVLALIAPPIDDMTPNAHVDDLTNEDSLKYNI
VVRHSKHHQAGNGTTEENVLSRYFKRFDEKLINEIRLEAPVEPVTDDSMLKRKRRDVDESHLPRESEEVDSHESF
VPLGGIDDFETPIKPREPHADITLKPTIGMKRAAPQHYQQPSVRYTPILVPVKNETLRQPSAFDTFASVALPIGT
ALALSAGATYMLAKRPRLAE


>U5
MLLIAAVTATVSYALTSVIGSLNIRKRCDGLREQMRSVDVKFYRRNNGKSVFIVWEQHQCAMWHNQKLFIYDHWH
GNEVFKENMLSADTFNLRSSELVLVYDFHEHTKNLQKNFPNEYMMRELFEVNGNCEEVLLNFRASPDDKFPALVK
QLYTVSDELNIIAVVYDALDKKILTC


>N-1
MNGDDDDDSRGSVGLLNYSEFDGGNMLQCEDDLEEEYNDEAENGGPEIYQTVSSDDELDGKLAISQSLPVVESMS
LECPSVPSLKKIVKRKPTAGGKNVLVGKVRKPDNTAPAGNAFSSEKKKSTIVSKERPAIETLSKGDSAAFAKKSS
VKKKKPVLKPKDPKEESSYVPAAAAEESRNKAKVAVKKSSKLVAMAAEEQPPVKKGRRDPKHQPEGVSLPPVTKD
GNSPSYRETQVQTFPQVNTSRKNSRTSGGTSTKSFLVLNSDIGDGRFWMETLAGDFASTGYNALLKNSQFTEQQA
YLLQHSFYQLCRKLSDISSYSDRFMKTHKCIECDFSISHQCMSIDFSQYVPASANITGVTNIAPQSTTMCVCQFA
FFHSHPTQPRVILANTLSWKSCSHMLELNRSLSLRCPRCHMNVVMKNRNNNVCSDFANWNSLEGINRRQFFKSLE
NRLQASKISRPHLDELYTCNRQCCQLFHRCAENALPDNTMPMRPH


>U2
METMDPNADTDGAASSDNNDDHTYEKLIELCHCTTLAIVAFTFQTEIELIVSDTEANSYEFSTFEDFQKFLVFVR
ELRAIVFHLRQNGDPAVYVSIASITVFRDLLPYSGSSMIQMLLKVHTLVEIIKSQWKLSLSEPPCKGLAHRMIRH
MTNSGTLAREAAALRQDLKFVHALLAEFHLSGSNHETVTEIYQTYIDYIEDRMTALGQSSKRQSNNQQHWSQKRQ
RFE
```

FIGURE 3   Sequence of additional IVSPER proteins identified by LC-MS/MS analysis of purified HdIV particles since the original report by Volkoff *et al.* (2010). For each of them, the peptide sequences that have been identified are underlined.

gene (Deng *et al.*, 2000). In *H. didymator*, it was shown that the complete region was amplified in the calyx cells and specifically in this tissue, at a level comparable to that measured for the viral segments. Amplification of IVSPERs might allow an increase in gene copy numbers required to produce viral structural proteins in amounts sufficient to generate large quantities of particles.

Among the three IVSPERs identified in *H. didymator*, two are located near a proviral segment (Fig. 2). This genomic proximity and the comparable amplification of IVSPERs and viral segments suggest that in these regions,

IVSPER and viral segment may belong to common viral replication units. However, the vicinity of a packaged segment is not required for IVSPER replication since the third *H. didymator* IVSPER is surrounded by wasp genes.

IVSPERS and Viral Segments Share Members of Multigenic Families

The fact that they are amplified together during virus replication is not the unique link between IVSPERs and packaged ichnoviral DNA. Both appear to share related sequences as

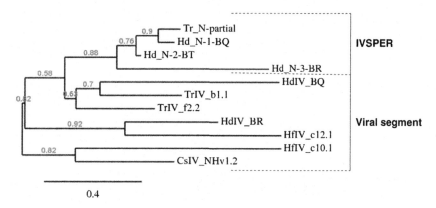

FIGURE 4   **PhyML-based tree using the alignment of a 165-amino-acid conserved C-terminal region of several proteins encoded by N-genes.** The tree was drawn using the tools available at http://www.phylogeny.fr/version2_cgi/simple_phylogeny.cgi. The tree indicates a clear separation between the *N*-genes found in packaged IV segments (HdIV_, CsIV_, TrIV_, and HfIV_) and the *N*-genes found in *H. didymator* IVSPERs (Hd_N-1, Hd_N-2, Hd_N-3) or in *T. rostrale* cDNA ovarian library (Tr_N-partial). Note that the sequence identified in *C. sonorensis* EST database was not used because of too many errors (leading to stop codons) in the sequence.

described below, which constitutes another strong argument for considering IVSPERs as part of the IV genome.

The N-Genes are Shared Between IVSPERS and Viral Segments

The IVSPERs characterized in *Hyposoter dydimator* contain members of a gene family, the *N*-gene family, also present in the encapsidated genome. The *N*-genes were first described in CsIV segment N and they are reported in all sequenced IV packaged genomes (Tanaka *et al.*, 2007; Webb *et al.*, 2006). In HdIV, several viral segments encode *N*-genes (Volkoff, personal data) including the two segments located in the vicinity of IVSPER-1 and IVSPER-2 (Fig. 2). In both *T. rostrale* and *C. sonorensis*, an *N*-gene not present in the packaged genome sequenced was found among the ESTs. Those additional *T. rostrale* and *C. sonorensis N*-genes are most likely located within an IVSPER. This suggests that in the three Campopleginae species, *N*-genes are present in the packaged genome and expressed in the parasitized host while other *N*-genes are present in the IVSPERs and expressed in the wasp ovaries.

In *H. didymator*, viral segments located in the vicinity of an IVSPER contain an *N*-gene. Presence of *N*-genes in both IVSPERs and viral segments suggest genomic exchanges and/or common ancestral sequences between HdIV and the *H. didymator* IVSPERs. When a phylogenetic tree is drawn using an alignment of IVSPER and packaged *N*-gene sequences (Fig. 4), the IVSPER *N*-genes group together and are separated from those located on the viral segment. This suggests that IVSPER *N*-genes have diverged from the packaged ones.

Similarities Between H. Didymator *IVSPER Sequences and a CsIV Viral Segment*

The *p12* gene was first described in *C. sonorensis* as a gene contained within a viral packaged molecule (Deng

and Webb, 1999). In *H. didymator*, three *p12* genes have been identified and all reside within IVSPERs, which are not packaged within the virus particle. The fact that all *H. didymator p12*-coding sequences are located within the wasp genome and are *a priori* absent from the packaged HdIV circular DNA could explain the fact that members of these gene families have not yet been described in other studied IVs. CsIV may thus constitute a rare example of modern polydnaviral segment still encoding a structural protein.

Interestingly, high similarity (65–77% identity) was found between the CsIV segment encoding the CsIV *p12* gene and a region of the *H. didymator* IVSPER-2, which contains two *p12* genes (Fig. 2). This similarity suggests that both derive from a common ancestor and, more generally, indicates a phylogenetic relationship between IVSPERs and IV genome segments. In this case, it appears that the CsIV segment has retained the ability to be encapsidated whereas the *H. didymator* IVSPER has lost this ability and now resides permanently in the wasp genome. This comparison gives insight into the dynamics of the system, where regions not involved in the control of host physiology might loose their ability to be packaged.

Conclusion: The Extended IV Genome

Altogether, results from *H. didymator* IV analysis prompted a revision of the definition of IV genomes. As illustrated in Fig. 5, the extended IV genome is therefore a multipartite genome comprising, on the one hand, the IVSPERs, which are regions actually remaining in the wasp that are specialized in the production of the viral particles and, on the other hand, the classical viral segments, which are regions packaged within the viral particles that mostly encode effectors of lepidopteran host regulation, some having being acquired from the wasp.

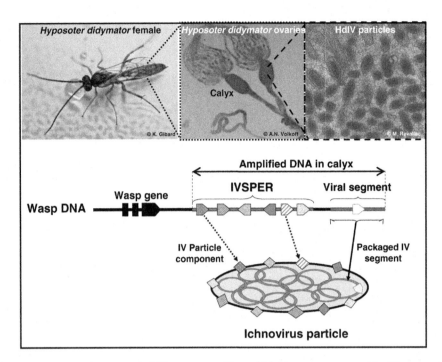

FIGURE 5   Schematic representation of the extended IV genome residing within the wasp genome. The IV genome is composed of (1) the IVSPERs, specialized regions containing genes encoding viral structural proteins and (2) the viral segments. Both regions are amplified in calyx cells but only the viral segments are incorporated into virus particles following DNA excision and circularization. Please see color plate section at the back of the book.

THE PROTEINS ASSOCIATED WITH IV PARTICLES

All IVSPER genes appear to be unrelated to known viral or eukaryotic sequences deposited in generalist databases such as GenBank, or known braconid or BV sequences. They constitute a specific set of genes conserved in IV-associated ichneumonid wasps.

The proteins identified as associated with HdIV particles included several related proteins. Indeed, members of seven gene families are distributed among the three IVSPERs and represent 40% of the genes. Three gene families are related to those previously described in CsIV, the *p12*, the *p53* and the *N* gene families. The other four IVSPER multigenic families had not been described before and were named *IVSP1* to *IVSP4* (*IVSP* for 'IchnoVirus Structural Protein'). All four gene families were composed of two sequences present in two different *H. didymator* IVSPERs.

Multiple p12 Proteins are Associated with IV Particles

Three sequences related to the CsIV *p12* were identified in *H. didymator* IVSPERs. Mass spectrometry analysis of purified HdIV particles allowed the detection of one of the corresponding proteins, thus demonstrating that the p12 proteins are components of HdIV virions, as described

for the CsIV p12 protein (Deng and Webb, 1999). The p12 proteins appear to be conserved proteins encoded by gene families in several PDV-associated Campopleginae wasps. Indeed, a second *C. sonorensis p12* gene was found by BLAST similarity searches against the GenBank EST databases. Similarly, two *p12*-related genes were found in a *T. rostrale* ovarian cDNA library. The p12 proteins are small proteins with sizes ranging between 8 and 11 kDa. Members of this family are quite divergent, sharing only a few amino acids (Fig. 6A).

Multiple p53 Proteins are Associated with IV Particles

Two sequences related to the CsIV p53 protein were identified in *H. didymator*, one being detected by mass spectrometry from purified HdIV particles. No additional sequences were found in *C. sonorensis* or *T. rostrale* cDNA libraries. The p53 proteins have predicted sizes ranging from 35.5 to 44.4 kDa and share a central motif of unknown function (VHAF-x-A-x-LRYPD-x(2)-VLA-x-I; Fig. 6B). Bigot *et al.* (2008) reported weak similarities in secondary structure between domains found in an ascovirus protein and CsIV p53, leading to the hypothesis that IVs might be related to ascoviruses. However, since no further similarity was found between IVSPER gene sequences and ascoviruses, there is at present no strong argument to support this hypothesis.

A

```
         ....|....| ....|....| ....|....| ....|....| ....|....| ....|....| ....|....|
                 10         20         30         40         50         60         70
Cs_p12-1  -MSYLPLGSS VISTGLSVFT CYQIYQLEEN --PTSSIDSH K----KNLGI AKFASVASAV SCTANGLLNA
Tr_p12-1  -MNYLPLGSS VLSTGLSVFT CYQVYQLGDG --SPNLVKDH -------LQT AKYASVASAV SSTANGLLNA
Hd_p12-1  -MSYLSLGSS VLGASLSVLT CYEVFQVEVT --EGN--KKN -------LLI AKVASVASAT SSLVSGGINV
CsIV_p12  -MSYYSLGTS VLSSGLSVLT AYHVYMLEKV NVENNPSDDD KGQYKQQLKV ARYGSLASAA SSTVNGALHA
Tr_p12-2  ----MSLAST VLSAGLSAIT CYQVYNLENT PIKSSMSQQD IDLYQSRLGI AQYLSLASAV TGSVNGLLEV
Hd_p12-2  MAGPLLAASS LFTAGVSVFN YYHVYQLGEM TISSDMNEQN LATYRSQIET VQWSSLAAAI TASADFVCRI
Hd_p12-3  MSGFLSVTPN LISAGISMYN YYHVQELAGI TVNNKMTNPD LEKYKQEMET VQWASLAAAV TTSINGVLSI
Consensus         .. :: :.:*  .  *.: :     .   .    :     .: *:*:*. :  .

          ....|....| ....|....| ....|....| ....|....| ....|....| ....|....| .
                 80         90        100        110        120        130
Cs_p12-1  YSTIQSEQND Y--------E AKQLD---MN PNMN------ ---------- ---------- -
Tr_p12-1  YSLIQPEQHY Y--------E PKVLD---MN PNMN------ ---------- ---------- -
Hd_p12-1  YDMIQSDDS- ---------S SKYLD---LN AS-------- ---------- ---------- -
CsIV_p12  YSAVRSRQND N--------D PMLLD---SD AGIN------ ---------- ---------- -
Tr_p12-2  YFSMPGGRYN T--------E PAHIEPCHMD PHIGPYHMDA HIGPYHVNPH IEGHHMNLDM H
Hd_p12-2  IHAAPPSHHH HGHPHWHHHH PGHVHHHMTD GLIN------ ---------- ---------- -
Hd_p12-3  FNAG--RHHH H-HPG-HIPP PPYTEHYLGD GHVH------ ---------- ---------- -
Consensus                        .   . :
```

B

```
         ....|....| ....|....| ....|....| ....|....| ....|....| ....|....| ....|....|
                 10         20         30         40         50         60         70
Hd_p53-1  MPTLTIYRPH APAYAAAWPL RNTISGVNSF LEYNEPREDK KRAIKTSVPE PDSVKKEIDL QVNTQTEITD
Cs_p53    MPLVTFQRPN TQIQGDGWVG SSECVSGDTL TGQYEHSNTD TQGSPTTHSE FVNDNEAGDD PIWQQKLIKI
Hd_p53-2  -----MSKPE HAVKSGAANE PAQVGAGGNA EDVDVATDEQ LNKELESHTY PMNCVKG--- --MRLKKIAK
Consensus      : :*.   .  .      .. .        :  .     :  .  .  :       . *

          ....|....| ....|....| ....|....| ....|....| ....|....| ....|....| ....|....|
                 80         90        100        110        120        130        140
Hd_p53-1  GRRTEADVDA TEAIIDKALV TKEFKVNCNK DMQLLKIMKY PNVRHDEPSA ILSHVLVKEQ DSKPFRMGAI
Cs_p53    WWKNRN---- RTRTVTSATT GEDFEISKDD ---------- ----KEEHSA ISSYVRLKKK MANLFVVAAT
Hd_p53-2  YPKH------ ---------- ---------- ---------- ----HDKGSP FESHIVVTEK HNKVIRVGAI
Consensus      :                                        ::: *. : *:: :.:: : : :.*

          ....|....| ....|....| ....|....| ....|....| ....|....| ....|....| ....|....|
                150        160        170        180        190        200        210
Hd_p53-1  VINDASIKGN DNITQWNVLS KYPEHSQALE AALKDSTPST VHAFRATLRY SDDFVLALIA PPIDDMTPN-
Cs_p53    IVNGAAVGKG GKIIHFQVLN KCPRYIKAME AAFRDSSENH VHAFHATLRY PDEYVLAVIT SPKDTWTPDG
Hd_p53-2  IVNGRAVKGS DKISHWKGLR LNPNYMDALL TSLGESIPNS VHAFQAKLRY PDDFVLAIIA PPNITMFPDE
Consensus ::*. ::** ..:* ::: *    *.: *: ::: :*  . ****:*.*** .*::***:*: .*   *:

          ....|....| ....|....| ....|....| ....|....| ....|....| ....|....| ....|....|
                220        230        240        250        260        270        280
Hd_p53-1  AHVDDLTNED SLKYNIVVRH SKHHQAGNGT TEENVLSRYF KRFDEKLINE IRLEAPVEPV TDDSMLKRKR
Cs_p53    SNVDSVTTDK --KYDIVIQR GKFPHKGDGR QLEIRLSSWF NQFNPKSISE SESAPSCDIV S---------
Hd_p53-2  PLPEGFVPDT SLQYEIVVQQ NKHDPDRLTS AKKSSIKRWF KDFTWKSVSE KVARAEQKCS H---------
Consensus  . :...:    :*:**::: .*.       : ..:* :* *  ..* *  :.*   .

          ....|....| ....|....| ....|....| ....|....| ....|....| ....|....| ....|....|
                290        300        310        320        330        340        350
Hd_p53-1  RDVDESHLPR ESEEVDSHES FVPLGGIDDF ETPIKPREPH ADITLKPTIG --------MK RAAPQHYQQP
Cs_p53    ---------- --QVSSSPNE YVSLHGNDGH RMPVKPTAPA MDAVQSHPII PQKLDDHDSR NGQPSHAVSA
Hd_p53-2  ---------- ----EDKTVG YVAMPESGHT HVPIRASAPR LDDEP----- -------YGF HHRLRKRMST
Consensus              ..   :*.:  .  . *::.  *  *          .   :  ..

          ....|....| ....|....| ....|....| ....|....| ....|....|
                360        370        380        390        400
Hd_p53-1  SVRYTPILVP VKNETLRQPS AFDTFAS-VA LPIGTALALS AGATYMLAKR PRLAE----
Cs_p53    TVYSPADNTS KRSNSETRPT NFIDFAVHPA LQIGSVCALS AGVTYLLVKK PKLAS----
Hd_p53-2  DSSHSNDVTG TKKPSEGIQA AMQTYGIPAV LAIGSAGAAI GGAVWLSKS- PRMKSYIGL
Consensus  .  .  :: .  : :     . . :  * **:. *  .*..:: .*   :  ..
```

FIGURE 6 **Protein sequence alignment of some IVSPER genes using ClustalW (http://www.phylogeny.fr/).** **A:** p12 proteins; the alignment includes the CsIV p12 (Cs_p12) and the translation of the EST sequences identified in *Tranosema rostrale* EST library (Tr_p12-1 and Tr_p12-2) and in *Campoletis sonorensis* EST library (Cs_p12-1). **B:** p53 proteins; the alignment includes the *C. sonorensis* p53 (Cs_p53). **C:** IVSP4 proteins; the alignment includes the translation of the EST sequences identified in *T. rostrale* EST library (Tr_IVSP4-1 and Tr_IVSP4-2; partial) and in *C. sonorensis* EST library (Cs_IVSP4-1; partial).

C

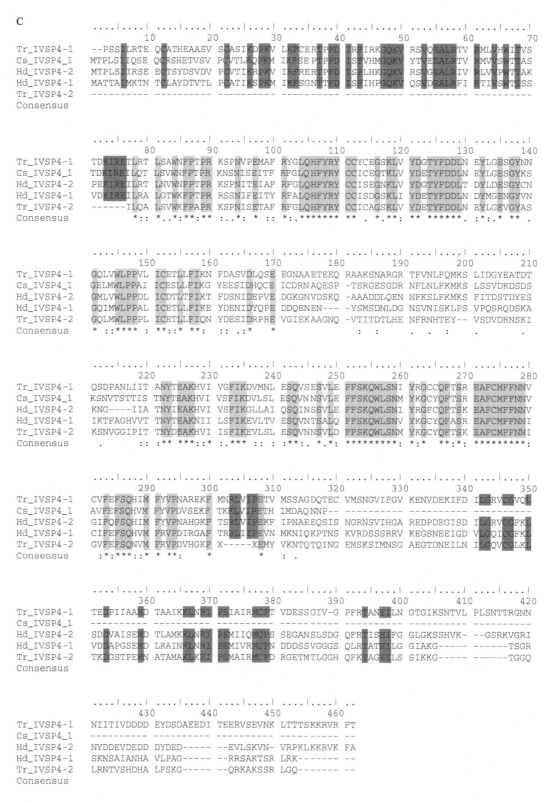

FIGURE 6   (Continued)

N-Gene Proteins are Associated with HdIV Particles

In *H. didymator*, two IVSPER members of the *N*-gene family, *N-1* and *N-2*, encode proteins shown to be associated with HdIV particles. The function of the *N*-genes is not known yet but since members of this family were first identified within the IV packaged genome, they were primarily hypothesized to be expressed in the parasitized lepidopteran larva and to participate in host physiology manipulation. Identification of *N*-gene proteins associated to virus particles suggests that they may constitute integral 'structural' proteins. On the other hand, *N*-genes may actually encode virulence factors that are delivered to the parasitized host. Based on the distribution of *N*-genes on both IVSPERs and packaged viral DNA, these factors may be simultaneously delivered as proteins, which would be produced in the calyx cells and incorporated within the virus particle, and as genes present in the encapsidated viral DNA. Functional analysis of the corresponding proteins would be necessary to confirm one or the other of the hypotheses.

Other Related Proteins Associated with IV Particles

In addition to the proteins encoded by the *p53*, *p12* and *N*-gene families, four other multigenic families, named IVSP1 to IVSP4, encode proteins associated with IV particles. In *H. didymator*, within a family, the corresponding proteins display 30–65% amino acid identity and over 70% similarity.

At least one of them, IVSP4, appears to be conserved in *Campopleginae* wasps. Indeed, sequences related to the IVSP4 genes were found in both *T. rostrale* cDNA library and *C. sonorensis* EST library. IVSP4 proteins have a predicted size of around 50 kDa and contain two central regions of about eighty amino acids that are 50% identical when the three species are compared (Fig. 6C).

Conclusion: Proteins Associated with IV Particles have no Clear Similarity with Known Viral or Eukaryotic Proteins

To conclude, none of the proteins shown to be associated with IV particles and/or encoded by IVSPERs show significant similarity with known sequences deposited in databases. This result differs from those obtained for bracoviruses, which were shown to employ a nudiviral-like machinery (Bézier *et al.*, 2009a, b).

The IVSPERs appear to contain primarily genes encoding virion components, as shown by mass spectrometry analysis of the HdIV particles. However their actual localization within the virion (nucleocapsid, envelopes, or matrix between the envelopes) remains to be investigated. Moreover, in the absence of a clear conserved functional domain, we have no indication of their function in virion formation or as to their function as virulence protein delivered to the host via the particles. In braconid species, subunits of the RNA polymerase were identified but no DNA polymerase. One hypothesis would be that modern IVs rely either on the wasp machinery to insure both viral DNA transcription and replication or on other viral genes not identified by the EST screen because they are expressed at a lower level.

IVSPERS REPRESENT FINGERPRINTS OF THE IV ANCESTOR

Discovery of large wasp genome regions specialized in production of virus particles that are amplified at the same time and in the same tissue where genome segments replicate raises the question of the link between IVSPERs and the IV ancestor. In addition to being dedicated to IV virion production, the IVSPERs are characterized by specific features. Firstly, IVSPERs display a higher density of coding sequences compared with the rest of the wasp genome. Secondly, none of the IVSPER sequences are predicted to be spliced, contrary to most of the wasp genes present in the flanking regions. These atypical genomic features of IVSPERs strongly suggest that such regions embedded in the wasp genome have an exogenous origin and may represent fingerprints of the IV ancestor that integrated its genome into the genome of an ancestral wasp.

An obvious hypothesis, based on what is known for BVs, would be that IVSPERs have a viral origin. The nature of the IV ancestor still remains unknown since none of the 40 IVSPER predicted proteins show similarity to other viral or eukaryotic proteins. A previous hypothesis suggested by Bigot *et al.* (2008) is that IVs have an ascovirus origin based on structural similarities between CsIV p53 and an ascovirus major capsid protein. However, no similarity was found between HdIV-associated proteins and ascovirus proteins. Therefore, IVs most probably do not derive from ascoviruses, although it cannot be excluded that they share a common ancestor but that sequences have considerably diverged during evolution. No evident similarity with other family of viral proteins was found either (Volkoff *et al.*, 2010). Using prediction tools such as homology detection and structure prediction (HHpred, at toolkit.tuebingen.mpg.de/hhpred) or the PHYRE automatic fold recognition server (at www.sbg.bio.ic.ac.uk/~phyre/), some IVSPER sequences displayed resemblance to proteins belonging to diverse families of DNA viruses, such as pox or orthopox viruses, or baculoviruses. However, similarity scores and matches frequency with a given virus species were not sufficient to conclude a homology between HdIV and another virus.

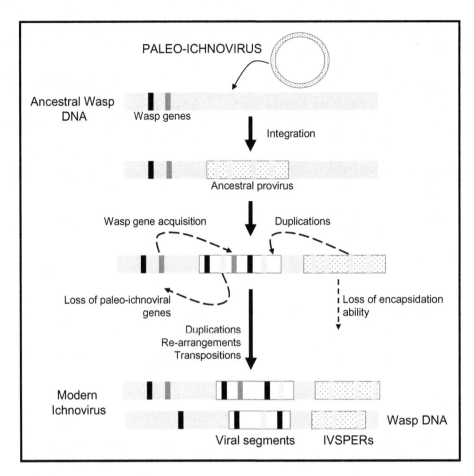

FIGURE 7 Hypothetical scenario of the origin of modern IVs. The genome of an ancestral virus, named 'paleo-ichnovirus', integrated into the genome of an ancestral ichneumonid wasp. Throughout evolution, the sequences of the ancestral provirus were duplicated, rearranged, and they probably acquired wasp genes. Some rearranged proviral regions, the modern IVSPERs, were selected for production of virus structural proteins and have lost their ability to be packaged into the virus particles. Other regions corresponding to the viral segments, which encode virulence factors, were selected for transmission to the lepidopteran host. Note that success of parasitoid development ensures vertical transmission of the viral genome.

One hypothesis is that the ancestral IV has not yet been sequenced. The other would be that IVs evolved rapidly, a fact illustrated by the elevated divergence observed even at the species level between the p12 and p53 proteins, which could lead to difficulties in detecting sequence similarities with other viruses.

In *H. didymator*, the three IVSPERs that were identified are combinations of related genes, as members of at least seven gene families were identified scattered amongst these regions. Although the noncoding sequences are divergent, these IVSPERs are nonetheless probably related to each other and the most plausible hypothesis is that they derive from an ancestral block of sequences (containing one representative of each of the modern gene families), followed by sequence duplication, transposition, and rearrangements (Fig. 7). During evolution, IVSPERs have probably lost their ability to be encapsidated whereas other regions kept this ability. The latter correspond

to the modern segments found within the virus particles. Concomitantly, these segments may have acquired sequences encoding virulence factors from the wasp as shown for BVs (Desjardins *et al.*, 2008). Note that coding sequence density on IV viral segments is more similar to that measured for the wasp genome than to IVSPERs (20–30% according to Tanaka *et al.*, 2007; Webb *et al.*, 2006). An as-yet unresolved issue is the question of whether there exists a larger region ('mega-cluster') of the wasp genome that would include IVSPERs and proviral segments and which could correspond to a highly rearranged ancestral virus genome.

In summary, results obtained so far for *H. didymator* demonstrate that the machinery of IVs associated with Campopleginae parasitic wasps originates (1) from an entity exogenous from the wasp and (2) from an ancestor that is different than that of BVs, although it still remains to be identified.

REFERENCES

Beckage, N.E., 1998. Modulation of immune responses to parasitoids by polydnaviruses. Parasitology 116, S57–S64.

Beckage, N.E., Gelman, D.B., 2004. Wasp parasitoid disruption of host development: implications for new biologically based strategies for insect control. Annu. Rev. Entomol. 49, 299–330.

Belle, E., Beckage, N.E., Rousselet, J., Poirie, M., Lemeunier, F., Drezen, J.-M., 2002. Visualization of polydnavirus sequences in a parasitoid wasp chromosome. J. Virol. 76, 5793–5796.

Bézier, A., Annaheim, M., Herbiniere, J., Wetterwald, C., Gyapay, G., Bernard-Samain, S., et al. 2009. Polydnaviruses of braconid wasps derive from an ancestral nudivirus. Science 323, 926–930.

Bézier, A., Herbiniere, J., Lanzrein, B., Drezen, J.-M., 2009. Polydnavirus hidden face: the genes producing virus particles of parasitic wasps. J. Invertebr. Pathol. 101, 194–203.

Bigot, Y., Samain, S., Auge-Gouillou, C., Federici, B.A., 2008. Molecular evidence for the evolution of ichnoviruses from ascoviruses by symbiogenesis. BMC Evol. Biol. 8, 253.

Deng, L., Stoltz, D.B., Webb, B.A., 2000. A gene encoding a polydnavirus structural polypeptide is not encapsidated. Virology 269, 440–450.

Deng, L., Webb, B.A., 1999. Cloning and expression of a gene encoding a Campoletis sonorensis polydnavirus structural protein. Arch. Insect Biochem. Physiol. 40, 30–40.

Desjardins, C.A., Gundersen-Rindal, D.E., Hostetler, J.B., Tallon, L.J., Fadrosh, D.W., Fuester, R.W., et al. 2008. Comparative genomics of mutualistic viruses of Glyptapanteles parasitic wasps. Genome Biol. 9, R183.

Espagne, E., Dupuy, C., Huguet, E., Cattolico, L., Provost, B., Martins, N., et al. 2004. Genome sequence of a polydnavirus: insights into symbiotic virus evolution. Science 306, 286–289.

Fleming, J.G., Summers, M.D., 1991. Polydnavirus DNA is integrated in the DNA of its parasitoid wasp host. Proc. Natl. Acad. Sci. U.S.A. 88, 9770–9774.

Glatz, R.V., Asgari, S., Schmidt, O., 2004. Evolution of polydnaviruses as insect immune suppressors. Trends Microbiol. 12, 545–554.

Krell, P.J., 1991. Polydnaviridae. In: Adams, J., Bonami, J.R. (Eds.), Atlas of Invertebrate Viruses (pp. 141–177). CRC Press, Boca Raton.

Krell, P.J., Stoltz, D.B., 1980. Virus-like particles in the ovary of an ichneumonid wasp: purification and preliminary characterization. Virology 101, 408–418.

Krell, P.J., Summers, M.D., Vinson, S.B., 1982. Virus with a multipartite superhelical DNA genome from the ichneumonid parasitoid Campoletis sonorensis. J. Virol. 43, 859–870.

Lapointe, R., Tanaka, K., Barney, W.E., Whitfield, J.B., Banks, J.C., Beliveau, C., et al. 2007. Genomic and morphological features of a banchine polydnavirus: comparison with bracoviruses and ichnoviruses. J. Virol. 81, 6491–6501.

Malva, C., Varricchio, P., Falabella, P., La Scaleia, R., Graziani, F., Pennacchio, F., 2004. Physiological and molecular interaction in the host–parasitoid system Heliothis virescens–Toxoneuron nigriceps: current status and future perspectives. Insect Biochem. Mol. Biol. 34, 177–183.

Murphy, N., Banks, J.C., Whitfield, J.B., Austin, A.D., 2008. Phylogeny of the parasitic microgastroid subfamilies (Hymenoptera: Braconidae) based on sequence data from seven genes, with an improved time estimate of the origin of the lineage. Mol. Phylogenet. Evol. 47, 378–395.

Pennacchio, F., Strand, M.R., 2006. Evolution of developmental strategies in parasitic hymenoptera. Annu. Rev. Entomol. 51, 233–258.

Stoltz, D.B., 1990. Evidence for chromosomal transmission of polydnavirus DNA. J. Gen. Virol. 71 (Pt 5), 1051–1056.

Stoltz, D.B., Guzo, D., Cook, D., 1986. Studies on polydnavirus transmission. Virology 155, 120–131.

Stoltz, D.B., Vinson, S.B., 1977. Baculovirus-like particles in the reproductive tracts of female parasitoid wasps. II. The genus Apanteles. Can. J. Microbiol. 23, 28–37.

Stoltz, D.B., Vinson, S.B., 1979. Penetration into caterpillar cells of virus-like particles injected during oviposition by parasitoid ichneumonid wasps. Can. J. Microbiol. 25, 207–216.

Tanaka, K., Lapointe, R., Barney, W.E., Makkay, A.M., Stoltz, D., Cusson, M., et al. 2007. Shared and species-specific features among ichnovirus genomes. Virology 363, 26–35.

Theilmann, D.A., Summers, M.D., 1986. Molecular analysis of Campoletis sonorensis virus DNA in the lepidopteran host Heliothis virescens. J. Gen. Virol. 67, 1961–1969.

Volkoff, A.N., Jouan, V., Urbach, S., Samain, S., Bergoin, M., Wincker, P., et al. 2010. Analysis of virion structural components reveals vestiges of the ancestral ichnovirus genome. PLoS Pathog. 6, e1000923.

Volkoff, A.N., Ravallec, M., Bossy, J.P., Cerutti, P., Rocher, J., Cerutti, M., et al. 1995. The replication of Hyposoter didymator polydnavirus: cytopathology of the calyx cells in the parasitoid. Biol. Cell 83, 1–13.

Webb, B.A., 1998. Polydnavirus biology, genome structure, and evolution. In: Miller, L.K., Ball, L.A. (Eds.), The Insect Viruses (pp. 105–139). Plenum Publishing Corporation, New York.

Webb, B.A., Strand, M.R., Dickey, S.E., Beck, M.H., Hilgarth, R.S., Barney, W.E., et al. 2006. Polydnavirus genomes reflect their dual roles as mutualists and pathogens. Virology 347, 160–174.

Webb, B.A., Summers, M.D., 1990. Venom and viral expression products of the endoparasitic wasp Campoletis sonorensis share epitopes and related sequences. Proc. Natl. Acad. Sci. U.S.A. 87, 4961–4965.

Wetterwald, C., Roth, T., Kaeslin, M., Annaheim, M., Wespi, G., Heller, M., et al. 2010. Identification of bracovirus particle proteins and analysis of their transcript levels at the stage of virion formation. J. Gen. Virol. 91, 2610–2619.

Whitfield, J.B., Asgari, S., 2003. Virus or not? Phylogenetics of polydnaviruses and their wasp carriers. J. Insect Physiol. 49, 397–405.

Chapter 4

Genomics and Replication of Polydnaviruses

Catherine Dupuy[*], Dawn Gundersen-Rindal[†] and Michel Cusson[‡]

[*]*Institut de Recherche sur la Biologie de l'Insecte, CNRS UMR 6035, Université François-Rabelais, Faculté des Sciences et Techniques, Parc Grandmond, 37200 Tours, France*

[†]*US Department of Agriculture, Agricultural Research Service, Invasive Insect Biocontrol and Behavior Laboratory, Room 267, 10300 Baltimore Avenue, Bldg 011A BARC-WEST Beltsville, MD 20705, U.S.A.*

[‡]*Laurentian Forestry Centre, Canadian Forest Service, Natural Resources Canada, 1055 rue du PEPS, Québec, QC, G1V 4C7, Canada*

SUMMARY

Most large DNA viruses have a pathogenic association with their hosts, with symbiosis being a rare exception. A striking example of the latter, however, is found among DNA viruses of the family Polydnaviridae (PDV), which have evolved a complex association with wasps that live as endoparasitoids of lepidopteran larvae. PDVs replicate asymptomatically in their wasp hosts but infect and cause severe disease in parasitized caterpillars. The two recognized PDV taxa, ichnoviruses (IVs) and bracoviruses (BVs), are associated with endoparasitic wasps of the families Ichneumonidae and Braconidae, respectively, and have distinct ancestors. Here, we survey the available data on the genome sequence and gene content of members of these two taxa. A comparison of the two groups shows that, despite their distinct origins, IV and BV genomes display similar organizational features. However, they share relatively few genes, which have diversified into multigene families in both taxa. We also review what is known about mechanisms of PDV replication in the wasp host. These viruses constitute a versatile, replication-defective system for delivery of virulence genes to parasitized host insects.

ABBREVIATIONS

BV	bracovirus
BEN	proteins with BEN domain
CrV	*Cotesia rubecula* V
DRJ	direct repeat junction
Egf-like	epidermal growth factor-like
EP1	'early-expressed protein 1'
Glc	glycosylated central domain proteins
H4	histone-4
Inx	innexins
IV	ichnovirus
IVSPERs	ichnovirus structural proteins encoding regions
MdBV	*Microplitis demolitor* BV
ORF	open reading frame
PDV	polydnavirus
PTP	protein tyrosine phosphatase
TE	transposable element
TrIV	*Tranosema rostrale* IV
vank	viral ankyrin

INTRODUCTION

The two recognized polydnavirus (PDV) taxa, ichnoviruses (IVs) and bracoviruses (BVs), share an obligate mutualistic association with endoparasitic wasps belonging to the families Ichneumonidae and Braconidae, respectively. Most studies on IVs have focused on the viral entities discovered in wasps of the subfamily Campopleginae. However, similar yet distinct viruses have also been described from wasps belonging to the subfamily Banchinae (Lapointe *et al.*, 2007), although it remains to be determined whether these two virus types have a common ancestor (Cusson *et al.*, Chapter 6 of this volume). At present, it has been clearly established that campoplegine IVs originated from a viral entity that is distinct from the recently identified nudiviral ancestor of BVs, demonstrating that the obligate mutualistic association between wasps and viruses arose at least twice during the evolution of parasitic wasps (Bézier *et al.*, 2009a, b; Volkoff *et al.*, 2010).

Despite their distinct origins, IVs and BVs share many features. In both groups, replication and virion assembly are restricted to specialized calyx cells of the parasitoid wasp ovaries. In addition, they share an atypical life-cycle distributed between two hosts, the primary parasitoid wasp host and the secondary host, typically a lepidopteran larva, in which virions are injected during oviposition and where viral genes are expressed to ensure successful parasitism (Dupuy *et al.*, 2006). In the primary wasp host, PDV genomes are present as integrated within the wasp chromosomes. Thus, PDV genomes exist in two forms: (1) integrated in the wasp genome ('provirus'); and (2) packaged in virions.

Interestingly, the extent of the integrated PDV genome has not yet been clearly defined, and has been a topic of much recent interest. One of the most unique features of these atypical viruses is that their genomes are made up of two types of genes, both of which reside permanently within the wasp chromosomes and are transmitted vertically. Together, they form the full PDV genome (Bézier *et al.*, 2009b). However, the genome that is packaged in the replication-deficient particles during virogenesis in wasp ovaries does not have the entire PDV gene complement. This packaged genome is made up of several segments, generated from the proviral genome through a process involving amplification, excision and circularization, and contains only virulence genes, none of which show similarity to genes involved in viral replication and synthesis of structural proteins (Dupuy *et al.*, 2006). These viral genes constitute the non-encapsidated, replication-associated components of BV and IV genomes and reside exclusively within the parasitoid genome. In BVs, these non-encapsidated genes have been identified as nudivirus-like genes and may represent remnants of the nudiviral ancestor (Bézier *et al.*, 2009a, b). Some of these genes are organized into large clusters in a macrolocus whereas others are dispersed in the wasp genome. In IVs, the non-encapsidated genes involved in particle production are clustered in specialized regions of the wasp genome that are amplified along with proviral DNA during virus particle replication, and are referred to as IVSPERs ('ichnovirus structural proteins encoding regions'; Volkoff *et al.*, 2010).

Genome sequencing and analyses undertaken to elucidate the genetic features of PDVs that relate to their unusual biology have tended to focus primarily on the genome packaged within the virions isolated from the reproductive tract of female wasps. These genome studies have been conducted for representative members from the two PDV taxa. The BV genomes sequenced to date are all from wasps of the subfamily Microgastrinae, one of the seven subfamilies that constitute the microgastroid complex, with which BVs are associated (Cheloninae, Dirrhoponae, Mendesellinae, Khoikhoiinae, Cardiochilinae, Miracinae and Microgastrinae) (Murphy *et al.*, 2008; Whitfield, 2002). BV genomes sequenced to date include those of *Cotesia congregata* BV (*Cc*BV) (Espagne *et al.*, 2004), *Microplitis demolitor* BV (*Md*BV) (Webb *et al.*, 2006), *Glyptapanteles flavicoxis* (*Gf*BV) and *Glyptapanteles indiensis* (*Gi*BV) (Desjardins *et al.*, 2008, 2007). In addition, the genome of *Cotesia vestalis* BV (*Cv*BV previously named *Cotesia plutellae* BV) has been almost completely sequenced (Choi *et al.*, 2009, 2005) and several other sequencing projects have allowed the acquisition of partial data on other BV genomes (see footnote 1 of Table 1). Campoplegine IV genomes sequenced to date include those of *Campoletis sonorensis* IV (*Cs*IV) (Webb *et al.*, 2006), *Hyposoter fugitivus* IV (*Hf*IV) and *Tranosema rostrale* IV

(*Tr*IV) (Tanaka *et al.*, 2007), whereas the only banchine IV genome sequenced is that of *Glypta fumiferanae* IV (*Gf*IV) (Lapointe *et al.*, 2007) (see Cusson *et al.,* Chapter 6 of this volume). These data allow comparative analyses of bracovirus and ichnovirus genomes, from which one can extract (1) features that are common to both groups and which likely reflect the parasitic and mutualistic lifestyle of these viruses, and (2) features that are specific to individual viruses. Comparative analyses of these genomes may increase understanding of the differential constraints imposed by host range and the pathological effects caused by individual PDVs.

GENERAL FEATURES OF PDV GENOMES

Common Genomic Organization Among PDV Genomes

PDV Genomes Contained Within Virion Particles are Highly Segmented

A hallmark of all PDV genomes, despite their different origins, is their large size and their segmentation, consisting of multiple dsDNA circular molecules. Although common in RNA viruses, genome segmentation is rare in DNA viruses, but reaches a high degree in both BVs and IVs.

The BV genomes sequenced to date have vastly different aggregate genome sizes, ranging from 189 kbp (*Md*BV) to 606 kbp (*Cc*BV), which are comprised of 15 and 32 segments, respectively (Fig. 1). Among BV genomes, those of *Cc*BV, *Gi*BV and *Gf*BV are most similar, with genomes averaging 550 kbp organized into 29–32 segments ranging in size from 3.611 bp to 50,691 bp. *Md*BV is the smallest BV genome sequenced to date and is organized in 15 segments that, together, have an aggregate size of 189 kbp (Webb *et al.*, 2006). Except TrIV, the two other campoplegine IV genomes fully sequenced to date have similar aggregate sizes of ~250 kbp. IV genomes contain greater than 20 segments, each ranging in size from 2.6 to 19.6 kbp. The *Hf*IV genome is highly segmented, composed of 56 unique segments. This degree of genome segmentation is higher than that assessed for *Tr*IV or *Cs*IV, making *Hf*IV the most highly segmented campoplegine IV genome characterized to date (Tanaka *et al.*, 2007). Its genome segments range in size from 2.6 to 8.9 kbp and its aggregate genome size is estimated at 246 kbp. Moreover, unlike BV genome segments, some IV segments exhibit a phenomenon referred to as segment nesting, with certain segments giving rise to smaller 'daughter' or 'nested' segments, which are generated by intramolecular recombination of larger genome segments (Kroemer and Webb, 2004; Tanaka *et al.*, 2007). For example, 11 of the 56 segments of the *Hf*IV genome are nested. These data reveal significant interspecific differences in the degree of

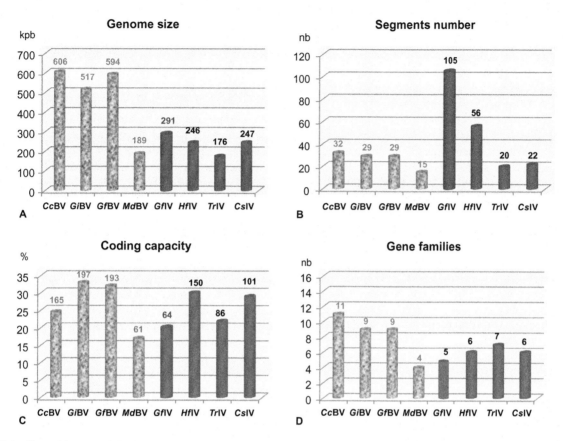

FIGURE 1 General features of the BV and IV genomes entirely sequenced to date. The characteristic of the PDV genome bearing by the banchine *Glypa fumiferana* (*Gf*V) is shown as indication. For more precision refer to Cusson *et al.*, Chapter 6 of this volume. **A:** Aggregate genome size. **B:** Number of genome segments per genome. **C:** Percentage of coding sequence in each genome. The number upper each column indicates the number of ORF. **D:** Number of gene families. Adapted from Tanaka *et al.*, 2007.

BV and IV segmentation. In BVs, there appears to be a relationship between genome size and the number of segments, with highly segmented genomes being the largest. In IVs, there are significant interspecific differences in the degree of genome segmentation even though aggregate genome sizes estimated are similar (Tanaka *et al.*, 2007) (Fig. 1). Thus, there is no clear relation between genome size and segmentation, nor is it clear which selective pressure drives PDV genome segmentation up or down. Indeed, the genomes of nudiviruses, from which BVs originated (Bézier *et al.*, 2009a) and of the related baculoviruses are non-segmented, although their genomes are large (~200 kb), indicating that a single capsid of such a virus is capable of holding up to 200 kbp. Moreover, extreme segmentation does not lead to enhanced coding capacity of the overall genome, as BVs contain approximately the same number of genes as described for AcMNPV (155 open reading frames (ORFs)) (Ayres *et al.*, 1994) and Hz_1 (154 ORFs) (Cheng *et al.*, 2002), versus 61–197 ORFs in BVs (Fig.1). Differences in the degree of segmentation could have a purely mechanistic origin, with variation in segmentation, within defined boundaries, being under little or no selective pressure

(Tanaka *et al.*, 2007). This could explain why, in a comparison of two BVs from wasps from a single genus (i.e., *Gi*BV and *Gf*BV), some segments appear to be unique to either *Gi*BV (segments 16 and 30) or *Gf*BV (segments 27 and 31) (Desjardins *et al.*, 2008).

Another genome property shared by BVs and IVs is the presence of specific segments in non-equimolar concentration in the calyx fluid, with some segments being present at higher copy number than others. The relative abundance of the 15 *Md*BV segments, as assessed by relative quantitative real-time PCR using segment specific primers and viral DNA from calyx fluid as template, revealed that five genomics segments (J, O, H, N, and B) accounted for more than 60% of the viral DNA (Beck *et al.*, 2007). In *Cv*BV, compared to the least abundant segment used as calibrator (segment C34), the other segments are 1.33–66.5-fold more abundant. The six more abundant segments (C1, C12, C14, C35, C15, C7) account for more than 35% of the viral DNA in calyx fluid (Chen *et al.*, 2011). In *Ci*BV, measurement of the relative abundance of six segments showed highest relative abundance for two segments (*Ci*V12 and *Ci*V16.8), the four other segments showing values between 7 and

25% compared with *Ci*V16.8 (Annaheim and Lanzrein, 2007). Interestingly, these authors suggested for the first time that, for each segment analyzed, abundance of proviral and excised forms vary in parallel, suggesting that abundant segments are present in multiple copies in the proviral genome. Hypermolar segments are also common in IV genomes. In *Cs*IV this was evident from random sequencing of genomic clones and prior electrophoretic and hybridization studies, which provided evidence for the presence of three high abundance segments (Blissard *et al.*, 1989; Cui *et al.*, 2000; Kroemer and Webb, 2004; Strand, 1994; Strand *et al.*, 1997). The organization of PDV genomes into multiple segments of variable abundance adds another layer of complexity that is also potentially important for function. In the absence of replication, the copy number of a given gene in a parasitized host is determined by the abundance of the corresponding viral genome segment within the viral inoculuum injected by the wasp at oviposition. Gene delivery to host cells could also be affected by how genomic segments are packaged into virions and which tissues virions infect (Beck *et al.*, 2007; Kroemer and Webb, 2004).

Encapsidated PDV Genomes Have Low Coding Density

Despite their large size, PDV genomes display a remarkably low coding density. Indeed, the second feature shared by both genomes is a strong A/T bias and gene density that is the lowest reported for any virus. IV genomes display coding densities of 22% (*Tr*IV) to 30% (*Hf*IV) and A/T biases (57–59%) that are similar to those observed in BVs, which display coding densities from 17% (*Md*BV) to 33% (*Gi*BV) and A/T biases from 65 to 66% overall (Desjardins *et al.*, 2008; Espagne *et al.*, 2004; Tanaka *et al.*, 2007; Webb *et al.*, 2006). Interestingly, the A/T bias is also a characteristic of hymenopteran genomes in general and of nudiviruses (Wang and Jehle 2009; Werren *et al.*, 2010). Except for *Hf*IV, in which coding density is higher among larger genome segments (>5 kbp: 38.9%) than among smaller ones (< 5 kbp: 26.8%) (Tanaka *et al.*, 2007), there is no direct link between the length of the segment and coding density. In IVs, a more extensive analysis allowed the identification of promoters and polyadenylation signals. Indeed, all predicted ORFs contain putative upstream promotor elements and polyadenylation signals. One and five promoters have been assigned to two ORFs in *Tr*IV and *Hf*IV, respectively. The predicted 5′ UTR lengths range from 1 to 2899 bp (mean of 302) and from 3 to 2315 bp (mean of 246) for *Hf*IV and *Tr*IV, respectively (Tanaka *et al.*, 2007). Among BVs, 58–69% of *Gi*BV and *Cc*BV genes have been predicted to contain introns (Desjardins *et al.*, 2008, 2007; Dupuy *et al.*, 2006; Espagne *et al.*, 2004; Webb *et al.*, 2006), whereas Webb *et al.* (2006), using sequence analysis combined with prior analysis of *Cs*IV and *Md*BV transcripts, predicted only 10% of *Cs*IV and 14% of *Md*BV genes to be spliced respectively.

Interestingly, due to the limited transcriptional data available for BVs, there is substantial disagreement on the structural complexity of *Cc*BV genes, particularly with regards to the percentage of PDV genes that contain introns. While Espagne *et al.* (2004) predicted that 69% of *Cc*BV protein-coding genes contain introns, Webb *et al.* (2006) re-annotated the *Cc*BV genome and predicted that only 6.8% of *Cc*BV genes contain introns, a 10-fold difference in intron content. Deep sequencing approaches should allow the resolution of these discrepancies using transcriptome analyses of PDV genomes expression in parasitized hosts.

Overall, PDV genomes are largely noncoding. Webb *et al.* (2006) analyzed these noncoding sequences and found that, in total, the genomes of *Md*BV and *Cs*IV contain 86 and 74 kb of repetitive DNA, encompassing approximately 15% of the genomic DNA in each virus. In *Wolbachia*, an intracellular symbiotic bacterium that infects a variety of arthropods and nematodes, the presence of a high percentage of repeats and noncoding sequence has also been observed and the authors suggested that this may reflect either very weak selection for its elimination or increased exposure to mobile elements as a consequence of a parasitic lifestyle (Moran, 2003; Wu *et al.*, 2004). In PDVs, 23 and 16 sequences showing homology to transposable elements have been identified in the noncoding sequences of *Cs*IV and *Md*BV, respectively; these TEs display no homology to one another, suggesting that their invasions occurred independently (Webb *et al.*, 2006). PDV TEs form a heterogenous group, as they belong to different classes: ACCORD1 retrotransposon from *Drosophila* (in *Cs*IV), *Drosophila* p-element like (in *Gi*BV), and *Maverick* element (in *Cc*BV) (Desjardins *et al.*, 2008; Drezen *et al.*, 2006; Webb *et al.*, 2006; Dupuy *et al.*, 2011). Interestingly, *Maverick* elements are also present in the flanking sequence of the *Gf*BV proviral locus 7 and in the genomes of the parasitoid wasps *C. congregata* and *C. sesamiae*, suggesting that *Cc*BV acquired this element from the wasp genome (Desjardins *et al.*, 2008; Drezen *et al.*, 2006; Webb *et al.*, 2006; Dupuy *et al.*, 2011). The presence of these TEs in both wasp and BV genomes suggests that, as hypothesized for *Wolbacchia*, the PDV genomes may have undergone extensive shuffling. Even if they are not complete and not functional, these elements (or their remnants) may be maintained in the PDV genome because there is no specific constraint in the length of the genome that might be encapsidated, or because they provide selective benefit by contributing to genome plasticity.

Characteristics of the Proviral Genome
Organization of the Integrated BV Proviral Genome

Braconid proviral genome segments were originally thought to be located at a single chromosomal locus in a tandem array. This was based on analyses of *C. congregata* and

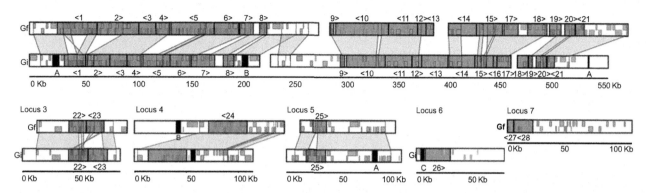

FIGURE 2 **Structure and organization of bracovirus proviral genome segments within the parasitoid genome: macrolocus and additional loci.**
Bracoviruses *Glyptapanteles flavicoxis* (Gf) and *G. indiensis* (Gi) proviral genome segments (gray) assembled from BAC clone data (Desjardins *et al.*, 2008) and their synteny are shown. For each, the corresponding segment number of the encapsidated viral genome segment form is given above the proviral genome segment, with the symbols > or < depicting the directionality of the segment excision regulatory signals. Protein coding genes within proviral genome segments are indicated by light blue boxes; genes encoded in parasitoid flanking DNA at each locus are indicated by light purple boxes. Regions of synteny between proviral segment and flanking DNA are shaded in light gray. Three different long tandem repeat classes (A, B, and C) are denoted by black boxes. Locus sizes are given in kilobase pairs. Figure modified from Desjardins *et al.*, 2008. Please see color plate section at the back of the book.

C. inanitus proviral genome segments, which were flanked on one or both sides by a different proviral segment (Savary *et al.*, 1997; Wyder *et al.*, 2002). Moreover, cytological evidence showed that probes from at least three unique *Cc*BV genome segments hybridized *in situ* on metaphase spreads to a common site and were likely clustered in a macrolocus on the short arm of subtelocentric wasp chromosome 5 (Belle *et al.*, 2002). Recently, linkage analyses of BV provirus structure and composition were undertaken through isolation and analysis of large parasitoid genome regions obtained either by using inverse amplification from known sequences within *C. inanitus* (Annaheim and Lanzrein, 2007) or through cloning, sequencing, and characterizing of numerous BAC clones containing long stretches of integrated provirus from *G. flavicoxis* and *G. indiensis* (Desjardins *et al.*, 2008, 2007). This has led to new discoveries as to the nature, structure, and organization of proviral genome segments within the parasitoid genome.

Based on a comprehensive analysis of proviral data, presented in Fig. 2 for *G. flavicoxis* and *G. indiensis*, most proviral genome segments found in the packaged genome are present in a macrolocus that is greater than 550 kb in size and which contains approximately 70% of the corresponding encapsidated viral form. These macrolocus-associated proviral genome segments are arranged in an imperfect tandem array, with each proviral genome segment being separated by small intersegmental regions of noncoding or spacer DNA, ranging in size from 117 bp to greater than 8 kb (Annaheim and Lanzrein, 2007; Desjardins *et al.*, 2008, 2007). Five additional proviral loci were identified, each containing one or two proviral genome segments; these loci likely reside in proximity to the macrolocus but could potentially reside in other region(s) of the parasitoid genome. Comparative sequence analyses (Fig. 2) among proviral

loci encoding the *Glyptapanteles* proviral genome segments revealed a high degree of synteny in provirus as well as flanking parasitoid DNA.

Conservation of the DRJ Excision Site Among BVs

Examination of proviral excision sites for BVs revealed a direct DNA sequence repeat (direct repeat junction, or DRJ) conserved sequence motif (GCTT) at the boundaries of proviral genome segments marking the site of proviral segment excision. There also appears to be directionality to the mechanism of circular segment excision from the site of proviral segment integration as indicated by differences between the extended sequence motifs at each end of a proviral segment (Desjardins *et al.*, 2008, 2007). The process of excision and circularization of viral genome segments from the parasitoid genome in calyx cells has been proposed to occur through juxtaposition of direct terminal repeats, 5′ and 3′ on each proviral genome segment, followed by recombination at the repeats involving a recombinase or recombination-associated enzyme (the process of BV replication is described in the secion on replication of pdvs below). A single copy of the repeat is retained within each corresponding circular genome segment excised from a progenitor and then encapsidated. The viral circular genome segment and the rejoined proviral DNA each retain a single DRJ repeat motif (Gruber *et al.*, 1996; Pasquier-Barre *et al.*, 2002; Savary *et al.*, 1997). Nearly every bracovirus genome segment sequenced to date contains the conserved DRJ excision site motif, including those from the most basal taxon, *Ci*BV, supporting the hypothesis of a highly conserved mechanism for BV proviral segment excision (Annaheim and Lanzrein, 2007; Desjardins *et al.*, 2008, 2007).

TABLE 1 Number of Genes Present in the Related Gene Families in BVs (A), Campoplegine IVs (B) and Banchine IV (C) Sequenced to Date[1]

A. BVs

Number of genes/family/genome

Present in:		5 BVs		4 BVs										3 BVs			2 BVs		1 BV	
Virus	# of ORFs	PTP[2]	Vank	BEN	BV4	Cys	CrV1-like	P94	Cystatin	RNase T2	EP1-like	C-type lectin	BV2	BV3	Duffy	BV1	ST	H4	Glc	Egf-like
CcBV[3]	165	27	8	12	2	4	1	2	3	2	6	1	6	2	1	2	0	1	0	0
CvBV[4]	125	36	8	11	3	0	1	4	3	3	7	1	3	1	1	5	0	1	0	0
GiBV[5]	197	42	9	9	2	1	3	1	1	2	2	2	6	2	1	0	3	1	0	0
GfBV[5]	193	32	8	6	2	2	2	1	2	2	2	5	5	4	0	0	5	1	0	0
MdBV[6]	61	13	12	1	1	2	0	0	0	0	0	0	0	0	0	0	0	0	2	6

B. Campoplegine IVs

Number of genes/family/genome

Present in:				3 IVs						1 IV	
Virus	# of ORFs	Rep	PTP	Vank	Cys	Inx	PRRP	N-gene		TrV	OSSP
CsIV[6]	101	30	23	7	10	4	5	2		0	0
HfIV[7]	150	38		9	5	11	11	3		0	0
TrIV[7,8]	86	17		2	1	3	1	4		7	4

C. Banchine IVs

Number of genes/family

Virus	# of ORFs	PTP	Vank	NTPase-like	BV-like	Recombinase-like (MULE)
GfV[9,10]	64	23	4	9	4	1

[1]Data presented in this table are restricted to those reported for fully or near-fully sequenced genomes. Additional data may be gleaned from various other studies and the NCBI database where individual or groups of genes were cloned and characterized for other polydnaviruses, including Toxoneuron nigriceps BV (TnBV), Chelonus inanitus BV (CiBV), Cotesia karyai BV (CkBV), Cotesia glomerata BV (CgBV), Cotesia rubicula BV (CrBV), and Hyposoter dydimator IV (HdIV) and Campoletis chlorideae IV (CcIV); see text for details.

[2]Gene family abbreviations: **PTP**, protein tyrosine phosphatase; **Vank**, viral ankyrin; **BEN**, proteins with BEN domain (CcBV hp2 in Espagne et al., 2004); **BV4**, BV family 4 (new family discovered since publication of Espagne et al., 2004); **Cys**, Cys-motif (or cystein-rich) proteins; **CrV1-like**, homologs of a gene first identified in CrBV; **P94**, related to P94 baculovirus protein; **EP1-like**, homologs of "early-expressed protein 1" of CcBV; **BV2**, BV family 2 (CcBVf2 in Espagne et al., 2004); **BV3**, BV family 3 (CcBV hp1 in Espagne et al., 2004); **BV1**, BV family 1 (CcBVf1 in Espagne et al., 2004); **ST**, sugar transporter; **H4**, histone-4-like; **Glc**, glycosylated central domain proteins; **Egf-like**, epidermal growth factor-like proteins; **Inx**, innexins; **PRRP**, polar-residue-rich proteins; **Recombinase-like**, homolog of a hypothetical protein from CiBV; **TrV**, homologs of TrV1 from TrIV; **OSSP**, ovary-specific secreted proteins; **Rep**, repeat-element; **Duffy**, proteins with Duffy-binding-like domain; **BV-like**, homologs of a CvBV-specific hypothetical protein. PTP, Vank and Cys proteins are color-coded to highlight their presence in two or three of the virus groups shown here.

[3]Cotesia congregata BV: Espagne et al., 2004; A. Bézier, personal communication.
[4]Cotesia plutellae BV (= C. vestalis BV): Choi et al., 2009; NCBI.
[5]Desjardins et al., 2008; NCBI.
[6]Webb et al., 2006.
[7]Tanaka et al., 2007.
[8]Rasoolizadeh et al., 2009.
[9]Lapointe et al., 2007.
[10]Cusson et al., Chapter 6 of this volume.

GENETIC CONTENT OF PDV GENOMES

Absence of Viral Genes

Unlike conventional viruses, which, within their respective families, share sets of conserved genes involved in DNA replication (polymerase or helicase, transcription machinery, virion structural proteins), the packaged PDV genome lacks such genes, explaining the absence of BV and IV replication in parasitized hosts (Dupuy *et al.*, 2006; Kroemer and Webb, 2004; Webb and Strand, 2005). These replication-associated genes identified recently in both BVs and IVs (Bézier *et al.*, 2009a, b; Volkoff *et al.*, 2010; Wetterwald *et al.*, 2010), reside permanently within the wasp genome and are not encapsidated. With respect to the genes encoding structural components, a clear dichotomy exists between BVs and IVs. The packaged genomes of BVs do not contain any of the structural nudivirus-like genes implicated in virion assembly and morphogenesis identified to date (Bézier *et al.*, 2009a, b; Volkoff *et al.*, 2010; Wetterwald *et al.*, 2010). In IVs, a relationship between genes contained in some IVSPERs and those packaged in virus particles has been observed. First, it has been demonstrated that some IVSPERs of *Hyposoter didymator* IV (*Hd*IV) encode for members of the N-gene family, previously described in the packaged genome of *Cs*IV (Webb *et al.*, 2006), *Hyposoter fugitivus* and *Tr*IV (Tanaka *et al.*, 2007). Second, synteny between *Hd*IV IVSPER and segment SH-C of *Cs*IV has been observed. Curiously, both the N-gene family and the corresponding region of the IVSPER-2 were not present in the encapsidated *Hd*IV genome. Volkoff *et al.* (2010) suggested that IV genomes may be at different stages of evolution, and that in the *H. didymator* lineage, IVSPER-2 (but not the corresponding region in *Cs*IV) may have lost the ability to be packaged.

Genes of PDV Packaged Genomes are Organized in Gene Families

A unique and distinctive feature of all PDV packaged genomes examined to date is the diversification of their genes into multigene families. This observation applies to both IVs and BVs, although the number of families identified so far in BVs exceeds that seen in IVs, irrespective of whether campoplegine and banchine IVs are considered as a single entity or as two separate groups (Table 1). Not surprisingly, given their distinct evolutionary origins, IV and BV genomes share relatively few gene families, the majority of which appear to have been acquired from the wasp host genome (as opposed to being of viral origin).

The viral ankyrins (vank) represent the only gene family that is shared by all PDV genomes sequenced to date (Table 1). Vank genes encode truncated homologs of inhibitor κβ (Iκβ; e.g. *Cactus* in *D. melanogaster*), a protein known to inhibit the translocation of the NFκβ transcription factor to cell nuclei. PDV vank proteins contain only the last four of the six ankyrin repeats seen in *Cactus* and lack both the N-terminal signal receiving domain and the PEST domain (polypeptide sequences enriched in proline (P), glutamate (E), serine (S) and threonine (T)) involved in protein turnover. As a result of such domain erosion, binding of vank proteins to NFκβ appears irreversible, thereby causing retention of the latter in the cytoplasm and preventing its regulatory action on the expression of genes involved in the immune response. The *Cs*IV vank-1 protein has a role as inhibitor of apoptosis in insect cells (Fath-Goodin *et al.*, 2009) and it has also been suggested that vank proteins could inhibit the expression of molecules involved in the encapsulation of parasitoid eggs or larvae and/or the antiviral responses directed at PDV virions (Thoetkiattikul *et al.*, 2005; Falabella *et al.*, 2007; Bae and Kim 2009).

Cys-motif (Cys) proteins are present in campoplegine IVs, but also in BVs, where they were originally referred to as cystein-rich proteins (crp; Espagne *et al.*, 2004) (Table 1). These gene products share a cystein motif known to confer a cystein-knot structural fold that imparts stability to the protein (Einerwold *et al.*, 2001). *Cys* genes typically contain introns and encode secreted proteins whose functions have been examined in the *Heliothis virescens–Campoletis sonorensis* host–parasitoid system. *Cs*IV Cys proteins bind to host hemocytes and expression of one of them, using a baculovirus expression system, was shown to reduce the encapsulation response towards wasp eggs (Li and Webb, 1994), suggesting a role in the inhibition of the cellular immune response. These proteins have also been shown to cause developmental delay and growth reduction in parasitized hosts (Fath-Goodin *et al.*, 2006), an effect that could result from their translation inhibitory properties (Kim, 2005).

Protein tyrosine phosphatases (PTPs) represent one of the most diversified PDV gene families studied to date, with members found in both BVs and banchine IVs. This is the family with the greatest number of genes recorded for a given PDV to date, with 42 members in *Gi*BV (Table 1). Among BVs, they have been found in viruses associated with both microgastrine and cardiochiline wasps (Provost *et al.*, 2004). The recombinant products of some of these genes have been shown to encode functional PTPs, which likely play a role in signal transduction. However, several polydnaviral PTPs are predicted to be inactive as phosphatases due to the lack of a critical cysteine residue in the active site, and it has been suggested that such products could be involved in trapping phosphorylated proteins (Provost *et al.*, 2004). Roles in both host developmental regulation and cellular immune response have been proposed for the active PTPs, namely through the dephosphorylation of proteins involved in ecdysone biosynthesis (therefore inhibiting its production and blocking the molt) and through the disruption of host signal transduction pathways controlling hemocyte cytoskeleton dynamics during encapsulation (Provost *et al.*, 2004). Interestingly, a *Cv*BV genome segment encoding seven PTPs has been shown to cause prolonged larval development following injection in *P. xylostella* larvae (Kwon *et al.*, 2010). Furthermore, transient expression of biochemically active *Md*BV two PTPs (PTP-H2 or PTP-H3) in *Drosophila* S2 cells led to reduction of phagocytosis of *Escherichia coli* by these cells (Pruijssers and Strand, 2007). In addition, PTPs could be involved in the apoptosis of granulocytes, therefore contributing to immunosuppression (Suderman *et al.*, 2008; Eum *et al.*, 2010).

Almost as diversified as the PTP family, the *repeat element* (rep) gene family constitutes the family with the greatest number of members among campoplegine IVs (Table 1), indicating that they likely play an important role in IV biology (Volkoff *et al.*, 2002). These genes owe their name to the presence of an imperfectly conserved ~540 bp repeat, first discovered in *Cs*IV (Theilmann and Summers, 1987). *Rep* genes are not spliced and encode proteins lacking a signal peptide, indicating that they are not secreted. Because these proteins display no similarity to known proteins, their functional analysis has proven difficult and their function remains unknown. On the basis that some *rep* genes are expressed primarily in the lepidopteran host while others are expressed more strongly in the wasp ovary, in the absence of any apparent pathology (Hilgarth and Webb 2002; Galibert *et al.*, 2006; Rasoolizadeh *et al.*, 2009a), it has been suggested that *rep* genes could play a role in the maintenance of homeostasis in IV-infected cells (Rasoolizadeh *et al.*, 2009b).

Among the remaining gene families that are shared by two or more PDV genomes, the bracoviral cystatins, C-type lectins, CrV1-like, EP1-like, BEN-domain and H4 proteins, as well as the ichnoviral innexins have been the focus of functional studies. The cystatins, found in all BV genomes sequenced to date except *Md*BV (Table 1), are functional cysteine protease inhibitors (Espagne *et al.*, 2004), and their rapid and strong expression shortly after parasitization suggests that they play an important role in host physiological manipulation. Compared to other PDV gene families such as the PTPs, cystatin genes from different *Cotesia* BVs show limited divergence, suggesting that they form a relatively young gene family (Serbielle *et al.*, 2008).

Bracoviral C-type lectins are secreted proteins encoded by intron-containing genes. In *Plutella xylostella* larvae parasitized by *C. plutellae*, C-type lectins are expressed shortly after oviposition and coat the wasp egg, contributing to its protection against encapsulation (Lee *et al.*, 2008; Glatz *et al.*, 2003).

The CrV1 gene was first identified in the *Cotesia rubecula* BV (*Cr*BV) and its expression characterized in the lepidopteran host *Pieris rapae* (Asgari *et al.*, 1996); homologs have since been identified in several other BVs (Table 1). CrV1 is a secreted protein whose transient expression takes place primarily in host hemocytes and fat body (Asgari *et al.*, 1996). The protein interacts with hemocytes, inducing

specific changes on their surface, including a rearrangement of lectin-binding sites that interferes with the encapsulation response (Asgari *et al.*, 1997). Binding of CrV1 to the surface of hemocytes and its uptake require the integrity of its coiled-coil domain (Asgari and Schmidt, 2002).

Early-expressed protein 1 (EP1) was the first PDV gene product to be isolated from a parasitized host insect (*Manduca sexta* parasitized by *C. congregata*; Harwood *et al.*, 1994). This glycosylated protein constitutes up to 5% of total *M. sexta* hemolymph protein by 24 h after parasitization. EP1-like genes have now been identified in other BVs, including *Cv*BV (Table 1), which encodes six EP1-like proteins that are expressed at varying levels in parasitized hosts, where they display hemolytic activity (Kwon and Kim, 2010).

BEN-domain gene products were originally described as hypothetical proteins (hp2) in *Cc*BV (Espagne *et al.*, 2004). The BEN domain has since been identified in many eukaryotic and viral proteins, some of which are known to mediate protein–DNA and protein–protein interactions during chromatin organization and transcription (Abhiman *et al.*, 2008). The *Cp*BV homologs have been shown to cause immunosuppression in parasitized hosts, namely through a reduction in hemocyte counts and an inhibition of nodule formation by hemocytes (Park and Kim, 2010). Because fusion of BEN domains with RNase T2 domains has been observed in bracoviruses, a role in RNA processing has been suggested (Abhiman *et al.*, 2008). Similarly, the histone 4 (H4) gene product is believed to affect chromatin conformation in the regulation of gene expression. *Cv*BV H4 displays high sequence identity with host *P. xylostella* H4, except for the N-terminal region. Expression of *Cv*BV H4 in host hemocytes compromised their ability to spread on an extracellular matrix, pointing again to a role in immune suppression (Gad and Kim, 2009).

The innexins (inx) are found in all campoplegine IVs examined to date, including *Hd*IV (Table 1). Innexins are proteins capable of forming functional gap junctions between cells. Although inx genes are expressed in several tissues of parasitized insects, their immunological detection is limited to hemocytes, where they are hypothesized to interfere with the function of endogenous innexins during capsule formation (Turnbull *et al.*, 2005).

Among the gene families that have so far been identified as specific to a single PDV, the glycosylated central domain (Glc) and epidermal growth factor-like (Egf-like) families of *Md*BV (Table 1) have received considerable attention and both have been shown to play a role in host immune suppression. Glc proteins are expressed in *Md*BV-infected host hemocytes and feature a glycosylated central core containing repeated elements flanked by hydrophobic N- and C-terminal domains (Trudeau *et al.*, 2000). The C terminus is an anchor sequence that is required for biological activity, which manifests itself by a loss of adhesion and phagocytic activity of infected cells. The adhesion-blocking properties of the protein are directly related to its number of glycosylated repeats (Beck and Strand 2003, 2005). Egf-like proteins are encoded by spliced genes expressed in both granulocytes and plasmatocytes (Strand *et al.*, 1997; Trudeau *et al.*, 2000). They contain a cystein-rich region at their N termini displaying similarity to epidermal growth factor-like motifs as well as trypsin-inhibitor-like domains of small serine protease inhibitors (smapins). As such, they have been shown to inhibit the prophenoloxidase-activating proteinase (PAP) and other enzymes of the prophenoloxidase cascade that are required for melanin formation (Beck and Strand, 2007).

Functional analyses for the other gene families so far observed in only one PDV are not yet available. These families include the TrV (Béliveau *et al.*, 2000, 2003) and ovary-specific secreted proteins (OSSP; Rasoolizadeh *et al.*, 2009a) of *TrIV*, the glycine-proline-rich (Volkoff *et al.*, 1999) and serine–threonine-rich (Galibert *et al.*, 2003) proteins of *HdIV*, and the recombinase-like proteins in CiBV (Johner and Lanzrein, 2002), although a homolog of the latter was identified in the banchine IV, *GfIV* (Table 1; Lapointe *et al.*, 2007). In as much as the latter virus remains the only banchine IV sequenced to date, the degree of uniqueness of its gene families cannot yet be assessed. Interestingly, however, unlike most other polydnaviral genes identified so far, its NTPase-like gene family appears to be of viral origin, presumably obtained through lateral gene transfer (Lapointe *et al.*, 2007; Cusson *et al.,* Chapter 6 of this volume). The bracoviral P94 protein (Table 1) and the ascoviral protein (SfAV1 orf14 gene homolog) are the only other PDV genes, in packaged genomes, that are clearly of viral origin (Drezen *et al.*, 2006).

Selective Pressure Driving the Evolution of PDV Gene Families

Although a good number of gene families are shared by several (e.g., PTPs, rep, Cys-motif) or all (vank) PDV genomes examined to date, important differences in both gene type and degree of within-family gene diversification are observed, even within the BV group, which has been shown to be monophyletic (Whitfield, 1997; Bézier *et al.,* 2009a; Wetterwald *et al.*, 2010). Strikingly, none of the genes identified to date in the partially sequenced genome of the chelonine virus *CiBV* (Annaheim and Lanzrein, 2007) have been found in the bracoviral genomes whose genes are listed in Table 1. This important difference has been attributed to the very distinct life histories—and, thus, requirements—of the parasitic wasps with which these viruses are associated. Indeed, chelonine wasps are egg–larval endoparasitoids whereas microgastrine wasps (hosts of *Cc*BV, *Cv*BV, *Gi*BV, *Gf*BV and *Md*BV) are larval endoparasitoids.

Other differences in life history strategies are likely to be the forces driving some of the other differences described in the previous section. For example, important differences in gene content between *Md*BV and other BVs (Table 1) are likely to result from the solitary versus gregarious life histories, respectively, of the two groups of wasp hosts. Similarly, although both campoplegine and banchine wasps are solitary endoparasitoids, the banchine parasitoid *Glypta fumiferanae*, unlike its campoplegine counterparts, lays its eggs in early instar larvae of its host a few days before they enter into winter diapause, a period during which development of the larval endoparasitoid is put on hold (Cusson *et al.,* Chapter 6 of this volume). Such differences in life history strategies are expected to have an impact on the sets of PDV genes a wasp needs for successful endoparasitic development.

The degree of diversification within individual polydnaviral gene families has been attributed to various factors including the need to use gene product variants to effectively overcome the immune response, either in different host species or within different tissues in a given host. The observation that there are important differences in tissue-specific expression of different genes within a single family (e.g., Provost *et al.*, 2004; Kroemer and Webb, 2005; Galibert *et al.*, 2006; Rasoolizadeh *et al.*, 2009a, b) tends to support the latter hypothesis. Whether different members of a gene family are selectively expressed in different host species has not yet been assessed. However, recent evidence on PDV genome polymorphism, within different populations of a wasp species, suggests that polymorphic variants of a single gene (Branca *et al.*, 2011), as opposed to different paralogs, may be responsible for local adaptations to different hosts, giving rise to a situation where a wasp species originally identified as a host generalist must now be seen as a collection of several sibling species forming a complex of cryptic host specialists (Smith *et al.,* in press).

REPLICATION OF PDVS

Bracovirus and ichnovirus particles are produced within the nuclei of epithelial cells located in a specialized region of the ovary called the calyx (de Buron and Beckage, 1992; Stoltz *et al.*, 1976; Volkoff *et al.*, 1995; Wyler and Lanzrein, 2003). Transmission electron microscopy analyses revealed that BV and IV virions display several morphological differences. The nucleocapsids of campoplegine IVs display an ellipsoid shape and are of relatively uniform size, with a single large nucleocapsid included in a double envelope, whereas BV virions contain one to several rod-shaped nucleocapsids of varying length within a single envelope (Krell and Stoltz, 1979). BV particles from the *Cotesia* genus contain several nucleocapsids while the particles of *Ci*BV contain only one (Albrecht *et al.*, 1994). These nucleocapsids are assembled in calyx cell nuclei and acquire envelopes during virion assembly within the nucleus (de Buron and Beckage, 1992; Stoltz *et al.*, 1976; Volkoff *et al.*, 1995; Wyler and Lanzrein, 2003). While IVs acquire their second membrane envelope by budding of virions through the plasma membrane into the oviduct lumen, BVs have a single envelope and are released into the oviduct lumen through cell lysis (Krell and Stoltz, 1979; Wyler and Lanzrein, 2003). Virions accumulate to high densities in the oviduct.

Despite the differences between the two PDV genera, replication for both follows a similar temporal specificity, possibly a consequence of convergent evolution. Females of these species have a short adult life and are ready to oviposit shortly after emergence. Indeed, for both genera, viral replication requires coordination between expression of viral structural genes (that are not encapsidated) and replication of the 'proviral' genome, which is encapsidated. The genes encoding structural components of PDVs as well as genes encoding the transcriptional machinery are present in the wasp genome and are expressed specifically in the calyx (Bézier *et al.*, 2009a; Volkoff *et al.*, 2010). Replication and assembly of both BVs and IVs is temporally associated with changes in ecdysteroid titers that allow melanization of adult cuticle late in pupal development. In BVs and IVs, initiation of virus segment replication begins between the last larval instar and the pharate adult stage of the wasp female though amplification and excision of proviral segment DNA (Albrecht *et al.*, 1994; Cui *et al.*, 2000; Gruber *et al.*, 1996; Marti *et al.*, 2003; Norton and Vinson, 1983; Pasquier-Barre *et al.*, 2002; Savary *et al.*, 1999; Webb and Summers, 1992; Wyler and Lanzrein, 2003). Characterization of pupal development allowed a precise analysis of the pattern of virus replication during this early time frame. In the course of pupal adult development, the pigmentation of pupae increased in a characteristic manner and ovarian development was tightly correlated with pigmentation patterns. For both *Ci*BV and *Cc*BV, six developmental markers for pupae and their ovaries were defined (Albrecht *et al.*, 1994; Pasquier-Barre *et al.*, 2002). These externally visible changes were accompanied by morphological and histological changes within the ovaries (Marti *et al.*, 2003; Pasquier-Barre *et al.*, 2002). More precisely, Wyler and Lanzrein (2003) provided details on the morphogenetic processes occurring during calyx cell differentiation and *Ci*BV replication in *C. inanitus*. During the first stages (continuing to pupal stage 3a), elongated peripheral cells (which will develop into virus-producing calyx cells) appear on the periphery of the ovarioles. The calyx cells increase massively in size. The nuclei of these cells contain many patches of dense heterochromatin and become polyploid. The viral genome is amplified to a similar extent as nonviral actin, indicating that calyx cells become polyploid. In *C. congregata*, a stage of calyx development with a highly lobulated nucleus and patches of heterochromatin was also observed (Pasquier-Barre *et al.*, 2002). During intermediate

stages, empty envelopes and envelopes containing some nucleocapsids appear in nuclei along with typical PDV ring-shaped and electron-dense virogenic stroma. At this stage, a selective amplification of viral DNA commences. In *C. congregata*, Southern blot and quantitative PCR data indicated that virus replication begins by day four of the pupal stage (Pasquier-Barre *et al.*, 2002). Gruber *et al.* (1996) and Marti *et al.* (2003) suggested that 20-hydroxyecdysone production, which peaks during early pupal stages (stages one and two), promotes differentiation of calyx cells prior to initiation of virus production. Male wasps lack calyx cells and therefore do not produce detectable levels of extrachromosomal DNA although a low level was detected by PCR in *Cotesia congregata* males (but not haploid males issued from a virgin female) suggesting some viral activity in diploid males (Savary *et al.*, 1999). In late stages, the nuclei are filled by mature BV virions and some mature calyx cells in the proximity of the oviduct begin to lyse. A phagocytic epithelial layer between lysed cells and the oviduct lumen functions to remove cell debris and transport mature virions into the concentrated calyx fluid, completing the process of virus production (Wyler and Lanzrein, 2003). In *Ci*BV adult stages, the excised circular form is predominant, suggesting that virus production is more intense in the second half of pupal development and diminishes in the adult.

As seen for BVs, replication and assembly of IVs are detected only in the ovarian cells of female parasitoids and are temporally associated with changes in ecdysteroid titers that induce synthesis and melanization of the adult cuticle late in pupal development. These data have been obtained by studying ovarian development and ichnovirus morphogenesis in two ichneumonids, *C. sonorensis* (Norton and Vinson, 1983) and *H. dydimator* (Volkoff *et al.*, 1995). As shown for BVs, excision of viral DNA is correlated with ovarian development and pigmentation patterns of the pupae, and calyx differentiation begins shortly after pupation. Calyx cell nuclei of *H. didymator* first increase in size, then pass through a lobulated stage with patches of heterochromatin followed by a stage with virogenic stroma and reduced levels of heterochromatin (Volkoff *et al.*, 1995). However, the final stages of virogenesis are different, as the release of ichnoviruses involves budding through the nuclear and plasma membranes (Norton and Vinson, 1983; Volkoff *et al.*, 1995). A major difference with *C. inanitus* is the finding that some calyx cells, in addition to producing PDV, appear to be secretory, releasing flocculent material into the oviduct (Volkoff *et al.*, 1995).

Little is known about the mechanism of PDV replication. However, the relationship between the integrated form of the virus present in the wasp chromosomes and the circular encapsidated form injected into parasitized hosts was analyzed for *C. congregata* and *C. inanitus*. For *Ci*BV, it was shown that viral DNA is amplified only in the proviral form and not after excision (Marti *et al.*, 2003). In

C. congregata, Pasquier-Barre *et al.* (2002) detected replication of the *Cc*BV EP1 segment within a large progenitor molecule containing both segments 8 (EP1) and 21 (circle A). This large molecule was amplified prior to excision of individual segments (Pasquier-Barre *et al.*, 2002; Savary *et al.*, 1997, 1999). As discussed earlier, this and further analysis of proviral genome segments revealed that all BV segments contained highly conserved DRJ located at 5′ and 3′ positions on segments. It has been proposed that these repeated sequences may act as excision motifs of the viral segments from the wasp proviral locus. The process of excision/ circularization is proposed to occur through juxtaposition of the DRJ followed by recombination across the repeats; as a consequence the circular segment and the rejoined DNA both contain one repeat (Annaheim and Lanzrein, 2007; Savary *et al.*, 1997) (Fig. 3). The questions raised to date concern the replication model for the production of the amplified molecule. Two models have been proposed. The first proposed that, as seen for the region containing chorion genes in *Drosophila* follicle cells (Calvi *et al.*, 1998; Osheim *et al.*, 1988), the entire chromosomal region of the proviral locus might be specifically amplified,

each segment being circularized later through a recombination event. However, this model predicts that the map of the amplified molecules corresponds to that of the chromosome, which was shown not to be the case for *Cc*BV 8–21 segments region (Pasquier-Barre *et al.*, 2002). A second model proposed that a large precursor molecule is excised from the chromosome and is amplified by a rolling circle mechanism in which DNA synthesis proceeds unidirectionally.

The recent finding of a nudiviral origin for BVs favors the latter model, given that this type of replication (rolling circle) is one of the most common mechanisms used by baculoviruses (Vanarsdall *et al.*, 2007) that are related to nudiviruses (for which the mode of replication is unknown). In such a replication model, the precursor molecule might contain several viral segments separated by spacers that are not encapsidated within viral particles (Annaheim and Lanzrein, 2007; Desjardins *et al.*, 2008; Pasquier-Barre *et al.*, 2002). The DRJ sites may act as encapsidation signals allowing the entry of only one molecule into a capsid.

For IVs, limited data suggest that IV proviral segments are dispersed within chromosomal DNA of their parasitoid hosts. The viral sequences are flanked by sequences that are

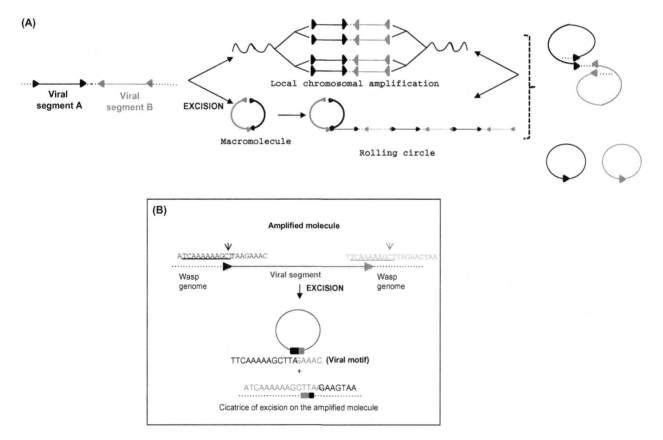

FIGURE 3 Model illustrating hypothetical mode of BV replication (adapted from Marti *et al.*, 2003 and Annaheim *et al.*, 2007). A: Two proviral segments physically linked may be co-amplified through local amplification or by rolling circles after excision of a macromolecule from the wasp chromosome. The arrows show the orientation of the DRJ in two segments. The orientation of the DRJ could contribute to the formation of a loop preceding excison of each circle. **B:** A model of how the segment may be excised from the amplified molecule due to the DRJ. The flanking sequence becomes continous following viral circle excision leaving a cicatrice of excision on the amplified molecule (from Savary *et al.*, 1997).

not encapsidated into virions, suggesting that, as seen for BVs, wasp genomic DNA separates IV genome segments. Some proviral DNA segments seem to have short imperfect DNA repeats (59 bp in segment B) or long perfect terminal repeats (1186 bp in segment W), a single copy of the repeat being retained after recombination to produce circular segments (Cui *et al.*, 2000; Fleming and Summers, 1991; Rattanadechakul and Webb, 2003). Moreover, smaller circles are generated from larger molecules through the alternative recombination pathway known as nesting (Cui *et al.*, 2000).

CONCLUDING REMARKS

Despite different origins, all PDV genomes share common structural features including large, highly segmented genomes with low coding densities. Many of the genes are organized into gene families that encode mainly virulence factors but none encode 'pure' viral genes. The PDV genome packaged in the particles presents evidence of genome expansion most probably by lateral transfer of virulence genes acquired from the wasp genomes and genome reduction in relation to genes regulating viral replication in wasps that reside permanently in the wasp genome. BVs and IVs have distinct origins and their shared features most likely reflect convergence driven by their similar role in parasitism. The diversification of virulence genes into families is likely a key adaptation that reflects the role of these viruses in parasitism. Variants within gene families might have different cellular and/or tissue tropism. Overall, the gene content of each packaged genome is unique, and only a few conserved gene families have been identified within different viruses. These observations allow us to conclude that the packaged genome is shaped by the physiological constraints with which the immature wasp is confronted within the lepidopteran host (e.g., parasitization of eggs versus larvae). In *Gi*BV and *Gf*BV the segments are primarily integrated into a macrolocus, each segment being flanked by conserved direct repeat junctions that may represent vestiges of the ancestral nudivirus integration. In *Cc*BV, the proviral sequence seems to be at some distance from the nudiviral cluster that represents vestiges of the ancestral nudivirus. In IVs the data currently available on integration sites within the wasp genome are limited but suggest that IV segments are dispersed in the wasp genome. However, a better understanding of the process that has driven the evolution of BVs and IVs will likely come from sequencing entire genomes of braconid wasps bearing BVs. This will make it possible to determine whether encapsidated genomes and ancestral nudivirus cluster genes are physically linked or not. At present we know the origins of both viruses and one of the questions that needs to be addressed is how they replicate. Have they conserved the ancestral viral replication machinery or do they replicate along with the cell chromosome under cell control? Thus, it seems likely that elucidating the replication mechanism of PDVs will require the isolation and analysis of genes directly involved in virus replication and will require that we determine how the genome segments are synthetized and encapsidated. Such work will shed new light on how these entities have evolved since the original integration of virus genomes into wasp genomes.

REFERENCES

Abhiman, S., Iyer, L.M., Aravind, L., 2008. BEN: a novel domain in chromatin factors and DNA viral proteins. Bioinformatics 24, 458–461.

Albrecht, U., Wyler, T., Pfister-Wilhelm, R., Gruber, A., Stettler, P., Heiniger, P., et al. 1994. Polydnavirus of the parasitic wasp *Chelonus inanitus* (Braconidae): characterization, genome organization and time point of replication. J. Gen. Virol. 75, 3353–3363.

Annaheim, M., Lanzrein, B., 2007. Genome organization of the *Chelonus inanitus* polydnavirus: excision sites, spacers and abundance of proviral and excised segments. J. Gen. Virol. 88, 450–457.

Asgari, S., Hellers, M., Schmidt, O., 1996. Host haemocyte inactivation by an insect parasitoid: transient expression of a polydnavirus gene. J. Gen. Virol. 77, 2653–2662.

Asgari, S., Schmidt, O., 2002. A coiled-coil region of an insect immune suppressor protein is involved in binding and uptake by hemocytes. Insect Biochem. Mol. Biol. 32, 497–504.

Asgari, S., Schmidt, O., Theopold, U., 1997. A polydnavirus-encoded protein of an endoparasitoid wasp is an immune suppressor. J. Gen. Virol. 78, 3061–3070.

Ayres, M.D., Howard, S.C., Kuzio, J., Lopez-Ferber, M., Possee, R.D., 1994. The complete DNA sequence of *Autographa californica* nuclear polyhedrosis virus. Virology 202, 586–605.

Bae, S., Kim, Y., 2009. Iκβ genes encoded in *Cotesia plutellae* bracovirus suppress an antiviral response and enhance baculovirus pathogenicity against the diamondback moth, *Plutella xylostella*. J. Invert. Pathol. 102, 79–87.

Beck, M., Strand, M.R., 2003. RNA interference silences *Microplitis demolitor* bracovirus genes and implicates glc1.8 in disruption of adhesion in infected host cells. Virology 314, 521–535.

Beck, M., Strand, M.R., 2005. Glc1.8 from *Microplitis demolitor* bracovirus induces a loss of adhesion and phagocytosis in insect high five and S2 cells. J. Virol. 79, 1861–1870.

Beck, M., Strand, M.R., 2007. A novel polydnavirus protein inhibits the insect prophenoloxidase activation pathway. Proc. Natl. Acad. Sci. U.S.A. 104, 19267–19272.

Beck, M.H., Inman, R.B., Strand, M.R., 2007. *Microplitis demolitor* bracovirus genome segments vary in abundance and are individually packaged in virions. Virology 359, 179–189.

Béliveau, C., Laforge, M., Cusson, M., Bellemare, G., 2000. Expression of a *Tranosema rostrale* polydnavirus gene in the spruce budworm, Choristoneura fumiferana. J. Gen. Virol. 81, 1871–1880.

Béliveau, C., Levasseur, A., Stoltz, D., Cusson, M., 2003. Three related TrIV genes: comparative sequence analysis and expression in host larvae and Cf-124T cells. J. Insect Physiol. 49, 501–511.

Belle, E., Beckage, N.E., Rousselet, J., Poirie, M., Lemeunier, F., Drezen, J.-M., 2002. Visualization of polydnavirus sequences in a parasitoid wasp chromosome. J. Virol. 76, 5793–5796.

Bézier, A., Annaheim, M., Herbiniere, J., Wetterwald, C., Gyapay, G., Bernard-Samain, S., et al. 2009. Polydnaviruses of braconid wasps derive from an ancestral nudivirus. Science 323, 926–930.

Bézier, A., Herbiniere, J., Lanzrein, B., Drezen, J.-M., 2009. Polydnavirus hidden face: the genes producing virus particles of parasitic wasps. J. Invertebr. Pathol. 101, 194–203.

Blissard, G.W., Theilmann, D.A., Summers, M.D., 1989. Segment W of Campoletis sonorensis virus: expression, gene products, and organization. Virology 169, 78–89.

Branca, A., Le Ru, B.P., Vavre, F., Silvain, J.F., Dupas, S., 2011. Intraspecific specialization of the generalist parasitoid Cotesia sesamiae revealed by polydnavirus polymorphism and associated with different Wolbachia infection. Mol. Ecol. 20, 959–971.

Calvi, B.R., Lilly, M.A., Spradling, A.C., 1998. Cell cycle control of chorion gene amplification. Genes Dev. 12, 734–744.

Chen, Y.F., Gao, F., Ye, X.Q., Wei, S.J., Shi, M., Zheng, H.J., et al., 2011. Deep sequencing of Cotesia vestalis bracovirus reveals the complexity of a polydnavirus genome. Virology. 414, 42–50.

Cheng, C.H., Liu, S.M., Chow, T.Y., Hsiao, Y.Y., Wang, D.P., Huang, J.J., et al. 2002. Analysis of the complete genome sequence of the Hz-1 virus suggests that it is related to members of the Baculoviridae. J. Virol. 76, 9024–9034.

Choi, J.Y., Kwon, S.J., Roh, J.Y., Yang, T.J., Yoon, S.H., Kim, H., et al. 2009. Sequence and gene organization of 24 circles from the Cotesia plutellae bracovirus genome. Arch. Virol. 154, 1313–1327.

Choi, J.Y., Roh, J.Y., Kang, J.N., Shim, H.J., Woo, S.D., Jin, B.R., et al. 2005. Genomic segments cloning and analysis of Cotesia plutellae polydnavirus using plasmid capture system. Biochem. Biophys. Res. Commun. 332, 487–493.

Cui, L., Soldevila, A.I., Webb, B.A., 2000. Relationships between polydnavirus gene expression and host range of the parasitoid wasp Campoletis sonorensis. J. Insect Physiol. 46, 1397–1407.

de Buron, I, B. N., Beckage, N.E., 1992. Characterization of a polydnavirus (PDV) and virus-like filamentous particles (VLFP) in the Braconid wasp Cotesia congregata (Hymenoptera: Braconidae). J. Invertebr. Pathol. 59, 440–450..

Desjardins, C.A., Gundersen-Rindal, D.E., Hostetler, J.B., Tallon, L.J., Fadrosh, D.W., Fuester, R.W., et al. 2008. Comparative genomics of mutualistic viruses of Glyptapanteles parasitic wasps. Genome Biol. 9, R183.

Desjardins, C.A., Gundersen-Rindal, D.E., Hostetler, J.B., Tallon, L.J., Fuester, R.W., Schatz, M.C., et al. 2007. Structure and evolution of a proviral locus of Glyptapanteles indiensis bracovirus. BMC Microbiol. 7, 61.

Drezen, J.-M., Bézier, A., Lesobre, J., Huguet, E., Cattolico, L., Periquet, G., et al. 2006. The few virus-like genes of Cotesia congregata bracovirus. Arch. Insect; Biochem. Physiol. 61, 110–122.

Dupuy, C., Huguet, E., Drezen, J.-M., 2006. Unfolding the evolutionary story of polydnaviruses. Virus Res. 117, 81–89.

Dupuy, C., Periquet, G., Serbielle, C., Bézier, A., Louis, F., Drezen, J.-M., 2011. Transfer of a chromosomal Maverick to endogenous bracovirus in a parasitoid wasp. Genetica 139, 489–496.

Einerwold, J., Jaseja, M., Hapner, K., Webb, B., Copié, V., 2001. Solution structure of the carboxyl-terminal cysteine-rich domain of the VHv1.1 polydnaviral gene product: comparison with other cysteine knot structural folds. Biochemistry 40, 14404–14412.

Espagne, E., Dupuy, C., Huguet, E., Cattolico, L., Provost, B., Martins, N., et al. 2004. Genome sequence of a polydnavirus: insights into symbiotic virus evolution. Science 306, 286–289.

Eum, J.H., Bottjen, R., Pruijssers, A.J., Clark, K.D., Strand, M.R., 2010. Characterization and kinetic analysis of protein tyrosine phosphatase-H2 from Microplitis demolitor bracovirus. Insect Biochem. Mol. Biol. 40, 690–698.

Falabella, P., Varricchio, P., Provost, B., Espagne, E., Ferrarese, R., Grimaldi, A., et al. 2007. Characterization of the IκB-like gene family in polydnaviruses associated with wasps belonging to different braconid subfamilies. J. Gen. Virol. 88, 92–104.

Fath-Goodin, A., Gill, T.A., Martin, S.B., Webb, B.A., 2006. Effect of Campoletis sonorensis ichnovirus cys-motif proteins on Heliothis virescens larval development. J. Insect Physiol. 52, 576–585.

Fath-Goodin, A., Kroemer, J.A., Webb, B.A., 2009. The Campoletis sonorensis ichnovirus vankyrin protein P-vank-1 inhibits apoptosis in insect Sf9 cells. Insect. Mol. Biol. 18, 497–506.

Fleming, J.G., Summers, M.D., 1991. Polydnavirus DNA is integrated in the DNA of its parasitoid wasp host. Proc. Natl. Acad. Sci. U.S.A. 88, 9770–9774.

Gad, W., Kim, Y., 2009. N-terminal tail of a viral histone H4 encoded in Cotesia plutellae bracovirus is essential to suppress gene expression of host histone H4. Insect Mol. Biol. 18, 111–118.

Galibert, L., Devauchelle, G., Cousserans, F., Rocher, J., Cerutti, P., Barat-Houari, M., et al. 2006. Members of the Hyposoter didymator ichnovirus repeat element gene family are differentially expressed in Spodoptera frugiperda. Virol. J. 3, 48.

Galibert, L., Rocher, J., Ravallec, M., Duonor-Cérutti, M., Webb, B.A., Volkoff, A.N., 2003. Two Hyposoter didymator ichnovirus genes expressed in the lepidopteran host encode secreted or membrane-associated serine and threonine rich proteins in segments that may be nested. J. Insect Physiol. 49, 441–451.

Glatz, R., Schmidt, O., Asgari, S., 2003. Characterization of a novel protein with homology to C-type lectins expressed by the Cotesia rubecula bracovirus in larvae of the lepidopteran host, Pieris rapae. J. Biol. Chem. 278, 19743–19750.

Gruber, A., Stettler, P., Heiniger, P., Schumperli, D., Lanzrein, B., 1996. Polydnavirus DNA of the braconid wasp Chelonus inanitus is integrated in the wasp's genome and excised only in later pupal and adult stages of the female. J. Gen. Virol. 77, 2873–2879.

Harwood, S.H., Grosovsky, A.J., Cowles, E.A., Davis, J.W., Beckage, N.E., 1994. An abundantly expressed hemolymph glycoprotein isolated from newly parasitized Manduca sexta larvae is a polydnavirus gene product. Virology 205, 381–392.

Hilgarth, R.S., Webb, B.A., 2002. Characterization of Campoletis sonorensis ichnovirus segment I genes as members of the repeat element gene family. J. Gen. Virol. 83, 2393–2402.

Johner, A., Lanzrein, B., 2002. Characterization of two genes of the polydnavirus of Chelonus inanitus and their stage-specific expression in the host Spodoptera littoralis. J. Gen. Virol. 83, 1075–1085.

Kim, Y., 2005. Identification of host translation inhibitory factor of Campoletis sonorensis ichnovirus on the tobacco budworm, Heliothis virescens. Arch. Insect Biochem. Physiol. 59, 230–244.

Krell, P.J., Stoltz, D.B., 1979. Unusual Baculovirus of the Parasitoid Wasp Apanteles melanoscelus: Isolation and Preliminary Characterization. J. Virol. 29, 1118–1130.

Kroemer, J.A., Webb, B.A., 2004. Polydnavirus genes and genomes: emerging gene families and new insights into polydnavirus replication. Annu. Rev. Entomol. 49, 431–456.

Kroemer, J.A., Webb, B.A., 2005. IκB-related vankyrin genes in the Campoletis sonorensis ichnovirus: temporal and tissue-specific patterns of expression in parasitized Heliothis virescens lepidopteran hosts. J. Virol. 79, 7617–7628.

Kwon, B., Kim, Y., 2010. Transient expression of an EP1-like gene encoded in *Cotesia plutellae* bracovirus suppresses the hemocyte population in the diamondback moth, *Plutella xylostella*. Dev. Comp. Immunol. 32, 932–942.

Kwon, B., Song, S., Choi, J.Y., Je, Y.H., Kim, Y., 2010. Transient expression of specific *Cotesia plutellae* bracoviral segments induces prolonged larval development of the diamondback moth, *Plutella xylostella*. J. Insect Physiol. 56, 650–658.

Lapointe, R., Tanaka, K., Barney, W.E., Whitfield, J.B., Banks, J.C., Beliveau, C., et al. 2007. Genomic and morphological features of a banchine polydnavirus: comparison with bracoviruses and ichnoviruses. J. Virol. 81, 6491–6501.

Lee, S., Nalini, M., Kim, Y., 2008. A viral lectin encoded in *Cotesia plutellae* bracovirus and its immunosuppressive effect on host hemocytes. Comp. Biochem. Physiol. A: Mol. Integr. Physiol. 149, 351–361.

Li, X., Webb, B.A., 1994. Apparent functional role for a cysteine-rich polydnavirus protein in suppression of the insect cellular immune response. J. Virol. 68, 7482–7489.

Marti, D., Grossniklaus-Burgin, C., Wyder, S., Wyler, T., Lanzrein, B., 2003. Ovary development and polydnavirus morphogenesis in the parasitic wasp *Chelonus inanitus*. I. Ovary morphogenesis, amplification of viral DNA and ecdysteroid titres. J. Gen. Virol. 84, 1141–1150.

Moran, N.A., 2003. Tracing the evolution of gene loss in obligate bacterial symbionts. Curr. Opin. Microbiol. 6, 512–518.

Murphy, N., Banks, J.C., Whitfield, J.B., Austin, A.D., 2008. Phylogeny of the parasitic microgastroid subfamilies (Hymenoptera: Braconidae) based on sequence data from seven genes, with an improved time estimate of the origin of the lineage. Mol. Phylogenet. Evol. 47, 378–395.

Norton, W.N., Vinson, S.B., 1983. Correlating the initiation of virus replication with a specific pupal developmental phase of an ichneumonid parasitoid. Cell. Tissue. Res. 231, 387–398.

Osheim, Y.N., Miller Jr., O.L., Beyer, A.L., 1988. Visualization of *Drosophila melanogaster* chorion genes undergoing amplification. Mol. Cell. Biol. 8, 2811–2821.

Park, B., Kim, Y., 2010. Transient transcription of a putative RNase containing BEN domain encoded in Cotesia plutellae bracovirus induces an immunosuppression of the diamondback month, Plutella xylostella. J. Invertebr. Pathol. 105, 156–163.

Pasquier-Barre, F., Dupuy, C., Huguet, E., Monteiro, F., Moreau, A., Poirie, M., et al. 2002. Polydnavirus replication: the EP1 segment of the parasitoid wasp *Cotesia congregata* is amplified within a larger precursor molecule. J. Gen. Virol. 83, 2035–2045.

Provost, B., Varricchio, P., Arana, E., Espagne, E., Falabella, P., Huguet, E., et al. 2004. Bracoviruses contain a large multigene family coding for protein tyrosine phosphatases. J. Virol. 78, 13090–13103.

Pruijssers, A.J., Strand, M.R., 2007. PTP-H2 and PTP-H3 from *Microplitis demolitor* bracovirus localize to focal adhesions and are antiphagocytic in insect immune cells. J. Virol. 81, 1209–1219.

Rasoolizadeh, A., Béliveau, C., Stewart, D., Cloutier, C., Cusson, M., 2009. *Tranosema rostrale* ichnovirus repeat element genes display distinct transcriptional patterns in caterpillar and wasp hosts. J. Gen. Virol. 90, 1505–1514.

Rasoolizadeh, A., Dallaire, F., Stewart, D., Béliveau, C., Lapointe, R., Cusson, M., 2009. Global transcriptional profile of *Tranosema*

rostrale ichnovirus genes in infected lepidopteran hosts and wasp ovaries. Virol. Sin. 24, 478–492.

Rattanadechakul, W., Webb, B.A., 2003. Characterization of *Campoletis sonorensis* ichnovirus unique segment B and excision locus structure. J. Insect. Physiol. 49, 523–532.

Savary, S., Beckage, N., Tan, F., Periquet, G., Drezen, J.-M., 1997. Excision of the polydnavirus chromosomal integrated EP1 sequence of the parasitoid wasp *Cotesia congregata* (Braconidae, Microgastinae) at potential recombinase binding sites. J. Gen. Virol. 78, 3125–3134.

Savary, S., Drezen, J.-M., Tan, F., Beckage, N.E., Periquet, G., 1999. The excision of polydnavirus sequences from the genome of the wasp *Cotesia congregata* (Braconidae, microgastrinae) is developmentally regulated but not strictly restricted to the ovaries in the adult. Insect. Mol. Biol. 8, 319–327.

Serbielle, C., Chowdhury, S., Pichon, S., Dupas, S., Lesobre, J., Purisima, E.O., et al. 2008. Viral cystatin evolution and three-dimensional structure modeling: a case of directional selection acting on a viral protein involved in a host–parasitoid interaction. BMC Biol. 6, 38.

Smith, M.A., Eveleigh, E.S., McCann, K.S., Merilo, M.T., McCarthy, P.C., Van Rooyen, K., 2011. Barcoding a quantified food web: crypsis, concepts, ecology and hypotheses. PLoS ONE. (In press.)

Stoltz, D.B., Vinson, S.B., MacKinnon, E.A., 1976. Baculovirus-like particles in the reproductive tracts of female parasitoid wasps. Can. J. Microbiol. 22, 1013–1023.

Strand, M.R., 1994. *Microplitis demolitor* polydnavirus infects and expresses in specific morphotypes of *Pseudoplusia includens* haemocytes. J. Gen. Virol. 75, 3007–3020.

Strand, M.R., Witherell, R.A., Trudeau, D., 1997. Two *Microplitis demolitor* polydnavirus mRNAs expressed in hemocytes of *Pseudoplusia includens* contain a common cysteine-rich domain. J. Virol. 71, 2146–2156.

Suderman, R.J., Pruijssers, A.J., Strand, M.R., 2008. Protein tyrosine phosphatase-H2 from a polydnavirus induces apoptosis of insect cells. J. Gen. Virol. 89, 1411–1420.

Tanaka, K., Lapointe, R., Barney, W.E., Makkay, A.M., Stoltz, D., Cusson, M., et al. 2007. Shared and species-specific features among ichnovirus genomes. Virology 363, 26–35.

Theilmann, D.A., Summers, M.D., 1987. Physical analysis of the campoletis sonorensis virus multipartite genome and identification of a family of tandemly repeated elements. J. Virol. 61, 2589–2598.

Thoetkiattikul, H., Beck, M.H., Strand, M.R., 2005. Inhibitor $\kappa\beta$-like proteins from a polydnavirus inhibit NF-$\kappa\beta$ activation and suppress the insect immune response. Proc. Natl. Acad. Sci. U.S.A. 102, 11426–11431.

Trudeau, D., Witherell, R.A., Strand, M.R., 2000. Characterization of two novel *Microplitis demolitor* polydnavirus mRNAs expressed in *Pseudoplusia includens* haemocytes. J. Gen. Virol. 81, 3049–3058.

Turnbull, M.W., Volkoff, A.N., Webb, B.A., Phelan, P., 2005. Functional gap junction genes are encoded by insect viruses. Curr. Biol. 15, R491–492.

Vanarsdall, A.L., Mikhailov, V.S., Rohrmann, G.F., 2007. Baculovirus DNA replication and processing. Curr. Drug Targets 8, 1096–1102.

Volkoff, A.N., Béliveau, C., Rocher, J., Hilgarth, R., Levasseur, A., Duonor-Cérutti, C., et al. 2002. Evidence for a conserved polydnavirus gene family: ichnovirus homologs of the CsIV repeat element genes. Virology 300, 316–331.

Volkoff, A.N., Cérutti, P., Rocher, J., Ohresser, M.C., Devauchelle, G., Duonor-Cérutti, M., 1999. Related RNAs in lepidopteran cells after in vitro infection with Hyposoter didymator virus define a new polydnavirus gene family. Virology 263, 349–363.

Volkoff, A.N., Jouan, V., Urbach, S., Samain, S., Bergoin, M., Wincker, P., et al. 2010. Analysis of virion structural components reveals vestiges of the ancestral ichnovirus genome. PLoS Pathog. 6, e1000923.

Volkoff, A.N., Ravallec, M., Bossy, J., Cerutti, P., Rocher, J., Cerutti, M., et al. 1995. The replication of *Hyposoter didymator* polydnavirus: cytopathology of the calyx cells in the parasitoid. Biol. Cell 83, 1–13.

Wang, Y., Jehle, J.A., 2009. Nudiviruses and other large, double-stranded circular DNA viruses of invertebrates: new insights on an old topic. J. Invertebr. Pathol. 101, 187–193.

Webb, B.A., Strand, M.R., 2005. The biology and genomic of polydnaviruses. In: Gilbert, L., Iatrou, K., Gill, S.S. (Eds.), Comprenhensive Molecular Insect Science (pp. 323–360). Elsevier Inc., San Diego.

Webb, B.A., Strand, M.R., Dickey, S.E., Beck, M.H., Hilgarth, R.S., Barney, W.E., et al. 2006. Polydnavirus genomes reflect their dual roles as mutualists and pathogens. Virology 347, 160–174.

Webb, B.A., Summers, M.D., 1992. Stimulation of polydnavirus replication by 20–hydroxyecdysone. Experientia 48, 1018–1022.

Werren, J.H., Richards, S., Desjardins, C.A., Niehuis, O., Gadau, J., Colbourne, J.K, et al. 2010. Functional and evolutionary insights from the genomes of three parasitoid *Nasonia species*. Science 327, 343–348.

Wetterwald, C., Roth, T., Kaeslin, M., Annaheim, M., Wespi, G., Heller, M., et al. 2010. Identification of bracovirus particle proteins and analysis of their transcript levels at the stage of virion formation. J. Gen. Virol. 91, 2610–2619.

Whitfield, J.B., 1997. Molecular and morphological data suggest a single origin of the polydnaviruses among braconid wasps. Naturwissenschaften 84, 502–507.

Whitfield, J.B., 2002. Estimating the age of the polydnavirus/braconid wasp symbiosis. Proc. Natl. Acad. Sci. U.S.A. 99, 7508–7513.

Wu, M., Sun, L.V., Vamathevan, J., Riegler, M., Deboy, R., Brownlie, J.C., et al. 2004. Phylogenomics of the reproductive parasite *Wolbachia pipientis* wMel: a streamlined genome overrun by mobile genetic elements. PLoS Biol. 2, e69.

Wyder, S., Tschannen, A., Hochuli, A., Gruber, A., Saladin, V., Zumbach, S., et al. 2002. Characterization of *Chelonus inanitus* polydnavirus segments: sequences and analysis, excision site and demonstration of clustering. J. Gen. Virol. 83, 247–256.

Wyler, T., Lanzrein, B., 2003. Ovary development and polydnavirus morphogenesis in the parasitic wasp *Chelonus inanitus*. II. Ultrastructural analysis of calyx cell development, virion formation and release. J. Gen. Virol. 84, 1151–1163.

Chapter 5

Evolution and Origin of Polydnavirus Virulence Genes

Elisabeth Huguet[*], Céline Serbielle[†] and Sébastien J.M. Moreau[*]

[*]Institut de Recherche sur la Biologie de l'Insecte, CNRS UMR6035 Université François-Rabelais, Faculté des Sciences et Techniques, Parc Grandmont, 37200 Tours, France

[†]GEMI, UMR CNRS IRD 2724, IRD, 911 avenue Agropolis, BP 64501, 34394 Montpellier CEDEX 5, France

SUMMARY

Parasitoid wasps and their associated polydnaviruses (PDVs) are likely to be under strong selection pressure to maintain efficient virulence genes in order to develop in caterpillar hosts. PDV genomes packaged in the particles harbor collectively hundreds of genes, many of which appear to be involved in lepidopteran host regulation. Here we focus on the PDV circular genomes that are injected in the lepidopteran hosts and highlight how selection pressures are likely to have shaped gene content, organization, and function. Another fascinating question concerns the origin and the way in which these 'virulence genes' were acquired. Many PDV genes resemble cellular genes, suggesting that significant gene transfer events have occurred betweens PDVs and the insects. We present the data that lend support to the different hypotheses concerning the origin of PDV packaged genes, and the possible mechanisms that led to such a complex viral entity.

ABBREVIATIONS

ank	ankyrin
ARD	ankyrin repeat domain
BV	bracovirus
IκB	inhibitor of nuclear factor-kappa B
IKK	IκB Kinase
IV	ichnovirus
NF-κB	nuclear factor-kappa B
NLS	nuclear localization signal
PDV	polydnavirus
PEST	domain protein domain rich in proline, glutamic acid, serine and threonine
SCF	SKP1/CUL-1/F-box protein
TE	transposable element
β-TRCP	β-transducin repeat-containing protein

INTRODUCTION

Polydnaviruses (PDVs) have a unique life-cycle, comprised of a mutualistic lifestyle with their associated parasitoid wasps and a parasitic interaction with the lepidopteran wasp hosts. They are present as proviruses in the parasitoid wasps that harbor them. The female wasps produce particles that contain circular double-stranded DNA versions of the viruses that are injected into the wasps' lepidopteran hosts. This 'injected circular form' is replication deficient but is absolutely essential for the physiological regulation of caterpillars that leads to parasitoid survival.

PDVs are divided into two genera, bracoviruses (BVs) and ichnoviruses (IVs) associated with tens of thousands of endoparasitoid braconid and ichneumonid wasps, respectively, belonging to the Ichneumonoidea superfamily. The absence of PDVs in basal lineages of Ichneumonoidea strongly suggests that the association of BVs with braconids and IVs with ichneumonids arose independently. In ichneumonids, PDVs have been identified in the Campopleginae and Banchinae subfamily. The unique genomic features of the first banchine virus examined to date also suggest they could have an origin distinct from those of IVs and BVs (Lapointe *et al.*, 2007).

The absence of genes involved in virus production in the injected circular form is likely to be a signature of the reductive evolution that these PDVs have been subjected to, due to their reliance on wasps for transmission and the absence of replication in host larvae. In this chapter, we will focus on the PDV circular genomes that are injected into the lepidopteran hosts and highlight how selection pressures are likely to have modeled their content, organization, gene function, and efficiency. These PDV genomes are atypical for viruses in the sense that they harbor numerous genes, many of which

Parasitoid Viruses: Symbionts and Pathogens. DOI: 10.1016/B978-0-12-384858-1.00005-9

TABLE 1 Features of Complete Genome Sequences of PDVs

	Genome size in kb	Circle number	Number of genes	Coding density	A + T ratio	Number of gene families	References
Bracoviruses							
CcBV	606.3	32	165	30%	66%	12	Espagne et al., 2004; Bézier et al., (in preparation)
GiBV	517	29	197	33%	64%	5	Desjardins et al., 2008
GfBV	594	29	193	32%	65%	6	Desjardins et al., 2008
MdBV	189	15	61	17%	66%	5	Webb et al., 2006
Ichnoviruses							
CsIV	246.7	24	101	29%	59%	6	Webb et al., 2006
HfIV	246	56	150	30%	56.9%	6	Tanaka et al., 2007
TrIV	Estimated 250	27	86	21.9%	57.8%	5	Tanaka et al., 2007
GfV	291.4	105	101	20.2%	63.3%	5	Lapointe et al., 2007

appear to be involved in the manipulation of the lepidopteran host's physiology. Parasitoid wasps, and thereby their associated PDVs, are likely to be under strong selection pressure to maintain efficient virulence genes in order to develop and perpetuate in caterpillar hosts. Here, we will describe how analysis of PDV genome content and the comparison of gene sequences between related wasp species can help us to understand how host–parasite interactions can shape gene evolution. Another fascinating question concerns the origin and the way in which these 'virulence genes' were acquired. We will present the data that lend support to the different hypotheses concerning the origin of PDV genes, and the possible mechanisms that led to such a complex viral entity.

EVOLUTION OF PDV VIRULENCE GENES

In recent years, completion of several sequencing projects has allowed the acquisition of a considerable amount of data on the DNA contained in PDV particles injected into caterpillars by parasitoid wasps. The sequences of eight PDV genomes have been published (Table 1). Partial data or near-complete sequence data have been obtained for several other viruses such as *Cotesia vestalis* bracovirus CvBV (ex-*Cotesia plutellae* bracovirus) (Choi et al., 2009; Choi et al., 2005) and *Chelonus inanitus* bracovirus CiBV (Annaheim and Lanzrein, 2007). Near-complete proviral sequences of PDVs are now also available for GiBV, GfBV and CcBV (Desjardins et al., 2008, 2007; Bézier et al., in preparation). This accumulation of sequence data from PDVs associated with either closely or distantly related wasps, enables comparisons to be made between the different genomes and allows us to draw scenarios of their evolution in relation to the life-history traits of the wasps and respective lepidopteran hosts.

All sequenced PDV viruses share common structural features: they possess large genomes (ranging from 189 to 606 kb) that are produced as highly segmented double-stranded DNA circles in varying proportions, with low coding densities (17–33%) and many genes organized into gene families (tables 1, 2). These characteristics may reflect both the lack of constraint on the size of the viral genome that is packaged, and the lower physiological cost associated with the production of these viruses by the wasps, compared to directly producing large quantities of venom proteins. Indeed parasitoid wasps also inject venom proteins during oviposition.

PDV Genome Content and Links with Host–Parasitoid Physiological Relationships

The gene content of each packaged PDV genome is unique and is composed of specific genes, half of which are organized into gene families. A few conserved gene families have been identified within all viruses (Table 2) and there is much conservation between closely related viruses. These observations have led to the hypothesis that the genome in the particles is shaped by the physiological constraints that the virus (or parasitoid) is confronted with in the lepidopteran host. For example, despite originating from the same ancestral nudivirus, to date no common genes have been found between *Chelonus inanitus* bracovirus (CiBV) and other bracovirus genomes (Weber et al., 2007). This could be explained by the fact that *C. inanitus* oviposit into the eggs of lepidopteran hosts and are therefore not confronted with the host immune system immediately. The genes maintained in CiBV are therefore predicted to be mainly involved in the control of development. In accordance, three CiBV genes were shown to be

TABLE 2 Examples of Common or Unique Gene Families Within Polydnavirus Families or Species

	Vank	Cys-motif	PTP	Cystatin	t-RNA	C-type lectin	Sugar-transporter	Glc	Egf-motif	References
Bracoviruses										
CcBV	9	4	27	3	7					Espagne et al., 2004; Brézier et al. (in preparation)
GiBV	9		42	1	3	2	3			Desjardins et al., 2008
GfBV	8		31	2	3	5	5			Desjardins et al., 2008
MdBV	12		13		7			2	6	Webb et al. (2006)

	Vank	Cys-motif	Rep	Vinnexin	N	PRRP	TrV	References
Ichnoviruses								
CsIV	7	10	30	4	2	1	0	Webb et al., 2006
HfIV	9	5	38	11	3	11	0	Tanaka et al., 2007
TrIV	2	1	17	3	4	1	7	Tanaka et al., 2007

	Vank	PTP	NTPase-like	BV-like	Recombinase-like	References
GfV	4	23	7	4	1	Lapointe et al., 2007

involved in the inhibition of development (Bonvin et al., 2005).

In contrast, bracoviruses associated with wasps having to face larval host defences have inhibitor of nuclear factor-kappa B (IκB)-like and cys-motif genes in common with ichnoviruses and have protein tyrosine phosphatase (PTP) genes in common with the *Glypta fumiferanae* virus (GfV, associated with a banchine wasp) suggesting that these factors constitute a 'basic arsenal' to regulate caterpillar host immune responses (Table 2). These striking similarities between IVs and BVs may be due to common ancestry or convergent evolution of these genes (see below). Globally, however, bracovirus genomes share more genes with other BVs than with IV genomes, and vice versa, which is consistent with the phylogeny of the associated wasps.

The specific genes or gene families within BVs or IVs may be a reflection of how different physiological interactions with different hosts have modeled these genomes. For instance, in contrast to HfIV and CsIV, the IV associated with *T. rostrale*, parasitoid wasp of the spruce budworm *Choristoneura fumiferana*, has been shown to have little or no apparent impact on the host cellular immune response, but it is involved in inhibition of caterpillar metamorphosis. The decrease, in the TrIV genome, in coding density and in the number of representatives within shared IV gene families has been interpreted as a consequence of the reduced requirement for depressing the host immune system (Tanaka et al., 2007) (Table 2). Conversely, the TrV gene family specifically found in TrIV may be involved in inhibition of host metamorphosis. Another fascinating

case is that of GfV, associated with the ichneumonid Banchinae wasp *G. fumiferanae* that is also a parasitoid of *C. fumiferana*. Despite the fact that banchinae wasps are more closely related to Campopleginae than Braconidae wasps, gene composition of the GfV genome has little in common with that of characterized IVs and GfV genes have more in common with BV genes (Lapointe et al., 2007). Whether this gene composition can be linked to the particular lifestyle of the associated wasp will be very interesting to investigate. Indeed, *G. fumiferanae* lay their eggs in prediapausing *C. fumiferana* hosts in late summer, but the larval progeny complete their development only in the following spring. It is conceivable that special PDV features may be necessary to ensure wasp larval development after the winter diapause.

Gene Expansion is a Remarkable Feature of PDV Genomes

One of the most striking features of PDV genomes is the organization of many genes into gene families. In certain cases, impressive levels of gene expansion have been attained with 17–38 Rep genes in IV genomes, 13–42 PTPs in BVs and 13 PTP genes in GfV (Table 2). The expansion of PTP genes in PDVs is particularly remarkable as it is comparable to PTP gene numbers in eukaryotic genomes. For example, the human genome (3×10^9 kb) contains 106 PTP genes, whereas the GiBV genome, which is 5×10^6 times smaller (517 kb), harbors 42 genes (Andersen et al., 2001).

Gene duplications have long been recognized as a source of evolutionary innovation and adaptation, but still relatively little is known on the evolutionary mechanisms involved in the emergence, maintenance, and evolution of gene duplications (Innan and Kondrashov, 2010). Classical evolutionary models assume that duplication does not affect fitness, so that the fixation of the duplicated copy is a neutral process and selection possibly leading to innovation occurs after fixation of the duplication event (Force *et al.*, 1999; Hughes, 1994; Ohno, 1970). Alternative models propose that the duplication itself is advantageous. This type of adaptation can be explained if an increase in dosage of a particular gene is beneficial then duplication of this particular gene may be fixed by positive selection. Genes that mediate interactions between the organism and its environment (i.e., a hostile host), or genes that have dosage-sensitive functions owing to protein–protein interaction properties of their products are predicted to be concerned by these models (Innan and Kondrashov, 2010). The process of gene duplication itself may create a new function, leading to fixation of the new copy and preservation by positive selection. For instance, partial gene duplications, or new genomic locations of the gene copy may introduce new functional aspects. Lastly, in the adaptive radiation model, Francino (2005) proposes that, in a population, gene copies appear that are preadapted to perform new functions while still performing the original function of the gene. The function of the preadapted duplicated copies are then 'improved' by positive selection. This model is particularly interesting because it could occur repeatedly, leading to substantial increases in gene copy numbers.

The growing information on PDV genomes (both proviral and injected circular genomes), combined with the phylogenetic analysis of wasp species will give us the opportunity to study evolutionary processes involved in gene duplications in PDVs (Friedman and Hughes, 2006). Comparative analyses between phylogenetically related PDV genomes should enable us to distinguish between original and new duplicated copies of PDV genes and to infer the evolutionary mechanisms involved in their emergence and maintenance. Analysis of two very different gene families—cystatins and PTPs—in PDVs associated with braconid *Cotesia* wasps have already generated interesting results which suggest that in these families duplication events are adaptive and have led to innovation (Serbielle *et al.*, in preparation; Serbielle *et al.*, 2008). We hypothesize that the diversification of PDV virulence genes into families reflects the pressure imposed by the caterpillar host on these genes with respect to their role in parasitism success (Bézier *et al.*, 2008; Serbielle *et al.*, 2008).

Cystatins constitute small multigene families that have only been identified in PDVs associated with *Cotesia* wasps and in the closely related PDVs, GfBV and GiBV, of *Glyptapanteles* wasps (three genes are present in CcBV, two

genes in GfBV and one gene in GiBV). The weak divergence between cystatin genes and their narrow phylogenetic distribution constitutes evidence of the recent acquisition of cystatin genes by BVs. As a consequence, studying cystatin gene evolution comes down to observing the preliminary evolutionary processes involved in the diversification of a 'young' gene family. By analyzing the nature of selection pressures acting on branches of a phylogenetic tree constructed with cystatin sequences from different *Cotesia* species, we could show that cystatin divergence is driven by positive selection, which has acted at different levels, before, during, or after the wasp speciation process (Serbielle *et al.*, 2008).

The study of PTP gene evolution constitutes a snapshot of a much older gene family, indeed PTP genes are common to most sequenced BVs, suggesting they were acquired early in the course of bracovirus–wasp associations. We used complete or partial available BV genome sequences to study the genomic organization and transmission of PTP duplications. This allowed us to infer the mechanisms at the basis of duplications, and to detect positive selection acting in certain lineages (Serbielle *et al.*, in preparation). We propose that PTP gene expansion could be a source of evolutionary innovations offering wasps adaptive properties. In accordance, different forms, expression patterns and functions have been described for PTPs in the context of host–parasite interactions (Moreau *et al.*, 2009). PTPs dephosphorylate phosphotyrosine residues on specific substrates, and given the high amino acid diversity of bracovirus PTPs they have the potential to each interact with a very specific and different substrate (Andersen *et al.*, 2001). Furthermore, BVs have been shown to encode both catalytically active and inactive PTPs. The inactive forms are believed to function by trapping phosphorylated tyrosine proteins (Ibrahim and Kim, 2008; Provost *et al.*, 2004). PDV PTPs have been implicated in a variety of functions involving immune deregulation such as antiphagocytic activity, apoptosis-inducing activity and alteration of hemocyte spreading and encapsulation (Ibrahim and Kim, 2008; Pruijssers and Strand, 2007; Suderman *et al.*, 2008). The diverse PTPs generated by adaptive gene duplications may also be implicated in other aspects of the host–parasite interaction. For instance, a baculovirus PTP was shown to be responsible for manipulating the behavior of caterpillars, inducing them to climb to the top of plants and thus enhancing viral dissemination (Kamita *et al.*, 2005). PTPs could also be implicated in the underphosphorylation of regulatory proteins that is observed in the prothoracicotropic hormone signaling pathway during parasitism by certain parasitoid wasps (Pennacchio *et al.*, 2001).

Further studies on duplications in PDV genomes will provide a theoretical framework to test different models of evolution, and will enable us to understand how PDV gene function is being innovated through duplications.

Diversifying Selection Acting on PDV Virulence Genes

In host–parasitoid interactions, reciprocal selective pressures involving host resistance and parasitoid virulence are likely to be running high. Indeed, parasitoids represent one of the major sources of biotic mortality on phytophagous insect populations (Hawkins *et al.*, 1997). Furthermore, the koinobiont parasitoid wasps that develop as larvae within caterpillars have to face the arsenal of a functional immune system produced by a living organism and the result of the interaction is the death of at least one of the partners. In consequence, parasitoid virulence factors are likely to be under strong evolutionary pressures. Selection pressures may act on virulence genes to radiate towards diverse targets and evolve to counteract defenses of novel hosts, thereby increasing wasp host range. Alternatively, pathogenicity factors could be under pressure to coevolve with already established hosts in a molecular arms race. To investigate these different scenarios, it is necessary to characterize and measure selection pressures acting on parasitoid/PDV virulence genes and on the suspected targets and resistance determinants in the host.

So far, in these systems, studies have focused on determining selection pressures acting on PDV virulence genes by measuring omega ratios. The ratio of non-synonymous (amino acid altering) versus synonymous (amino acid conserving) substitution rates ($\omega = dN/dS$) is indeed an indicator of selection pressure at the protein level. Omega ratios equal to one indicate absence of selection, ω ratios less than one indicate purifying selection (that acts to conserve protein function), whereas ω ratios greater than one are an indication of positive or diversifying selection (that acts to modify protein function) (Yang and Bielawski, 2000; Yang *et al.*, 2000a, b). This latter type of selection is expected to be observed for genes encoding products that are involved in interactions with hosts.

In a global analysis, Desjardins *et al.* (2008) examined the ω ratio for 72 homologous pairs of genes present in the proviral segments of GiBV and GfBV and compared them to ω ratios obtained for 41 homologous pairs of flanking wasp cellular genes. Both sets of genes were shown to be evolving at significantly different rates. Flanking wasp genes predominantly have dN/dS values of zero, indicating they are under strong negative selection. Proviral genes had dN/dS values closer to one suggesting that viral genes are mainly evolving under neutral or positive selection.

A closer examination of specific PDV gene families in different host–parasitoid systems has revealed evidence of positive selection, notably in the cystatin and PTP gene families in BVs and in the Cys-motif family in IVs (Serbielle *et al.*, 2008; Desjardins *et al.*, 2008, Dupas *et al.*, 2003). Here we shall present the published data on cystatins and Cys-motif genes.

Positive Selection Acting on Cystatin Genes

Cystatins are tight-binding reversible inhibitors of cysteine proteases of the C1 family (Rawlings *et al.*, 2004). Cystatins are characterized by three conserved domains forming the inhibitory sites of interaction with proteases: an N-terminal glycine, a glutamine-X-valine-X-glycine motif and a C-terminal proline–tryptophan amino acid pair (Bode *et al.*, 1988; Stubbs *et al.*, 1990). CcBV cystatins are suspected to play an important role in the host–parasitoid interaction owing to their high level of expression and the fact that a recombinant CcBV cystatin1 was shown to be a functional and specific cysteine protease inhibitor (Espagne *et al.*, 2005). Furthermore, considering that, in insects, C1 cysteine proteases are implicated in physiological processes such as development and immunity, which are two functions targeted by PDVs, PDV cystatins could act by inhibiting target proteases implicated in these functions (Serbielle *et al.*, 2009). CcBV cystatins therefore appeared to be a good candidate gene family to study the potential influence of host selection pressures.

The molecular evolution of viral cystatin genes was studied by isolating 48 sequences from BVs associated with nine species of *Cotesia* wasps. Cystatin gene evolution was shown to be driven by strong positive selection with a global ω value of 1.2 over all sites. Furthermore, a generated three-dimensional model of CcBV cystatin 1 provided a powerful framework to position positively selected residues and revealed that they are concentrated in the vicinity of active sites that interact with cysteine proteases directly (Serbielle *et al.*, 2008). These results suggest that positive selection is acting presumably to modulate viral cystatin affinity for caterpillar protease targets and that cystatins have to continuously evolve in order to adapt to evolving targets in the same host, or to new targets in different hosts.

Positive Selection Acting on Cys-Motif Genes

Genes encoding cysteine motifs have been identified in IVs and BVs (Dupuy *et al.*, 2006). These proteins are characterized by one or more cysteine-knot structural motifs consisting of six invariable cysteine residues in a stretch of approximately 50 amino acids (Dib-Hajj *et al.*, 1993). The VHv1.1 cys-motif protein in CsIV is involved in the reduction of the host encapsulation response and this effect is suggested to be mediated by host surface receptors (Li and Webb, 1994). Cys-motif genes from CsIV were analyzed for synonymous and non-synonymous substitution patterns. Positive selection was shown to act on 8 out of 51 codons in aligned cysteine-rich regions. In this system, the solved structure of VHv1.1 allowed the overlaying of the data on the tertiary structure (Dupas *et al.*, 2003). Residues under positive selection occurred primarily in accessible, exposed regions of the tertiary structure, whereas residues under strong negative selection were situated in B-sheets

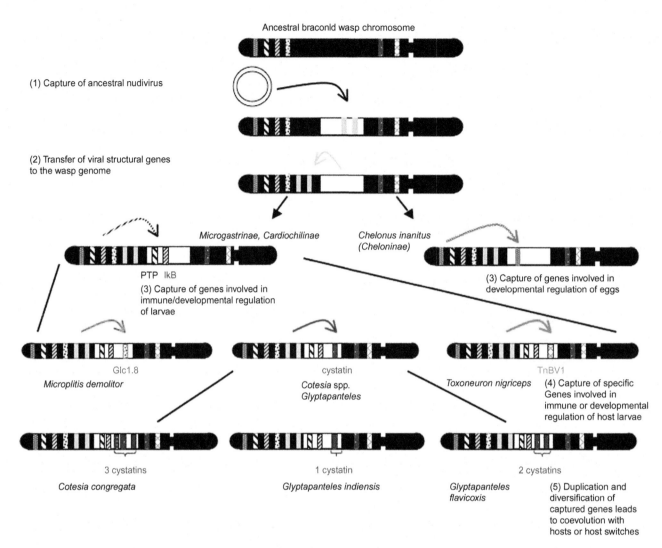

Ancestral braconid wasp chromosome

(1) Capture of ancestral nudivirus

(2) Transfer of viral structural genes
to the wasp genome

Microgastrinae, Cardiochilinae   Chelonus inanitus
                                        (Cheloninae)

PTP IkB
(3) Capture of genes involved in
immune/developmental regulation
of larvae

(3) Capture of genes involved in
developmental regulation of eggs

Glc1.8   cystatin   TnBV1
Microplitis demolitor   Cotesia spp.   Toxoneuron nigriceps   (4) Capture of specific
                                Glyptapanteles   Genes involved in
                                                                              immune or developmental
                                                                              regulation of host larvae

3 cystatins   1 cystatin   2 cystatins
Cotesia congregata   Glyptapanteles indiensis   Glyptapanteles   (5) Duplication and
                                                            flavicoxis   diversification of
                                                                               captured genes leads
                                                                               to coevolution with
                                                                               hosts or host switches

FIGURE 1   **Proposed evolutionary scenario for the origin and evolution of bracoviruses.** Glc1.8, cystatins and TnBV1 are genes specific to each bracovirus associated with *M. demolitor*, *Cotesia* and *Glyptapanteles*, and *Toxoneuron nigriceps*, respectively. Please see color plate section at the back of the book.

and disulfide bridges (Dupas *et al.*, 2003). Cys-motif proteins therefore appear to be under evolutionary pressure to simultaneously encode stable proteins with certain positions under positive selection, probably to evolve in response to host receptors. In this molecular interaction, however, the nature of the host receptor is not known.

The directional selection acting on key sites in PDV virulence factors both in BVs and IVs suggest that these factors could be coevolving with their respective targets in host caterpillars. Alternatively, diversifying evolutionary patterns could also be explained by wasp host switches. The challenge is now to also study the molecular evolution of the lepidopteran targets to be able to distinguish between the two hypotheses. This approach could be taken for example in the *Cotesia/Melitaea* parasite–host complex in which the natural histories of the host–parasite relationships are well characterized and wasps and hosts,

respectively, display close phylogenetic relationships, facilitating gene isolation (Kankare and Shaw, 2004).

In conclusion, studying the evolution of PDV genes present in the injected circular form of the virus not only gives insight on how these genes have participated in the construction of this complex entity (Fig. 1), but can also be a means to pinpoint genes of interest that are potentially actively involved in the suppression of the host immune response.

ORIGIN OF PDV VIRULENCE GENES

A fascinating question concerns the origins of PDV virulence genes and the mechanisms of acquisition. Early on, it was proposed that PDV virulence genes could originate from genes encoding wasp venom products (Webb and Summers, 1990). Indeed, the injection and expression of

TABLE 3 Number of Ankyrin-Like Genes Present in all Available Bracovirus or Ichnovirus Sequences

	Number of ankyrin-like genes	References
Bracoviruses		
CcBV	6 + 3	Falabella et al., 2007; Drezen et al. (in preparation)
CpBV (CvBV)	8	Choi et al., 2005; Chen et al., 2008; Shi et al., 2008
GiBV	9	Desjardins et al., 2008
GfBV	8	Desjardins et al., 2008
MdBV	12	Thoetkiattikul et al., 2005; Webb et al., 2006
TnBV	3	Falabella et al., 2007
Ichnoviruses		
CcIV	2	Tian et al., 2007
CsIV	7	Kroemer and Webb, 2005; Webb et al., 2006
GfIV	4	Lapointe et al. (2007)
HdIV	1 + 8	Webb et al., 2006; Volkoff et al. (in preparation)
HfIV	9	Tanaka et al., 2007
TrIV	2	Tanaka et al., 2007

CcIV, Campoletis chloridae ichnovirus.

virulence genes by the means of a virus may represent a less costly delivery mechanism of virulence products within the caterpillar host for the wasp. In effect, many PDV genes present in the injected circular form resemble cellular genes, suggesting that significant gene transfer events have occurred betweens PDVs and the insects (wasps and lepidopteran hosts) with which they are confronted. However, so far it has been difficult to formally demonstrate gene transfer between insects and PDVs due to a high divergence of PDV genes when compared to their insect counterparts. For instance, for bracovirus PTP and ankyrin proteins, no clear phylogenetic link can be established with corresponding insect proteins. The high level of divergence appears to be a hallmark of PDV genes as the analysis of the content of a braconid wasp (*C. congregata*) cDNA library revealed that most sequences displayed significant similarities with those of the honeybee (Bézier *et al.*, 2008).

Here, we will present the evidence for the capture of different insect genes by PDVs. Clearly, as shown by the genome compositions of the different viruses (Table 2), types of genes transferred and timing of transfers have differed, and specific scenarios can be envisaged for different genes. Ankyrin-like genes, for example, that are present both in IVs and BVs, are likely to have been acquired at a very early stage whereas genes present only in certain species (i.e., cystatins) certainly represent recent gene acquisitions. Furthermore, there is also evidence that certain PDV genes were acquired via horizontal transfer from other

viruses, mainly lepidopteran viruses that are likely to coinfect parasitized caterpillar hosts. Transposable elements have also been shown to be a source of gene acquisition in PDVs, and they are suspected to play a role in the shuffling of genes towards PDVs.

The Mysterious Case of the Ankyrin-Like Proteins from PDVs

Distribution of ank Genes Among PDV Genomes

Genes encoding ankyrin-like proteins have been reported from all polydnavirus genomes that have been completely sequenced so far. Along with genes encoding cysteine-motifs, they constitute one of the few gene families identified both in bracoviruses and ichnoviruses (tables 2, 3). Members of this viral gene family share weak but significant sequence identities with the *Drosophila* dorsal/inhibitor of nuclear factor kappa-B known as cactus and have been referred to as 'vankyrins', 'IκB-like', 'cactus-like', 'viral ankyrin' or 'ank' genes according to authors. The definitive number of ankyrin-like genes contained in these genomes is likely to exceed what has been published so far. For example, three additional new genes were recently discovered in CcBV (Drezen *et al.*, in preparation) and at least eight ankyrin-like genes could be present in HdIV (Volkoff *et al.*, in preparation).

Structural Features and Roles of the Nuclear Factor-Kappa B/IκB Interactions in Insects

The presence of ankyrin-like genes in PDVs raised the attractive hypothesis that nuclear factor kappa-B (NF-κB)-mediated activities may be targeted for disruption by PDV infections. NF-κB is an eukaryotic transcription factor that is present in the cytoplasm of cells in a dimeric inactive form bound to the inhibitor IκB (Ghosh *et al.*, 1998). The ankyrin repeat domain (ARD), shared by all IκBs and composed of several ankyrin (ank) repeats, plays a critical role in this interaction, in mediating the binding to NF-κB dimer and the masking of their nuclear localization signal (NLS) (Karin, 1999). In response to an appropriate signal, IκB is phosphorylated by an IκB, Kinase (IKK), undergoes immediate polyubiquitination, and is degraded by the 26S proteasome. Once released from its inhibitor, NF-κB translocates to the nucleus where it can upregulate transcription of specific genes (Ghosh *et al.*, 1998). For instance, in *Drosophila*, NF-κB-like transcription factors Relish, Dorsal, and/or DIF activate the specific transcription of antimicrobial peptides dependent on IMD (immune deficiency) or Toll pathways (Lemaitre, 2004). Since NF-κB transcription factors are key regulators of development, innate and cellular immunity pathways in insects, it has been proposed that alterations of their activity could benefit parasitization by preventing the triggering of essential aspects of host immune response and development (Kroemer and Webb, 2005). The initial observation that CsIV vankyrins lacked regulatory elements associated with signal-induced and basal degradation of typical IκB proteins (Kroemer and Webb, 2004) raised the idea that viral ankyrin-like proteins could act as irreversible inhibitors of NF-κB transcription factors in parasitized hosts.

The N-terminal ends of most known PDV ankyrin-like proteins do not align with the N-terminal regulatory domain of typical cactus/IκB proteins that contain two highly conserved serine residues in the motif DSGXXS (Fuchs *et al.*, 2004; Ghosh *et al.*, 1998; Shirane *et al.*, 1999). However, such a DSGXXS motif can be found in the C-terminal end of a viral ankyrin from GiBV (ACE75486.1). In human cells, the serine residues contained in this motif are phosphorylated by an inducible IκB kinase complex (IKK) in response to immune or developmental stimuli. Phosphorylated IκB proteins can then bind SCF$^\beta$-TrCP ubiquitin ligase which mediates the ubiquitination of lysins and the subsequent proteolysis of IκB by the 26S proteasome. The absence of this N-terminal motif may thus increase the stability of viral ankyrins in insect host cells. Interestingly, some viral proteins, like the HIV type 1 Vpu protein, are known to inhibit NF-κB activation by interfering with β-TrCP-mediated degradation of IκB (Bour *et al.*, 2001). In this case, Vpu uses a DSGXXS pseudosubstrate to interact with the human β-TrCP but is not targeted to degradation by proteasome. This causes the depletion of the stock of available ubiquitin ligase and consequently the inhibition of IκB degradation and the reduction of NF-κB activity. It can not be ruled out that the C-terminal DSGXXS motif found in the GiBV viral ankyrin may similarly act as a pseudosubstrate in host cells. This idea is reinforced by the absence of lysine residues downstream of this motif, that would prevent any possibility of ubiquitination of the viral protein. This hypothetical complementary mechanism for NF-kB inhibition may thus be worth further investigations.

C-terminal ends of PDV ankyrin-like proteins also do not align with the C-terminal PEST domains of cactus/IκB proteins, which led to the idea, largely diffused, that viral proteins lacked another important regulatory element of these inhibitors (Kroemer and Webb, 2005; Thoetkiattikul *et al.*, 2005). In fact, this domain does not display a conserved consensus sequence but is rather defined by its richness in proline, glutamic acid, serine, threonine and to a lesser extent aspartic acid residues (Rogers *et al.*, 1986). In addition, its role in the functional regulation of IκB proteins varies from one protein to another. For instance, the PEST domain is absent from the IκBε human protein (Ghosh *et al.*, 1998). Phosphorylation of the PEST domain is essential for stability of the human IκBα protein (Lin *et al.*, 1996), is required for the degradation of the cactus protein (Liu *et al.*, 1997), or is involved in inhibition of Rel-type proteins (Chu *et al.*, 1996). A careful examination of 73 PDV ankyrin-like proteins using the EMBOSS program *epestfind* with the default cut-off pest score of +5.0 has revealed the presence of potential PEST domains in 10 PDV proteins in MdBV, GiBV, GfBV, HfIV and GfIV. All the predicted PEST domains were systematically found to be positioned at the C-terminal end of the proteins. This finding might be of real biological interest given that a recent study on 9827 human PEST domain-containing proteins using the same program, showed the absence of any preferential localization of PEST sequences in the analyzed proteins (Singh *et al.*, 2006).

Functional analysis of certain PDV ankyrin-like proteins have revealed that these proteins have the potential to disrupt NF-κB signaling in lepidopteran hosts (Falabella *et al.*, 2007; Thoetkiattikul *et al.*, 2005). The consequences of such disruptions on host development and immunity have not been well established in most cases, but a number of PDV ankyrin-like proteins appear to have the ability to disrupt NF-κB-mediated processes such as antimicrobial peptide production (Shrestha *et al.*, 2009), elicitation of antiviral responses (Bae and Kim, 2009; Rivkin *et al.*, 2006), and regulation of apoptosis (Fath-Goodin *et al.*, 2009; Kroemer and Webb, 2006). Globally, the data indicate that the ankyrin-like gene family is not a homogenous entity but gathers members that have evolved differently to acquire various types of tissue expression, subcellular localization, and activity (Gundersen-Rindal and Pedroni, 2006; Kroemer and Webb, 2006). The detailed analysis of the structural features of their ank repeats highlights the heterogeneity of this gene family and helps to better appreciate their evolutionary history.

Structural Features of ank Repeats as a Tool to Apprehend the Origin and Evolution of PDV Ankyrin-Like Genes

From a structural point of view, ankyrin-like proteins are primarily defined by the occurrence of multiple ank repeats in their central region corresponding to repeated motifs of 33 residues (Bennett, 1992; Bork, 1993). Each ank repeat folds into a helix-loop-helix structure with a β-hairpin/loop region projecting outward from the helices. Actual methods of prediction of the occurrence of ank repeats in proteins are mainly based on the recognition of the ability of amino acids to form such structures. In addition, consensus-derived structural determinants of ank repeat motifs have been established through a statistical approach and confirmed by structural studies (Mosavi et al., 2002).

A striking feature of polydnaviral ankyrin-like proteins, with respect to cactus/IκB proteins to which they are often compared, is the variation observed in the number of their predicted ank repeats. This number ranges from one to eight with the majority of proteins containing three or four repeats in BV and IV, respectively. In comparison, IκBα and its Drosophila homolog cactus would contain six or seven repeats (Baldwin, 1996; Geisler et al., 1992; Jacobs and Harrison, 1998) although only five of them can be predicted with certainty with the available annotation tools (see annotations at the UniProtKB database). Such variations might have arisen from loss or duplication events of ank repeats during the course of PDV gene evolution and may constitute a first cause of functional diversification within this family.

Several genes bear witness to this evolution. For example, in CcBV, three ank genes have been identified that are either interrupted by a stop codon (CcBV ank5), or encode proteins containing a single ank repeat (CcBV ank8) or devoid of any ank repeat (CcBV ank9, initially named CcBV ank 26.5) (Falabella et al., 2007). Given that these genes are expressed in parasitized M. sexta larvae, it is not yet clearly established whether they are undergoing pseudogenization or if they fulfil new functions.

In contrast, some viral ankyrin-like proteins are characterized by sequences containing additional ank repeats. In HfIV, for example, the vankyrin-d8.2 gene encodes a protein containing eight ank repeats. Three nearly identical regions in the protein are highly similar to the region encompassing ank repeats one to three of the vankyrin-d8.1 protein. Such sequence identities suggest that vankyrin-d8.2 originates from vankyrin-d8.1, located on the same d8 viral circle (Tanaka et al., 2007). A parsimonious hypothesis explaining the formation of the vankyrin-d8.2 gene is described in Fig. 2 and relies on one initial duplication event followed by two unequal crossing over events.

Sequence alignments of known polydnaviral ankyrin-like proteins show that some residues of ank repeats are highly conserved while others are variable both within and between proteins. Sequence variations are likely to constitute another cause of functional diversity among these proteins. It has been suggested for IκB-like proteins that these variable residues could confer different binding specificities to each ank repeat and consequently different functions (Geisler et al., 1992). For instance, ank repeats one and two of human IκBα are likely to participate in the masking of the NLS of NF-κB while repeat three has relatively little contact with NF-κB (Jacobs and Harrison, 1998). Interestingly, corresponding repeats one and two of IκBα are lacking in PDV proteins, which suggests that they may incompletely mask the NLS of NF-κB-like ligands. In CsIV, it has been proposed that vankyrins may hence actively inhibit and sequester NF-κB complexes in the host nucleus which would lead to a significant reduction of cytoplasmic pools of NF-κB (Kroemer and Webb, 2005).

The two most conserved ank repeats within PDV proteins and between these proteins and cactus/IκB proteins, are located centrally in the sequences while N-terminal or C-terminal repeats are more divergent (Thoetkiattikul et al., 2005). In particular, the penultimate ank repeat, which aligns with ank repeat five of cactus and IκBα, possesses a consensus sequence shared by all BV and IV ankyrin-like proteins with few variations (Fig. 3). This consensus sequence is relatively close to the general consensus sequence of ank repeats defined by Mosavi et al. (2002) and is always conserved even in PDV proteins that only contain a single predicted ank repeat. Functionally, it suggests that this core ank repeat may play a key role in the molecular interactions between PDV ankyrin-like proteins and their targets. From an evolutionary point of view, such a conserved repeat is likely to have been under strong conservative selection since its acquisition by PDV genomes. Interestingly, Falabella et al. (2007) have noticed the presence in this conserved repeat of a short Trp-Leu-Cys (WLC) sequence only found in ankyrin-like proteins of PDVs. The existence of this short molecular signature, along with the overall conservation of repeat organization in BV and IV ankyrin-like proteins, led to exciting hypotheses on the origin of the related genes. Indeed, it suggests a common origin despite the fact that BVs and IVs originated and evolved independently in the braconid and ichneumonid lineages (Whitfield, 2002). Phylogenetic analyses not only confirm this common origin but may also help define the main steps of the diversification of PDV ankyrin-like genes.

Phylogenetic Analysis of PDV Ankyrin-Like Proteins

The phylogenetic analysis displayed in Fig. 4 shows that BV and IV ankyrin domain-containing proteins constitute a monophyletic group. It confirms, in a more exhaustive manner, previous analyses performed by Falabella et al., (2007), Tian et al. (2007), Lapointe et al. (2007), Chen et al. (2008), Choi et al. (2009) and Bae and Kim (2009).

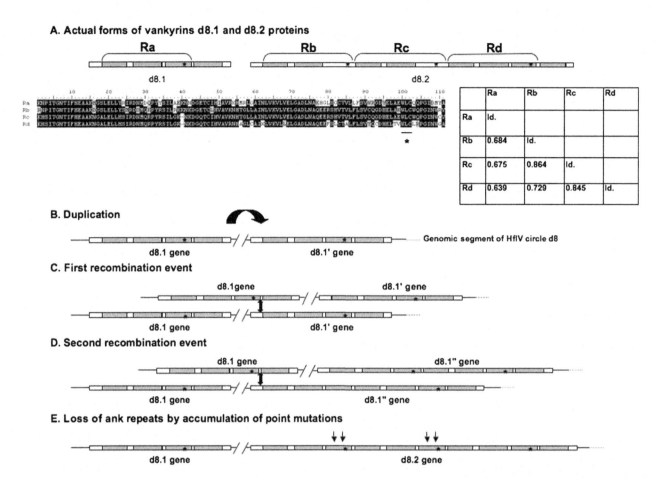

FIGURE 2   Hypothesis on the origin of the *vankyrin*-d8.2 gene of HfIV. A: Comparison between the actual forms of vankyrins d8.1 and d8.2 of HfIV. The *vankyrin-d8.2* gene encodes a protein containing eight ank repeats clustered into three nearly identical regions (regions Rb, Rc and Rd, respectively). Each region is highly similar to the region encompassing ank repeats one to three of the vankyrin-d8.1 protein (Ra). The amino acid sequence alignment of regions Ra, Rb, Rc and Rd is shown at the left of the figure and the corresponding sequence identity matrix is shown at the right. Stars indicate the WLC signature, found in most BV and IV ankyrin-like proteins, and predicted ank repeats are shown in gray. **B:** The most probable hypothesis explaining the formation of the *vankyrin-d8.2* gene relies on one initial duplication event by which a copy of the *vankyrin-d8.1* gene has been duplicated downstream of the original gene. **C:** An unequal crossing-over has occurred afterwards between the region coding for the beginning of the fourth ank repeat of *vankyrin-d8.1* gene and the 5′ end of the copied gene. This event could have led to an intermediate recombinant gene that possessed seven ank repeats, as found in a sequence encoded by the closely related viral genome HdIV (accession number AAR99844.1). **D:** Another unequal crossing-over has probably occurred in the same conditions as the previous one, between the recombinant gene and *vankyrin-d8.1*. **E:** Subsequent accumulation of point mutations has modified the regions encoding the third and sixth ank repeats leading to the actual form of the *vankyrin-d8.2* gene.

Within this group of PDV ankyrin genes, our analysis reveals the existence of two well-supported clades (A and B), while a third group (named C) is not well-supported. Clade A is basal and exclusively composed of sequences from BV species associated to braconid wasps of the microgastrine subfamily. Within clade A, subclades are well supported and their relatedness mirrors wasps phylogeny (Murphy *et al.*, 2008). Clade B is more diversified and also formed by BV products, 50% of BV proteins studied in this analysis belong to this clade. Again, proteins are grouped in different subclades mostly well supported and in accordance with wasp phylogeny. All other BV sequences and ankyrin-like proteins from IVs are found in group C. Although

some external nodes are robust, this group presents internal nodes globally weakly supported, which constitutes a limit for the interpretation of its organization. However, it can be observed that all BV sequences figure in a basal position with respect to the IV sequences. Interestingly, sequences from TnBV, which is associated with a wasp from an ancestral subfamily (Cardiochilinae) of the microgastroid complex were found in clades B and C but not in the basal clade A.

This phylogenetic analysis allows us to draw a possible scenario for the origin and evolution of ankyrin-like genes in PDVs. The first capture of an ankyrin gene probably occurred in a BV associated with an ancestor

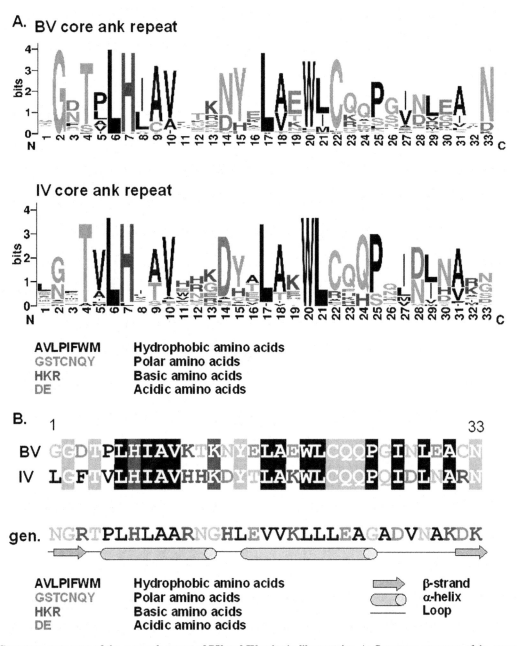

FIGURE 3 Consensus sequences of the core ank repeat of BV and IV ankyrin-like proteins. A: Consensus sequences of the core ank repeats of BV and IV ankyrin-like proteins were drawn using the WebLogo application (http://weblogo.berkeley.edu) after multiple sequence alignment of 45 BV and 23 IV ankyrin-like proteins. B: Comparison between the consensus sequences of BV and IV core ank repeats. Identical amino acids are highlighted. The general consensus sequence of ank repeats, adapted from Mosavi *et al.* (2002) is shown for comparison. Please see color plate section at the back of the book.

wasp of the microgastroid complex. Due to the sequence similarities observed between contemporary PDV ankyrin-like proteins and insect cactus/IκB proteins, the captured ankyrin gene probably originated from an insect genome and gave rise to the gene products present in basal clade A. However, the WLC signature of PDV genes is not present in hymenopteran and more generally in insect cactus-related protein sequences available suggesting that

the filiation was not direct and that PDV ankyrin genes may have undergone an initial step of evolution within an ancestral PDV genome or a pathogenic virus. Identifying a larger set of PDV, virus and insect genes would allow us to test this hypothesis.

The second important point concerns the diversification of BV ankyrins that occurred afterwards and which resulted in a large multigene family whose products

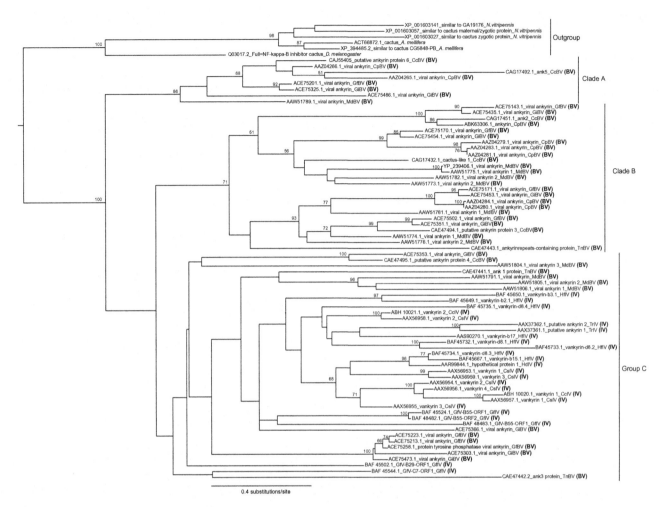

FIGURE 4   Ankyrin protein tree obtained by maximum likelihood on Phyml 3.0 (Guindon and Gascuel, 2003) under the LG model. One hundred bootstraps were performed and only node supports higher than 50 are represented.

compose clade B. Finally, the basal position of BV proteins in group C, which contains all IV proteins, and the presence of the WLC signature in the core ank repeat of most known IV and BV ankyrin-like proteins strongly suggests an acquisition of ankyrin-like genes by horizontal transfer(s) from BV to IV genomes. However, due to the lack of robustness of the organization of group C, it cannot be determined with precision which BV ankyrin-like gene could be at the origin of IV genes. The definitive position of TnBV sequences, which are very divergent compared to other BV and IV sequences also remains to be established.

Hence, PDV ankyrin genes constitute a multigene family with divergent functions that could intervene in various aspects of regulation of host physiology through their ability to interact with NF-κB-like ligands. Some, but not all, lack regulatory elements of cactus/IκB proteins. The structural organization of the ank repeats that mediate such interactions is variable between proteins and is the result of duplication and recombination events associated with accumulation of point mutations. These structural

modifications have contributed to the diversification of the family from an ancestral pattern probably inherited from a wasp gene captured by BVs and transferred horizontally to IVs. Future studies should aim at elucidating their phylogenetic link with NF-κB inhibitors and at delineating more precisely their functions in parasitoid–host relationships.

Evidence for Gene Transfer from Wasps to PDV Sequences

Since PDVs are present as an integrated provirus in wasp genomes, the potential for genetic transfer between wasp and virus exists. The fact that genes encoding structural components of the particles have mainly been found in wasp genomic loci at some distance from proviruses, shows that transfer of genetic material from the provirus to the rest of the wasp genome has occurred (Bézier *et al.*, 2009; Deng *et al.*, 2000).

The most robust evidence of gene transfer from wasp to PDVs has been provided by phylogenetic analyses performed on bracovirus genes (Desjardins *et al.*, 2008; Webb *et al.*, 2006). In GiBV and GfBV, a novel family of genes encoding products with similarities to insect sugar transporters was identified. Bayesian phylogenetic analysis conducted on the 8 GiBV and GfBV sugar transporter sequences and the 16 most similar transporters in insects revealed that the GiBV and GfBV sequences group strongly with sugar transporters from the parasitoid wasp *Nasonia vitripennis* and the honeybee *Apis mellifera*, but do not group with that of the silk moth, *Bombyx mori*. The results therefore suggest that these PDV genes share a common hymenopteran ancestry, and that they are likely to have been acquired from the wasp genome and not secondarily from the host caterpillar genome (Desjardins *et al.*, 2008). The fact that these genes have not been described within other PDV genomes sequenced to date, suggests that their presence results from a relatively recent acquisition event, which could explain why, in this case, the phylogenetic signal has not been lost with related insect genes. Another example of a phylogenetic link between BV sequences and hymenopteran sequences was reported for the MdBV EGF proteins. The egf-motif genes of MdBV code for proteins that contain a cysteine-rich domain. The MdBV 1.5 and 1.0 EGF proteins were shown to cluster with high bootstrap values to the cys-motif, TSP14 protein, of the braconid parasitoid wasp, *Microplitis croceipes*, pointing again towards a transfer between wasp and provirus (Webb *et al.*, 2006). Finally, histone genes present in PDV genomes evidently originate from gene transfer; however, in these cases the lack of sequence variation in histone genes of different organisms makes it impossible to conduct phylogenetic analysis.

For the other PDV virulence genes that have sequence similarity to known eukaryotic genes (i.e., PTPs and anks; see also Table 2), but for which there is no phylogenetic link, the divergence of these proteins can be interpreted by the high selection pressures imposed by the host–parasite relationship, which shapes the evolution of PDVs (see above).

Role of Other Viruses and Transposable Elements in PDV Gene Origin and Shuffling

Another common feature of described PDV genomes is the rare presence (and sometimes total absence) of coding sequences with significant similarity to sequences from known viruses. In CcBV, for example, three gene products share significant similarity with lepidopteran viral proteins (Drezen *et al.*, 2006). The presence of these lepidopteran viral genes in PDVs suggest that horizontal transfers between PDVs and lepidopteran viruses are possible, even though the phenomenon is likely to be quite rare. In the case of endoparasitoid wasps, penetration of lepidopteran virus particles in the wasps during their development in

an infected host caterpillar, could result in integration of a gene in the provirus locus in the wasp germline. Whether these genes are maintained or not would then depend on whether these genes are beneficial to PDV 'fitness'.

A remarkable feature of PDV genomes is the presence of mobile elements (Drezen *et al.*, 2006). This is not surprising in the sense that these viruses have been integrated for millions of years as proviruses in parasitoid wasp genomes. The age of the microgastroid wasp lineage, for example, has been estimated to be approximately 100 million years, therefore the chromosomally integrated bracovirus sequences have been a potential target of mobile elements for a considerably long time. In MdBV and CsIV, for example, 16 and 23 transposable element (TE) sequences, respectively, were identified mainly in noncoding domains of the viral genome. In CsIV, two sequences that are 71% similar to a *Drosophila* retrotransposon flank members of the rep gene family (Webb *et al.*, 2006). In GiBV, Desjardins *et al.* (2008) reported what appears to be a recent acquisition of a p-element-like transposable element. This element is indeed present at low frequency in GiBV genomes from different wasp populations.

An interesting observation has been the recent identification of *Maverick* TEs in wasp genomes and both in flanking wasp sequences next to GfBV proviral locus 7 and within the injected viral sequences of CcBV (Desjardins *et al.*, 2008, Dupuy *et al.*, 2011). *Mavericks* are a novel class of giant TEs also named Polintons that display characteristics of both TEs but also of Nucleo Cytoplasmic Large DNA viruses (Filee *et al.*, 2008; Kapitonov and Jurka, 2006; Pritham *et al.*, 2007). The CcBV *Maverick* element is the first example of such a TE in the genome of a virus and it has already been the siege of rearrangements with secondary insertions of retro-elements. Very interestingly, *Maverick* sequences have been identified in insect genomes and in particular in certain parasitoid wasp genomes, allowing phylogenetic analyses to be performed that strongly suggest that the CcBV *Maverick* originated by transposition of a copy present in the wasp genome (Dupuy *et al.*, 2011).

Altogether, the data show that TEs can enter PDV genomes and become packaged successfully. Transposable elements represent a plausible mechanism by which virulence or other beneficial genes could be acquired by PDVs from the host wasp. They may also have served to move core viral genes out of the ancestral proviral genomes. It could also be envisaged that PDVs play the roles of TE shuttles, allowing horizontal gene transfers between Hymenoptera and Lepidoptera, or even between Lepidoptera. A clear example of a PDV gene that originates either from a retrotransposon or a retrovirus is the aspartyl protease identified in the TnBV genome (Falabella *et al.*, 2003). This gene is highly expressed in the parasitized host suggesting it has a physiological function in the course of the interaction. Other potential cases of gene transfer via retroelements concern the BV cystatins and

IV vinnexins. The corresponding cellular genes contain characteristic introns that are notably absent in the viral forms suggesting that they may have been acquired via integration of cDNA (Espagne *et al.*, 2005; Turnbull and Webb, 2002). Human long interspersed element retransposons have been shown to be able to mobilize transcribed DNA sequences, which can result in the generation of processed (i.e., intronless) genes that fulfil physiological functions (Esnault *et al.*, 2000). Similar mechanisms could explain the presence of certain genes such as vinnexins and cystatins in PDV genomes.

CONCLUSION

Symbiotic associations have provided key biological innovations conferring selective advantages for the colonization of new ecological niches and subsequent specialization (Moran, 2007). The plataspid stinkbug insects, for example, show plant specializations that are in fact mediated by the associated bacterial symbiont which potentially induces host diversification (Hosokawa *et al.*, 2007).

In the case of the symbiotic associations implicating PDV and parasitoid wasps, it has also been inferred that the capture of these viruses represents an evolutionary novelty predicted to confer a selective advantage to the wasp, thereby playing an active part in wasp radiation. For example, braconid wasps associated with PDVs form the Microgastroid complex diverged from an unique ancestral association estimated to have occurred 103 million years ago. The capture of an ancestral nudivirus by the ancestor braconid wasp is thought to have been followed by a rapid wasp diversification, probably in response to lepidopteran host speciation (Bézier *et al.* 2009; Banks and Whitfield, 2006; Mardulyn and Whitfield, 1999; Murphy *et al.*, 2008). Genomic analyses of several polydnavirus genomes are now starting to provide keys to understanding how this association is maintained and what the physiological functions supplied by the wasp to ensure parasitism success are. Ongoing functional studies are unraveling the role of PDV virulence genes during parasitism. However, their divergence during symbiosis evolution and their contribution to wasp adaptation and diversification are only starting to be investigated (Fig. 1).

Studies based on well-characterized natural populations will certainly give strong impetus to this field of research. In this respect, recent studies conducted by Dupas and co-workers have revealed for the first time that variations in a PDV gene (CrV1) can be correlated to the parasitoid's specificity to hosts in Kenya. This study shows that parasitoids (and their associated PDVs) can be more or less well-adapted to hosts and that hosts are likely playing an important role in parasitoid diversification (Dupas *et al.*, 2008; Gitau *et al.*, 2007; Branca *et al.*, Chapter 10 of this volume). Furthermore, with the advent of next generation sequencing, it is now theoretically possible to address questions concerning the molecular ecology of host–parasitoid interactions. High coverage sequence data of PDVs harbored by wasps with different life history traits will conceivably open the door to investigating how ecological conditions (i.e., different caterpillar hosts) are shaping PDV genomes.

REFERENCES

Andersen, J.N., Mortensen, O.H., Peters, G.H., Drake, P.G., Iversen, L.F., Olsen, O.H., et al. 2001. Structural and evolutionary relationships among protein tyrosine phosphatase domains. Mol. Cell. Biol. 21, 7117–7136.

Annaheim, M., Lanzrein, B., 2007. Genome organization of the *Chelonus inanitus* polydnavirus: excision sites, spacers and abundance of proviral and excised segments. J. Gen. Virol. 88, 450–457.

Bae, S., Kim, Y., 2009. IκB genes encoded in *Cotesia plutellae* bracovirus suppress an antiviral response and enhance baculovirus pathogenicity against the diamondback moth, *Plutella xylostella*. J. Invertebr. Pathol. 102, 79–87.

Baldwin Jr., A.S., 1996. The NF-κB and IκB proteins: new discoveries and insights. Annu. Rev. Immunol. 14, 649–683.

Banks, J.C., Whitfield, J.B., 2006. Dissecting the ancient rapid radiation of microgastrine wasp genera using additional nuclear genes. Mol. Phylogenet. Evol. 41, 690–703.

Bennett, V., 1992. Ankyrins. Adaptors between diverse plasma membrane proteins and the cytoplasm. J. Biol. Chem. 267, 8703–8706.

Bézier, A., Annaheim, M., Herbiniere, J., Wetterwald, C., Gyapay, G., Bernard-Samain, S., et al. 2009. Polydnaviruses of braconid wasps derive from an ancestral nudivirus. Science 323, 926–930.

Bézier, A., Herbiniere, J., Serbielle, C., Lesobre, J., Wincker, P., Huguet, E., et al. 2008. Bracovirus gene products are highly divergent from insect proteins. Arch. Insect Biochem. Physiol. 67, 172–187.

Bode, W., Engh, R., Musil, D., Thiele, U., Huber, R., Karshikov, A., et al. 1988. The 2.0 A X-ray crystal structure of chicken egg white cystatin and its possible mode of interaction with cysteine proteinases. Embo. J. 7, 2593–2599.

Bonvin, M., Marti, D., Wyder, S., Kojic, D., Annaheim, M., Lanzrein, B., 2005. Cloning, characterization and analysis by RNA interference of various genes of the *Chelonus inanitus* polydnavirus. J. Gen. Virol. 86, 973–983.

Bork, P., 1993. Hundreds of ankyrin-like repeats in functionally diverse proteins: mobile modules that cross phyla horizontally? Proteins 17, 363–374.

Bour, S., Perrin, C., Akari, H., Strebel, K., 2001. The human immunodeficiency virus type 1 Vpu protein inhibits NF-κB activation by interfering with beta TrCP-mediated degradation of IκB. J. Biol. Chem. 276, 15920–15928.

Chen, Y.F., Shi, M., Liu, P.C., Huang, F., Chen, X.X., 2008. Characterization of an IκB-like gene in *Cotesia vestalis* polydnavirus. Arch. Insect Biochem. Physiol. 68, 71–78.

Choi, J.Y., Kwon, S.J., Roh, J.Y., Yang, T.J., Yoon, S.H., Kim, H., et al. 2009. Sequence and gene organization of 24 circles from the *Cotesia plutellae* bracovirus genome. Arch. Virol. 154, 1313–1327.

Choi, J.Y., Roh, J.Y., Kang, J.N., Shim, H.J., Woo, S.D., Jin, B.R., et al. 2005. Genomic segments cloning and analysis of *Cotesia plutellae* polydnavirus using plasmid capture system. Biochem. Biophys. Res. Commun. 332, 487–493.

Chu, Z.L., McKinsey, T.A., Liu, L., Qi, X., Ballard, D.W., 1996. Basal phosphorylation of the PEST domain in the IκBβ regulates its functional interaction with the c-rel proto-oncogene product. Mol. Cell. Biol. 16, 5974–5984.

Deng, L., Stoltz, D.B., Webb, B.A., 2000. A gene encoding a polydnavirus structural polypeptide is not encapsidated. Virology 269, 440–450.

Desjardins, C.A., Gundersen-Rindal, D.E., Hostetler, J.B., Tallon, L.J., Fadrosh, D.W., Fuester, R.W., et al. 2008. Comparative genomics of mutualistic viruses of *Glyptapanteles* parasitic wasps. Genome Biol. 9, R183.

Desjardins, C.A., Gundersen-Rindal, D.E., Hostetler, J.B., Tallon, L.J., Fuester, R.W., Schatz, M.C., et al. 2007. Structure and evolution of a proviral locus of *Glyptapanteles indiensis* bracovirus. BMC Microbiol. 7, 61.

Dib-Hajj, S.D., Webb, B.A., Summers, M.D., 1993. Structure and evolutionary implications of a 'cysteine-rich' *Campoletis sonorensis* polydnavirus gene family. Proc. Natl. Acad. Sci. U.S.A. 90, 3765–3769.

Drezen, J.-M., Bézier, A., Lesobre, J., Huguet, E., Cattolico, L., Periquet, G., et al. 2006. The few virus-like genes of *Cotesia congregata* bracovirus. Arch. Insect Biochem. Physiol. 61, 110–122.

Dupas, S., Gitau, C.W., Branca, A., Le Ru, B.P., Silvain, J.F., 2008. Evolution of a polydnavirus gene in relation to parasitoid–host species immune resistance. J. Hered. 99, 491–499.

Dupuy, C., Huguet, E., Drezen, J.-M., 2006. Unfolding the evolutionary story of polydnaviruses. Virus Res. 117, 81–89.

Dupuy, C., Periquet, G., Serbielle, S., Bézier, A., Louis, F., Drezen, J.-M., 2011. Transfer of a chromosomal Maverick to endogenous bracovirus in a parasitoid wasp. Genetica 139, 489–496.

Dupas, S., Turnbull, M.W., Webb, B.A., 2003. Diversifying selection in a parasitoid's symbiotic virus among genes involved in inhibiting host immunity. Immunogenetics 55, 351–361.

Esnault, C., Maestre, J., Heidmann, T., 2000. Human LINE retrotransposons generate processed pseudogenes. Nat. Genet. 24, 363–367.

Espagne, E., Douris, V., Lalmanach, G., Provost, B., Cattolico, L., Lesobre, J., et al. 2005. A virus essential for insect host–parasite interactions encodes cystatins. J. Virol. 79, 9765–9776.

Espagne, E., Dupuy, C., Huguet, E., Cattolico, L., Provost, B., Martins, N., et al. 2004. Genome sequence of a polydnavirus: insights into symbiotic virus evolution. Science 306, 286–289.

Falabella, P., Varricchio, P., Gigliotti, S., Tranfaglia, A., Pennacchio, F., Malva, C., 2003. *Toxoneuron nigriceps* polydnavirus encodes a putative aspartyl protease highly expressed in parasitized host larvae. Insect Mol. Biol. 12, 9–17.

Falabella, P., Varricchio, P., Provost, B., Espagne, E., Ferrarese, R., Grimaldi, A., et al. 2007. Characterization of the IκB-like gene family in polydnaviruses associated with wasps belonging to different braconid subfamilies. J. Gen. Virol. 88, 92–104.

Fath-Goodin, A., Kroemer, J.A., Webb, B.A., 2009. The *Campoletis sonorensis* ichnovirus vankyrin protein P-vank-1 inhibits apoptosis in insect Sf9 cells. Insect Mol. Biol. 18, 497–506.

Filee, J., Pouget, N., Chandler, M., 2008. Phylogenetic evidence for extensive lateral acquisition of cellular genes by Nucleocytoplasmic large DNA viruses. BMC Evol. Biol. 8, 320.

Force, A., Lynch, M., Pickett, F.B., Amores, A., Yan, Y.L., Postlethwait, J., 1999. Preservation of duplicate genes by complementary, degenerative mutations. Genetics 151, 1531–1545.

Francino, M.P., 2005. An adaptive radiation model for the origin of new gene functions. Nat. Genet. 37, 573–577.

Friedman, R., Hughes, A.L., 2006. Pattern of gene duplication in the *Cotesia congregata* bracovirus. Infect. Genet. Evol. 6, 315–322.

Fuchs, S.Y., Spiegelman, V.S., Kumar, K.G., 2004. The many faces of beta-TrCP E3 ubiquitin ligases: reflections in the magic mirror of cancer. Oncogene 23, 2028–2036.

Geisler, R., Bergmann, A., Hiromi, Y., Nusslein-Volhard, C., 1992. Cactus, a gene involved in dorsoventral pattern formation of Drosophila, is related to the IκB gene family of vertebrates. Cell 71, 613–621.

Ghosh, S., May, M.J., Kopp, E.B., 1998. NF-κ B and Rel proteins: evolutionarily conserved mediators of immune responses. Annu. Rev. Immunol. 16, 225–260.

Gitau, C.W., Gundersen-Rindal, D., Pedroni, M., Mbugi, P.J., Dupas, S., 2007. Differential expression of the CrV1 haemocyte inactivation-associated polydnavirus gene in the African maize stem borer *Busseola fusca* (Fuller) parasitized by two biotypes of the endoparasitoid *Cotesia sesamiae* (Cameron). J. Insect Physiol. 53, 676–684.

Guindon, S., Gascuel, O., 2003. A simple, fast and accurate algorithm to estimate large phylogenies by maximum likelihood. Syst. Biol. 52, 696–704.

Gundersen-Rindal, D.E., Pedroni, M.J., 2006. Characterization and transcriptional analysis of protein tyrosine phosphatase genes and an ankyrin repeat gene of the parasitoid *Glyptapanteles indiensis* polydnavirus in the parasitized host. J. Gen. Virol. 87, 311–322.

Hawkins, B.A., Cornel, H.V., Hochberg, M.E., 1997. Predators, parasitoids, and pathogens as motality agents in phytophagous insect populations. Ecology 78, 2145–2152.

Hosokawa, T., Kikuchi, Y., Shimada, M., Fukatsu, T., 2007. Obligate symbiont involved in pest status of host insect. Proc. Biol. Sci. 274, 1979–1984.

Hughes, A.L., 1994. The evolution of functionally novel proteins after gene duplication. Proc. Biol. Sci. 256, 119–124.

Ibrahim, A.M., Kim, Y., 2008. Transient expression of protein tyrosine phosphatases encoded in *Cotesia plutellae* bracovirus inhibits insect cellular immune responses. Naturwissenschaften 95, 25–32.

Innan, H., Kondrashov, F., 2010. The evolution of gene duplications: classifying and distinguishing between models. Nat. Rev. Genet. 11, 97–108.

Jacobs, M.D., Harrison, S.C., 1998. Structure of an IκBα/ F − κB complex. Cell 95, 749–758.

Kamita, S.G., Nagasaka, K., Chua, J.W., Shimada, T., Mita, K., Kobayashi, M., et al. 2005. A baculovirus-encoded protein tyrosine phosphatase gene induces enhanced locomotory activity in a lepidopteran host. Proc. Natl. Acad. Sci. U.S.A. 102, 2584–2589.

Kankare, M., Shaw, M.R., 2004. Molecular phylogeny of *Cotesia* Cameron, 1891 (Insecta: Hymenoptera: Braconidae: Microgastrinae) parasitoids associated with *Melitaeini* butterflies (Insecta: Lepidoptera: Nymphalidae: Melitaeini). Mol. Phylogenet. Evol. 32, 207–220.

Kapitonov, V.V., Jurka, J., 2006. Self-synthesizing DNA transposons in eukaryotes. Proc. Natl. Acad. Sci. U.S.A. 103, 4540–4545.

Karin, M., 1999. The beginning of the end: IκB kinase (IKK) and NF-κB activation. J. Biol. Chem. 274, 27339–27342.

Kroemer, J.A., Webb, B.A., 2004. Polydnavirus genes and genomes: emerging gene families and new insights into polydnavirus replication. Annu. Rev. Entomol. 49, 431–456.

Kroemer, J.A., Webb, B.A., 2005. IκB-related vankyrin genes in the *Campoletis sonorensis* ichnovirus: temporal and tissue-specific patterns of expression in parasitized *Heliothis virescens* lepidopteran hosts. J. Virol. 79, 7617–7628.

Kroemer, J.A., Webb, B.A., 2006. Divergences in protein activity and cellular localization within the *Campoletis sonorensis* ichnovirus vankyrin family. J. Virol. 80, 12219–12228.

Lapointe, R., Tanaka, K., Barney, W.E., Whitfield, J.B., Banks, J.C., Beliveau, C., et al. 2007. Genomic and morphological features of a banchine polydnavirus: comparison with bracoviruses and ichnoviruses. J. Virol. 81, 6491–6501.

Lemaitre, B., 2004. The road to Toll. Nat. Rev. Immunol. 4, 521–527.

Li, X., Webb, B.A., 1994. Apparent functional role for a cysteine-rich polydnavirus protein in suppression of the insect cellular immune response. J. Virol. 68, 7482–7489.

Lin, R., Beauparlant, P., Makris, C., Meloche, S., Hiscott, J., 1996. Phosphorylation of IκBα in the C-terminal PEST domain by casein kinase II affects intrinsic protein stability. Mol. Cell. Biol. 16, 1401–1409.

Liu, Z.P., Galindo, R.L., Wasserman, S.A., 1997. A role for CKII phosphorylation of the cactus PEST domain in dorsoventral patterning of the Drosophila embryo. Genes Dev. 11, 3413–3422.

Mardulyn, P., Whitfield, J.B., 1999. Phylogenetic signal in the COI, 16S, and 28S genes for inferring relationships among genera of Microgastrinae (Hymenoptera: Braconidae): Evidence of a high diversification rate in this group of parasitoids. Mol. Phylogenet. Evol. 12, 282–294.

Moran, N.A., 2007. Symbiosis as an adaptive process and source of phenotypic complexity. Proc. Natl. Acad. Sci. U.S.A. 104, 8627–8633.

Moreau, S., Huguet, E., Drezen, J.-M., 2009. Polydnaviruses as tools to deliver wasp virulence factors to impair lepidopteran host immunity. In: Reynolds, S., Rolff, J. (Eds.), Insect Infection and Immunity: Evolution, Ecology and Mechanisms (pp. 137–158). Oxford University Press, Oxford.

Mosavi, L.K., Minor Jr., D.L., Peng, Z.Y., 2002. Consensus-derived structural determinants of the ankyrin repeat motif. Proc. Natl. Acad. Sci. U.S.A. 99, 16029–16034.

Murphy, N., Banks, J.C., Whitfield, J.B., Austin, A.D., 2008. Phylogeny of the parasitic microgastroid subfamilies (Hymenoptera: Braconidae) based on sequence data from seven genes, with an improved time estimate of the origin of the lineage. Mol. Phylogenet. Evol. 47, 378–395.

Ohno, S., 1970. Evolution by Gene Duplication. Springer, New York.

Pennacchio, F., Malva, C., Vinson, S.B., 2001. Regulation of host endocrine system by the endophagous braconid *Cardiochiles nigriceps* and its polydnavirus. In: Edwards, J.P., Weaver, R.J. (Eds.), Endocrine Interactions of Insect Parasites and Pathogens (pp. 123–132). BIOS, Oxford.

Pritham, E.J., Putliwala, T., Feschotte, C., 2007. Mavericks, a novel class of giant transposable elements widespread in eukaryotes and related to DNA viruses. Gene 390, 3–17.

Provost, B., Varricchio, P., Arana, E., Espagne, E., Falabella, P., Huguet, E., et al. 2004. Bracoviruses contain a large multigene family coding for protein tyrosine phosphatases. J. Virol. 78, 13090–13103.

Pruijssers, A.J., Strand, M.R., 2007. PTP-H2 and PTP-H3 from *Microplitis demolitor* bracovirus localize to focal adhesions and are antiphagocytic in insect immune cells. J. Virol. 81, 1209–1219.

Rawlings, N.D., Tolle, D.P., Barrett, A.J., 2004. Evolutionary families of peptidase inhibitors. Biochem. J. 378, 705–716.

Rivkin, H., Kroemer, J.A., Bronshtein, A., Belausov, E., Webb, B.A., Chejanovsky, N., 2006. Response of immunocompetent and immunosuppressed *Spodoptera littoralis* larvae to baculovirus infection. J. Gen. Virol. 87, 2217–2225.

Rogers, S., Wells, R., Rechsteiner, M., 1986. Amino acid sequences common to rapidly degraded proteins: the PEST hypothesis. Science 234, 364–368.

Serbielle, C., Chowdhury, S., Pichon, S., Dupas, S., Lesobre, J., Purisima, E.O., et al. 2008. Viral cystatin evolution and three-dimensional structure modelling: a case of directional selection acting on a viral protein involved in a host–parasitoid interaction. BMC Biol. 6, 38.

Serbielle, C., Moreau, S., Veillard, F., Voldoire, E., Bézier, A., Mannucci, M.A., et al. 2009. Identification of parasite-responsive cysteine proteases in *Manduca sexta*. Biol. Chem. 390, 493–502.

Shi, M., Chen, Y.F., Huang, F., Liu, P.C., Zhou, X.P., 2008. Characterization of a novel gene encoding ankyrin repeat domain from Cotesia vestalis polydnavirus (CvBV). Virology 375, 374–382.

Shirane, M., Hatakeyama, S., Hattori, K., Nakayama, K., Nakayama, K., 1999. Common pathway for the ubiquitination of IκBα, IκBβ, and IκBε mediated by the F-box protein FWD1. J. Biol. Chem. 274, 28169–28174.

Shrestha, S., Kim, H.H., Kim, Y., 2009. An inhibitor of NF-κB encoded in *Cotesia plutella* bracovirus inhibits expression of antimicrobial peptides and enhances pathogenicity of *Bacillus thuringiensis*. J. Asia Pac. Entomol. 12, 277–283.

Singh, G.P., Ganapathi, M., Sandhu, K.S., Dash, D., 2006. Intrinsic unstructuredness and abundance of PEST motifs in eukaryotic proteomes. Proteins 62, 309–315.

Stubbs, M.T., Laber, B., Bode, W., Huber, R., Jerala, R., Lenarcic, B., et al. 1990. The refined 2.4 A X-ray crystal structure of recombinant human stefin B in complex with the cysteine proteinase papain: a novel type of proteinase inhibitor interaction. Embo J. 9, 1939–1947.

Suderman, R.J., Pruijssers, A.J., Strand, M.R., 2008. Protein tyrosine phosphatase-H2 from a polydnavirus induces apoptosis of insect cells. J. Gen. Virol. 89, 1411–1420.

Tanaka, K., Lapointe, R., Barney, W.E., Makkay, A.M., Stoltz, D., Cusson, M., et al. 2007. Shared and species-specific features among ichnovirus genomes. Virology 363, 26–35.

Thoetkiattikul, H., Beck, M.H., Strand, M.R., 2005. Inhibitor κB-like proteins from a polydnavirus inhibit NF-κB activation and suppress the insect immune response. Proc. Natl. Acad. Sci. U.S.A. 102, 11426–11431.

Tian, S.P., Zhang, J.H., Wang, C.Z., 2007. Cloning and characterization of two *Campoletis chlorideae* ichnovirus vankyrin genes expressed in parasitized host *Helicoverpa armigera*. J. Insect Physiol. 53, 699–707.

Turnbull, M., Webb, B., 2002. Perspectives on polydnavirus origins and evolution. Adv. Virus Res. 58, 203–254.

Webb, B.A., Strand, M.R., Dickey, S.E., Beck, M.H., Hilgarth, R.S., Barney, W.E., et al. 2006. Polydnavirus genomes reflect their dual roles as mutualists and pathogens. Virology 347, 160–174.

Webb, B.A., Summers, M.D., 1990. Venom and viral expression products of the endoparasitic wasp *Campoletis sonorensis* share epitopes and related sequences. Proc. Natl. Acad. Sci. U.S.A. 87, 4961–4965.

Weber, B., Annaheim, M., Lanzrein, B., 2007. Transcriptional analysis of polydnaviral genes in the course of parasitization reveals segment-specific patterns. Arch. Insect. Biochem. Physiol. 66, 9–22.

Whitfield, J.B., 2002. Estimating the age of the polydnavirus/braconid wasp symbiosis. Proc. Natl. Acad. Sci. U.S.A. 21, 7508–7513.

Yang, Z., Bielawski, J.P., 2000. Statistical methods for detecting molecular adaptation. Trends Ecol. Evol. 15, 496–503.

Yang, Z., Nielsen, R., Goldman, N., Pedersen, A.M., 2000. Codon-substitution models for heterogeneous selection pressure at amino acid sites. Genetics 155, 431–449.

Yang, Z., Swanson, W.J., Vacquier, V.D., 2000. Maximum-likelihood analysis of molecular adaptation in abalone sperm lysin reveals variable selective pressures among lineages and sites. Mol. Biol. Evol. 17, 1446–1455.

Chapter 6

Genomics of Banchine Ichnoviruses:[1] Insights into their Relationship to Bracoviruses and Campoplegine Ichnoviruses

Michel Cusson[*], Don Stoltz[†], Renée Lapointe[‡], Catherine Béliveau[*], Audrey Nisole[*], Anne-Nathalie Volkoff[§], Jean-Michel Drezen[¶¶], Halim Maaroufi[¶] and Roger C. Levesque[¶]

[*]Laurentian Forestry Centre, Natural Resources Canada, 1055 du P.E.P.S, Quebec, G1V 4C7, Canada

[†]Department of Microbiology and Immunology, Dalhousie University, Halifax, NS B3H 1X5, Canada

[‡]Sylvar Technologies Inc., P.O. Box 636, Stn. "A", Fredericton, NB, E3B 5A6, Canada

[§]UMR 1333 INRA – Université Montpellier 2, Diversité, Génomes et Interactions Microorganismes–Insectes, Place Eugène Bataillon, cc101, 34095 Montpellier Cedex 5, France

[¶¶]Institut de Recherche sur la Biologie de l'Insecte, CNRS, UMR 6035, Faculté des Sciences et Techniques, Université François Rabelais, Parc Grandmont, 37200 Tours, France

[¶]Institut de Biologie Intégrative et des Systèmes, Université Laval, 1030 Avenue de la Médecine, Québec, QC, G1V 0A6, Canada

SUMMARY

Ichneumonid polydnaviruses, referred to as ichnoviruses (IVs), have so far been observed in only two parasitic wasp subfamilies, the Campopleginae and Banchinae. The IVs carried by banchine wasps have received limited attention, and most of what we know about them is based on the characterization of a single virus, the *Glypta fumiferanae* ichnovirus (GfIV). The latter differs considerably from the more extensively studied campoplegine IVs, both in terms of virion morphology and features of the packaged genome. These differences have in fact raised the question as to whether campoplegine and banchine IVs have a common ancestor. The present chapter provides a brief review of the current state of knowledge on banchine IVs, including results from recent genomic analyses. It also provides suggestions as to how to address, in future research, the question of whether the campoplegine and banchine IVs have a common origin.

---

[1] In an earlier publication that provided the first detailed description of a banchine polydnavirus, the results presented raised questions as to whether this virus could be considered an ichnovirus (Lapointe *et al.*, 2007). For this reason, the polydnavirus isolated from the banchine wasp *Glypta fumiferanae* was referred to as 'GfV' (as opposed to 'GfIV', for *G. fumiferanae* ichnovirus). Since then, the Polydnavirus Study Group, a body within the International Committee on the Taxonomy of Viruses (ICTV), has provided an update on the classification of polydnaviruses, in which the virus of *G. fumiferanae* is classified as an ichnovirus. Thus, for now, we use the expression 'banchine ichnoviruses' and the abbreviation 'GfIV', recognising that the issue may not as yet be settled.

ABBREVIATIONS

BV	bracovirus
IV	ichnovirus
ORF	open reading frame
PTP	protein tyrosine phosphatase
TE	transposable elements
NCLDV	nucleo-cytoplasmic large DNA viruses

INTRODUCTION

Of the two taxa currently recognized in the family Polydnaviridae, namely the ichnoviruses (IV) and the bracoviruses (BV), the latter appear monophyletic (Whitfield, 1997), with all members presumed to have a common nudiviral ancestor (Bézier *et al.* 2009; Wetterwald *et al.*, 2010). Monophyly in the case of the former, however, is not as readily apparent. All known BVs are associated with wasps belonging to a group of closely related braconid subfamilies, the microgastroid lineage (Murphy *et al.*, 2008), and share various features including homologous virion structural proteins (Bézier *et al.*, 2009; Wetterwald *et al.*, 2010) and similar genome segment excision/circularization sequences (Desjardins *et al.*, 2007). The IVs, however, have so far been found in only two ichneumonid subfamilies, the Campopleginae and the Banchinae (Stoltz *et al.*, 1981; Krell, 1991; Lapointe *et al.*, 2007) and the viruses associated

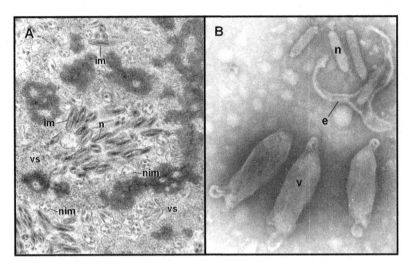

FIGURE 1   **Morphological features of a banchine IV (*Glypta fumiferanae* IV or GfIV). A:** Micrograph of section through an ovarian calyx epithelial cell nucleus. **B:** Negatively stained virions and nucleocapsids. In B, the envelope of one of the four virions was disrupted by a detergent, releasing four nucleocapsids. Abbreviations: e, envelope remnant; im, inner membrane of virion; nim, nascent inner membrane at the periphery of the virogenic stroma; n, nucleocapsid; v, virion; vs, virogenic stroma.

with these two taxa display several morphological and genomic differences (Lapointe *et al.*, 2007). In addition, although they belong to a common ichneumonid lineage, the Ophioniformes, Campopleginae and Banchinae appear to be phylogenetically separated by several intervening subfamilies (Quicke *et al.*, 2009). Thus, the question as to whether banchine and campoplegine IVs have a common ancestor has yet to be resolved. Recent work, however, has clearly established that the ancestor of campoplegine IVs is not a nudivirus (Volkoff *et al.*, 2010), providing unequivocal support for the hypothesis that BVs and IVs have distinct evolutionary origins (Whitfield, 1997, 2002; Whitfield and Asgari, 2003; Bézier *et al.*, 2009). Although the identity of the viral ancestor of campoplegine IVs was not elucidated in that study, available data on virion structural proteins do not support the hypothesis suggested earlier (Federici and Bigot, 2003; Bigot *et al.*, 2008), that IVs have evolved from ascoviruses.

The present chapter provides an overview of the limited amount of work that has been conducted on the genomics of banchine IVs, all of which is based on the analysis of a single genome, i.e. that of the *Glypta fumiferanae* ichnovirus (GfIV; Lapointe *et al.*, 2007). We also provide more recent analyses and preliminary findings from ongoing studies aimed at characterizing GfIV and elucidating the question of whether campoplegine and banchine IVs have a common origin.

EARLY STUDIES

The first report of a putative banchine polydnavirus was provided by Stoltz *et al.* (1981), who observed virions in the calyx epithelial cells of a wasp identified as *Glypta* sp. These ovarian particles displayed morphological features

that were seen as distinct from those of campoplegine IVs, including the presence of several smaller nucleocapsids per virion (see Fig. 1B in Stoltz *et al.*, 1981). The segmented nature of the dsDNA genome of this virus was documented for the first time by Krell (1991), who compared the electrophoretic profile of viral DNAs extracted from the ovaries of *Glypta* sp. and several campoplegine wasps. This analysis indicated that the genome segments of the *Glypta* virus were, on average, smaller than those found in the viruses of campoplegine wasps (see Fig. 8B in Krell, 1991). This same study also indicated that *Glypta* viral DNA did not hybridize with that of the virus isolated from the campoplegine parasitoid *Hyposoter fugitivus* (HfIV), pointing to limited, if any, relatedness between the two viruses. The only other mention of a banchine polydnavirus in the early literature is that of Stoltz and Whitfield (1992), who reported the presence of a similar virus in *Lissonota* sp.

UNIQUE MORPHOLOGICAL FEATURES OF THE GfIV VIRION

The first detailed characterization of a banchine polydnavirus virion was provided several years later (Lapointe *et al.*, 2007), when the virus isolated from the spruce budworm (*Choristoneura fumiferana*) parasitoid, *G. fumiferanae*, was examined by EM of thin sections of ovarian calyces as well as by negative staining of virions. This study pointed to a process of viriogenesis similar to that described for campoplegine IVs, which involves nucleocapsid assembly and envelopment by a unit membrane (referred to as the inner membrane) at the periphery of virogenic stroma in nuclei of calyx epithelial cells (Fig. 1A), followed by migration of the 'subvirions' to the cytoplasm through the nuclear envelope,

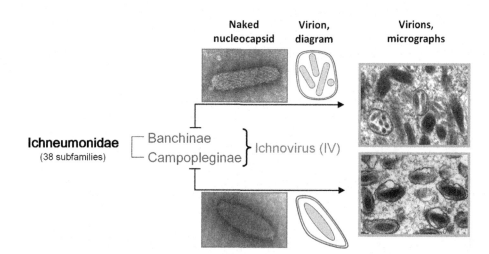

FIGURE 2 The Banchinae and the Campopleginae are the only two ichneumonid subfamilies with members known to harbor polydnaviruses in their ovaries. Virions found in these two subfamilies (both at this point considered ichnoviruses) display distinct morphological features. Banchine nucleocapsids are rod-shaped, with rounded ends, and show a distinctly reticulated surface, whereas campoplegine nucleocapsids are often more lenticular in shape and display a somewhat smoother surface. Both vary in length. Campoplegine virions contain a single nucleocapsid whereas those of the few banchine wasps examined thus far typically contain several. Both have a double-membrane envelope (each unit membrane is represented by a single line in the diagram); the inner membrane is produced *de novo* in nuclei of calyx cells whereas the outer membrane is acquired during virion exocytosis and is derived from the plasma membrane of calyx epithelial cells. Note that the scale differs among the images shown here; nucleocapsids of banchine IV are in reality smaller than those of campoplegine IVs (~35 × 170 nm versus ~85 × 330 nm).

and exocytosis into the lumen of the ovary. This is the process during which the virion acquires a second, outer membrane derived from the plasma membrane of calyx cells (Lapointe *et al.*, 2007). As seen in the campoplegine IVs (Stoltz and Vinson, 1979), the two additional membranes transiently acquired by passage through the nuclear envelope are, in some as yet undetermined manner, sloughed off in the cytoplasm prior to exit from calyx epithelial cells (Stoltz, unpublished observations). Although this process is shared by campoplegine IVs and GfIV, the latter displays morphological features that are clearly distinct from those described for the former, including the presence of several nucleocapsids in each virion (Fig. 1A) and the different shape, appearance, and size of nucleocapsids: whereas campoplegine IV virions contain a single, large, lenticular nucleocapsid displaying a relatively smooth surface, GfIV virions contain several rod-shaped/rounded-end nucleocapsids displaying a distinctly reticulated surface (Fig. 1A, B). The latter are also smaller (~35 × 170 nm) than those of campoplegine IVs (~85 × 330 nm). Whether each nucleocapsid contains one or more GfIV genome segments is not known. Some of the differences reported here are summarized in Fig. 2.

GENERAL FEATURES OF THE ENCAPSIDATED GfIV GENOME AND COMPARISONS WITH OTHER POLYDNAVIRUSES

Although we do not know yet whether the genome of GfIV is representative of a typical banchine IV, it clearly

displays a number of features that set it apart from the genomes of BVs and campoplegine IVs. With the exception of a few variables that are of limited diagnostic value for GfIV (e.g., aggregate genome size, G + C content, coding density), some features are clearly distinct, including (to this point in time) a higher degree of genome segmentation, a lower size of genome segments, and a higher proportion of genome segments apparently devoid of coding regions (Table 1). With respect to the number of predicted genes, GfIV falls at the lower end of the range established for campoplegine IV genomes (Table 1), and putative open reading frames (ORFs) tend to cluster into multigene families, as shown for other polydnaviruses. However, GfIV gene families have little in common with those reported for campoplegine viruses.

GENE FAMILIES

Of the 118 ORFs that were predicted from the sequence of the GfIV genome, 27 can be assigned to previously described polydnavirus gene families, namely the viral ankyrins (*ank*), with four members, and the protein tyrosine phosphatases (PTPs), with 23 members. In addition to these two families, a group of four genes, designated BV-like, were observed to encode proteins related to a hypothetical gene product from the BV of *Cotesia plutellae*, and one ORF was found to encode a protein related to a putative recombinase from the BV of *Chelonus inanitus*. Finally, the GfIV genome featured a seven-member gene family not previously described for any polydnavirus and encoding

TABLE 1 Genomic Features of a Banchine Ichnovirus (GfIV), Compared with Those of Campoplegine Ichnoviruses and Bracoviruses

	Banchine IVs*	Campoplegine IVs†	BVs‡
Genome size	~300 kb	~250 kb	189–594 kb
No. GSs§	106	24–56	15–30
Size range of GSs	1.5–5.2 kb	3–20 kb	4–50 kb
Median size of GSs	2.5 kb	4–9 kb	12–16 kb
Coding density	18.6%	22–30%	17–33%
G + C content	37%	41–43%	34–36%
% GS without ORFs	43%	0–2%	0–3%
Predicted ORFs	82	86–150	61–197
Assigned ORFs¶¶	36	35–77	39–57

*From Lapointe et al. (2007; GfIV).
†From Tanaka et al. (2007; HfIV and TrIV) and Webb et al. (2006; CsIV).
‡From Espagne et al. (2004; CcBV), Webb et al. (2006; MdBV) and Desjardins et al. (2008; GiBV and GfBV).
§GS: genome segment.
¶¶Number of ORFs assigned to a known gene family.

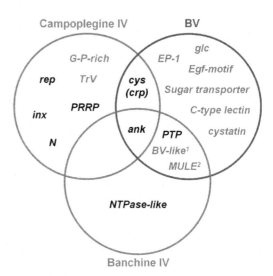

FIGURE 3 **Venn diagram illustrating the degree of gene family sharing among bracoviruses (BV), and campoplegine and banchine ichnoviruses (IV) whose genomes have been completely or partially sequenced.** Gene families shown in black letters have been identified in two or more representatives of the group(s), except for 'NTPase-like' genes, which are found in multiple copies in the genome of GfIV, the only banchine IV sequenced to date. Gene families shown in gray letters have been identified in only one polydnavirus. (1) 'BV-like' refers to a hypothetical protein originally identified in CpBV (CpBV ORF1103; AAZ04291) and four homologs found in the genome of GfIV. (2) 'MULE' refers to two proteins originally identified in the genome of CiBV (CAA91234 and CAC82100) as putative recombinases (named 'recombinase-like' in Lapointe et al., 2007) and a homolog identified in the genome of GfIV (GfIV-C17-ORF1; YP_001029420). All three proteins display a motif that is common to all Mutator-like element (MULE) transposases (pfam1055). After Strand (2010).

proteins related to the D5 NTPases of nucleo-cytoplasmic large DNA viruses (NCLDV; a recent re-analysis of GfIV putative ORFs indicates that two additional genes encode products related to this group; see below). All other putative ORFs encode proteins showing no or little similarity to any known protein, although some were seen to form families of up to five members (Lapointe et al., 2007). Interestingly, GfIV was found to share far more genes with BVs than with campoplegine IVs (Fig. 3).

The viral ankyrins form the only family shared by all three groups of viruses shown in Fig. 3. Phylogenetic analyses and BLASTP searches indicated that two of the four GfIV *ank* proteins are more closely related to their BV counterparts than to those of campoplegine IVs, while the opposite is true for the remaining two (Lapointe et al., 2007). More recent BLASTP analyses[2] indicate that the closest non-polydnaviral relatives of GfIV *ank* proteins are hymenopteran orthologs of *cactus*, providing further support for the hypothesis that polydnavirus *ank* proteins have all evolved from a wasp *cactus* gene, albeit independently (at least in the case of the IV–BV dichotomy), and thus pointing to their convergent evolution. The same appears to be true of the PTPs, a gene family that GfIV shares with BVs (Fig. 3); for example, the non-polydnaviral

[2]The power of these analyses, when using polydnavirus genes as query sequences, is now enhanced by the recent addition of hymenopteran genome data to the NCBI database, including those of *Apis mellifera* (HGSC, 2006), *Nasonia* spp. (Werren et al., 2010), and *Camponotus floridanus* and *Harpegnathos saltator* (Bonasio et al., 2010).

protein most closely related to the PTP encoded by GfIV-C8-ORF1 is a PTP from the jumping ant *Harpegnathos saltator* (E-value: 6e-40), which is followed closely by a *Nasonia* PTP (E-value: 2e-38). This suggests that the apparent relatedness between some GfIV gene families and those of BVs does not reflect a common viral origin. Rather, it appears to reflect a shared necessity to subjugate the lepidopteran host using similar molecular strategies.

The NTPase-like proteins of GfIV constitute the most interesting gene family identified in the genome of this virus. Not only is it a novel polydnavirus gene family, but some of its members are also the most highly expressed GfIV genes in spruce budworm larvae parasitized by *G. fumiferanae* (unpublished data). Although this family contains nine interrelated proteins, only seven of them display sequences showing similarity to domains that have been described for D5 NTPases (Fig. 4). First identified in vaccinia virus, the D5 protein was initially shown to display NTPase activity and to be required for viral DNA replication (Evans et al., 1995). The vaccinia D5 protein has since been found to contain both an N-terminal primase domain and a C-terminal NTPase/helicase domain, each of which can be inactivated independently (De Silva et al.,

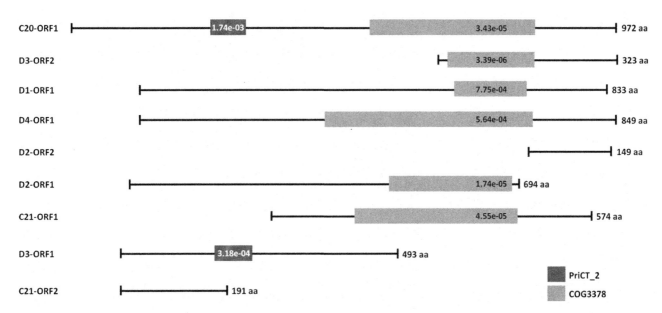

FIGURE 4 **Diagrammatical representation of the nine predicted GfIV proteins that form the NTPase-like family.** Proteins are each represented by a line whose length is proportional to the number of amino acid residues (indicated at the right end of each line), and their horizontal positioning reflects their alignment. These proteins show similarity to nucleo-cytoplasmic large DNA viruses (NCLDV) primases, which contain an N-terminal primase domain and a C-terminal NTPase/helicase domain. However, domain erosion has apparently resulted in the disappearance of the primase function, although two proteins, C20-ORF1 and D3-ORF1, still display a primase-C terminal-2 domain (PriCT_2), which is an alpha-helical domain found at the C terminus of primases. Six of the proteins contain a sequence that is recognized as belonging to the COG3378 family (NTPase/helicase), but these domains typically appear eroded and do not appear to represent functional helicases. Light and dark boxes contain E values for the level of similarity to the domain detected by the BLASTP program.

2007). The D5 protein has orthologs in other NCLDVs and is now considered to be part of a large superfamily of primases known as the archaeo-eukaryotic primases (AEPs), the eukaryotic members of which are believed to be involved in DNA repair pathways (Iyer *et al.*, 2005). The GfIV NTPase-like proteins are most closely related to the NCLDV primases, including one from an insect ascovirus (Bigot *et al.*, 2008). However, none of them contain a detectable primase domain and the NTPase/helicase domain, when present, has undergone substantial erosion, to the point where the proteins are unlikely to possess any helicase activity (Fig. 4). Thus, the GfIV NTPase-like proteins do not appear to be functional enzymes. They could, however, form complexes with endogenous primases, causing their inactivation, as shown recently for a bacteriophage primase (Sturino and Klaenhammer, 2007). How such inactivation would benefit the immature wasp is not clear; one possibility is that it may block the replication of competing NCLDVs such as entomopoxviruses and ascoviruses.

LATERAL GENE TRANSFER FROM AND INTO THE GfIV GENOME

The number of documented cases of lateral gene transfer in the eukaryotes is limited, and most of them involve transposable elements (TE; Schaack *et al.*, 2010).

However, the intimate relationship between a parasite and its host is believed to be conducive to such lateral transfer of genetic material between unrelated organisms. Evidence for the transfer of a TE from a caterpillar to its hymenopteran endoparasite has been presented (Yoshiyama *et al.*, 2001) and a mechanism for the transfer of DNA from polydnaviruses to the genomes of lepidopteran hosts has been described. Indeed, instances of spontaneous integration of polydnaviral genome segments into lepidopteran chromosomal DNA *in vitro* have been reported for both BVs (Gundersen-Rindal and Lynn, 2003; Gundersen-Rindal *et al.*, this volume) and IVs (Doucet *et al.*, 2007). Given that caterpillars stung by a polydnavirus-carrying wasp in many cases survive the attack, transmission of the foreign DNA to the next generation is possible if genome segment integration takes place in germ cells.

To assess the possibility of lateral gene transfer from the GfIV genome to that of the spruce budworm, whose genome is currently being sequenced, we searched the available sequence data (~6 × genome coverage of 454 random sequences) using all GfIV genome segments as query sequences. A portion of a GfIV genome segment (B42) encoding a PTP displayed a high degree of sequence identity to a group of overlapping spruce budworm 454 reads (Fig. 5). The latter sequences encode an N-terminal truncated PTP. To assess the likelihood that the spruce budworm sequence is the result of a lateral gene transfer from

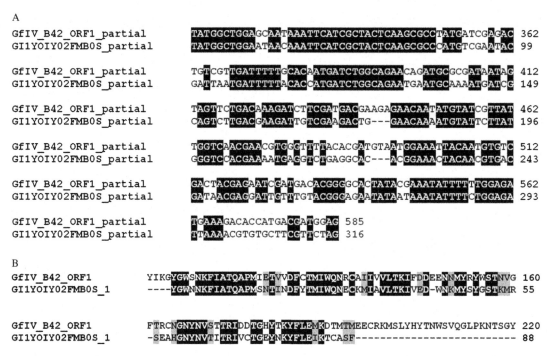

FIGURE 5   Lateral gene transfer from GfIV to the genome of *Choristoneura fumiferana*. A: ClustalW alignment of a portion of the coding region found on GfIV genome segment B42 (GfIV-B42-ORF1; GenBank BAF45510) and a portion of a sequence obtained through 454 random sequencing of the *C. fumiferana* genome. **B:** ClustalW alignment of the deduced amino acid sequences. E-value for BLASTX search using full *C. fumiferana* DNA sequence as query (453 nt): 8e-22.

GfIV, the longest 454 read was submitted to a BLASTX analysis against the NCBI database. The eight proteins showing the highest degree of similarity to the query were all GfIV PTPs, with B42-ORF1 producing an E value of 8e-22. Among the remaining 'hits' were other polydnaviral PTPs but no insect homologs, suggesting very strongly that the spruce budworm sequence was acquired from the GfIV genome. Whether this truncated PTP is expressed and plays a role in spruce budworm biology is not known, but the present example of lateral gene transfer from a polydnavirus to its lepidopteran host points to the potential impact of this process on the evolution of lepidopteran genomes.

Another case of lateral gene transfer involving GfIV was documented by another group (Bigot *et al.*, 2008), although this particular instance involved the transfer of DNA from the genome of an ascovirus (DpAV4) known to infect a parasitic wasp to that of a putative GfIV ancestor. As in the example given above, the six best 'hits' of a BLAST search against the NCBI database, using the DpAV4 sequence (which encodes a D5 NTPase protein) as the query, were all GfIV proteins (NTPase-like), with levels of significance much higher than those obtained for other NCLDV D5 NTPases, including those of other ascoviruses. Bigot *et al.* (2008) argued that the presence of a sequence derived from an ascovirus in the GfIV genome supported the hypothesis that IVs (both banchine and campoplegine) have an ascoviral ancestor (Federici and Bigot,

2003). However, current evidence obtained for campoplegine IVs does not support this hypothesis (Volkoff *et al.*, 2010, this volume); likewise, preliminary data pertaining to GfIV do not support an ascoviral origin for the IVs.

ON THE ORIGIN OF BANCHINE ICHNOVIRUSES

Although campoplegine and banchine IVs share some virion structural features and other characteristics, they also display important morphological and genomic differences. But do these differences necessarily imply that they have different ancestors? In light of the evidence recently presented in support of a common origin for chelonine and microgastrine BVs, which show differences with respect to both virion morphology (e.g., number of nucleocapsids per virion) and gene content (Bézier *et al.*, 2009; Wetterwald *et al.*, 2010), the answer would have to be no. Of course, it can be argued that the differences observed between banchine and campoplegine IVs are more pronounced than those reported within the BV lineage (e.g., the two groups of IV nucleocapsids differ with respect to shape, size, and surface texture, which is not the case within the BV lineage). In addition, a recent phylogenetic analysis of the family Ichneumonidae, based on both morphological and molecular data, suggests that there are eight or nine subfamilies separating the Banchinae from the Campopleginae

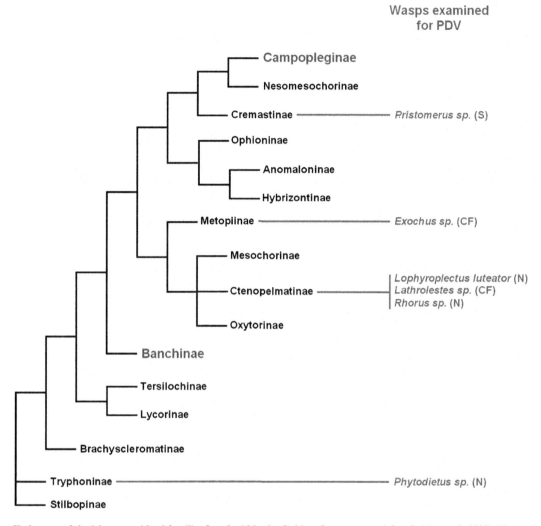

FIGURE 6 Cladogram of the ichneumonid subfamilies found within the Ophionoformes group (after Quicke *et al.*, 2009), illustrating the phylogenetic relationship between Banchinae and Campopleginae, which are separated by eight or nine subfamilies. A few wasp species belonging to these intervening subfamilies, and one belonging to a subfamily that is basal to the Banchinae, have been examined for the presence of polydnavirus (by negative staining of calyx fluid (N), TEM of ovarian sections (S), or simple visual examination of the calyx fluid (CF), which is opalescent when it contains polydnavirus virions), but none showed clear signs of possessing polydnaviruses.

(Quicke *et al.*, 2009). Although only a few wasps belonging to these intervening taxa have been examined for the presence of polydnaviruses, the latter were not found in those that have thus far been inspected (Fig. 6). However, this observation could equally be taken to indicate that the two IVs have a distinct origin, because the virus is apparently not found in the intervening subfamilies, or that they are the outcome of divergence following a common integration event in an ancestor of banchine and campoplegine wasps. In the latter process, the virus could have been lost by wasps belonging to the intervening subfamilies. Notwithstanding the possibility that more extensive sampling of the latter taxa could reveal the presence of polydnavirus in some of them, it should be pointed out that some campoplegine wasps are apparently devoid of IV particles in their ovaries (Stoltz, unpublished data), which

would support the 'polydnavirus loss hypothesis', unless the Campopleginae are polyphyletic.

Ultimately, a convincing answer to the question of whether banchine and campoplegine IVs have a common ancestor will likely come from an analysis of GfIV virion structural proteins and the genes that encode them, as well as other genes that may be involved in viral genome replication and that are not present in the assembled virus particle. An analysis of transcripts isolated from the ovaries of *G. fumiferanae* is currently underway and initial results indicate that some transcripts are homologous to genes identified as involved in particle production in the campoplegine wasps *H. didymator* and *Tranosema rostrale* (Volkoff *et al.*, 2010). More extensive analyses will be required to determine whether viral proteins typical of another virus family contribute to GfIV particle production.

FUTURE DIRECTIONS

It is clear from the foregoing discussion that priority in going forward must be placed on the proteomics analysis of GfIV virion structural proteins and on the sequencing of the wasp genomic DNA that encodes these proteins and, possibly, others involved in viriogenesis. Should these genes and proteins all prove to be homologous to those characterized in campoplegine wasps (Volkoff *et al.,* 2010), a strong argument could be made in favor of a common origin of banchine and campoplegine IVs. In parallel, additional banchine wasps must be examined for the presence of polydnavirus and the latter must be compared with GfIV so as to determine whether GfIV features are typical of the group. Similarly, many additional wasps belonging to the subfamilies separating the Banchinae from the Campopleginae should be examined for the presence of polydnavirus in their ovarian tissue; should the two known groups of IVs have a common ancestor, such a survey could lead to the identification of intermediate forms of ichnoviruses. Finally, identification of GfIV genome segment excision/junction sequences could also help clarify the evolutionary origin of banchine IVs; analyses conducted to date suggest that these sequences, in GfIV, are distinct from those of campoplegine IVs and BVs. Sequencing of the GfIV provirus would provide a definitive answer to this question.

With respect to the role(s) played by GfIV in the success of parasitism by *G. fumiferanae,* the uncommon seasonal relationship between this parasitoid and its spruce budworm host must be taken into consideration in formulating hypotheses. Indeed, *G. fumiferanae* eggs are laid in the late summer, in prediapause first or second instars, and hatch shortly after oviposition. The wasp larva is then believed to enter diapause, in synchrony with its host, and to remain quiescent throughout the winter until host development resumes in the spring. The parasitoid larva then completes its development and egresses from a fifth or sixth (final) instar host. Recent qPCR analyses indicate that GfIV remains transcriptionally active throughout host diapause, with some NTPase-like genes generating the most abundant transcripts (unpublished observations). Could this gene family represent an adaptation to the prolonged period of endoparasitic development made necessary as a consequence of winter diapause? A partial answer to this question could be provided by a genomic analysis of additional banchine IVs obtained from wasps that do not undergo such a larval winter diapause. Certainly, research aimed at understanding the role(s) played by NTPase-like proteins in host–parasitoid relationships, and elucidating the molecular events associated with their action, should be most rewarding.

ACKNOWLEDGMENTS

We wish to thank Eugene Koonin (NCBI) and Donald L.J. Quicke (Imperial College) for sharing their insights on the D5 NTPases and ichneumonid phylogeny respectively. The research described in this chapter was funded by Genome Canada, the Natural Sciences and Engineering Research Council of Canada, Genoscope and the Canadian Forest Service.

REFERENCES

Bézier, A., Annaheim, M., Herbiniere, J., Wetterwald, C., Gyapay, G., Bernard-Samain, S., et al. 2009. Polydnaviruses of braconid wasps derive from an ancestral nudivirus. Science 323, 926–930.

Bigot, Y., Samain, S., Auge-Gouillou, C., Federici, B.A., 2008. Molecular evidence for the evolution of ichnoviruses from ascoviruses by symbiogenesis. BMC Evol. Biol. 8, 253.

Bonasio, R., Bonasio, R., Zhang, G., Ye, C., Mutti, N.S., Fang, X., et al. 2010. Genomic comparison of the ants *Camponotus floridanus* and *Harpegnathos saltator.* Science 329, 1068–1071.

De Silva, F.S., Lewis, W., Berglund, P., Koonin, E.V., Moss, B., 2007. Poxvirus DNA primase. Proc. Natl. Acad. Sci. U.S.A. 104, 18724–18729.

Desjardins, C.A., Gundersen-Rindal, D.E., Hostetler, J.B., Tallon, L.J., Fadrosh, D.W., Fuester, R.W., et al. 2008. Comparative genomics of mutualistic viruses of *Glyptapanteles* parasitic wasps. Genome Biol. 9, R183.

Desjardins, C.A., Gundersen-Rindal, D.E., Hostetler, J.B., Tallon, L.J., Fuester, R.W., Schatz, M.C., et al. 2007. Structure and evolution of a proviral locus of *Glyptapanteles indiensis* bracovirus. BMC Microbiol. 7, 61.

Doucet, D., Levasseur, A., Beliveau, C., Lapointe, R., Stoltz, D., Cusson, M., 2007. In vitro integration of an ichnovirus genome segment into the genomic DNA of lepidopteran cells. J. Gen. Virol. 88, 105–113.

Espagne, E., Dupuy, C., Huguet, E., Cattolico, L., Provost, B., Martins, N., et al. 2004. Genome sequence of a polydnavirus: insights into symbiotic virus evolution. Science 306, 286–289.

Evans, E., Klemperer, N., Ghosh, R., Traktman, P., 1995. The vaccinia virus D5 protein, which is required for DNA replication, is a nucleic acid-independent nucleoside triphosphatase. J. Virol. 69, 5353–5361.

Federici, B.A., Bigot, Y., 2003. Origin and evolution of polydnaviruses by symbiogenesis of insect DNA viruses in endoparasitic wasps. J. Insect Physiol. 49, 419–432.

Gundersen-Rindal, D.E., Lynn, D.E., 2003. Polydnavirus integration in lepidopteran host cells in vitro. J. Insect Physiol. 49, 453–462.

HGSC (Honeybee Genome Sequencing Consortium), 2006. Insights into social insects from the genome of the honeybee *Apis mellifera.* Nature 443, 931–949.

Iyer, L.M., Koonin, E.V., Leipe, D.D., Aravind, L., 2005. Origin and evolution of the archaeo-eukaryotic primase superfamily and related palm-domain proteins: structural insights and new members. Nucleic Acids Res. 33, 3875–3896.

Krell, P.J., 1991. The polydnavirus: multipartite DNA viruses from parasitic hymenoptera. In: Kurstak, E. (Ed.), Viruses of Invertebrates (pp. 141–177). Dekker, New York.

Lapointe, R., Tanaka, K., Barney, W.E., Whitfield, J.B., Banks, J.C., Beliveau, C., et al. 2007. Genomic and morphological features of a banchine polydnavirus: comparison with bracoviruses and ichnoviruses. J. Virol. 81, 6491–6501.

Murphy, N., Banks, J.C., Whitfield, J.B., Austin, A.D., 2008. Phylogeny of the parasitic microgastroid subfamilies (Hymenoptera: Braconidae) based on sequence data from seven genes, with an improved time estimate of the origin of the lineage. Mol. Phylogenet. Evol. 47, 378–395.

Quicke, D.L.J., Laurenne, N.M., Fitton, M.G., Broad, G.R., 2009. A thousand and one wasps: a 28S rDNA and morphological phylogeny of the Ichneumonidae (Insecta: Hymenoptera) with an investigation into alignment parameter space and elision. J. Nat. Hist. 43, 1305–1421.

Schaack, S., Gilbert, C., Feschotte, C., 2010. Promiscuous DNA: horizontal transfer of transposable elements and why it matters for eukaryotic evolution. Trends Ecol. Evol. 25, 537–546.

Stoltz, D.B., Krell, P.J., Vinson, S.B., 1981. Polydisperse viral DNA's in ichneumonid ovaries: a survey. Can. J. Microbiol. 27, 123–130.

Stoltz, D.B., Vinson, S.B., 1979. Viruses and parasitism in insects. Adv. Virus Res. 24, 125–170.

Stoltz, D.B., Whitfield, J.B., 1992. Viruses and virus-like entities in the parasitic hymenoptera. J. Hym. Res. 1, 125–139.

Strand, M.R., 2010. The interactions between polydnavirus-carrying parasitoids and their lepidopteran hosts. In: Goldsmith, M.R., Marec, F. (Eds.), Molecular Biology and Genetics of the Lepidoptera (pp. 321–336). Boca Raton: CRC Press/ Taylor & Francis.

Sturino, J.M., Klaenhammer, T.R., 2007. Inhibition of bacteriophage replication in Streptococcus thermophilus by subunit poisoning of primase. Microbiology 153, 3295–3302.

Tanaka, K., Lapointe, R., Barney, W., Makkay, A., Stoltz, D., Cusson, M., et al. 2007. Shared and species-specific features among ichnovirus genomes. Virology 363, 26–35.

Volkoff, A.N., Jouan, V., Urbach, S., Samain, S., Bergoin, M., Wincker, P., et al. 2010. Analysis of virion structural components reveals vestiges of the ancestral ichnovirus genome. PLoS Pathog. 6, e1000923.

Webb, B.A., Strand, M.R., Dickey, S.E., Beck, M.H., Hilgarth, R.S., Barney, W.E., et al. 2006. Polydnavirus genomes reflect their dual roles as mutualists and pathogens. Virology 347, 160–174.

Werren, J.H., Richards, S., Desjardins, C.A. Niehuis, O., Gadau, J., Colbourne, J.K., et al. 2010. Functional and evolutionary insights from the genomes of three parasitoid Nasonia species. Science 327, 343–348.

Wetterwald, C., Roth, T., Kaeslin, M., Annaheim, M., Wespi, G., Heller, M., et al. 2010. Identification of bracovirus particle proteins and analysis of their transcript levels at the stage of virion formation. J. Gen. Virol. 91, 2610–2619.

Whitfield, J.B., 1997. Molecular and morphological data suggest a single origin of the polydnaviruses among braconid wasps. Naturwissenschaften 84, 502–507.

Whitfield, J.B., 2002. Estimating the age of the polydnavirus/braconid wasp symbiosis. Proc. Natl. Acad. Sci. U.S.A. 99, 7508–7513.

Whitfield, J.B., Asgari, S., 2003. Virus or not? Phylogenetics of polydnaviruses and their wasp carriers. J. Insect Physiol. 49, 397–405.

Yoshiyama, M., Tu, Z., Kainoh, Y., Honda, H., Shono, T., Kimura, K., 2001. Possible horizontal transfer of a transposable element from host to parasitoid. Mol. Biol. Evol. 18, 1952–1958.

Chapter 7

Molecular Systematics of Wasp and Polydnavirus Genomes and their Coevolution

James B. Whitfield and Jaqueline M. O'Connor,

Department of Entomology, 320 Morrill Hall, 505 S. Goodwin Avenue, University of Illinois, Urbana, IL 62801, U.S.A

SUMMARY

Polydnaviruses (PDVs) are integrated, inherited DNA viral endosymbionts of parasitoid wasps found within some lineages of Braconidae and Ichneumonidae that attack caterpillars (larval Lepidoptera). Due to their chromosomal integration and mode of inheritance, phylogenomic relationships among the PDVs are closely tied to phylogenetic relationships among the host wasps that carry them. We have inferred that PDVs must have at least three independent ancient origins in the wasps: bracoviruses (BVs) from ancestral nudiviruses associated with braconid wasps, ichnoviruses, and 'banchoviruses' from still unidentified progenitor viruses and associated with unrelated families of Ichneumonidae. Phylogenetic studies of the BV-carrying lineage of braconid wasps are now guiding comparative studies of BV genomes and advancing our understanding of the roles that BV genes may play in the evolution of wasp/caterpillar parasitic relationships. Similar studies of IVs and 'banchoviruses' await more extensive and well-supported phylogenies of ichneumonid wasps.

ABBREVIATIONS

PDV   polydnavirus
BV   bracovirus
IV   ichnovirus

INTRODUCTION

From the early days of polydnavirus (PDV) research, the taxonomy of the wasps that carry these endosymbiotic viruses has been seen to be of high relevance. The 1979 review of parasitoid viruses by Stoltz and Vinson (before the PDVs were even named as a separate viral family) had already made note of the fact that the PDVs associated with braconid and ichneumonid wasps, respectively, showed striking morphological differences, and that more subtle morphological differences were expressed even among PDVs from different braconids. Cook and Stoltz (1983) showed, through comparative serological studies, that ichneumonid PDVs from more distantly related ichneumonids appeared to be less genetically similar. It eventually was revealed that at least some PDVs were inherited in Mendelian fashion and in fact were integrated into the wasp chromosomal DNA (Stoltz, 1990; Belle *et al.*, 2002); the underlying cause for these correlations between wasp and viral relationships became clear and it became possible to make explicit predictions about PDV evolutionary relationships (Whitfield, 1990; Stoltz and Whitfield, 1992). Among these were the predictions that *all* members of several large lineages of wasps (already found to contain species that possessed PDVs) would carry PDVs, and that PDV genetic and phylogenetic relationships would mirror those of their wasp carriers. These predictions assumed that the current distributions of PDVs among wasp carriers were inherited from unique and very ancient colonization events.

While these predictions have, for the most part, borne out with further study (especially in the case of braconid PDVs), it was some years before the molecular systematic tools were in place to properly test them. Even now we are rapidly accumulating additional tools to investigate relationships at the genomic level that will likely continue to illuminate PDV evolution.

Below, we provide a brief tour through the insights gained so far into PDV evolution from phylogenetic analysis of wasp and PDV genes, and discuss some newer developments that promise to deepen our understanding of this remarkable symbiotic relationship.

Parasitoid Viruses: Symbionts and Pathogens. DOI: 10.1016/B978-0-12-384858-1.00007-2

TABLE 1 Taxonomic Distribution of PDVs Among Wasp Groups

PDV subgroup	Wasp family	Wasp subfamilies	Representative wasp genera
Bracovirus (BV)	Braconidae	Cardiochilinae	*Cardiochiles, Toxoneuron*
		Cheloninae	*Adelius, Chelonus, Ascogaster, Phanerotoma*
		Khoikhoiinae	*Sania, Khoikhoia*
		Mendesellinae	*Epsilogaster, Mendesella*
		Microgastrinae	*Apanteles, Cotesia, Diolcogaster, Dolichogenidea, Glyptapanteles, Hypomicrogaster, Microgaster, Microplitis, Pholetesor*
		Miracinae	*Mirax*
Ichnovirus	Ichneumonidae	Campopleginae	*Campoletis, Campoplex, Dusona, Hyposoter, Sinophorus, Venturia*
Banchovirus	Ichneumonidae	Banchinae	*Banchus, Glypta*

The braconid subfamilies that carry BVs form a monophyletic lineage (originating from a common ancestor). The two ichneumonid subfamilies that carry IVs and 'banchoviruses', respectively, appear not to be each others' closest relatives.

TAXONOMIC DISTRIBUTION OF PDVS

As summarized by Stoltz and Whitfield (1992), PDVs are known from two families of parasitoid wasps: Braconidae and Ichneumonidae. The two families are sister groups (at least among extant families; Sharkey and Wahl, 1992; Whitfield, 1998, 2003), and both contain subfamilies whose species are endoparasitoids of larval Lepidoptera (caterpillars) and carry PDVs. Surprisingly, however, the subfamilies that have been found to carry PDVs are *not* all closely related. Instead, at least three separate lineages of these wasps are now known to carry highly divergent groups of PDVs: bracoviruses (BVs) in members of the subfamilies of the relatively recently derived microgastroid lineage of braconid wasps, ichnoviruses (IVs) in some groups of campoplegine ichneumonid wasps, and the more recently characterized 'banchoviruses' (Lapointe *et al.*, 2007) in at least some banchine ichneumonid wasps. The distribution of the three PDV groups among the subfamilies of ichneumonoid wasps is summarized in Table 1.

It should be pointed out that other symbiotic viruses have been described in other groups of both these two wasp families and in other wasp families (Stoltz and Whitfield, 1992; Lawrence, 2005), but they appear to be only distantly related, or unrelated, to the PDVs (with the possible exception of ascoviruses, see below).

MONOPHYLY OF THE WASP LINEAGE BEARING BRACOVIRUSES

In order to demonstrate that the current distribution of PDVs among wasps could be explained by inheritance from unique early acquisition of viruses by the wasps, it was necessary to first investigate how the wasps that carry the PDVs are themselves related. For various reasons, both biological and historical, progress has been more rapid on this issue with braconid wasps than with Ichneumonidae. All known BVs are found in wasps belonging to a long-recognized set of putatively related subfamilies informally referred to as the microgastroid assemblage or complex (summarized in Whitfield and Mason, 1994). The group was not formally analyzed phylogenetically in any detailed sense until Whitfield and Mason (1994) used morphological data in the recognition of a new subfamily. Subsequently, Whitfield (1997) analyzed 16S rDNA sequences, combined with earlier morphological data, to show that the microgastroid assemblage was likely monophyletic; several studies incorporating 28S rDNA (Belshaw *et al.*, 1998; Dowton and Austin, 1998) quickly followed and came to the same result. Whitfield (2002) estimated the age of this monophyletic clade to be 74 million years (plus or minus about 11 million years) using fossil-calibrated relaxed molecular clock estimates from three genes (16S, COI and 28S). A more taxonomically comprehensive and more thoroughly fossil-calibrated study based on seven genes (Murphy *et al.*, 2008) later revised this age estimate up to roughly 100 million years. Zaldivar-Riveron *et al.* (2008) have suggested that the Baltic amber ages used in calibration for this study might be overestimated by about 6 million years; if true (the dating of these fossils is still controversial), the 100 million year estimate might be reduced to 85–90 million years, closer to the original estimate. In either case, the wasp lineage carrying BVs is mid Cretaceous in age and has had a long time to evolve into a diverse lineage. Our current phylogenetic understanding of the relationship between braconid wasps and BVs is summarized in Fig. 1.

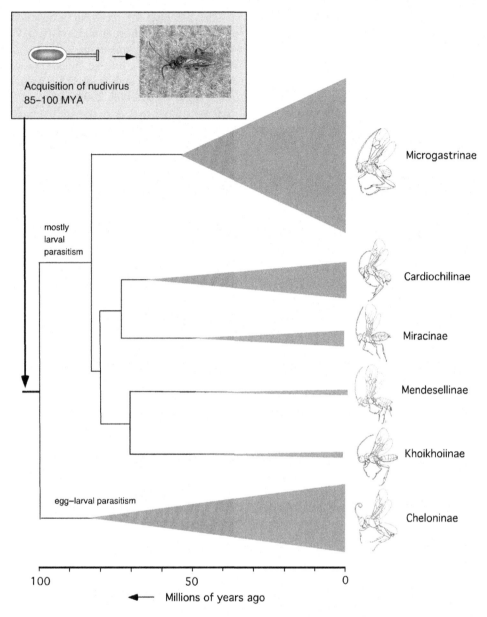

FIGURE 1　Schematic overview of current understanding of the origin and taxonomic distribution of BVs. Wasp phylogeny and timescale from Murphy *et al.* (2008); nudivirus origin of BVs from Bézier *et al.* (2009) and Stoltz and Whitfield (2009). Photograph of *Microplitis* adult by Won Young Choi; line drawings of subfamily representatives modified from versions earlier published in Goulet and Huber (1993) and Wharton *et al.* (1997); used with permission. Width of triangular 'lineages' corresponds roughly to relative species diversity. MYA, millions of years ago.

POSSIBLE POLYPHYLY OF ICHNEUMONID WASPS BEARING PDVS

The phylogenetic relationships among the Ichneumonidae that carry IVs are less resolved for several reasons. At this point we know that some, but not all, Campopleginae carry IVs, but what absence really means is not always clear. Some genera have not been seriously surveyed for IVs, and in one case, *Venturia canescens*, virus-like particles (VLPs) morphologically similar to IVs (Norton *et al.*, 1975) are present but do not contain DNA. Without

screening of suspect genera using IV-specific molecular probes, we cannot entirely rule out the existence of IV genes residing in wasps with no obvious viral particles.

Initially, it was puzzling that IVs also were known to occur in banchine ichmeumonids, which were not considered to be the closest relatives of Campopleginae (discussed in Stoltz and Whitfield, 1992). Once the genome of the 'IV' associated with *Glypta fumiferanae* (Banchinae) was fully sequenced and characterized, it was clear that the banchine-associated viruses were not related (i.e. showed very little similarity even in gene content) to the IVs in campoplegines, in some respects they more closely resembled BVs, and actually represented a third PDV group (Lapointe *et al.*, 2007).

Recent ichneumonid molecular phylogenies (Belshaw *et al.*, 1998; Quicke *et al.*, 2009) are still inconclusive with respect to many deeper evolutionary relationships, largely due to insufficient molecular data being yet applied; detailed molecular phylogenetic studies within Banchinae and Campopleginae remain to be conducted. This is clearly an area with some potential for fruitful investigation using a battery of well-conserved nuclear genes.

ORIGINS OF POLYDNAVIRUSES

Despite the formal description of PDVs as a viral family (Stoltz *et al.*, 1984), for a long time there has been dispute about whether PDVs actually qualify as viruses, or whether indeed they even originated from viruses (Whitfield, 1990; Stoltz and Whitfield 1992, 2009; Whitfield and Asgari, 2003). The recent demonstration that BVs were derived from nudiviruses or very nudivirus-like progenitors (Bézier *et al.*, 2009), and the identification of a conserved battery of viral assembly genes in IVs (Volkoff *et al.*, 2010), seem to have settled the issue that at least as far as we know, the basic PDV assembly and delivery systems have derived from ancestral associations between wasps and viruses. There is of course still the issue that PDVs do not fully match the usual definition of a virus (Stoltz, 1993; Federici and Bigot, 2003; Whitfield and Asgari, 2003; Stoltz and Whitfield, 2009), and are perhaps best viewed from a functional standpoint as a viral endosymbiosis with functional gene contributions from the wasps. The elucidation of the viral origin of PDVs is nevertheless important in our understanding of how PDVs (and parasitism in some wasps) came to be the way they are.

The identities of the viral progenitors of the PDVs associated with ichneumonid wasps (IVs and 'banchoviruses') are still uncertain, despite some suggestive studies implicating a lineage including ascoviruses and iridoviruses (Federici and Bigot, 2003; Whitfield and Asgari, 2003; Bigot *et al.*, 2008, 2009). Detailed study of core IV viral assembly genes does not obviously implicate any known viral group as closely related (Volkoff *et al.*, 2010). What appears clear so far, however, is that the origins of IVs and 'banchoviruses' from progenitor viruses were separate evolutionary events from that of the BVs, and probably also from each other.

COPHYLOGENY OF PDVS WITH WASPS

Demonstration that PDVs appear to have unique and ancient origins in wasp lineages suggests that some sort of coevolution between the wasp and viral genomes might be occurring. It is another thing, however, to show that the diversifications of the wasps and viruses are linked, or to show that changes in wasp/host caterpillar biology might be linked with specific genetic responses in the PDVs (and vice versa).

To show that diversification in the two partner entities are linked, it is necessary to first have estimates of the phylogenetic relationships among the wasps, and independent estimates of the phylogenetic relationships among the viruses they carry. Correspondence between the two, or congruence between the phylogeny estimated from wasp genes and that estimated from genes from the PDVs they carry, would indicate a long-term evolutionary relationship, as expected from the vertical transmission of the PDVs and their integration into wasp DNA (Fig. 2). Such correspondence has already been demonstrated between braconid wasps in the genus *Cotesia*, as estimated from

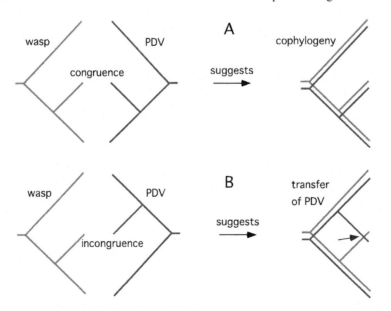

FIGURE 2   Simple schematic showing logic of co-phylogeny inferences from examining congruence among wasp and virus gene trees. Please see color plate section at the back of the book.

several wasp genes, and the BVs they carry (Fig. 3), as estimated from the *CrV1* homologs packaged in the BV particles of the various wasps (Whitfield, 2000; Whitfield and Asgari, 2003). Continuing work on populations within wasps of the *Cotesia flavipes* complex show similar patterns, although local incongruences can arise at the population level (Muirhead, personal communication). Efforts to demonstrate the same pattern further back in the phylogeny, i.e. at the generic and subfamily level within the bracovirus-bearing lineage of braconid wasps, has been complicated by several factors. First, the relationships among the genera of Microgastrinae, and to some extent also the relationships among subfamilies within the microgastroid assemblage, have been notoriously difficult to resolve, even after employment of batteries of up to seven genes and a variety of analytical approaches (Mardulyn

and Whitfield, 1999; Whitfield *et al.*, 2002; Banks and Whitfield, 2006; Huson *et al.*, 2006; Grünewald *et al.*, 2008; Holland *et al.*, 2007; Murphy *et al.*, 2008; Whitfield *et al.*, 2008). Second, orthology among genes from BVs of wasps in different genera is not well established yet, as BVs from only a few genera have been fully sequenced, and a number of the BV genes are members of large multigene families (Kroemer and Webb, 2004). Thirdly, the genomes of BVs in different subfamilies of wasps can be so extremely divergent even in terms of gene content as to be virtually incomparable, especially the BVs from Cheloninae (Bézier *et al.*, 2009).

To complicate matters, it is now clear that many of the BV genes appear not to have been inherited from a progenitor virus (Espagne *et al.*, 2004; Bézier *et al.*, 2009). This uncertainty about specific gene origins clouds the interpretation of what the correspondence between wasp and 'BV' gene trees actually means. It is no accident that the studies of BV and IV origins have had to focus on the difficult task of identifying the core viral genes involved in particle assembly. Many (IVs) if not all (BVs) of these genes are no longer packaged in the particles (Bézier *et al.*, 2009; Volkoff *et al.*, 2010). Disappointingly for coevolutionary analysis, it is precisely the genes involved in the wasp/ caterpillar interaction that are at the same time of highest evolutionary interest and of the least certain (and probably nonviral) origin.

No doubt somewhat similar correspondences between wasp and PDV gene trees will be found in the IV- and 'banchovirus'-bearing lineages, once these are studied in more detail.

WHAT CAN WASP PHYLOGENY TELL US ABOUT PDV EVOLUTION?

It is possible that a detailed PDV phylogeny will not show full and direct co-diversification with wasp gene trees. However, the generation of PDV-carrying wasp phylogenies still has the potential to be informative, and indeed already has proven fruitful in elucidating and providing insight into many important factors regarding PDV evolutionary relationships.

As mentioned previously, it was possible to use wasp phylogenetics to infer the possible single origin of BVs within braconid wasps and its approximate timing, and further to confirm the cophylogeny of the viral genes with wasp genes. This work established the history and extreme antiquity of BV/braconid wasp interactions, but the exact ages are increasingly difficult to pin down due to the rapid diversification of wasps within a few million years of the acquisition of PDVs, making the backbone of the phylogeny nearly impossible to resolve (Mardulyn and Whitfield, 1999; Banks and Whitfield, 2006; Whitfield and Lockhart, 2007; Murphy *et al.*, 2008).

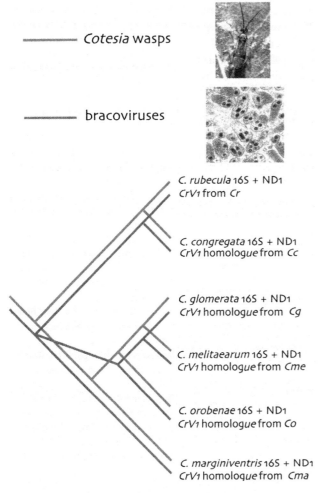

Cotesia wasps

bracoviruses

C. rubecula 16S + ND1
CrV1 from Cr

C. congregata 16S + ND1
CrV1 homologue from Cc

C. glomerata 16S + ND1
CrV1 homologue from Cg

C. melitaearum 16S + ND1
CrV1 homologue from Cme

C. orobenae 16S + ND1
CrV1 homologue from Co

C. marginiventris 16S + ND1
CrV1 homologue from Cma

FIGURE 3 Inferred co-phylogeny of BV genes with genes of wasps in the genus *Cotesia* (from Whitfield and Asgari, 2003, re-used with permission from Elsevier). The wasp phylogeny was estimated based on partial sequences from the 16S and ND1 mtDNA genes; BV relationships were estimated based on partial sequences from the CrV1 gene and its orthologs in other *Cotesia* species. Please see color plate section at the back of the book.

Molecular phylogenies have provided an interpretive framework upon which to study PDV genomic evolution, and it is possible that inclusion of more complete PDV genomic data, including the original viral functional genes, in turn will actually help resolve the remaining uncertainties in the wasp phylogenies. Moreover, knowledge of the approximate age of the symbiosis has allowed us to pair the PDV endosymbiosis with examination of an explosive radiation of wasp diversification.

When combined with well-sampled natural history data including host records, wasp phylogenies provide a time-lined ecological framework with which to analyse PDV evolution. Phylogenies will allow us to use the unique historical setting within which each PDV has evolved in comparative analyses of PDV genomic data to examine whether, or which, specific similarities and/or differences between PDV genomes arise in association with given ecological or host contexts. Provisional analyses of wasp diversity have shown that 94% of Microgastrinae attack only one or two hosts, at least within a given geographic area (Smith *et al.*, 2008). Exactly how this level of host specificity is determined or otherwise influenced by variation in sequence and gene content of PDVs is not yet known, but is now open to direct study with the integration of host relationships, wasp phylogeny, and comparative genomics of the PDVs. Some preliminary steps in this direction (cross-protection experiments exploring the effects of novel BV/wasp combinations) were taken for several *Microplitis* species by Kadash *et al.* (2003), but we are now able to approach the problem more comprehensively.

The key factor which has held back construction of this detailed wasp biological phylogeny is the difficulty of collecting and assembling enough high-quality life history information from PDV-carrying parasitoids to populate a coherent phylogeny. Assembly of such a biologically informative phylogeny requires that groups be taxonomically well sampled and paired with reliable ecological information including, but not limited to, whether they exhibit solitary or gregarious development, mode of parasitism (larval or in some cases egg–larval parasitism), identity of the lepidopteran host, breadth of host range, and host feeding plant. In order to obtain such detailed information in association with a large number of parasitoids, data must be collected in areas of high parasitoid diversity, in large-scale rearing operations, employing experienced staff, using systematic sampling methods, and be conducted over an extended period of time.

Over a period of more than thirty years, a rearing inventory of the Area de Conservacion Guanacaste (ACG) in northwestern Costa Rica has put together the largest data set ever assembled of wild-caught caterpillars and their reared parasitoids (Janzen *et al.*, 2009). All data collected include detailed life history information, COI barcoding sequences to aid in species identification, and are linked with storage of specimens in 95% EtOH for later genomic and morphological analysis. To date, a database of over 400,000 caught and reared caterpillars (and their rearing results, be they adult moths, or butterflies, or parasites) has been compiled (Smith *et al.*, 2008), totaling thousands of attacked caterpillars and nearly 500 provisional identified microgastrine wasp species. The unparalleled size and quality of this database (http://janzen.sas.upenn.edu/), mean that, to date, it is the only (albeit still regional) data set suitable to produce a number of large wasp phylogenies, which will be paired with PDV genomic data.

In association with this rearing inventory, we have produced a number of detailed Microgastrinae phylogenies incorporating molecular, morphological and fossil record information (Whitfield *et al.*, 2002, Banks and Whitfield, 2006; Murphy *et al.*, 2008). To date we have chosen to target six braconid genera of the extensively sampled Microgastrinae to estimate further phylogenies including detailed natural history data, upon which PDV genomic traits can be superimposed.

PHYLOGENOMICS OF WASPS AND PDVS

Currently, only a small number of PDV genomes have been sequenced, each coming from a relatively unrelated taxon compared to the others. Due to the high level of difference, both at the sequence level and in terms of gene content between PDVs, genome comparisons/analyses between such disparate taxa cannot easily result in phylogenetically informative analyses of PDV genes common within subsets of taxa, nor is it currently straightforward to detect hierarchically distributed genes among related taxa. In order to perform such broad studies, it will be necessary to build a database of PDV genomic data, including groups of both closely and distantly related species genera, but until now two key factors have held back assembly of more than but a few PDV genomes.

The first hurdle in the study of PDVs is a factor of small body size: the difficulty of obtaining sufficient quantities of PDV DNA from a single female wasp to serve as a sequencing template. On average, a female wasp contains 40–100 ng of viral DNA (Webb *et al.*, 2006; Beck *et al.*, 2007), whereas conventional sequencing requires many micrograms of intact DNA. In solitary parasitoid wasp species, only one wasp emerges from the parasitized host, and thus there is only a single chance to extract the required DNA. It has also been shown that when comparing genomes for detecting signatures of selection, it is ideal that the genomic DNA be assembled from a single individual (Pool *et al.*, 2010). In parasitoids, this avoids obtaining an unrealistic representation of single nucleotide polymorphisms (SNP's) and other variations occurring in individuals of a species. We have the advantage in Hymenoptera that male wasps are haploid; the disadvantage to this is that females contain many more copies of PDV genomic material in their ovaries.

Despite the latter limitation, it is now becoming possible to use rolling circle DNA amplification techniques and pyrosequencing to obtain complete PDV genomes efficiently and cheaply from individual wasps (Strand, personal communication). It is now possible to design comparative projects employing PDV genomes obtained at the cost of only a few thousand dollars per genome, and this cost is likely to continue to go down in coming years. With these reduced costs, it will soon be possible to assemble a large 'PDV genome library' including wasps distributed through known phylogenies with detailed life history information. Once we have this comparative phylogenomic framework, we will have the ability to ask a wide range of questions, and to design analyses which could identify features of the wasp–PDV–caterpillar interaction that are functionally significant in determining the wasps' host ranges. We discuss a few of the possibilities below.

HOW AND WHY DO PDV GENE FAMILIES DIVERSIFY?

A significant proportion of the PDV genes that appear to be functional in the host–parasitoid relationship belong to large multigene families (e.g., in BVs, protein tyrosine phosphatases, ankyrins and cystatins; Espagne et al., 2004; Kroemer and Webb, 2004; Webb et al., 2006; Lapointe et al., 2007; Serbielle et al., 2008). How is this high rate of gene duplication and diversification achieved? And how much of the diversity of resulting genes is significant in the host–parasitoid relationship? As a start, it is straightforward enough to infer the history of gene duplications and losses using tree reconciliation analysis (e.g., Durand et al., 2006) of PDV gene family trees in the context of wasp backbone trees, but evaluation of how much of the actual genetic change has functional significance depends on its association with biological traits such as host switches, adaptation to host defenses, etc. The phylogenies can then provide a framework for informative experimental studies.

WHICH GENES UNDERGO POSITIVE SELECTION IN ASSOCIATION WITH A HOST SWITCH?

Of obvious interest would be the identification of which genes actually play key roles in host specificity of the wasps. Crucial to any study of this would be the ability to identify those genes that have changed most significantly when wasps have switched hosts during their evolutionary history. Armed with a wasp phylogeny for which we have associated host information, we should be able to use the ratio of nonsynonymous to synonymous substitutions in viral genes in phylogenetically independent contrasts of wasp lineages that have undergone recent host switches versus those that have not.

Previous selection studies have been conducted on PDV genes of known virulence. As an example, Serbielle et al. (2008) conducted a study on PDV cystatin gene evolution in a subset of the wasp genus Cotesia. This study showed that particular BV genes were under strong positive selection, with variations being most strongly selected for in amino acids within the active site of the modeled protein structure. Yet, studies such as this have predominantly focused on pre-specified genes; by superimposing PDV 'genomes of interest' onto the wasp phylogeny, it should be possible to detect genes of interest and then investigate their function, essentially reversing the process.

ARE PARTICULAR GENES OR MODIFICATIONS 'REQUIRED' FOR ATTACK OF A GIVEN HOST GROUP?

This question can be approached by testing contrasting evolutionarily independent species pairs, e.g., distantly and closely related species pairs which parasitize closely related hosts or the same host versus closely and distantly related species parasitizing distantly related host species. In such comparative analyses, it will be important to factor out the level of background variation in PDV sequence expected in ecologically and genetically similar species at various hierarchical levels, in order to unambiguously associate specific genetic changes in PDVs with putative causative factors. Of special interest would be any genes or gene motifs uniquely found in unrelated wasps attacking the same host.

AVAILABILITY OF WASP GENOMES

So far, we have focused on PDVs genomes, not wasp genomes. Currently, for parasitoid wasps only three Nasonia spp. (Chalcidoidea, non PDV bearing wasps) have had their genomes completely sequenced (Nasonia Genome Working Group, 2010), along with the honey bee among the Hymenoptera. However, several more are currently underway, including Microplitis demolitor, a BV-bearing microgastrine braconid (Robertson et al., unpublished data). Given the increase in speed, and decrease in costs, of obtaining whole genome sequences, it is likely that within the next ten years or so a wide variety of parasitoid wasp genomes will be available, and comparatively we will be able to say quite a bit more about the molecular mechanisms of parasitism in Hymenoptera. With respect to PDV evolution, comparison of PDV-bearing and non PDV-bearing groups could especially provide insight into wasp, viral, or other origins of PDV genes that function within caterpillar hosts.

POSSIBLE APPLICATIONS IN BIOPESTICIDE DESIGN

Despite the current lack of understanding of the physiological and genetic mechanisms by which PDVs affect the immune system and development of their host caterpillar, hundreds of microgastrine wasps have been tested to date as possible biological control agents in agriculture. The apparent specificity of host–parasitoid–PDV interaction implies that it is likely that we will eventually be able to pinpoint key genes involved in PDV virulence. It may even be possible to identify the genes responsible for compromising the immune defenses of a specific caterpillar species. Such advances in the specificity of our functional understanding might not only lead to improvements in selection of biological control agents, but also to the development of highly specific biopesticides for management of caterpillar pests (Beckage and Gelman, 2004).

ACKNOWLEDGMENTS

We would especially like to thank Nancy Beckage and Jean-Michel Drezen for inviting us to contribute to this volume, the first truly extensive treatment of PDVs. Both of us have benefited from discussions of recent and current work not only with Nancy and Jean-Michel but also with many of the other authors in this book. The members of the Whitfield and Cameron insect systematics lab groups at the University of Illinois provided many useful comments on the manuscript. JBW's and JMO's work on PDVs is supported by USDA award no. 2009-35302-05250 and research on microgastrine wasps phylogenetics by NSF grant DEB 1020510.

REFERENCES

Banks, J.C., Whitfield, J.B., 2006. Dissecting the ancient rapid radiation of microgastrine wasp genera using additional nuclear genes. Mol. Phylogen. Evol. 41, 690–703.

Beck, M.H., Inman, R.B., Strand, M.R., 2007. *Microplitis demolitor* bracovirus segments vary in abundance and are individually packaged in virions. Virology 359, 179–189.

Beckage, N.E., Gelman, D.B., 2004. Wasp parasitoid disruption of host development: implications for new biologically based strategies for insect control. Ann. Rev. Entomol. 49, 299–330.

Belle, E., Beckage, N.E., Rousselet, J., Poirie, M., Lemeunier, F., Drezen, J.-M., 2002. Visualization of polydnavirus sequences in a parasitoid wasp chromosome. J. Virol. 76, 5793–5796.

Belshaw, R., Fitton, M., Herniou, E., Gimeno, C., Quicke, D.L.J., 1998. A phylogenetic reconstruction of the Ichneumonoidea (Hymenoptera) based on the D2 variable region of 28S ribosomal RNA. Syst. Entomol. 23, 109–123.

Bézier, A., Annaheim, M., Herbiniere, J., Wetterwald, C., Gyapay, G., Bernard-Samain, S., et al. 2009. Polydnaviruses of braconid wasps derive from an ancestral nudivirus. Science 323, 926–930.

Bigot, Y., Renault, S., Nicolas, J., Moundras, C., Demattei, M.-V., Samain, S., et al. 2009. Symbiotic virus at the evolutionary intersection of three types of large DNA viruses; iridoviruses, ascoviruses and ichnoviruses. PLoS One 4 (7), e6397.

Bigot, Y., Samain, S., Auge-Gouillou, C., Federici, B.A., 2008. Molecular evidence for the evolution of ichnoviruses from ascoviruses by symbiogenesis. BMC Evol. Biol. 8, 253.

Cook, D.J., Stoltz, D.B., 1983. Comparative serology of viruses isolated from ichneumonid parasitoids. Virology 130, 215–220.

Dowton, M., Austin, A.D., 1998. Phylogenetic relationships among the microgastroid wasps (Hymenoptera: Braconidae); combined analysis of 16S and 28S rDNA genes and morphological data. Mol. Phylogenet. Evol. 10, 354–366.

Durand, D., Halldorsson, B., Vernot, B., 2006. A hybrid micro-macroevolutionary approach to gene tree reconstruction. J. Comput. Biol. 13, 320–335.

Espagne, E., Dupuy, C., Huguet, E., Cattolico, L., Provost, B., Martins, N., et al. 2004. Genome sequence of a polydnavirus: insights into symbiotic virus evolution. Science 306, 286–289.

Federici, B.A., Bigot, Y., 2003. Origin and evolution of polydnaviruses by symbiogenesis of insect DNA viruses in endoparasitic wasps. J. Insect Physiol. 49, 419–432.

Goulet, H., Huber, J.T. (Eds.), 1993. Hymenoptera of the World: An Identification Guide to Families. Res. Branch Agric. Canada Publication 1894/E.

Grünewald, S., Spillner, A., Forslund, K., Moulton, M., 2008. Constructing phylogenetic supernetworks from quartets. Workshop on Algorithms in Bioinformatics 2008 (Lecture Notes in Bioinformatics) 5251, 284–295, Springer, Berlin.

Holland, B.R., Conner, G., Huber, K.T., Moulton, V., 2007. Imputing supertrees and supernetworks from quartets. Syst. Biol. 56, 57–67.

Huson, D.H., Steel, M., Whitfield, J.B., 2006. Reducing distortion in phylogenetic networks. Workshop on Algorithms in Bioinformatics 2006 (Lecture Notes in Bioinformatics) 4175, 150–161.

Janzen, D.H., Hallwachs, W., Blandin, P., Burns, J.M., Cadiou, J.-M., Chacon, I., 2009. Integration of DNA barcoding into an ongoing inventory of tropical complex biodiversity. Mol. Ecol. Res. 9 (suppl.1), 1–26.

Kadash, K., Harvey, J.A., Strand, M.R., 2003. Cross-protection experiments with parasitoids in the genus *Microplitis* (Hymenoptera: Braconidae) suggest a high level of specificity in their associated polydnaviruses. J. Insect. Physiol. 49, 473–482.

Kroemer, J.A., Webb, B.A., 2004. Polydnavirus genes and genomes: emerging gene families and new insights into polydnavirus replication. Ann. Rev. Entomol. 49, 431–456.

Lapointe, R., Tanaka, K., Barney, W.E., Whitfield, J.B., Banks, J.C., Beliveau, C., et al. 2007. Genomic and morphological features of a banchine polydnavirus: comparison with bracoviruses and ichnoviruses. J. Virol. 81, 6491–6501.

Lawrence, P.O., 2005. Non-poly-DNA viruses, their parasitic wasps, and hosts. J. Insect Physiol. 51, 99–101.

Mardulyn, P., Whitfield, J.B., 1999. Phylogenetic signal in the COI, 16S and 28S genes for inferring relationships among genera of Microgastrinae (Hymenoptera: Braconidae); evidence of a high diversification rate in this group of parasitoids. Mol. Phylogenet. Evol. 12, 282–294.

Murphy, N., Banks, J.C., Whitfield, J.B., Austin, A.D., 2008. Phylogeny of the microgastroid complex of subfamilies of braconid parasitoid wasps (Hymenoptera) based on sequence data from seven genes, with an improved estimate of the time of origin of the lineage. Mol. Phylogen. Evol. 47, 378–395.

Nasonia Genome Working Group, 2010. Functional and evolutionary insights from the genomes of three parasitoid *Nasonia* species. Science 327, 343–348.

Norton, W.N., Vinson, S.B., Stoltz, D.B., 1975. Nuclear secretory particles associated with the calyx cells of the ichneumonid parasitoid *Campoletis sonorensis* (Cameron). Cell Tissue Res. 162, 195–208.

Pool, J.E., Hellmann, I., Jenson, J.D., Nielsen, R., 2010. Population genetic inference from genomic sequence variation. Genome Res. 20, 291–300.

Quicke, D.L.J., Laurenne, N.M., Fitton, M.G., Broad, G.R., 2009. A thousand and one wasps: a 28S rDNA and morphological phylogeny of the Ichneumonidae (Insecta: Hymenoptera) with an investigation into alignment parameter space and elision. J. Nat. Hist. 43, 1305–1421.

Serbielle, C., Chowdhury, S., Pichon, S., Dupas, S., Lesobre, J., Purisima, E.O., et al. 2008. Viral cystatin evolution and three-dimensional structure modeling: a case of directional selection acting on a viral protein involved in a host–parasitoid interaction. BMC Biol. 6 (38), 1–16.

Sharkey, M.J., Wahl, D.B., 1992. Cladistics of the Ichneumonoidea. J. Hymenopt. Res. 1, 15–24.

Smith, M.A., Rodriguez, J.J., Whitfield, J.B., Janzen, D.H., Hallwachs, W., Deans, A.R., et al. 2008. Extreme diversity of tropical parasitoid wasps exposed by iterative integration of natural history, DNA barcoding, morphology and collections. PNAS 105, 12359–12364.

Stoltz, D.B., 1990. Evidence for chromosomal transmission of polydnavirus DNA. J. Gen. Virol. 71, 1051–1056.

Stoltz, D.B., 1993. The polydnavirus life cycle. In: Beckage, N.E., Thompson, S.N., Federici, B.A. (Eds.), Parasites and Pathogens of Insects, Vol. 1: Parasites (pp. 167–187). Academic Press, San Diego.

Stoltz, D.B., Krell, P., Summers, M.D., Vinson, S.B., 1984. Polydnaviridae – a proposed family of insect viruses with segmented, double-stranded, circular DNA genomes. Intervirology 21, 1–4.

Stoltz, D.B., Vinson, S.B., 1979. Viruses and parasitism in insects. Adv. Virus Res. 24, 125–171.

Stoltz, D.B., Whitfield, J.B., 1992. Viruses and virus-like entities in the parasitic Hymenoptera. J. Hymenopt. Res. 1, 125–139.

Stoltz, D.B., Whitfield, J.B., 2009. Making nice with viruses. Science 323, 884–885.

Volkoff, A.-N., Véronique, J., Urbach, S., Samain, S., Bergoin, M., Wincker, P., et al. 2010. Analysis of virion structural components reveals vestiges of the ancestral ichnovirus genome. PLoS Pathog. 6 (5), e1000923.

Webb, B.A., Strand, M.R., Deborde, S.E., Beck, M., Hilgarth, R.S., Kadash, K., et al. 2006. Polydnavirus genomes reflect their dual roles as mutualists and pathogens. Virology 347, 160–174.

Wharton, R.A., Marsh, P.M., Sharkey, M.J., 1997. Manual of the New World Genera of the Family Braconidae (Hymenoptera). International Society of Hymenopterists, Washington, DC.

Whitfield, J.B., 1990. Parasitoids, polydnaviruses and endosymbioses. Parasitol. Today 6, 381–384.

Whitfield, J.B., 1997. Molecular and morphological data suggest a common origin for the polydnaviruses among braconid wasps. Naturwissenschaften 84, 502–507.

Whitfield, J.B., 1998. Phylogeny and evolution of the host/parasitoid relationship in the Hymenoptera. Ann. Rev. Entomol. 43, 129–151.

Whitfield, J.B., 2000. Phylogeny of microgastroid braconid wasps, and what it tells us about polydnavirus evolution. In: Austin, A.D., Dowton, M. (Eds.), The Hymenoptera: Evolution, Biodiversity and Biological Control (pp. 97–105). CSIRO Publishing, Melbourne.

Whitfield, J.B., 2002. Estimating the age of the polydnavirus/braconid wasp symbiosis. PNAS 99, 7508–7513.

Whitfield, J.B., 2003. Phylogenetic insights into the evolution of parasitism in Hymenoptera. In: Littlewood, T.J. (Ed.), The Evolution of Parasitism – A Phylogenetic Approach, Advances in Parasitology 54, pp. 69–100). Elsevier, Amsterdam.

Whitfield, J.B., Asgari, S., 2003. Virus or not? Phylogenetics or polydnaviruses and their wasp carriers. J. Ins. Physiol. 49, 397–405.

Whitfield, J.B., Cameron, S.A., Huson, D.H., Steel, M.A., 2008. Filtered Z-closure supernetworks for extracting and visualizing recurrent signals from incongruent gene trees. Syst. Biol. 57, 939–947.

Whitfield, J.B., Lockhart, P.J., 2007. Deciphering ancient rapid radiations. Trends Ecol. Evol. 22, 258–265.

Whitfield, J.B., Mason, W.R.M., 1994. Mendesellinae, a new subfamily of braconid wasps (Hymenoptera, Braconidae) with a review of relationships within the microgastroid assemblage. Syst. Entomol. 19, 61–76.

Whitfield, J.B., Mardulyn, P., Austin, A.D., Dowton, M., 2002. Phylogenetic analysis of relationships among microgastrine braconid wasp genera based on data from the 16S, COI and 28S genes and morphology. Syst. Entomol. 27, 337–359.

Zalidivar-Riveron, A., Shaw, M.R., Saez, A.G., Mori, M., Belokobylskij, S.A., Shaw, S.R., et al. 2008. Evolution of the parasitic wasp subfamily Rogadinae (Braconidae): phylogeny and evolution of lepidopteran host ranges and mummy characteristics. BMC Evol. Biol. 8, 329.

Chapter 8

Integration of Polydnavirus DNA into Host Cellular Genomic DNA

Dawn Gundersen-Rindal

US Department of Agriculture, Agricultural Research Service, Invasive Insect Biocontrol and Behavior Laboratory, Room 267, 10300 Baltimore Avenue, Bldg 011A BARC WEST, Beltsville, Maryland, MD 20705, U.S.A.

SUMMARY

Polydnaviruses are unique insect viruses that are obligately associated with thousands of parasitoid wasp species in intimate mutualistic symbioses. Polydnaviruses have evolved by atypical life-cycle and replication strategies that require two separate insect hosts. In addition, they exist in two distinct forms, both as linear integrated provirus and as double-strand circular forms of genome segments encapsidated into virions. The linear integrated provirus form is transferred vertically through the parasitoid germline. The encapsidated circular form packaged in virions is produced from the integrated proviral form in the female parasitoid and is transferred horizontally by injection into the secondary insect host during oviposition along with egg(s) and ovarian and other proteins, where upon infection viral genes are expressed and cause immune suppression, developmental arrest, and other negative effects. Integration, the insertion of viral DNA into host-cell DNA, is a central theme in polydnavirus biology for both provirus and encapsidated virion. In addition to analysis of vertical provirus integration, several studies have examined integration of the encapsidated form polydnavirus virion episomal DNAs in infected lepidopteran host cells, finding that some to all of the virus genome became integrated permanently into the secondary host cellular DNA, which appears also to occur *in vivo*. Analyses of proviral and encapsidated polydnavirus genomes suggest unique sequence features of polydnaviruses themselves. Both forms also have an association with a diversity of transposable elements. The intimacy in parasitism, unique replication and life strategies involving two hosts, novel structural features, and organization of polydnavirus integrated proviral and circular encapsidated forms, and even associations with transposable elements, suggest numerous opportunities for genetic exchange and sequence integration within the primary host parasitoid, polydnavirus, and secondary lepidopteran hosts. Cellular integration of proviral and encapsidated polydnaviral genomes may confer advantages to parasitoid as well as secondary insect host genomes, and may have played roles in their mutual evolution.

ABBREVIATIONS

IVSPERs	Ichnovirus structural protein encoding regions
LTR	long terminal repeat
NCLDV	nucleocytoplasmic large DNA viruses

Polydnaviruses (<u>poly</u>disperse-<u>DNA</u>-<u>viruses</u>) are unique insect viruses that are obligately associated with thousands of parasitoid wasp species in intimate mutualistic symbioses. The virus family *Polydnaviridae* contains two taxa. These are ichnoviruses and bracoviruses, associated with members of distinct lineages within *Campopleginae* and the microgastroid complex (*Microgastrinae, Cardiochilinae, Miracinae, Khoikholiinae,* and *Cheloninae*), respectively. A third taxon was recently proposed based on a virus associated with an ichneumonid belonging to the subfamily *Banchinae* (Lapointe *et al.*, 2007). Bracoviruses and ichnoviruses differ in particle morphology, physiology of viral particle production, molecular characteristics, and the families of the associated parasitoid wasps (Stoltz and Whitfield, 1992; Webb *et al.*, 2005; reviewed in Webb and Strand, 2005). Banchine polydnaviruses share some morphological and genetic properties with ichnoviruses, and a common gene family with bracoviruses.

In contrast to typical viruses, polydnaviruses exist in symbiotic relationships with their primary parasitoid wasp hosts in which they are formed but display few perceptible negative effects. Their morphogenesis and active replication is limited to a specific population of specialized calyx cells within the female parasitoid oviduct. Like most viruses, polydnaviruses have replication cycles characterized by attachment, penetration, and entry into host cells, uncoating of virus nucleocapsid(s), replication of viral nucleic acid in the nucleus, virus gene expression, synthesis of virus-specific proteins, and the packaging of infectious progeny

Parasitoid Viruses: Symbionts and Pathogens. DOI: 10.1016/B978-0-12-384858-1.00008-4

virions. Unlike typical viruses, polydnavirus life-cycles and replication are distributed between and require two distinct insect hosts. Polydnavirus pathogenic effects are exerted only within the secondary insect host, in most cases lepidopteran larva, into which they are introduced during oviposition and whose cells they infect following parasitization. However, infection of the secondary host cells does not lead to active replication and production of infectious progeny virions, largely because genes for viral structural proteins have been shown to reside integrated within the chromosomes of the parasitoid host and are not packaged into virions (Bézier et al., 2009a; Volkoff et al., 2010).

Virus integration, the insertion of viral DNA into host-cell DNA, plays a central role in the life-cycle of polydnaviruses. They are integrated within their primary parasitoid wasp host cellular chromosomal DNA as provirus and vertically transmitted to progeny within the germline as integrated provirus. Within the unique life-cycle of polydnaviruses, infectious virions are formed in a specialized population of female ovarian calyx cells, involving excision and packaging into virions of circular dsDNA segments excised from replicative form(s) amplified from a linear integrated proviral template. Several polydnaviruses, both ichnoviruses and bracoviruses, have also exhibited the ability of circular encapsidated episomal viral genome segments to become integrated into their lepidopteran secondary host cellular DNA after horizontal transfer from the primary host parasitoid.

INTEGRATION OF POLYDNAVIRUS (PROVIRAL FORM) WITHIN THE PRIMARY PARASITOID HOST CELLULAR DNA: VERTICAL TRANSFER

Many lines of evidence have shown that sequences of the segmented polydnavirus genome are present within the respective host parasitoid wasp carrier in two distinct forms: a provirus form (linear) that is integrated within the parasitoid wasp chromosomes, and a circular episomal form in which circular viral dsDNA segments are encapsidated and packaged into infectious virus particles that are subsequently delivered to the secondary host. The transmission of polydnaviruses, both bracoviruses and ichnoviruses, is accomplished vertically via the integrated provirus form, which is present in all cells of the parasitoid wasp host, including those of male parasitoids that do not produce infectious virions (Theilmann and Summers 1986, 1987; Fleming and Summers, 1986, 1990, 1991; Fleming, 1991; Fleming and Krell, 1993; Gruber et al., 1996; Savary et al., 1997; Stoltz and Xu, 1990; Stoltz, 1993; Stoltz et al., 1986; Wyder et al., 2002; Xu and Stoltz, 1991).

The integration of polydnavirus as a provirus was recognized in early studies of their transmission in parasitism

that suggested the virus was maintained by incorporation into the parasitoid genome (Stoltz and Vinson, 1979). Later, extensive genetic studies showed both ichnoviruses and bracoviruses share the property of vertical transmission through the germ line (Stoltz et al., 1986), through integration into their respective wasp host genomes (Fleming and Summers, 1990, 1991; Fleming, 1991; Gruber et al., 1996; Savary et al., 1997; Stoltz, 1990, 1993; Stoltz et al., 1986; Wyder et al., 2002). Fleming and Summers (1986) initially demonstrated the existence of the provirus form for ichnoviruses by showing sequences were present in both male and female Campoletis sonorensis parasitoid wasps. Later, they verified the truly integrated provirus form by cloning a C. sonorensis ichnovirus (CsIV) fragment flanked by parasitoid wasp genomic DNA (1991). Further studies confirmed parasitoid proviral DNA integration by demonstration of off-size polydnavirus restriction fragments within parasitoid DNAs through Southern hybridization to blots containing restricted male and female parasitoid and polydnavirus genomic DNAs (Fleming and Krell, 1993; Fleming and Summers, 1990; Xu and Stoltz, 1991). For bracoviruses, Gruber et al. (1996) confirmed the existence of the proviral form by showing integrated Chelonus inanitus BV (CiBV) sequence within female parasitoid genomic DNA. This was accomplished by molecular genetic demonstration that a 12-kb DNA genome segment integrated in the C. inanitus parasitoid genome was excised and circularized in female parasitoids in later pupal and early adult stages. Savary et al. (1997) showed sequences of both circular and integrated forms of Cotesia congregata bracovirus (CcBV) containing viral early protein 1 (EP1) were present in both male and female parasitoid wasps, and that the circular form in female parasitoids was produced by excision of the linear integrated form. Numerous studies since that time have confirmed integrated provirus status for both ichnoviruses, bracoviruses, and recently the integration into parasitoid wasp chromosomes was also demonstrated for the polydnavirus associated with the banchine Glypta fumiferanae (Lapointe et al., 2007).

The mechanism of vertical transmission as integrated provirus has been the subject of investigation. During vertical transmission, germline integrated proviral sequences are replicated as part of parasitoid chromosomes in the course of cell division. The integrated provirus is passed from parent to progeny with chromosomal inheritance of the provirus subjected to constraints of Mendelian genetics (Stoltz, 1993) in a transmission mode that is similar to the vertical transmission of lysogenic viruses (Webb et al., 2009). Within the braconid parasitoids, proviral segment sequences were originally thought to be located at a single chromosomal locus in a tandem array based on the analyses of CcBV and CiBV proviral segments flanked on one or both sides by a different proviral segment (Savary et al., 1997; Wyder et al., 2002) and the cytological evidence from Belle et al. (2002), which showed probes from at least 3 (of >30) unique CcBV genome segments hybridized to a common site and were likely clustered in a macrolocus on the short arm of subtelocentric wasp chromosome 5, as observed on metaphase spreads. The structure and organization of the integrated braconid provirus was later characterized in detail through analyses of *C. inanitus* amplicons containing intersegmental DNA (Annaheim and Lanzrein, 2007) and large cloned contiguous regions of *Glyptapanteles indiensis* and *G. flavicoxis* chromosomal DNAs. Desjardins *et al.* (2007, 2008) showed that bracovirus genome segments within the parasitoid chromosome were organized in greater than seven loci, with each proviral locus containing one to many viral genome segments. The two largest loci are linked and contain greater than 70% of the segments, thus constituting a viral macrolocus. It is not known whether the other loci are located in the same vicinity of the macrolocus. The segments were generally arranged in tandem within a locus, but were separated by short (most shorter than 1 kb) spans of non coding DNA. Each major locus was flanked and separated by parasitoid chromosomal gene-encoding DNA. Three types of repeated regions were noted in parasitoid chromosomal sequences adjacent to some proviral loci.

Early studies focused on ichnoviruses and showed excision sites marking proviral genome segment integration within parasitoid chromosomes were characterized by terminal repeats, which varied in length and homology among polydnaviruses (Fleming and Summers, 1991; Cui and Webb, 1997; Gruber *et al.*, 1996; Wyder *et al.*, 2002). Later studies on bracoviruses showed each proviral genome segment end contained highly conserved 5′- and 3′- sequence motifs that defined sites on the provirus for excision and circularization, possibly via conservative site-specific recombination, of the bracovirus circular form segments. Once excised from a progenitor replicative form, circularized, and encapsidated into virus particles, each circular virus genome segment retained one of these sequence motifs (Savary *et al.*, 1997; Annaheim and Lanzrein, 2007; Desjardins *et al.*, 2008). There was directionality to the mechanism for individual circular segment excision from an amplified progenitor suggested by the 5′ or 3′ strand arrangement of these motif sites within proviral segments (Desjardins *et al.*, 2007, 2008). Within the microgastroid complex the Cheloninae are the most basal phylogenetic subgroup (Murphy *et al.*, 2008) and have the least viral genome sequence similarity with other bracoviruses (Weber *et al.*, 2007; Wetterwald *et al.*, 2010), yet their terminal excision motif sequences (GCTT motif) for production of the circular form were highly conserved and were shared among the entire microgastroid complex, supporting a highly conserved mechanism for bracovirus proviral genome segment excision (Annaheim and Lanzrein, 2007; Desjardins *et al.*, 2008; Wetterwald *et al.*, 2010).

Because of structural and morphological similarities to insect baculoviruses, the ancestor bracovirus was suspected to be a baculovirus (Stoltz and Vinson, 1979; reviewed in Federici and Bigot, 2003). Analyses of vertical transmission of polydnavirus as integrated provirus led to speculations on their obligate associations within certain parasitoid lineages. Early phylogentic analysis within microgastroid complex representatives suggested the association of integrated bracovirus as provirus with its primary host parasitoid wasp originated with the integration of an ancestral virus within the ancestor braconid parasitoid genome (Whitfield and Asgari, 2003). In detailed studies of the phylogeny of the microgastroid complex, Murphy *et al.* (2008) placed the association of bracovirus with the braconid parasitoid ancestor at approximately 103 million years ago and hypothesized a single integration event, creating a provirus, within the microgastroid lineage. These studies suggested the presence of an integrated virus conferred a selective advantage to the parasitoid during parasitism and was therefore maintained through selection pressure and coevolved with the parasitoid host. They also predicted that the presence of vertically transmitted integrated provirus DNA maintained for its contribution to successful parasitism may have contributed to the diversification of the microgastroid complex into at least 17,000 species (Webb, 1998 and others; Murphy *et al.*, 2008; Bézier *et al.*, 2009b). The ancestral virus remained unknown until Bézier *et al.* (2009a, 2009b) demonstrated that bracoviruses derive from an ancestral nudivirus, an ancient sister group to baculoviruses, as 24 nudivirus-related genes were identified in ovaries of distantly related braconid parasitoids, *C. congregata* and *C. inanitus*, at the time of virus formation. In further analyses of braconid ovarian proteins, additional nudiviral proteins were identified similar to those of another braconid, modified version of cellular proteins, and lineage- or species-specific proteins possibly involved in parasitoid–host interaction (Wetterwald *et al.*, 2010).

In contrast to members of the microgastroid complex, less is known about ichnovirus-possessing ichneumonid parasitoids of the *Campopleginae*, which because of morphological similarity were at one time thought to originate from an ancestral ascovirus (Bigot *et al.*, 2008), though this was not supported by recent work (Volkoff *et al.*, 2010). Less is understood about ichnovirus integrated provirus form and the formation of the circular virus genome segments encapsidated in virion particles, which appear to be produced by excision at repeats from ichneumonid integrated provirus (reviewed in Webb, 1998). Early studies on *C. sonorensis* ichnovirus showed several proviral segments were flanked by genomic DNA that was not encapsidated, suggesting CsIV proviral segment sequences were dispersed within the parasitoid genome. Moreover, an excision site locus was characterized for one circular CsIV segment (Fleming and Summers, 1991; Cui and Webb, 1997; Rattanadechakul and Webb, 2003). Recently, Volkoff *et al.* (2010) characterized numerous parasitoid genes that appeared to be involved in the production of ichnovirus circular dsDNA encapsidated virion particles, and analyzed their organization integrated within the parasitoid genome. They found a class of genes clustered in specialized regions of the parasitoid genome, which were amplified along with proviral DNA during the virion replication in the ovaries but were not packaged into virion particles. The ichnovirus particle products produced from these ichnovirus structural protein encoding regions (IVSPERs) were conserved among ichnovirus-bearing parasitoids and stongly suggested ichnoviruses originated from an ancestral virus belonging to a group that has not yet been characterized that integrated its genome into that of an ichneumonid wasp ancestor (Volkoff *et al.*, 2010). Thus, ichnoviruses and bracoviruses originated from different viral entities in a remarkable example of convergent evolution where two distinct lineages of parasitoid wasps independently integrated and domesticated distinct viruses into their genomes, ultimately for the delivery of pathogenic genes into their hosts (Volkoff *et al.*, 2010).

INTEGRATION OF POLYDNAVIRUS (CIRCULAR ENCAPSIDATED FORM) WITHIN THE SECONDARY INSECT HOST CELLULAR DNA: HORIZONTAL TRANSFER

During oviposition, one or more parasitoid eggs are injected into a secondary host (usually a lepidopteran larva) along with parasitoid calyx fluid containing the circular episomal form of polydnavirus, as well as venom and ovarian-associated proteins. Polydnaviruses then infect larval host tissues but they do not replicate (Stoltz and Vinson, 1979; Theilmann and Summers, 1986; Wyder *et al.*, 2002). For bracoviruses, genes associated with replication and structural proteins have recently been shown to have been lost and now reside in the parasitoid chromosome, no longer packaged within the virion particle (Bézier *et al.* 2009a; Volkoff *et al.*, 2010). Although virions do not replicate within the secondary host, polydnavirus genes are transcribed, translated, and expressed transiently or sustained in host cells in temporal and tissue-specific fashion (Stoltz *et al.*, 1986; Theilmann and Summers, 1986, Strand *et al.*, 1992; Asgari *et al.*, 1996; Béliveau *et al.*, 2000). The polydnavirus gene expression products are important in regulation of the larval host and act in interrelated ways, which differ depending on the wasp and on the parasitized host, to enable parasitoid development and ensure survival. The most pronounced pathogenic effects exerted by polydnaviruses on the larval host result in suppression of immunity, such as alteration of haemocyte behavior, reduction of haemocyte population, inhibition of haemolymph phenoloxidase activity, and inhibition of melanization (Edson *et al.*, 1981; Blissard *et al.*, 1986; Theilman and Summers, 1986; Li and Webb, 1994;

Strand and Pech, 1995; Summers and Dib-Hajj, 1995; Lavine and Beckage, 1995, 1996; Asgari *et al.*, 1997; Webb and Cui, 1998; Shelby and Webb, 1999). Polydnavirus gene products are also responsible for inhibition of host protein synthesis (Shelby and Webb, 1994, 1997; Pennacchio *et al.*, 1998) and regulation of host development and physiology, including developmental arrest, disruption of hormone balance, and inhibition of growth of the host (Vinson *et al.*, 1979; Beckage, 1985; Davies *et al.*, 1987; Dover *et al.*, 1987, 1988; Tanaka *et al.*, 1987; Lawrence and Lanzrein, 1993; Stoltz, 1993; Strand and Pech, 1995; Yamanaka *et al.*, 1996; Béliveau *et al.*, 2000; Cusson *et al.*, 2000). It was recently shown that bracoviruses also alter host metabolic physiology, preventing the host from achieving critical weight, which inhibits the endocrine events that trigger metamorphosis (Pruijssers *et al.*, 2009).

Early attempts to propagate ichnoviruses and bracoviruses in infected lepidopteran host cells in culture led to laboratory observations of viral infection pathologies (Fig. 1). Studies of the infected cell lines resulted in detection of polydnaviral persistence in association with host cellular DNA, the nature of which was further examined *in vitro* and *in vivo*.

In Vitro Integration of Polydnavirus DNA into Insect Host Cellular Genomic DNA

Early studies examined the apparent long-term persistence of certain detectable ichnovirus and bracovirus DNAs in high-molecular-weight DNA associated with infected cultured insect cells (Kim *et al.*, 1996; McKelvey *et al.*, 1996; Volkoff *et al.*, 2001; Gundersen-Rindal and Lynn, 2003; Rocher *et al.*, 2004; Doucet *et al.*, 2007). Although polydnaviruses do not actively replicate to produce progeny virions in infected cultured cells, circular viral genome segments were demonstrated to persist in chromosomally integrated form in host-derived (and non host) cultured cells, where, after the passage of time, the circular episomal form of the virus was no longer detected (Kim *et al.*, 1996; Gundersen-Rindal *et al.*, 1999; Gundersen-Rindal and Dougherty, 2000; Doucet *et al.*, 2007).

In Vitro Persistence of Encapsidated Form Polydnavirus Genome DNA Within Lepidopteran Host and Non Host Insect Cells in Culture

Certain insect cells are highly permissive to polydnavirus infection *in vitro* even if they are not derived from the parasitoid host. Other insect cells types may be refractory to polydnavirus infection (Johnson *et al.*, 2010). Upon polydnavirus infection, many cultured insect cells display significant cytopathic effects immediately, which continue

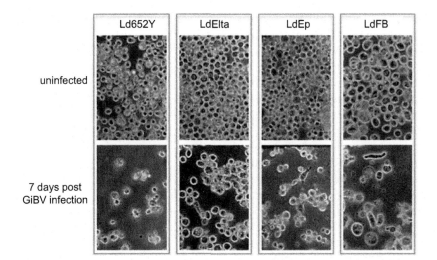

FIGURE 1 Four different *Lymantria dispar* (gypsy moth) cultured cell lines (IPLB-Ld652Y, IPLB-LdEIta, IPLB-LdEp, and IPLB-LdFB) shown uninfected, and displaying cytopathic effects seven days post infection with *G. indiensis* bracovirus. Partly modified from *Biochemical and Biophysical Research Communications* Vol. 225, T.A. McKelvey, D.E. Lynn, D. Gundersen-Rindal, D. Guzo, D.A. Stoltz, K.P. Guthrie, P.B. Taylor, and E.M. Dougherty, 1996. Transformation of gypsy moth (*Lymantria dispar*) cell lines by infection with *Glyptapanteles indiensis* polydnavirus pp. 764–770, with permission from Elsevier.

either transiently or consistently for approximately three weeks thereafter. This phase is followed by a period of cellular recovery in which numbers, morphologies, and growth patterns of surviving cells return to states comparable to those of their uninfected counterparts (Fig. 1). Both ichnovirus and bracovirus DNAs were shown to persist over time maintained *in vitro* within susceptible insect cell populations, including ichnoviruses associated with *Hyposoter fugitivus* (HfPV), *H. didymator* (HdIV), *Tranosema rostrale* (TrIV), and bracoviruses associated with *G. indiensis* (GiBV) and *G. flavicoxis* (GfBV). HfPV was used to infect a pupal ovary cell line derived from *L. dispar* (Ld652Y), which is not the host of the parasitoid (Kim *et al.*, 1996); HdIV was used to infect several cell lines derived from host *Spodoptera littoralis* hemocytes, as well as nonpreferred hosts of the parasitoid *S. frugiperda* (SF9) and *Trichoplusia ni* (High Five) (Volkoff *et al.*, 1999, 2001); TrIV was used to infect a natural host *Choristoneura fumiferana* (CF-124T) cells (Doucet *et al.*, 2007) and GiBV was used to infect several cell lines (LdEp, LdEIta, and LdFB) derived from various tissue types of natural host *L. dispar* (McKelvey *et al.*, 1996; Gundersen-Rindal and Dougherty, 2000) and also non preferred hosts of the parasitoid, including lepidopterans (*T. ni* (TN-R²), *S. frugiperda* (SF-21), *Plodia interpunctella* (PID2), and *Heliothis virescens* (HvT1)) and a coleopteran (*Diabrotica undecimpunctata* (DU182E)) cell line (Gundersen-Rindal *et al.*, 1999). In these examples, Southern analyses of cellular DNAs from the recovered cell populations showed viral sequences were maintained in newly transformed host cells in association with high molecular weight cellular DNAs of chromosomal origin (Kim *et al.*, 1996; McKelvey *et al.*, 1996; Volkoff *et al.*,

2001; Gundersen-Rindal *et al.*, 1999; Doucet *et al.*, 2007). In addition, for those examined, viral sequences were readily detectable within cellular DNAs and viral genes were expressed even after long-term (>60 months for HdIV-infected SF9 and GiBV-infected LdEp cells) routine weekly passage, signifying ichnovirus and brachovirus persistence in integrated form. For HdIV, Volkoff *et al.* (2001) described a K-gene of unknown function and containing 13 imperfectly repeated sequences, in HdIV that was stably maintained, transcribed, and expressed in transformed lepidopteran cell lines from Spodoptera (SF9 and *S. littoralis* haemocytes) and *Trichoplusia* for greater than three years post infection. Because no transcripts related to seven additional HdIV-encoded genes were detected in the same cell lines, it was proposed the viral genome segments associated with K-gene DNAs were selectively retained *in vitro* long-term (Volkoff *et al.*, 2001). Similarly, the *rep* gene (TrFrep1) carried by TrIV genome segment displayed long-term transcription in infected cultured cells (CF-124T) (Doucet *et al.*, 2007).

In Vitro *Polydnavirus DNA Integration Confirmation by Isolation of Clones Containing Integration Junctions from Infected Cultured Cellular DNAs*

The nature of viral persistence *in vitro* was analyzed in detail, where isolation of clones from virus-infected cellular DNAs containing viral:cellular junctions definitively demonstrated *in vitro* integration of certain viral segments. Integration of a bracovirus circular genome segment *in vitro* appeared to occur, in at least one example, at a consistent locus, as the GiBV genome segment F (GiPDVsegF = segment 25 of the GiBV full genome) was shown to be covalently linked to *L. dispar* DNA at the same position on this virus genome segment locus within different cell lines, with one identified cellular DNA site having sequence similarity to a lepidopteran retrotransposon (Gundersen-Rindal and Lynn, 2003). A CATG palindrome repeat was noted at the junction site. More recently, Doucet *et al.* (2007) examined the persistence of a viral genome segment from ichnovirus TrIV cloned directly from genomic DNA extracted from infected cultured natural host *C. fumiferana* (CF-124T) cells. Junction regions on either side of the viral DNA sequence were sequenced and integration border sequences analyzed, showing TrIV genome segment F lost an (ATTCT)$_2$ repeat, potentially associated with DNA unpairing, within a 33-nucleotide region through the process of integration. This was the first definitive demonstration of integration of an ichnovirus genome segment within infected lepidopteran cells. The circular episomal form of this genome segment did not persist in these cells; yet, a gene (TrFrep1) carried by this genome segment displayed long-term transcription in infected cultured cells (Doucet *et al.*, 2007). Both GiBV and TrIV *in vitro* integrations were associated with polyA, polyT, or AT runs in host cellular DNA near the site(s) of integration, which sequences may favor DNA unpairing and instability and facilitate integration (Doucet *et al.*, 2007; Gundersen-Rindal and Lynn, 2003).

In Vivo Integration of Polydnavirus DNA into Insect Host Cellular Genomic DNA

Studies of polydnavirus persistence *in vivo* within cells of parasitized or infected lepidopteran larvae over the course of time have been limited, and to date, *in vivo* analyses have examined only bracovirus-infected host larvae. Le *et al.* (2003) conducted Southern hybridization analyses of *Manduca sexta* DNAs extracted from different tissues of naturally parasitized larvae infected with CcBV where viral DNAs were detected in *M. sexta* fat body cells and brain (faint binding) collected at two time points: the day of parasitoid emergence and six days post emergence (Fig. 2). A smear of hybridization, as opposed to discrete bands that would represent CcBV dsDNA genome segments or fragments from their restriction digestion, was observed, which the authors suggested represented viral sequences which had integrated into the genomic cellular DNA of the caterpillar host (Le *et al.*, 2003). In both tissues, CcBV hybridization continued to be detected in the parasitized larvae a considerable time (six days) post parasitoid egress from the host. This study, which utilized labeled total CcBV genome viral DNA as a probe, indicated possible host genomic integration of CcBV sequences though it was not possible to fully differentiate whether the CcBV DNAs persisted *in vivo* in episomal form, in a chromosomally integrated form, or both

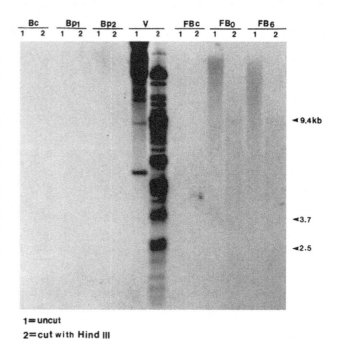

1 = uncut
2 = cut with Hind III

FIGURE 2   Southern blot of *Manduca sexta* (tobacco hornworm) larval DNAs hybridized to a *Cotesia congregata* bracovirus total DNA labeled probe. Agarose gel electrophoresis of uncut (lanes 1) and cut (2) genomic DNA isolated from brains (B) or fat body (FB) of larvae dissected on the day the wasps began emerging (Bo) or six days later (B6). Tissues from control unparasitized larvae were also included (Bc, FBc). The central lanes (V) represent the CcBV genome which is uncut (lane 1) or cut with HindIII (lane 2). Hybridization with total viral DNA as a probe for the presence of viral sequences in the brains (B) or fat body (FB) of parasitized larvae. Note there was no hybridization with control unparasitized tissues (Bc, FBc), but there is extensive hybridization with the genomic DNAs (V uncut and cut) and also faint binding to sequences from genomic fat body DNA, which is cut with HindIII. Reprinted from *Journal of Insect Physiology* Vol. 49/5, Le, N.T., Asgari, S., Amaya, K., Tan, F.F., and Beckage, N.E., 2003. Persistence and expression of *Cotesia congregata* polydnavirus in host larvae of the tobacco hornworm, *Manduca sexta*. 533–543, with permission from Elsevier.

(Fig. 2), or whether faint hybridization could have been due to parasitoid cells present in the parasitized fat body.

Similar investigation of *in vivo* persistence and potential integration within *L. dispar* larvae host genomic DNA of *G. indiensis* bracovirus (GiBV) strongly suggest *in vivo* chromosomal integration (Fig. 3). Southern hybridization was conducted to analyze DNAs extracted from *L. dispar* larvae that had been manually injected with GiBV-containing calyx fluid into the haemocoel for evidence of *in vivo* viral persistence. The *L. dispar* larvae were manually injected in order to introduce a greater viral load than would normally enter the host during natural parasitization. Here, instead of total virus, the unique circular genome segment GiPDVsegF (=GiBV25) was utilized as a probe. This probe was selected because DNA from GiPDVsegF (=GiBV25) had been shown to integrate chromosomal cellular DNA band, with faint circular episomal virus still detectable at day 14 post infection (Fig. 3). The associated GiBV genes encoded on this genome segment were also expressed *in vivo* long term, as all but one of nine genes (eight protein tyrosine phosphatase (PTP), one ankyrin (ank)) encoded on this GiBV circular genome segment continued to be expressed 13+ days post parasitization (Gundersen-Rindal and Pedroni, 2006), even several days after emergence (day 10 post parasitization) of the larval endoparasitoid.

VERTICAL AND HORIZONTAL TRANSFER OF POLYDNAVIRUS DNAs WITH INSECT HOST(S)

Vertical and horizontal transfers within the unique lifestyle and the intimacy of relationships in parasitism are important to the associations that may have potential to result in genetic interchange between polydnavirus and insect hosts: polydnavirus acquisition of foreign genes, polydnavirus loss or transfer of replication-associated genes to the primary parasitoid host, or integration of polydnavirus gene sequence in the primary or the secondary host. Instances of polydnavirus, transposable element, or foreign DNA genetic exchange with and integration into cellular DNA of primary or secondary insect host could result from passive or random horizontal transfers, homologous recombination, or non-homologous end joining. Integration could be facilitated by inherent polydnavirus sequence and structure, or by parasitoid or secondary lepidopteran host genome-encoded replication, repair, and/or integration-related proteins.

The intimate eukaryotic cellular environment in parasitism containing bacteria or other viruses may sometimes result in a passive, horizontal, or reciprocal transfer of genetic information, where polydnaviral genetic material could become integrated within or exchanged with host cellular DNA. In fact, polydnaviral or any other DNA reaching the nucleus has a chance of becoming integrated into the host nuclear DNA. An example of passive horizontal transfer in insects has been described for nucleocytoplasmic large DNA viruses (NCLDV), where the intimate environment in the insect host allows reciprocal passive horizontal transfers between viral genomes, host chromosomes, and other existing microbial DNA, resulting in genes being acquired in the NCLDV or the insect host (Bigot *et al.*, 2008).

The integration and physical strand invasion joining polydnavirus DNAs to those of the secondary host might also be facilitated by both the inherent sequence and structure of the integrating polydnavirus genome segment(s), in conjunction with host sequence features favoring DNA uncoupling or instability, such as AT-rich genome sequences. Polydnavirus dsDNA circular genome segment sequences commonly contain internal repeated sequences, direct

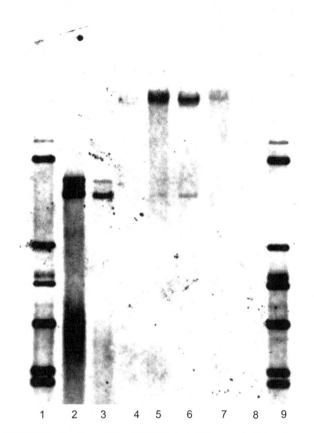

FIGURE 3   **Southern blot of GiBV-injected** *Lymantria dispar* **(gypsy moth) larval DNAs hybridized to GiPDVsegF (=GiBV25) digoxigenin-labeled probe.** Field inversion gel electrophoresis was conducted using 0.8% pulsed filed grade agarose (BioRad) in 0.5 × TBE buffer for 20h at 4°C; with switch times 180v 0.1s, 120v 0.1s. Numerous early third-instar larvae were manually injected with *G. indiensis* parasitoid calyx fluid (1, 2, or 3 μl, injecting greater virus dose than would be delivered during natural parasitization). Haemolymph was removed and extracted separately to remove episomal virions that may have been still circulating in haemolymph. The remaining body was ground and extracted at specified days post injection (pi). DNA extraction was conducted by standard methods; 5 μg DNA was loaded per lane. **Lane 1**: dig-labeled marker II (Roche); **lane 2**: GiBV 10 ng; **lane 3**: day 0; **lane 4**: day 5 pi; **lane 5**: day 14 pi; **lane 6**: day 14 pi; **lane 7**: day 21 pi; **lane 8**: non-injected control; **lane 9**: dig-labeled marker II (Roche).

in vitro into *L. dispar* cellular DNAs following exposure to calyx fluid containing virions in culture (Gundersen-Rindal and Dougherty, 2000; Gundersen-Rindal and Lynn, 2003). DNAs were extracted from the injected whole larvae, after haemolymph removal, at several time points post injection. They were then separated by field inversion gel electrophoresis, and probed in Southern hybridization with the labeled integrative genome segment GiPDVsegF (=GiBV25). Results (Fig. 3) strongly supported *in vivo* chromosomal integration. The episomal pattern indicative of the circular form GiBV viral segment dsDNA was seen immediately post viral injection; however, with increasing time post injection, strong hybridization of GiPDVsegF was noted with the high-molecular-weight *L. dispar* larvae repeats, inverted repeats, sub-terminal inverted repeats, or other sequence features, distinct from the repeated and highly conserved direct repeat excision site motifs of braconid proviral genome segments. Perhaps polydnavirus genome segments carry binding site(s), replication origins, or recognition sequences that facilitate their replication or act for recognition for integration in host DNA. These have potential to contribute to mobility and integration of viral sequences *in vitro* or *in vivo*, as may recipient host sequence features facilitating DNA breakage and instability.

Integrations may also result from host-directed horizontal transfer, via self- or transposable element-encoded replication/integration genes. Interactions with host-encoded genes for replication and integration may have been utilized in passive or active horizontal transfers of genetic sequence among polydnavirus forms and insect host(s) and may have had effects in shaping gene content. Genes abundant in polydnavirus genomes and having sequence similarity to eukaryotic genes are thought to be of eukaryotic origin (Dupuy *et al.*, 2006). Insects carry a wide variety of transposable elements encoding replication or integration genes that may have had roles in exchange; there are many examples regarding insect viruses. A mutant baculovirus contained a functional long terminal repeat (LTR) Gypsy-like retrotransposon (Errantivirus), TED, that may have been acquired by horizontal transfer between lepidopteran species via virus infection (Miller and Miller, 1982). Likewise, the TTAA-specific piggyback transposon originally discovered in association within a baculovirus was believed to have been transferred horizontally from the lepidopteran cell line TN-368 in which the virus was growing (Fraser *et al.*, 1983). A gene expressed during parasitism of *Heliothis virescens* likely involved in alteration of host physiology, an aspartyl protease of retroviral origin, was identified within the *Toxoneuron nigriceps* bracovirus genome (Falabella *et al.*, 2003). As Drezen *et al.* (2006) pointed out, genes acquired from the domestication of retrotransposable elements could have acquired a cellular function secondarily (Nouaud *et al.*, 1999) or could have become fixed and no longer autonomous. Horizontal transfers between baculovirus and polydnavirus have been described in CcBV, GiBV, and other bracovirus genomes, which contain genes encoding *Autographa californica* multinucleocapsid nuclear polyhedrosis virus (AcMNPV) P94-related proteins. Additional horizontal transfers of insect genes to virus have likely occurred, such genes may have conferred a benefit and been retained. Hughes and Friedman (2003) described evidence for horizontal transfer of IAP and UDP-glucosyltransferase genes from insect to baculovirus at a time after infection of the ancestor baculovirus, which were likely retained in the baculovirus because they improved adaptation to insect hosts. Two *Glyptapanteles* bracovirus genome segment homologs were shown to encode a novel family of eight genes of major

facilitator superfamily (MFS) transporter proteins, which had high-level sequence homology to insect sugar transporter genes, but which had not been described within other bracovirus genomes (Desjardins *et al.*, 2008). These genes were phylogenetically related to hymenopteran sugar transporter genes and did not group with those of lepidopteran *Bombyx mori*, thus it is likely these were acquired into the bracovirus proviral form from the parasitoid.

JUMPING GENES: A WIDE ARRAY OF TRANSPOSABLE ELEMENT ASSOCIATIONS WITH POLYDNAVIRUSES

Transposable elements are found in all kingdoms and, like some viruses, are parasitic and integrating. Transposable elements (or transposons) in eukaryotes have been divided into two major classes: retrotransposons (class 1, mobilize via RNA intermediates) and DNA transposons (class 2, mobilize via DNA intermediates), with retrotransposons further divided into non-LTR and LTR retrotransposons (Kapitonov and Jurka, 2008). LTR retrotransposons include Copia, BEL/pao, Gypsy, endogenous retroviruses and DIRS elements, while DNA transposons include 'cut and paste' transposons (Mutator), self-replicating transposons (Polinton/Maverick), rolling circle transposons (Helitron), and tyrosine recombinase transposons (Crypton). In eukaryotes, LTR retrotransposons such as Copia, BEL/pao and Gypsy integrate their DNA copies into the host genome using the transposase integrase (Bao *et al.*, 2010) and some produce virus-like particles. The Gypsy integrase-like transposase is also conserved in the Polinton/Maverick self-synthesizing DNA transposable elements and in 'cut and paste' DNA transposable elements (Kapitonov and Jurka, 2006).

A wide array of distinctive transposable elements has recently been identified in association with both integrated parasitoid proviral and circular encapsidated polydnavirus forms. The relationship of polydnaviruses with transposable elements is not new, and their proviral genomes have been associated with a diversity of transposable elements over the course of evolution. Through greater than 103 million years (Murphy *et al.*, 2008), there has been a great deal of opportunity for exposure of integrated bracovirus and ichnovirus proviral forms to genetically mobile transposable elements carried in connection with both primary parasitoid and secondary host insects, or with their associated viruses and other microbes. Horizontally transferred elements and/or the genetic material they carried that were acquired in an ancestor polydnavirus may have been selectively retained, maintained, and transmitted vertically through the course of parasitoid evolution, contributing to evolution and diversification in polydnavirus-bearing taxa. Prior to the Bézier *et al.* (2009a) studies demonstrating bracovirus origin in ancestral nudiviruses, researchers hypothesized heterodisperse polydnaviruses (both bracoviruses and ichnoviruses) could have arisen in association with their parasitoid from wasp genes and mobile elements that evolved viral form during horizontal movement to another insect. (Summers and Dib-Hajj, 1995; Cui and Webb, 1997; Drezen et al., 2003, 2006; Dupuy et al., 2006). This was supported by limited data, although specific features of polydnaviruses, for example the terminal repeated sequences and conserved excision and circularization sites associated with integrated bracoviral genome segments, and the loss of one excision site as the ds circular genome segment was encapsidated, were reminiscent of transposable elements which often result in target site duplication post insertion in host chromosomes.

Both bracovirus and ichnovirus genomes show evidence for transposable elements that suggest they have received multiple insertions, deletions, and rearrangements that may have affected proviral organization or gene content, and could have been retained in the circular encapsidated form. Significantly, 3.8% of the CcBV encapsidated genome is represented by genes that resemble retrovirus-like elements (Espagne et al., 2004), though very few GiBV or GfBV segments encode retrovirus-like elements (Desjardins et al., 2008). The comparative genomics of available integrated provirus and circular encapsidated virus data suggests both forms of polydnavirus have been subjected to numerous integration, excision, and transposition events. Transposable elements thus have potential to have played roles in acquisition of pathogenicity genes into the proviral genome or in reduction within provirus genomes, where replication-associated and structural genes may have been transferred to the parasitoid host (Bézier et al., 2009a). Recently, in addition to transposable elements from both major classes, new classes of transposable elements have been identified in insect and proviral genomes. In this regard, as suggested by Drezen et al. (2006), domestications of transposable elements and retroelements may have had roles in horizontal transfer of transposable elements between hymenopteran/parasitoid species and lepidopteran species via polydnavirus infection. Alternatively, these transposable elements could simply represent selfish elements that are randomly integrated.

Maverick DNA transposable elements are one novel class that has been identified in association with proviral genomes. Mavericks encode their own cellular-integrase, create a target site duplication of 5–6 bp upon integration, display long terminal-inverted repeats, and encode a number of genes with homology to replication and packaging proteins of some bacteriophages and eukaryotic dsDNA viruses (Pritham et al., 2006). Drezen et al. (2006) described the presence in CcBV of a Maverick element into which there was insertion of both a DIRS LTR-retrotransposon and a Dong non-LTR-retrotranposon. Just upstream of the direct repeat of the DIRS element, an N-terminal region of a second distinct Maverick was observed, suggesting the original

Maverick element had been subjected to secondary element insertions that had potentially truncated or rearranged the original (Dupuy et al., 2011). Similarly, Desjardins et al. (2008) noted within integrated G. flavicoxis proviral locus 7 (containing two GiBV proviral genome segments: segment 28 and partial segment 27 flanked on one side by approximately 95 kbp of non-GfBV) a large Maverick element that contained both a Gypsy LTR-retrotransposon and Mariner-like element. Desjardins et al. (2008) further hypothesized that a Maverick element present within a parasitoid wasp ancestor containing the ancestral bracovirus genome may have captured virulence genes from the parasitoid genome and transferred them to the proviral genome. Whether Maverick elements play a role in bracovirus biology is unknown, but there are similarities between hypothesized methods of replication in Mavericks (Kapitonov and Jurka, 2006) and polydnaviruses, as both are thought to replicate extrachromosomally from a circular or stem-loop molecule. These and other transposable elements could have served to move core viral genes out of the ancestral bracovirus genome to the host genome and/or virulence genes into it; alternatively genes could have moved out of the proviral genome by degradation of proviral genome segment excision conserved motifs, creating integrated polydnaviral genes no longer encapsidated (Desjardins et al., 2008). It is also possible that Mavericks are simply integrating randomly within the parasitoid genome as selfish elements and have rested within proviral regions.

In addition to Maverick elements, Bézier et al. (2009a) showed many remnants of transposable elements within the C. congregata integrated nudiviral gene clusters containing structural and CcBV-replication-associated genes in the parasitoid that were not packaged in virion particles, suggesting transposable elements presence in the vicinity of genes potentially lost to the host. This was also supported by Desjardins et al. (2008). Recently, Phantom/Mutator transposable elements were identified in association with polydnavirus genomes (Marquez and Pritham, 2010), where associations with Phantom-like element transposase sequences were described after analyses of both C. inanitus bracovirus and the banchine G. fumiferana polydnavirus sequences. Phantom and Mutator elements are widespread in eukaryotes and have significant terminal inverted repeats in various arrangements. The Phantom-like element transposases associated with C. inanitus and G. fumiferana bore homology and were closely related to hymenopteran wasp (N. vitripennis) phantom-like element transposase, suggesting a potential horizontal transfer of Phantom-like element to polydnavirus from the parasitoid host.

Recently, Thomas et al. (2010) identified yet another novel class (non class 1 or 2) of DNA transposable elements called Helitrons associated with bracovirus proviral genomes. They are apparently related to genetic elements that replicate through a rolling-circle mechanism, including single-strand DNA plant geminiviruses that may have originated from or given rise to them (Feschotte and Wessler, 2001; Kapitonov and Jurka, 2001; Murad et al., 2004). In addition, Helitrons lack terminal repeats, do not duplicate host sequence upon insertion, and do not use transposase. Helitrons function as 'exon shuffling machines' as they can recombine exons from multiple genes and shuttle gene fragments between genomes (Feschotte and Wessler 2001; Pritham and Freschotte, 2007). Horizontally transferred Helitrons are frequently found in insect genomes, and Helitron horizontal transfer in insect viruses could potentially act as shuttle systems for the delivery of DNA between species (Loreto et al., 2008). Helitrons can have major roles in the evolution of host genomes as they frequently capture diverse host genes, some of which can evolve into novel genes or become essential for Helitron transposition (reviewed in Kapitonov and Jurka, 2007). Schaak et al. (2010) recently identified the presence of Helitron elements within Cotesia sesamiae genomic regions containing integrated CsBV proviral genome segments originally sequenced in the Desjardins et al. study (2008). C. sesamiae is a parasitoid found in sub-Saharan Africa which exists in at least two biotypes and parasitizes two distinct lepidopteran stem borer species (Gitau et al., 2007; Dupas et al., 2008). Specifically, Helitrons belonging to families Heligloria and Helisimi were identified within the braconid C. sesamiae Mombasa (coastal biotype) and Kitale (inland biotype) integrated provirus genome; one copy intact with captured host genomic sequence and a secondary Tc1 mariner element insertion, and one copy truncated at the 5′ end were present within the Mombasa biotype. Helitron elements belonging to these same two Helitron families, Heligloria and Helisimi, were also identified in the analyses of Thomas et al. (2010), within the public full genome sequences for B. mori, currently the only substantially sequenced lepidopteran genome. Recently, Coates et al. (2010) showed that Helitron and Helitron-like transposable elements appear to be common in lepidopterans. The similar transposable elements within parasitoid and lepidopteran insect genomic sequences support the notion that the lepidopteran secondary host they infect may act as a means for horizontal transfer between the two. In this regard, analyses of full-genome sequence data for lepidopteran secondary hosts involved in some of the well-characterized parasitoid/polydnavirus/lepidopteran interactions will become valuable for studying this potential route of lateral exchange.

REVISITING DEMONSTRATED POLYDNAVIRUS CELLULAR DNA INTEGRATION

Doucet et al. (2007) ichnovirus DNA integration junction sequences in Cf-124T cells in vitro showed no association of integrated TrIV with any identifiable transposable element in host sequence, though an $(ATTCT)_2$ repeat was lost from the integrated TrIV genome segment post integration. Data for GiPDVsegF (=GiBV25), shown to integrate into host L. dispar cells in vitro (and in vivo as described; Fig. 3), did show an association with a transposable element (Gundersen-Rindal and Lynn, 2003). The L. dispar cellular DNA sequence at the junction site of integration contained evidence for a host retrotransposable element sequences with homology to B. mori LTR retrotransposons, pao-like, and mag-like (GenBank AB126055), a close relative of Gypsy, encoding tranposase and integrase, which could potentially have facilitated polydnavirus sequence integration in vitro. Further, comparison of the DNA sequences at the site of integration and linearization on this dsDNA circular genome segment GiPDVsegF (=GiBV25) with the conserved excision and circularization highly conserved motif site (GCTT) that is conserved for most bracovirus genome segments (Annaheim and Lanzrein, 2007; Desjardins et al., 2007, 2008) showed they were not the same sequences. This was in contrast to a prior hypothesis suggesting they may be, conserved, at least for this circular segment and this site of in vitro integration into lepidopteran cellular DNA.

Transposing 'jumping genes' acquired into the integrated provirus from the primary parasitoid, or as suggested earlier, from the secondary lepidopteran host, sometimes mediated by transposable element insertion, may have become stabilized as part of the circular encapsidated form of polydnaviral genomes that were then injected to deliver pathogenicity genes to the secondary host during parasitization. There is an example of this potential within the G. indiensis parasitoid provirus BAC clone sequences analyzed in the Desjardins et al. (2008) study of integrated provirus. Two G. indiensis BAC clones were isolated that overlapped each other for a large stretch of chromosome. In the overlap region, one BAC (clone 1C20) contained a transposable element (class 2, Drosophila p-element related, though larger and having distinct direct repeat sequences) while the other BAC clone (clone 2C5) from the same chromosomal region did not contain any element. The p-like transposable element encoded a transposase but no other genes. The consensus GiBV circular encapsidated viral genome segment 11 sequence excised from this chromosomal region did not contain the p-like transposable element sequence; however, the sequence for this transposable element did appear within whole genome shotgun reads used for the assembly of GiBV circular encapsidated genome segment sequences. The presence of this transposable element in the G. indiensis provirus strongly suggests that transposable element(s) could enter a bracovirus genome, become stabilized, and become packaged successfully into the circular encapsidated form (Desjardins et al., 2008).

Of note, the large GiBV genome segment sequence described by Gundersen-Rindal and Dougherty (2000) as clone p384, for which integration in two different *L. dispar* host cell lines *in vitro* was illustrated by the presence of off-size fragments in the host cellular DNA on Southern hybridization, represents the aforementioned GiBV segment 11, which encodes numerous hypothetical proteins. It is possible that a transposable element(s) was associated with this viral genome segment within the population of *G. indiensis* parasitoids virus that was used to infect the host cell lines, which may have facilitated cellular integration of at least part of the GiBV genome segment 11 *in vitro*.

Associations with transposable elements may have played roles in polydnavirus evolution, particularly in directing gene content, especially considering some characteristics of the newly identified DNA transposable elements described above. For example, 'cut and paste' transposable elements have a simple structure in which autonomous elements carry a single transposase gene flanked on either side by transposase binding sites, which may be embedded within terminal inverted repeats (Craig *et al.*, 2002; Feschotte and Pritham, 2007). Nonautonomous elements do not encode transposase but instead carry only the binding sites. They are then able to move *in trans* using transposase encoded by another transposable element located elsewhere in the genome (Craig *et al.* 2002; Feschotte and Pritham, 2007; Marquez and Pritham, 2010). In addition to other similarities, as described earlier, perhaps certain polydnavirus genome segments carry binding site, replication origins, or recognition sequences that facilitate their replication or act for recognition for integration in host DNA. These types of repeated or inverted repeated sequences are common in polydnavirus genome segment sequences. In examining GiPDVsegF (=GiBV25), the observation that the site of this genome segment integration *in vitro* was not the same sequence nor in the same circular segment location as the conserved excision motif of bracovirus genome segments is consistent with the hypothesis there could be internal genome segment binding or recognition sequences encoded within certain polydnavirus dsDNA genome segments. In this case, integration could have been facilitated *in trans* by the transposase or integrase in the vicinity. Alternatively, bracoviruses could carry their own integrase within viral particles since two identified nudivirus-related genes encode products having integrase domains (Bézier *et al.*, 2009a).

FUNCTIONAL SIGNIFICANCE FOR POLYDNAVIRUS INTEGRATION AND MAINTENANCE IN HOST CELLULAR DNA

Many viruses actively integrate their genomes into host chromatin during replication, such as bacteriophages, adenoviruses, papillomaviruses, and many retroviruses. Some viruses can be detected within host chromatin but do not actively integrate their genomes, either because integration

is not required for their replication or they do not encode integration enzymes. There are several examples of such viruses whose DNA sequences can be found integrated within dsDNA genomes of plants and arthropods. Complete or fragmented pararetroviruses and geminiviruses have been found integrated in several plant genomes. Some endogenous pararetroviruses genome integrated forms are activable, and have structural similarities with polydnaviruses in that they are found in episomal and integrated forms in the host, have been observed in clusters, and have preferences for certain chromosomal locations as integration sites (reviewed in Hohn *et al.*, 2008). DNA sequences originating from flaviviruses, nonretro RNA viruses that replicate without a recognized DNA intermediate, have been found integrated in the dsDNA genome of *Aedes albopictus*, *Aedes aegypti*, and other *Aedes* species mosquito populations (Crochu *et al.*, 2004; Roiz *et al.*, 2009). Integration of virus sequences sometimes confers an advantage to the host. For example, integrated DNA sequences from non-retro RNA Israeli acute paralysis virus (IAPV) Dicistrovirus were shown within genomes of certain *Apis mellifera* honeybees, which integration conferred resistance to further IAPV infection (Maori *et al.*, 2007). The dsDNA polydnaviruses are integrated as proviruses in their primary parasitoid hosts as part of their unique lifestyle, but only replicate or become activated within certain ovarian calyx cells. Polydnaviral DNAs can clearly become integrated within dsDNA of their secondary insect host genome (cultured cells as well as whole larvae), yet functional significance for their integration has not been clear.

The integration of polydnavirus sequences in both primary parasitoid host and the secondary insect host (*in vitro* and *in vivo*) suggests an advantage is conferred in both hosts. The simplest explanation for integration and maintenance of polydnavirus as provirus within the parasitoid host is that virus sequences have been subverted and incorporated into the parasitoid genome because expression of those viral gene sequences upon parasitization of the host insect are essential for survival of the parasitoid offspring, and thus confer benefit to the parasitoid. Integration of polydnavirus in provirus form may have contributed this evolutionary advantage in protecting the parasitoid: integration and maintenance could protect the parasitoid from infection by related viruses or transposable elements by an RNA interference (RNAi) type resistance, as it is known to function in the examples above and in many living systems (reviewed in Ding and Voinnett, 2007; and Goic and Saleh, 2010). Integrated polydnavirus sequences may also benefit parasitoid gene expression by acting as promoters or in other aspects of gene regulation, or by contributing to the ability of the parasitoid to modify its host range.

The functional significance for polydnavirus integration *in vitro* (and *in vivo* at least for bracovirus) within secondary host cellular genomic DNA is not as clear. In many cases, the integration event leads to persistent sustained expression of the virus-encoded chromosomally integrated gene(s) in host cells. Polydnavirus distinctive transmission and replication, requiring two separate hosts and resulting in the formation of non-replicative virion particles for introduction into the secondary host, necessitates viral genes be transcribed and continue to be expressed during parasitization for sufficient time to suppress host immunity, impair development, and ensure parasitoid survival. After introduction into the secondary host, polydnavirus circular genome segments persist as episomes for variable periods of time allowing viral gene transcription and expression, whether sustained or transient. Integration may allow the persistence of the appropriate dosage of polydnavirus sequences in the absence of replication in the growing host while viral genome segments are progressively diluted. Note the earlier examples describing *in vitro* integration for ichnovirus (HdIV, K-gene, transcribed and expressed in long-term transformed *Spodoptera* and *Trichoplusia* cells (Volkoff *et al.*, 2001); and TrIV rep gene carried by TrIV genome segment displaying long-term transcription in *C. fumiferana* CF-124T cells (Doucet *et al.*, 2007)) and bracovirus (GiPDVsegF (=GiBVseg25) (Gundersen-Rindal and Lynn, 2003) with long-term transcribed PTP and other genes in *L. dispar* cells) where integrated viral genome segments encoded genes belonging to the numerically most abundant gene families (rep genes in ichnoviruses and PTPs in bracoviruses) (Doucet *et al.*, 2007), which also have key roles in pathogenicity. Though rep gene functions have not been fully defined, they are expressed in tissue-specific fashion and may have tissue-specific functions, including a role in regulation of transcript level in gene dosage (Rasoolizadeh *et al.*, 2009). With integration of the TrIV genome segment F1, the most abundantly expressed rep gene transcript in infected *C. fumiferana* larvae *in vivo*, TrFrep1, also had sustained expression *in vitro* (Doucet *et al.*, 2007; Rasoolizadeh *et al.*, 2009). Likewise, bracovirus PTP genes are expressed in tissue-specific fashion and serve a wide variety of important functions in the disruption of signaling pathways involved in hormone biosynthesis or haemocyte cytoskeleton dynamics (reviewed in Moreau *et al.*, 2009) and PTPs are expressed *in vitro* in long term Ld cells containing integrated GiBV DNA. Thus, as suggested by Doucet *et al.* (2007), the fact that integrated genome segments involve numerically and functionally significant genes may be a reason for integration, for continued sustained expression of important horizontally transferred and subsequently integrated polydnavirus genes within the secondary host after episomal copies have decreased.

For bracoviruses, the integration and maintenance of virus in cellular DNA of the lepidopteran secondary host cells may reflect the evolutionary relationship to ancestral nudiviruses as well as antiviral evolution of the host. The *Heliothis zea* nudivirus 1 (HzNV-1), which encodes replication machinery and can be reactivated to produce infection, is able to integrate its DNA into host chromosomes and establish persistent infection within numerous lepidopteran cell lines, which are subsequently resistant to superinfection with the same virus (Lin *et al.*, 1999; Wu *et al.*, 2010). Integration of polydnavirus sequences within the secondary host cellular genome might activate an RNAi-based defense (or counterdefense) to prevent superinfection by related viruses. Of the many defense systems in insects, RNA interference is the major mechanism evolved to combat and confer protection against pathogens, particularly against viruses and transposable elements, which are abundant in these insect systems. RNA interference in insects can also regulate certain aspects of host innate immunity by modulating gene expression (Goic and Saleh, 2010). Immune system genes and the viruses and transposable elements interact closely and may evolve rapidly, which is thought to result from a co evolutionary arms race in which there are reciprocal processes of adaptation and counter-adaptation between pathogens and host (Ding and Voinnet, 2007). Integration may be an important part of these processes.

The close relationships among polydnavirus, primary parasitoid host and secondary parasitized insect host and the impact of polydnaviral integration on the evolution of their genomes is complex. Clearly there is much to discover with regard to polydnavirus integration within host cellular DNAs and the many possibilities for gene exchange in parasitoid/viral/lepidopteran host relationships among all parties involved in their mutual evolution.

REFERENCES

Annaheim, M., Lanzrein, B., 2007. Genome organization of the *Chelonus inanitus* polydnavirus: excision sites, spacers and abundance of proviral and excised segments. J. Gen. Virol. 88, 450–457.

Asgari, S., Hellers, M., Schmidt, O., 1996. Host haemocyte inactivation by an insect parasitoid: transient expression of a polydnavirus gene. J. Gen. Virol. 77, 2653–2662.

Asgari, S., Schmidt, O., Theopold, U., 1997. A polydnavirus-encoded protein of an endoparasitoid wasp is an immune suppressor. J. Gen. Virol. 78, 3061–3070.

Bao, W., Kapitonov, V., Jurka, J., 2010. Ginger DNA transposons in eukaryotes and their evolutionary relationships with long terminal repeat retrotransposons. Mobile DNA 1, 3.

Beckage, N.E., 1985. Endocrine interactions between endoparasitic insects and their hosts. Annu. Rev. Entomol. 30, 371–413.

Béliveau, C., Laforge, M., Cusson, M., Bellemare, G., 2000. Expression of a *Tranosema rostrale* polydnavirus gene in the spruce budworm, *Choristoneura fumiferana*. J. Gen. Virol. 81, 1871–1880.

Belle, E., Beckage, N.E., Rousselet, J., Poirié, M., Lemeunier, F., Drezen, J.-M., 2002. Visualization of polydnavirus sequences in a parasitoid wasp chromosome. J. Virol. 76, 5793–5796.

Bézier, A., Annaheim, M., Herbinière, J., Wetterwald, C., Gyapay, G., et al. 2009. Polydnaviruses of braconid wasps derive from an ancestral nudivirus. Science 323, 926–930.

Bézier, A., Herbinière, J., Lanzrein, B., Drezen, J.-M., 2009. Polydnavirus hidden face: the genes producing virus particles of parasitic wasps. J. Invertebr. Pathol. 101, 194–203.

Bigot, Y., Samain, S., Augé-Gouillou, C., Federici, B.A., 2008. Molecular evidence for the evolution of ichnoviruses from ascoviruses by symbiogenesis. BMC Evol. Biol. 8, 253.

Blissard, G.W., Vinson, S.B., Summers, M.D., 1986. Identification, mapping, and in vitro translation of Campoletis sonorensis virus mRNAs from parasitized Heliothis virescens larvae. J. Virol. 57, 318–327.

Coates, B.S., Sumerford, D.V., Hellmich, R.L., Lewis, L.C., 2010. A helitron-like transposon superfamily from lepidoptera disrupts (GAAA)n microsatellites and is responsible for flanking sequence similarity within a microsatellite family. J. Mol. Evol. 70, 275–288.

Craig, N.L., Craigie, R., Gellert, M., Lambowitz, A.M., 2002. Mobile DNA II. American Society for Microbiology Press, Washington, D.C.

Crochu, S., Cook, S., Attoui, H., Charrel, R.N., De Chesse, R., Belhouchet, M., et al. 2004. Sequences of flavivirus-related RNA viruses persist in DNA form integrated in the genome of Aedes spp. mosquitoes. Gen. Virol. 85, 1971–1980.

Cui, L., Webb, B.A., 1997. Homologous sequences in the *Campoletis sonorensis* PDV genome are implicated in replication and nesting of the W segment family. J. Virol. 71, 8504–8513.

Cusson, M., Laforge, M., Miller, D., Cloutier, C., Stolz, D., 2000. Functional significance of parasitism-induced suppression of juvenile hormone esterase activity in developmentally delayed Choristoneura fumiferana larvae. Gen. Comp. Endocrinol. 117, 343–354.

Davies, D.H., Strand, M.R., Vinson, S.B., 1987. Changes in differential haemocyte count and in vitro behaviour of plasmatocytes from host Heliothis virescens caused by Campoletis sonorensis polydnavirus. J. Insect Physiol. 33, 143–153.

Desjardins, C.A., Gundersen-Rindal, D.E., Hostetler, J.B., Tallon, L.J., Fadrosh, D.W., Fuester, R.W., et al. 2008. Comparative genomics of mutualistic viruses of Glyptapanteles parasitic wasps. Genome Biol. 9, R183.

Desjardins, C.A., Gundersen-Rindal, D.E., Hostetler, J.B., Tallon, L.J., Fuester, R.W., Schatz, M.C., et al. 2007. Structure and evolution of a proviral locus of Glyptapanteles indiensis bracovirus. BMC Microbiol. 7, 61.

Ding, S.W., Voinnet, O., 2007. Antiviral immunity directed by small RNAs. Cell 130 (3), 413–426.

Doucet, D., Levasseur, A., Béliveau, C., Lapointe, R., Stoltz, D., Cusson, M., 2007. In vitro integration of an ichnovirus genome segment into the genomic DNA of lepidopteran cells. J. Gen. Virol. 88, 105–113.

Dover, B.A., Davies, D.H., Strand, M.R., Gary, R.S., Keeley, L.L., Vinson, S.B., 1987. Ecdysteroid-titer reduction and developmental arrest of last-instar Heliothis virescens larvae by calyx fluid from the parasitoid Campoletis sonorensis. J. Insect Physiol. 33, 333–338.

Dover, B.A., Davies, D.H., Vinson, S.B., 1988. Degeneration of last-instar Heliothis virescens prothoracic glands by Campoletis sonorensis polydnavirus. J. Invert Pathol. 51, 80–91.

Drezen, J.-M., Bézier, A., Lesobre, J., Huguet, E., Cattolico, L., Périquet, G., et al. 2006. The few virus-like genes of *Cotesia congregata*. Arch. Insect Biochem. Phys. 61, 110–122.

Drezen, J.-M., Provost, B., Espagne, E., Cattolico, L., Dupuy, C., Poirié, M., et al. 2003. Polydnavirus genome: integrated vs. free virus. J. Insect Physiol. 49, 407–417.

Dupas, S., Gitau, C.W., Branca, A., Le Rü, B., Silvain, J.-F., 2008. Evolution of a polydnavirus gene in relation to parasitoid–host species immune resistance. J. Hered. 99, 491–499.

Dupuy, C., Huguet, E., Drezen, J.-M., 2006. Unfolding the evolutionary story of polydnaviruses. Virus Res. 117, 81–89.

Dupuy, C., Periquet, G., Serbielle, C., Bézier, A., Louis, F., Drezen, J.-M., 2011. Transfer of a chromosomal Maverick to endogenous bracovirus in a parasitoid wasp. Genetica 139, 489–496.

Edson, K.M., Vinson, S.B., Stoltz, D.B., Summers, M.D., 1981. Virus in a parasitoid wasp: suppression of the cellular immune response in the parasitoid's host. Science 211, 582–583.

Espagne, E., Dupuy, C., Huguet, E., Cattolico, L., Provost, B., Martins, N., et al. 2004. Genome sequence of a polydnavirus: insights into symbiotic virus evolution. Science 306, 286–289.

Falabella, P., Varricchio, P., Gigliotti, S., Tranfaglia, A., Pennacchio, F., Malva, C., 2003. Toxoneuron nigriceps polydnavirus encodes a putative aspartyl protease highly expressed in parasitized host larvae. Insect Mol. Biol. 12, 9–17.

Federici, B.A., Bigot, Y., 2003. Origin and evolution of polydnaviruses by symbiogenesis of insect DNA viruses in endoparasitic wasps. J. Insect Physiol. 49, 419–432.

Feschotte, C., Pritham, E.J., 2007. DNA Transposons and the evolution of eukaryotic genomes. Annu. Rev. Genet. 41, 331–368.

Feschotte, C., Wessler, S.R., 2001. Treasures in the attic: rolling circle transposons discovered in eukaryotic genomes. Proc. Natl. Acad. Sci. U.S.A. 98, 8923–8924.

Fleming, J.A.G.W., 1991. The integration of polydnavirus genomes in parasitoid genomes: implications for biocontrol and genetic analyses of parasitoid wasps. Biol. Control 1, 127–135.

Fleming, J.A.G.W., Krell, P.J., 1993. Polydnavirus genome organization. In: Beckage, N.E., Thompson, S.N., Federici, B.A. (Eds.), Parasites and Pathogens of Insects. Vol. 1: Parasites (pp. 189–225). Academic Press, San Diego.

Fleming, J.A.G.W., Summers, M.D., 1986. *Campoletis sonorensis* endoparasitic wasps contain forms of *C. sonorensis* virus DNA suggestive of integrated and extrachromosomal polydnavirus DNAs. J. Virol. 57, 552–562.

Fleming, J.A.G.W., Summers, M.D., 1990. The integration of the genome of a segmented DNA virus in the host insect's genome. In: Hagedorn, H.H., Hildebrand, J.G., Kidwell, M.G., Law, J.H. (Eds.), Molecular Insect Science (pp. 99–105). Plenum Press, New York.

Fleming, J.A.G.W., Summers, M.D., 1991. Polydnavirus DNA is integrated in the DNA of its parasitoid wasp host. Proc. Natl. Acad. Sci. U.S.A. 88, 9770–9774.

Fraser, M.J., Smith, G.E., Summers, M.D., 1983. The acquisition of host cell DNA sequences by Baculoviruses: Relation between host DNA insertions and FP mutants of *Autographa californica* and *Galleria mellonella* NPVs. J. Virol. 47, 287–300.

Gitau, C.W., Gundersen-Rindal, D., Pedroni, M., Mbugi, P.J., Dupas, S., 2007. Differential expression of the CrV1 haemocyte inactivation-associated polydnavirus gene in the African maize stem borer *Busseola fusca* (Fuller) parasitized by two biotypes of the endoparasitoid *Cotesia sesamiae* (Cameron). J. Insect Physiol. 53, 676–684.

Goic, B., Saleh, M.-C., 2010. RNAi: an antiviral defence system in insects. In: Martinez., M.A. (Ed.), RNA Interference and Viruses: Current Innovations and Future Trends (pp. 1–24). Caister Academic Press, Norfolk.

Gruber, A., Stettler, P., Heiniger, P., Schümperli, D., Lanzrein, B., 1996. Polydnavirus DNA of the braconid wasp Chelonus inanitus is integrated in the wasp's genome and excised only in later pupal and adult stages of the female. J. Gen. Virol. 77, 2873–2879.

Gundersen-Rindal, D., Dougherty, E.M., 2000. Evidence for integration of *Glyptapanteles indiensis* polydnavirus DNA into the chromosome of *Lymantria dispar* in vitro. Virus Res. 66, 27–37.

Gundersen-Rindal, D.E., Lynn, D.E., 2003. Polydnavirus integration in lepidopteran host cells *in vitro*. J. Insect Physiol. 49, 453–462.

Gundersen-Rindal, D., Lynn, D.E., Dougherty, E.M., 1999. Transformation of lepidopteran and coleopteran insect cell lines by *Glyptapanteles indiensis* polydnavirus DNA. In Vitro Cell. Dev. Biol. 35, 111–114.

Gundersen-Rindal, D.E., Pedroni, M.J., 2006. Characterization and transcriptional analysis of protein tyrosine phosphatase genes and an ankyrin repeat gene of the parasitoid *Glyptapanteles indiensis* polydnavirus in the parasitized host. J. Gen. Virol. 87, 311–322.

Hohn, T., Richert–Poggeler, K.R., Staginnus, C., Harper, G., Schwarzacher, T., Teo, C.H., et al. 2008. Evolution of integrated plant viruses. In: Roossinck, M.J. (Ed.), Plant Virus Evolution. Springer-Verlag, Berlin.

Hughes, A.L., Friedman, R., 2003. Genome-wide survey for genes horizontally transferred from cellular organisms to baculoviruses. Mol. Biol. Evol. 20, 979–987.

Johnson, J.A., Bitra, K., Zhang, S., Wang, L., Lynn, D.E., Strand, M.R., 2010. The UGA-CiE1 cell line from Chrysodeixis includens exhibits characteristics of granulocytes and is permissive to infection by two viruses. Insect Biochem. Mol. Biol. 40, 394–404.

Kapitonov, V.V., Jurka, J., 2001. Rolling-circle transposons in eukaryotes. Proc. Natl. Acad. Sci. U.S.A. 98, 8714–8719.

Kapitonov, V.V., Jurka, J., 2006. Self-synthesizing DNA transposons in eukaryotes. Proc. Natl. Acad. Sci. U.S.A. 103, 4540–4545.

Kapitonov, V.V., Jurka, J., 2007. Helitrons on a roll: eukaryotic rolling-circle transposons. Trends Genet. 23, 521–529.

Kapitonov, V.V., Jurka, J., 2008. A universal classification of eukaryotic transposable elements implemented in Repbase. Nat. Rev. Genet. 9, 411–412.

Kim, M.-K., Sisson, G., Stoltz, D.B., 1996. Ichnovirus infection of an established gypsy moth cell line. J. Gen. Virol. 77, 2321–2328.

Krell, P.J., Stoltz, D.B., 1979. An unusual baculovirus of the parasitoid wasp *Apanteles melanoscelus*: isolation and characterization. J. Virol. 29, 1118–1130.

Lapointe, R., Tanaka, K., Barney, W., Whitfield, J., Banks, J., Béliveau, C., et al. 2007. Genomic and morphological features of a banchine polydnavirus: comparison with bracoviruses and ichnoviruses. J. Virol. 81, 6491–6501.

Lavine, M.D., Beckage, N.E., 1995. Polydnaviruses: potent mediators of host insect immune dysfunction. Parasitol. Today 11, 368–378.

Lavine, M.D., Beckage, N.E., 1996. Temporal pattern of parasitism-induced immunosuppression in Manduca sexta larvae parasitized by Cotesia congregata. J. Insect Physiol. 42, 41–51.

Lawrence, P.O., Lanzrein, B., 1993. Hormonal interactions between insect endoparasites and their host insects. In: Beckage, N.E., Thompson, S.N., Federici, B.A. (Eds.), Parasites and Pathogens of Insects, Vol. 1: Parasites (pp. 59–86). Academic Press, San Diego.

Le, N.T., Asgari, S., Amaya, K., Tan, F.F., Beckage, N.E., 2003. Persistence and expression of *Cotesia congregata* polydnavirus in host larvae of the tobacco hornworm, *Manduca sexta*. J. Insect Physiol. 49, 533–543.

Li, X., Webb, B.A., 1994. Apparent functional role for a cysteine-rich polydnavirus protein in suppression of the insect cellular immune response. J. Virol. 68, 7482–7489.

Lin, C.L., Lee, J.C., Chen, S.S., Wood, H.A., Li, M.L., Li, C.F., et al. 1999. Persistent Hz-1 virus infection in insect cells: evidence for insertion of viral DNA into host chromosomes and viral infection in a latent status. J. Virol. 73, 128–139.

Loreto, E.L.S., Carareto, C.M.A., Capy, P., 2008. Revisiting horizontal transfer of transposable elements in *Drosophila*. Heredity 100, 545–554.

McKelvey, T.A., Lynn, D.E., Gundersen-Rindal, D., Guzo, D., Stoltz, D.A., Guthrie, K.P., et al. 1996. Transformation of gypsy moth (*Lymantria dispar*) cell lines by infection with *Glyptapanteles indiensis* polydnavirus. Biochem. Biophys. Res. Comm. 225, 764–770.

Maori, E., Tanne, E., Sela, I., 2007. Reciprocal sequence exchange between non-retro viruses and hosts leading to the appearance of new host phenotypes. Virology 362, 342–349.

Marquez, C.P., Pritham E.J., 2010. Phantom, a new subclass of Mutator DNA transposons found in insect viruses and widely distributed in animals. E-print ahead of pub 10.1534/genetics.110.116673.

Miller, D.W., Miller, L.K., 1982. A virus mutant with an insertion of copia-like transposable element. Nature (London) 299, 562–564.

Moreau, S., Huguet, E., Drezen, J.-M., 2009. Polydnaviruses as tools to deliver wasp virulence factors to impair lepidopteran host immunity. In: Rolff, J., Reynolds., S. (Eds.), Insect Infection and Immunity (pp. 137–158). Oxford University Press, Oxford.

Murad, L., Bielawski, J.P., Matyasek, R., Kovařík, A., Nichols, R.A., Leitch, A.R., 2004. The origin and evolution of geminivirus-related DNA sequences in Nicotiana. Heredity 92, 352–358.

Murphy, N., Banks, J.C., Whitfield, J.B., Austin, A.D., 2008. Phylogeny of the parasitic microgastroid subfamilies (Hymenoptera: Braconidae) based on sequence data from seven genes, with an improved time estimate of the origin of the lineage. Mol. Phylogenet. Evol. 47, 378–395.

Nouaud, D., Boeda, B., Levy, L., Anxolabehere, D., 1999. A P element has induced intron formation in Drosophila. Mol. Biol. Evol. 16, 1503–1510.

Pennacchio, F., Falabella, P., Vinson, S.B., 1998. Regulation of Heliothis virescens prothoracic glands by Cardiochiles nigriceps polydnavirus. Arch. Insect Biochem. Physiol. 38, 1–10.

Pritham, E.J., Feschotte, C., 2007. Massive amplification of rolling-circle transposons in the lineage of the bat *Myotis lucifugus*. Proc. Natl. Acad. Sci. U.S.A. 104, 1895–1900.

Pritham, E.J., Putliwala, T., Feschotte, C., 2006. Mavericks, a novel class of giant transposable elements widespread in eukaryotes and related to DNA viruses. Gene 390, 3–17.

Pruijssers, A.J., Falabella, P., Eum, J.H., Pennacchio, F., Brown, M.R., Strand, M.R., 2009. Infection by a symbiotic polydnavirus induces wasting and inhibits metamorphosis of the moth *Pseudoplusia includens*. J. Exp. Biol. 212 (18), 2998–3006.

Rasoolizadeh, A., Beliveau, C., Stewart, D., Cloutier, C., Cusson, M., 2009. *Tranosema rostrale* ichnovirus repeat element genes display distinct transcriptional patterns in caterpillar and wasp hosts. J. Gen. Virol. 90, 1505–1514.

Rattanadechakul, W., Webb, B.A., 2003. Characterization of *Campoletis sonorensis* ichnovirus unique segment B and excision locus structure. J. Insect Physiol. 49, 523–532.

Rocher, J., Ravallec, M., Barry, P., Volkoff, A.-N., Ray, D., Devauchelle, G., et al. 2004. Establishment of cell lines from the wasp *Hyposoter didymator* (Hym., *Ichneumonidae*) containing the symbiotic polydnavirus *H. didymator* ichnovirus. J. Gen. Virol. 85, 863–868.

Roiz, D., Vázquez, A., Sánchez Seco, M.P., Tenorio, A., Rizzoli, A., 2009. Detection of novel insect flavivirus sequences integrated in *Aedes albopictus* (Diptera: Culicidae) in Northern Italy. Virol. J. 6, 93.

Savary, S., Beckage, N., Tan, F., Periquet, G., Drezen, J.-M., 1997. Excision of the polydnavirus chromosomal integrated EP1 sequence of the parasitoid wasp *Cotesia congregata* (Braconidae, Microgastrinae) at potential recombinase binding sites. J. Gen. Virol. 78, 3125–3134.

Schaack, S., Gilbert, C., Feschotte, C., 2010. Promiscuous DNA: horizontal transfer of transposable elements and why it matters for eukaryotic evolution. Trends Ecol. Evol. 25, 537–546.

Shelby, K.S., Webb, B.A., 1994. Polydnavirus infection inhibits synthesis of an insect plasma protein, arylphorin. J. Gen. Virol. 75, 2285–2292.

Shelby, K.S., Webb, B.A., 1997. Polydnavirus infection inhibits translation of specific growth-associated host proteins. Insect Biochem. Mol. Biol. 27, 263–270.

Shelby, K.S., Webb, B.A., 1999. Polydnavirus-mediated suppression of insect immunity. J. Insect Physiol. 45, 507–514.

Stoltz, D.B., 1993. The polydnavirus life cycle. In: Beckage, N.E., Thompson, S.N., Federici, B.A. (Eds.), Parasites and Pathogens of Insects. Vol. 1: Parasites (pp. 167–187). Academic Press, San Diego.

Stoltz, D.B., Vinson, S.B., 1979. Viruses and parasitism in insects. Adv. Virus Res. 24, 125–171.

Stoltz, D.B., Guzo, D., Cook, D., 1986. Studies on polydnavirus transmission. Virology 155, 120–131.

Stoltz, D., Whitfield, J.B., 1992. Viruses and virus-like entities in the parasitic hymenoptera. J. Hymenoptera. Res. 1, 125–139.

Stoltz, D.B., Xu, D., 1990. Polymorphism in polydnavirus genomes. Can. J. Microbiol. 36, 538–543.

Strand, M.R., McKenzie, D.I., Grassl, V., Dover, B.A., Aiken, J.M., 1992. Persistence and expression of *Microplitis demolitor* polydnavirus in *Pseudoplusia includens*. J. Gen. Virol. 73, 1627–1635.

Strand, M.R., Pech, L.L., 1995. Microplitis demolitor polydnavirus induced apoptosis of a specific haemocyte morphotype in Pseudoplusia includents. J. Gen. Virol. 76, 283–291.

Summers, M.D., Dib-Hajj, S., 1995. Polydnavirus-facilitated endoparasite protection against host immune defenses. Proc. Natl. Acad. Sci. U.S.A. 92 (1), 29–36.

Tanaka, T., Agui, N., Hiruma, K., 1987. The parasitoid Apanteles kariyai inhibits pupation of its host, Pseudaletia separata, via disruption of prothoracicotropic hormone release. Gen. Compar. Endocrinol. 67, 364–374.

Theilmann, D.A., Summers, M.D., 1986. Molecular analysis of *Campoletis sonorensis* virus DNA in the lepidopteran host *Heliothis virescens*. J. Gen. Virol. 67, 1961–1969.

Theilmann, D.A., Summers, M.D., 1987. Physical analysis of the *Campoletis sonorensis* virus multipartite genome and identification of a family of tandemly repeated elements. J. Virol. 61 (8), 2589–2598.

Thomas, J., Schaack, S., Pritham, E.J., 2010. Pervasive horizontal transfer of rolling-circle transposons among animals. Genome Biol. Evol. 2, 656–664.

Vinson, S.B., Edson, K.M., Stoltz, D.B., 1979. Effect of a virus associated with the reproduction system of the parasitoid wasp, Campoletis sonorensis, on host weight gain. J. Invertebr. Pathol. 34, 133–137.

Volkoff, A.-N., Cérutti P., Rocher J., Ohresser M., Devauchelle G., Duonor-Cérutti M., 1999. Related RNAs in lepidopteran cells after in vitro infection with *Hyposoter didymator* virus define a new polydnavirus gene family. Virology 263, 349–363.

Volkoff, A.-N., Jouan, V., Urbach, S., Samain, S., Bergoin, M., Wincker, P., 2010. Analysis of virion structural components reveals vestiges of the ancestral ichnovirus genome. PLoS Pathog. 6 (5), e1000923.

Volkoff, A.-N., Rocher, J., Cerutti, P., Ohresser, M.C.P., d'Aubenton-Carafa, Y., Devauchelle, G., et al. 2001. Persistent expression of a newly characterized *Hyposoter didymator* polydnavirus gene in long-term infected lepidopteran cell lines. J. Gen. Virol. 82, 963–969.

Webb, B.A., 1998. Polydnavirus biology, genome structure, and evolution. In: Miller, L.K., Ball., L.A. (Eds.), The Insect Viruses (pp. 105–139). Plenum Press, New York.

Webb, B.A., Cui, L., 1998. Relationships between PDV genomes and viral gene expression. J. Insect Physiol. 44, 785–793.

Webb B.A., Beckage N.E., Harayakawa Y., Krell P.J., Lanzrein B., Stoltz D.B., et al. 2005. Family Polydnaviridae. In: Fauquet C.M., Mayot M.A., Maniloff J., Dessselberger U., Ball L.A., (Eds.). In: Virus Taxonomy. Eighth report of the International Commitee on Taxonomy of Viruses (pp. 253–260). Academic Press, London.

Webb, B., Fisher, T., Nusawardani, T., 2009. The natural genetic engineering of polydnaviruses. Ann. N. Y. Acad. Sci. 1178 (1), 146–156.

Webb, B.A., Strand, M.R., 2005. The Biology and Genomics of Polydnaviruses. In: Gilbert, L.I., Iatrou, Gill, S.S., (Eds.), Comprehensive Molecular Insect Science. Vol. 6 (pp. 323–360). Elsevier, San Diego.

Weber, B., Annaheim, M., Lanzrein, B., 2007. Transcriptional analysis of polydnaviral genes in the course of parasitization reveals segment-specific patterns. Arch. Insect. Biochem. Physiol. 66, 9–22.

Wetterwald, C., Roth, T., Kaeslin, M., Annaheim, M., Wespi, G., Heller, M., et al. 2010. Identification of bracovirus particle proteins and analysis of their transcript levels at the stage of virion formation. J. Gen. Virol. 91, 2610–2619.

Whitfield, J.B., Asgari, S., 2003. Virus or not? Phylogenetics of polydnaviruses and their wasp carriers. J. Insect Physiol. 49, 397–405.

Wu, Y.-L., Wu, C., Lee, S.-T., Tang, H., Chang, C.-H., Chen, H.-W., et al. 2010. The early gene hhi1 reactivates *Heliothis zea* nudivirus 1 in latently infected cells. J. Virol. 84, 1057–1065.

Wyder, S., Tschannen, A., Hochuli, A., Gruber, A., Saladin, V., Zumbach, S., et al. 2002. Characterization of *Chelonus inanitus* polydnavirus segments: sequences and analysis, excision site and demonstration of clustering. J. Gen. Virol. 83, 247–256.

Xu, D., Stoltz, D.B., 1991. Evidence for a chromosomal location of polydnavirus DNA in the ichneumonid parasitoid, *Hyposoter fugitivus*. J. Virol. 65, 6693–6704.

Yamanaka, A., Hayakawa, Y., Noda, H., Nakashima, N., Watanabe, H., 1996. Characterization of polydnavirus-encoded RNA in parasitized armyworm larvae. Insect Biochem. Mol. Biol. 5, 529–536.

Chapter 9

Unusual Viral Genomes: Mimivirus and the Polydnaviruses

Christopher A. Desjardins,

The Broad Institute, 320 Charles Street, Cambridge, MA 02141, U.S.A.

SUMMARY

Recent publication of some large dsDNA viral genome sequences has challenged the ways in which we view viral classification and evolution. Here, I focus on comparative analysis of the genomes of two of these viruses. The first is the amoebal virus Mimivirus, which has the largest viral genome sequenced to date and shows complexity comparable to some small parasitic prokaryotes. The second virus is the polydnavirus, which shares a unique symbiotic relationship with parasitoid wasps, in which the virus is entirely dependent on the wasp for replication and in return viral gene expression promotes parasitoid survival. While the genomes of these two viruses share few homologous features, in parallel ways they have challenged our definition of viruses and illuminated new avenues in which viruses interact with their hosts and the environment over an evolutionary timescale.

INTRODUCTION

Over the past decade, a number of viral genome sequences have been published which have altered the way in which we view the definition and evolution of viruses. Large dsDNA viral genomes have shown complexity comparable to some small parasitic prokaryotes, and the biologies of these viruses have illuminated new ways in which viruses interact with their hosts and the environment over an evolutionary timescale. In 2005, I coauthored a review (Desjardins *et al.*, 2005) in which we briefly discussed the genomes of two viruses that are drastically changing the way we look at viruses.

One of these viruses was the polydnavirus, which shares a symbiotic relationship with parasitoid wasps. Virions are manufactured solely in the female wasp's reproductive tract from proviral DNA, and virions themselves do not encode the capacity for replication or packaging. When polydnavirus-containing wasps inject eggs into their caterpillar host, they co-inject virions which, through suppression of the caterpillar immune system, act to ensure the survival of the young parasitoid. These viruses persist only by being transmitted vertically in their primary wasp host as proviral DNA, calling into question even the classification of polydnaviruses as viruses. Polydnavirus genomes are in many ways more eukaryote-like than any other virus, as the heritable form of the virus lies within a eukaryotic genome.

The other virus we examined was the amoebal virus Mimivirus, and at 1.2 Mbp, the Mimivirus genome is the largest viral genome sequenced to date (Raoult *et al.*, 2004). The Mimivirus genome rivals in both size and complexity the genomes of many small parasitic prokaryotes. This has even led scientists to propose the term 'girus' for giant viruses which show complexity comparable to or greater than these prokaryotes (Claverie *et al.*, 2006). While the Mimivirus genome lacks some of the eukaryote-like features of polydnavirus genomes, it does encode a level of translational machinery never before seen in viruses. Furthermore, Mimivirus relatives are the only viruses known to have their own viruses (virophages), opening up new avenues for virus–virus interactions. The polydnavirus and Mimivirus genomes, in different ways, have pushed the envelope of both viral complexity and the relationships of viruses with their host and host environment.

In part, the purpose of this chapter is to compare the genomes and biology of Mimivirus and the polydnaviruses. However, the other purpose of this chapter is to bring the discussions centered around Mimivirus to the polydnavirus researcher. Unfortunately, it seems that polydnaviruses are often left out of discussions on 'giant' viruses, despite the fact that many polydnavirus genomes fall well above the most stringent 280–300-Mbp minimum size required

Parasitoid Viruses: Symbionts and Pathogens. DOI: 10.1016/B978-0-12-384858-1.00009-6

for classification as giant viruses (Claverie *et al.*, 2006; Van Etten *et al.*, 2010). In many articles on giant viruses, the existence of polydnaviruses is mentioned initially, and then it is stated that they will 'not [be] discussed further' (e.g., Claverie *et al.*, 2006). This appears to be due to the polydnavirus genome's eukaryote-like low coding density, which is argued to make comparisons to other viruses difficult. Regardless, many of the questions and debates Mimivirus has brought to the field of viral evolution are also quite relevant to polydnavirus research. First, I will introduce the genomes of Mimivirus and its virophage Sputnik, followed by the polydnavirus genomes. I will then compare these genomes and discuss the implications of these genome sequences for our understanding of the definition and evolution of viruses.

MIMIVIRUS: THE LARGEST SEQUENCED VIRUS

Mimivirus is a virus of the amoeba *Acanthamoeba polyphaga* and was first isolated from a cooling tower in England in 1992 (La Scola *et al.*, 2003). Gram-positive staining and the large size of Mimivirus led its discoverers to believe it was a bacteria for over ten years. In 2003, it was renamed Mimivirus, meaning 'mimicking microbe'. The virus particle is surrounded by an isocahedral capsid with a diameter of 500nM, which is then covered by a thick layer of fibrils giving Mimivirus a 'hairy' appearance and an overall diameter of 750nM (Xiao *et al.*, 2005). Due to their enormous size, Mimivirus particles do not fit though the 0.2 μM filter typically used to isolate viruses.

The replication cycle of Mimivirus is shown in Fig. 1. It begins when a virion is phagocytized by an amoeba. The binding of lysosomes to the Mimivirus-containing phagosome triggers the opening of the capsid, probably through lysosomal activity, which then fuses to the phagosome membrane and releases viral DNA in the cytoplasm (Zauberman *et al.*, 2008). It is then hypothesized that Mimivirus DNA enters the amoebal nucleus, as fluorescent staining of this time-point shows an increase in nuclear AT content, which could correspond to the AT-rich Mimivirus DNA (Suzan-Monti *et al.*, 2007). This may be indicative of a transient nuclear DNA replication phase as seen in asfarviruses. Subsequent to the entry of Mimivirus DNA into the nucleus, large virus factories develop in the cytoplasm which subsequently package new Mimivirus particles. These viral factories appear centered around a compartment the size of a Mimivirus virion, which could indicate replication and packaging machinery similar to poxviruses, where transcription of early genes occurs immediately following infection from within the virion (Claverie and Abergel, 2009). Mimivirus particles are then released through cell lysis.

The Mimivirus genome is the largest viral genome sequenced to date, composed of a single linear dsDNA molecule of ~1.18Mbp and predicted to encode 911 protein-coding genes (Raoult *et al.*, 2004). Additional transcript sequencing increased the number of predicted genes to 960 and identified 26 non coding RNAs (Legendre *et al.*, 2010). The Mimivirus genome encodes a number of genes never before seen in viruses, such as an unprecedented number of genes involved in protein translation: four aminoacyl-tRNA synthetases, five translation factors, and the first identified viral homolog of a tRNA modification enzyme (Raoult *et al.*, 2004). Furthermore, Mimivirus encodes five DNA repair genes, three topoisomerases, three protein chaperones, and at least eight genes involved in the synthesis and modification of proteins and polysaccharides. None of these functional classes of genes has been identified in this number in any previously sequenced viral genome. However, like all viruses, the Mimivirus genome does not encode any ribosomal proteins and cannot replicate without host replication machinery.

Mimivirus genes show an unprecedented level of promoter conservation—45% of all Mimivirus genes have a perfectly conserved AAAATTGA motif within 150bp of their start site (Suhre *et al.*, 2005). It is hypothesized that this motif is functionally equivalent to the eukaryotic TATA box. This appears to be a unique feature of Mimivirus, as related viruses do not show a similar level of conservation of this or any other promoter element.

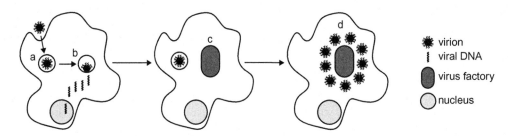

FIGURE 1   Replication cycle of Mimivirus. A: Mimivirus is phagocytized by the amoeba and sequestered in a phagosome. **B:** The Mimivirus virion releases DNA into amoebal cytoplasm, and may enter the nucleus. **C:** Virus factories appear in the cytoplasm of the infected amoeba. **D:** Mature Mimivirus virions are produced by the virus factory, and subsequently released by lysis of the amoebal cell.

Furthermore, most Mimivirus mRNAs have a 3′ polyadenylation signal which forms a perfectly conserved hairpin structure (Byrne *et al.*, 2009). This degree of transcriptional complexity has not been found in any other virus.

Mimivirus is classified as a nucleo-cytoplasmic large DNA virus (NCLDV) (La Scola *et al.*, 2003) and encodes all nine of the class I core genes assigned to NCLDVs (Iyer *et al.*, 2001; Raoult *et al.*, 2004). NCLDVs were named as such because they either replicate entirely in the cytoplasm (e.g., asfarviruses, iridoviruses, poxviruses) or begin replication in the nucleus and end replication in the cytoplasm (e.g., phycodnaviruses) (Iyer *et al.*, 2001). Also, as the name implies, NCLDVs have large genomes, generally over 100 kbp, although inclusion in the group is phylogenetic rather than phenotypic. NCLDVs attack a wide range of eukaryotic hosts ranging from algae (phycodnaviruses) to insects and vertebrates (e.g., poxviruses). Based on phylogenetic analysis of the NCLDV core genes, Mimivirus was placed in its own family Mimiviridae (La Scola *et al.*, 2003), and further phylogenetic analysis has supported this classification (Iyer *et al.*, 2006; Raoult *et al.*, 2004).

Numerous Mimivirus relatives have been identified through environmental sequencing of ocean samples (Ghedin and Claverie, 2005; Monier *et al.*, 2008) and targeted culture from fresh water environmental samples (La Scola *et al.*, 2010). Mamavirus, which was isolated from amoebas in France and has a slightly larger capsid than Mimivirus, has a similar genome size of ~1.2 Mbp (La Scola *et al.*, 2008). It is closely related to Mimivirus, as 99% of Mamavirus open reading frames (ORFs) have 75–100% sequence identity with orthologs in Mimivirus. Mimivirus' ocean-dwelling relatives certainly attack a range of hosts beyond amoebas, and Claverie *et al.*, (2009) hypothesize that this host range includes such disparate organisms as algae, corals, and sponges. These relatives may have even larger genomes than Mimivirus, as viruses with larger capsid sizes to Mimivirus have been isolated, and capsid size is often reflective of genome size (La Scola *et al.*, 2010). However, it remains to be seen how much overlap exists between the coding capacity of the Mimivirus genome and the genomes of these currently uncharacterized viruses, and whether or not these new viruses will stretch our definition of viruses.

SPUTNIK: A VIRUS OF A VIRUS

Mimivirus and its relatives are even complex enough to have their own viruses. Sputnik, which infects the Mimivirus relative Mamavirus, in addition to the Mimivirus itself, has been termed a 'virophage' (La Scola *et al.*, 2008). Sputnik is much smaller than Mimivirus, only 50 nM in diameter, and has a small circular dsDNA genome of 18.3 kbp, encoding only 21 proteins, and is 73% AT. Only eight of the proteins have detectable homologs in viruses, bacteria, or eukaryotes; three of these genes have identifiable homologs in

Mimivirus. Given the chimeric nature of this small genome, it is currently unknown what other viruses Sputnik may be related to.

Sputnik can only replicate in virus factories produced during an amoeba's infection with Mimivirus, and replicating is deleterious to the replication of Mimivirus by causing defective capsid formation, resulting in a significant decrease in the rate of amoebal lysis. In some ways, Sputnik is similar to satellite viruses, which utilize the factories of co-infected viruses to assist with their own packaging (Desnues and Raoult, 2010). However, while some satellite viruses may reduce the replication rate of their host virus, Sputnik is the first virus described to cause it's host virus to create defective particles. Also, satellite viruses generally replicate their DNA in the host cell nucleus whereas Sputnik replicates within the virus factory. Furthermore, the presence of Mimivirus-like hairpin structures in 14 intergenic regions in the Sputnik genome suggest that Sputnik is using Mimivirus machinery for replication and is therefore more than a satellite virus (Claverie and Abergel, 2009). Together, these observations provide convincing evidence that Sputnik is more than a satellite and is a virus of a virus.

POLYDNAVIRUSES: VIRAL SYMBIONTS

Mimivirus and Sputnik showed us new levels of viral genome complexity and new types of virus–virus interactions, but the sequencing of the first polydnavirus genome opened up a different set of questions regarding the limits of viral complexity and virus–eukaryote interactions. The polydnavirus replication cycle is shown in Fig. 2. The polydnavirus genome is integrated into the parasitoid wasp genome. Replication of proviral segments occurs only in the nuclei of calyx cells in the reproductive tract of the female wasp, followed by excision and circularization of viral genome segments from the amplified proviral DNA. Virions are subsequently produced utilizing a virion packaging system present in flanking wasp DNA (Bézier *et al.*, 2009a). During packaging, nuclear pores increase in abundance and the cytoplasm of the calyx cells fill with ribosomes, suggesting that viral structural proteins are synthesized in the cytoplasm and then imported into the nucleus (Wyler and Lanzrein, 2003). In bracoviruses, virions are released by lysis of the calyx cells into the oviduct lumen (Wyler and Lanzrein, 2003), while in ichnoviruses virions are released by budding (Volkoff *et al.*, 1995). The female wasps then co-inject polydnavirus virions along with eggs into their caterpillar hosts. The viruses then enter caterpillar host cells where they act to suppress the host immune response and alter host physiology, promoting survival of the young parasitoid.

The first genome sequence of a packaged polydnavirus was that of *Cotesia congregata* bracovirus (CcBV) (Espagne *et al.*, 2004). This genome is composed of 30

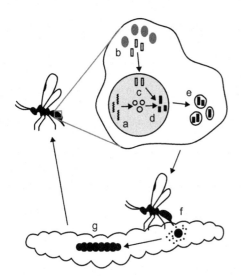

FIGURE 2   Replication cycle of a polydnavirus (adapted from (Desjardins et al., 2005)). A: Replication of proviral DNA occurs only in the nuclei of calyx cells in the reproductive tract of the female wasp. **B:** Simultaneously, viral structural proteins are generated in the cytoplasm. **C:** Viral DNA is excised and circularized, and viral structural proteins are imported into the nucleus. **D:** Virus DNA is encapsidated. **E:** Virions are generated by budding from the nucleus and released by lysis (in bracoviruses) or a second budding (in ichnoviruses). **F:** Female wasps co-inject virions along with eggs into their caterpillar hosts. Virions enter caterpillar host cells where they act to suppress the host immune response and alter host physiology, promoting survival of the growing young parasitoid and therefore survival of proviral DNA (G).

circular dsDNA segments totaling ~560 kbp and is predicted to encode 156 protein-coding genes. Polydnaviruses are unique among viruses in having their genome broken into multiple segments (hence the name 'poly-DNA-virus'). A number of packaged polydnavirus genome sequences from both the bracovirus and ichnovirus lineages followed: MdBV and CsIV (Webb et al., 2006), HfIV and TrIV (Tanaka et al., 2007), GfIV (Lapointe et al., 2007), and GiBV and GfBV (Desjardins et al., 2008). These genomes range in segment number from 15 in MdBV to 105 in GfIV and in size from 189 kbp in MdBV to 589 kbp in GfBV. Ichnoviruses typically encode a larger number of smaller circular segments than bracoviruses. The polydnavirus proviral genome is integrated into the parasitoid wasp genome at multiple loci, each containing one to many proviral genome segments (Annaheim and Lanzrein, 2007; Desjardins et al., 2008).

Polydnavirus genomes are unusual among viruses in that they encode large gene families, and the CcBV genome encodes a number of these previously identified as virulence factor in non viral pathogens, including protein tyrosine phosphatases, inhibitors of NF-κB (also called vankyrins (Kroemer and Webb, 2005)), cysteine-rich proteins, EP1 (early protein 1) genes, and cysteine protease inhibitors

(cystatins) (Espagne et al., 2004). Webb et al. (2006) identified a number of gene families unique to ichnoviruses, including rep genes, viral innexins (vinnexins), plus two unique gene families in MdBV: egf-motif (epidermal growth factor) genes and glc genes (mucins). Further polydnavirus genome sequencing identified new and sometimes taxon-specific genes families with functional predictions, e.g., sugar transporters in GiBV and GfBV (Desjardins et al., 2008). In fact, polydnavirus gene families are remarkably taxon-specific, and the only gene family that appears to be shared across bracoviruses and ichnoviruses is the vankyrins (Falabella et al., 2007; Webb et al., 2006).

Conservation of gene families within polydnavirus lineages is more variable. Campoplegine ichnovirus genomes appear to have a large overlap between gene family content (Tanaka et al., 2007), while bracoviruses do not (Webb et al., 2006). The gene family content of the banchine ichnovirus GfIV includes the ubiquitous vankyrins, the bracovirus protein tyrosine phosphatases, and a previously undescribed NTPase-like gene family with similarity to NCLDV NTPases (Lapointe et al., 2007). However, only having a single banchine ichnovirus genome sequence available makes drawing generalities about the group's gene content difficult. Some features appear to be more conserved within polydnavirus lineages than packaged genes, including the packaging machinery (Bézier et al., 2009a) and motifs governing segment excision (Desjardins et al., 2007).

Polydnavirus genomes have a remarkably low coding density, ranging from 20–30% (see Table 1). In fact, some ichnoviruses have entire genome segments which are not predicted to encode any proteins; the function of these segments is currently unknown (Tanaka et al., 2007). This situation is most extreme in the banchine ichnovirus GfIV, where 30% of the genome segments contain no identifiable gene (Lapointe et al., 2007). Despite this eukaryotic-like low coding density, polydnavirus genomes do have some properties which may be indicative of an external viral origin, including atypical nucleotide composition and simpler gene structures compared to flanking wasp DNA (Desjardins et al., 2008, 2007).

Perhaps the most unusual feature of polydnavirus genomes, that which calls their classification as viruses into question, is their complete lack of replication and packaging genes. The exception to this is the CsIV structural protein p12, which is encoded in the packaged ichnovirus genome (Deng and Webb, 1999), although a single structural protein is certainly not enough to replicate and package an entire virus. Recently, Bézier et al. (2009a) discovered nudivirus-derived packaging machinery within the genomes of the polydnavirus-carrying wasps Cotesia congregata and Chelonus inanitus. This included genes involved in transcription, packaging and assembly, and envelope components. In total, genes representing 22

TABLE 1 Morphological and Genomic Features of Mimivirus, its Virophage 'Sputnik', and Polydnaviruses from the Three Major Lineages: Bracoviruses (*Cotesia Congregata* Bracovirus, CcBV), Camplopegine Ichnoviruses (*Hyposoter Fugitivus* Ichnovirus, HfIV), and Banchine Ichnoviruses (*Glypta Fumiferanae* Ichnovirus, GfIV)

Feature	Mimivirus	Sputnik	CcBV	HfIV	GfIV
Capsid size (nM)	500	50	35 × 30–150	85 × 330	30 × 125
Capsid shape	Isocahedral	Isocahedral	Rod	Lenticular	Rounded rod[1]
Capsids per virion	One	One	Multiple	One	Multiple
Envelopes	One	None	One	Two	Two
Replication site	Cytoplasm	Virus factory	Nucleus	Nucleus	Nucleus
Genome size (kbp)	1181	18	568	246	291
Genome segments	1	1	30	56[2]	105
Genome orientation	Linear	Circular	Circular	Circular	Circular
% AT	72	73	66	57	63
% coding density	91	80	27	30	20
Packaged genes	960	21	155	143	103
Unpackaged genes	0	0	19	?[3]	?
tRNA genes	6	0	7	0	0
Non coding RNAs	26	0	0	0	0

Unless otherwise stated, features refer to the packaged viral genome.
[1]*GfIV capsids appear intermediate in shape between bracovirus and campoplegine ichnovirus virions, forming a rounded rod (Lapointe et al., 2007).*
[2]*Eleven of the 56 genome segments in HfIV can generate smaller nested segments, further adding to genome complexity (Tanaka et al., 2007).*
[3]*Twenty unpackaged genes were identified in the related Hyposoter didymator ichnovirus (Volkoff et al., 2010).*

nudivirus genes were identified; 19 in *Cotesia congregata* and 18 in the distantly related *Chelonus inanitus*. Through mass spectrometry and Q-PCR experiments, Bézier *et al.* provided substantial evidence that the nudivirus-related genes in the wasp genome are the packaging machinery for their associated bracoviruses. While Bézier *et al.* 2009a identified numerous genes involved in virion packaging and transcription, they did not find any viral DNA replication genes, suggesting that host genes perform this function. Viral packaging machinery has also been recently identified for *Hyposoter didymator* ichnovirus (Volkoff *et al.*, 2010). As with the bracovirus packing machinery, the genes involved in viral particle production were clustered in the wasp genome. Unlike the bracovirus machinery, however, the genes involved shared no similarity with any known viruses.

GENOME FEATURES SHARED ACROSS MIMIVIRUS AND POLYDNAVIRUSES

A comparison of the genomes of Mimivirus, Sputnik, and representatives of the three lineages of polydnaviruses can be seen in Table 1. At first glance, it would appear that little is shared between Mimivirus and the polydnaviruses other than both being encoded by large, AT-rich dsDNA genomes. Comparative genome analysis of NCLDVs found that polydnaviruses and NCLDVs 'hardly share any homologous proteins' (Iyer *et al.*, 2006). One rare example is a DNA Polymerase B2 domain in CcBV, although it appears that this domain is a component of a Maverick transposable element rather than of viral origin (Drezen *et al.*, 2006). However, one might not expect to find genes in common given that the most conserved viral genes belong to functional categories absent from the packaged polydnavirus genomes published at the time, such as those involved in viral replication and packaging.

Utilizing recently published transcriptional data for Mimivirus (Legendre *et al.*, 2010), and polydnavirus packaging machinery identified for both CcBV and HdIV (Bézier *et al.*, 2009a; Volkoff *et al.*, 2010), I queried the Mimivirus proteome with the polydnavirus packaging genes using BLASTP with a cutoff of 1e-5. No hits were identified, reinforcing the idea that neither bracoviruses nor ichnoviruses are closely related to Mimivirus. Querying environmental metagenomic data with Mimivirus genes has revealed a diversity of Mimivirus-like sequences (Ghedin and Claverie, 2005), so I similarly queried Genbank's environmental protein sequence database with the polydnavirus packaging genes as above. No significant matches were found, although the marine environment, from which a large quantity of environmental sequence data originates, is less ideal for searches of polydnavirus relatives than it is the relatives of protist and algae viruses.

One genome feature shared by Mimivirus and some polydnaviruses is the encoding of tRNA genes, although the exact function of these genes has not been proven in either virus. All four sequenced bracovirus genomes are predicted to encode from three to seven tRNAs each, while only one ichnovirus genome, that of TrIV, was predicted to encode a single tRNA. In Mimivirus, tRNAs appear to be transcribed in a unique way. In eukaryotes, tRNA transcripts are generated in the nucleus using RNA polymerase III. However, Mimivirus both lacks RNA polymerase III and also replicates entirely in the cytoplasm. This led to Byrne et al.'s (2009) discovery that Mimivirus tRNA transcripts are polyadenylated, sometimes in pairs and in one case paired with a protein-coding gene, suggesting that Mimivirus tRNAs are likely transcribed by the same machinery as protein-coding genes. The most obvious purpose of these viral tRNAs would be to adjust the difference in codon and/or amino acid usage from that of the virus to that of the host, as has been hypothesized for phages (Bailly-Bechet et al., 2007), but thus far no convincing evidence has been put forth to support this idea for either Mimivirus or polydnaviruses.

A second feature shared by Mimivirus and polydnaviruses which is unusual for viruses in general is the presence of introns. Mimivirus contains a small number of self-splicing introns and inteins, which are self-splicing regions of a protein. The related phycodnaviruses also encode both self-splicing introns and inteins (Wilson et al., 2009), suggesting that both of these viruses may have acquired them prior to a common ancestor. While transcriptional data have verified the existence of introns in at least some polydnavirus genes (Desjardins et al., 2007; Webb et al., 2006), the exact degree of introns present in polydnaviruses is still up for debate. Espagne et al. (2004) predicted that 69% of CcBV genes contained introns, while Webb et al. (2006) reannotated the same data and predicted 7% contained introns. Given the limited transcriptional data available for polydnavirus genes, this difference is largely the result of whether the gene predictor used incorporates introns or not. However, even when widespread introns are predicted in polydnavirus genes, such as for GiBV and GfBV, polydnavirus genes appear to have fewer introns on average than flanking wasp genes, implying in some sense that even intron-containing polydnavirus genes are simpler than their wasp counterparts (Desjardins et al., 2008).

Perhaps one of the most intriguing elements of both the Mimivirus and polydnavirus genomes is what they don't tell us—in total these genomes encode an enormous number of proteins of completely unknown function, with no similarities to any proteins in unrelated viruses. Here, I used BLAST2GO (Conesa and Gotz, 2008) with a relaxed 1e-5 cutoff to functionally annotate the proteomes of Mimivirus and CcBV with Gene Ontology (Ashburner et al., 2000) terms. Functional annotations were assigned to only 42% of Mimivirus genes and 24% of CcBV genes. Extensive laboratory studies will need to be conducted in order to begin to understand the functions of this wide array of currently unannotated viral genes.

MIMIVIRUS ORIGINS

There has been much discussion and debate over the origins of unusual large dsDNA viruses such as Mimivirus and the polydnaviruses. A good starting point for discussion is a phylogeny of large dsDNA viruses estimated by Iyer et al. (2006) by comparative analysis of a large number of viral genomes (Fig. 3). They divide NCLDVs into two major lineages: one containing poxviruses and asfarviruses, and one containing mimiviruses, phycodnaviruses, and iridoviruses. They argue that many of the unique features of Mimivirus are derived and that it likely evolved from a smaller ancestral NCLDV. They further hypothesize that this ancestral NCLDV and other large DNA viruses evolved from prokaryotic viruses shortly after the emergence of the eukaryotic cell, based on phylogenetic analyses suggesting eukaryotic origins of a large number of core viral genes.

Iyer et al., 2006 cannot determine if all large dsDNA viruses share a common ancestry, or evolved independently and converged when faced with similar selective pressures, although they favor the former hypothesis. They hypothesize that NCLDVs and baculoviruses have nonhomologous virion packaging systems, which suggests that any baculovirus-related polydnavirus packaging machinery cannot be phylogenetically compared to that of Mimivirus. Furthermore, they hypothesize that while that NCLDVs and baculoviruses share a homologous DNA replication system, the systems are not orthologous, i.e., they did not originate from a common ancestor, and therefore the replication systems of Mimivirus and polydnaviruses may also not be comparable. Alternate hypotheses for phylogenetic relationships between large dsDNA viruses have been proposed, such as all viruses with isocahedral capsids forming a monophyletic group (see Krupovic and Bamford (2008) for a review of this hypothesis). However, given the different capsid shapes of Mimivirus, bracoviruses, and ichnoviruses, this hypothesis also fails to provide phylogenetic relationships between these viruses.

An ancestral NCLDV is not the only origin hypothesized for Mimivirus. Using phylogenetic analysis, the authors of the Mimivirus genome sequence hypothesize that Mimivirus represents a branch of life basal to the eukaryotes, the product of reductive evolution from a more complex ancestor, and should be considered the fourth domain of life (Raoult

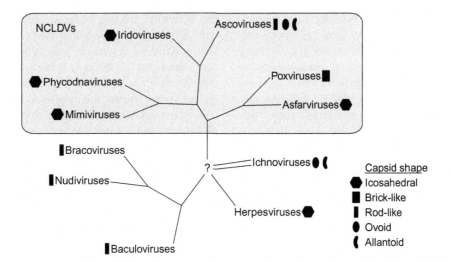

FIGURE 3 **Phylogeny of large dsDNA viruses showing the relative placement of Mimivirus and polydnaviruses (adapted from Iyer *et al.*, 2006 to include the supported placement of bracoviruses (Bézier *et al.*, 2009a)).** While it was originally hypothesized that ichnoviruses are related to ascoviruses (Bigot *et al.*, 2008; Federici and Bigot, 2003), ichnovirus structural proteins show no similarity to structural proteins of ascoviruses or any other viruses, leaving the origins of ichnoviruses unknown (Volkoff *et al.*, 2010). Two lines are shown leading to the ichnovirus lineage, representing the two hypothesized origins of ichnoviruses (Lapointe *et al.*, 2007). Diversity of capsid shapes within each group is also depicted.

et al., 2004). Further arguing this hypothesis, Claverie (2006) stated that the presence of incomplete translational machinery implies the ancestor of Mimivirus had a complete machinery and lost components. Conserved transcriptional elements found in Mimivirus but no other virus, such as the promoter element (Suhre *et al.*, 2005), have also been used to support this hypothesis. It is not explicitly clear in this hypothesis how the other NCLDVs are phylogenetically related to Mimivirus, although presumably if Mimivirus represents the most ancestral NCLDV, the remaining NCLDVs represent derived lineages with further reduced genomes. Yet a third hypothesis for the origin of Mimivirus is that it is largely chimeric and its evolution has been driven by the acquisition of genes through horizontal transfer (Moreira and Brochier-Armanet, 2008).

These arguments over the origin of Mimivirus reflect arguments on the origins of viruses as a whole. The two primary hypotheses for viral origins are the reduction hypothesis and the escape hypothesis. In the reduction hypothesis, viruses began as complex organisms capable of self-replication, and over time evolved a parasitic lifestyle which caused them to lose the ability to self-replicate. This is similar to the hypothesis of the Mimivirus origin by Raoult *et al.* (2004). Alternatively, the escape hypothesis postulates that viruses originated from a set of viral components escaping from a more complex organism. One main difference in the predictions of these hypotheses is that in the reduction hypothesis, many viral genes have a specifically viral origin, wherein the escape hypothesis viral genes have a cellular origin. While many authors claim the lack of similarity between a large fraction of viral genes and cellular genes

implies that the two do not share common ancestry (e.g., Claverie (2006)), the vast evolutionary timescale over which this has occurred makes this conclusion difficult to draw. A third hypothesis is that viruses originated during the emergence of cells from DNA replicons ancestral to prokaryotes and potentially played a role in the origin of cellular life (Forterre, 2006; Iyer *et al.*, 2006).

ORIGINS OF PARASITOID–VIRUS SYMBIOSES

The current phylogenetic placement of polydnaviruses relative to other large dsDNA viruses is shown in Fig. 3. The wasp–polydnavirus association is believed to have evolved multiple times independently, as the wasp hosts of bracoviruses and ichnoviruses lack a common ancestor (Stoltz and Whitfield, 1992). A variety of morphological differences also exist between the capsids and virions of bracoviruses and ichnoviruses (see Table 1). Relationships between polydnaviruses and other viruses were proposed based on the structure of their virions. Bracoviruses were proposed to be related to baculoviruses while ichnoviruses were proposed to be related to ascoviruses (Federici and Bigot, 2003; Whitfield and Asgari, 2003). Bézier *et al.* (2009a) uncovered the packaging machinery of bracoviruses and identified their relatives as nudiviruses, which are closely related to their previously proposed relatives, the baculoviruses. Identification of ichnovirus packaging machinery did not provide the same resolution: none of the identified packaging genes were similar to ascovirus genes or genes of any known virus (Volkoff *et al.*, 2010).

It has been further proposed that the campoplegine and banchine ichnovirus lineages arose independently (Lapointe *et al.*, 2007). This is based on the lack of a common ancestor between the polydnavirus-containing clades of campoplegine and banchine wasps, differences in virion size and structure between campoplegine and banchine ichnoviruses (see Table 1), and a gene family representation in the banchine ichnovirus HfIV that shares more with bracoviruses than ichnoviruses. If correct, this hypothesis implies three independent origins of a remarkably convergent polydnavirus system.

The lack of core viral machinery in the packaged genome sequences of polydnaviruses led to multiple hypotheses on their origins. One hypothesis was that polydnaviruses originated from a large virus which was integrated into the wasp's genome, which then underwent reductive genome evolution, having its replication and packaging machinery transferred to the wasp genome (Whitfield and Asgari, 2003). The second hypothesis was that polydnaviruses derived from wasp DNA which had captured viral structural and virulence proteins and over time accumulated additional virulence components from the wasp genome (Federici and Bigot, 2003). An alternate version of this second hypothesis specifically involves circular DNA mobile elements capturing viral structural proteins (Espagne *et al.*, 2004). This is not unprecedented as a group of ssDNA viruses known as geminiviruses are hypothesized to have evolved from a bacterial plasmid which captured a capsid-coding gene (Koonin and Ilyina, 1992; Krupovic *et al.*, 2009). A third hypothesis was that polydnavirus structural components were evolved by the wasp *de novo* rather than acquired from a virus (Federici and Bigot, 2003). However, given the extensive nudivirus-related packaging machinery discovered by Bézier *et al.* (2009a) the initial hypothesis of viral genome capture and reduction seems the most likely.

ARE MIMIVIRUS AND POLYDNAVIRUSES REDEFINING VIRUSES?

The Mimivirus genome exceeds in both sheer genome size and number of encoded proteins a large number of parasitic prokaryotes (a good comparison can be found in Ward and Fraser, (2005)). This unprecedented complexity has led much debate on the definition of viruses and their relationship to cellular life (e.g., see Moreira and Lopez-Garcia (2009) and associated correspondence for a debate over whether or not viruses should be included in the tree of life). Several new definitions of the term virus have been proposed (reviewed in Forterre (2010)). One such proposition is that viruses could be defined by encoding capsids while bacteria, archaea, and eukaryotes be defined by encoding ribosomes (Raoult and Forterre, 2008). While at first glance this definition may seem sensible, attempts to classify polydnaviruses and their wasp hosts using this definition produce strange results. Under this definition, whether or not polydnaviruses would be considered viruses depends on whether non-packaged components are considered part of the polydnavirus or wasp genome. If the polydnavirus packaging machinery is considered part of the wasp genome, then polydnavirus-associated wasps could be considered both alive (by nature of encoding ribosomes) and viruses (by nature of encoding capsids). Many parasitoid wasps also encode the ability to produce non-polydnavirus encapsidated particles, termed virus-like particles, furthering the idea that the mere ability to encode a capsid is not specific to viruses. While it may be possible to define bacteria, archaea, and eukaryotes by the presence of ribosomes, it seems that viruses cannot be defined by the presence of capsids alone. It is possible that both encoding a capsid and being packaged entirely by that capsid are required to be considered a virus—this definition would exclude parasitoid wasps from viruses.

Other definitions of viruses which have been recently proposed include the idea of viruses as viral factories (Claverie, 2006) and viruses as cellular organisms: the virocell (Forterre, 2010). Claverie (2006) argues that viewing a virus solely as a virion ignores much of the complexity of large dsDNA viruses. The viral factory is a complex intracellular structure where viral genes are transcribed and translated into proteins, and viral DNA is replicated and packaged, and that this structure shows remarkable similarity to an intracellular parasitic bacterium. However, Forterre (2010) points out that this concept only encompasses viruses which replicate in the cytoplasm, excluding a large fraction of viruses which replicate in the nucleus (including polydnaviruses). Using an extension of Claverie's logic, in which the entire cell is taken over by the virus, Forterre argues that the infected cell, or 'virocell' is the definitive form of the virus. At this point, however, the debate veers away from hypotheses based on the Mimivirus and into speculation on what should be considered life; therefore, I will not further explore this argument.

The definition debate within the polydnavirus community seems not to center around how to redefine viruses to properly account for polydnaviruses, but rather to decide if polydnaviruses should be considered viruses at all. According to the International Committee on Taxonomy of Viruses (www.ictvdb.org, accessed 11/13/2010), polydnaviruses are currently defined as viruses. However, the absence of polydnaviruses from virtually all discussion on the evolution and classification of large dsDNA viruses (e.g., Claverie *et al.* (2006), Van Etten *et al.* (2010)) implies that many viral researchers may not consider them true viruses. Already, it has been proposed that polydnaviruses should not be considered viruses but rather immunosuppressive organelles (Federici and Bigot, 2003). Certainly, being derived from a nudivirus means that polydnaviruses are viruses in a phylogenetic sense, whereas if the structural components were not of viral origin, then polydnaviruses would be excluded from viruses. Given this, whether or not a polydnavirus is considered a virus or an organelle-like component of the wasp genome is dependent on whether or not viral packaging machinery must be included with the packaged viral DNA in order for a virion to be considered a virus.

THE OVIPOSITOR AND AMOEBA AS EVOLUTIONARY ENVIRONMENTS

Stoltz and Whitfield (2009) point out that regardless of whether or not polydnaviruses expand our definition of viruses, they should expand our view of viral evolution, in that viruses are not always antagonistic and can evolve symbiotic mutualistic relationships with eukaryotes. Symbiotic relationships between parasitoid wasps and 'free-living' viruses have already been identified, such as the relationship between the DpAV4a ascovirus and its ichneumonid wasp host, *Diadromus puchellus* (Bigot *et al.*, 2009). All polydnavirus-associated wasps are koinobiont endoparasitoids, where the host continues to develop subsequent to the injection of the parasitoid egg—this provides an ideal environment for parasitoid–virus coevolution. In this situation, the wasp, under attack from the host immune system, can benefit from viral immune suppression. The parasitoid ovipositor, in a sense nature's 'dirty needle', provides an ideal mechanism for the spread of viruses between multiple hosts.

While the parasitoid–host environment may provide an ideal situation for the development of symbiotic relations between parasitoid and virus, the amoeba provides an ideal environment for the evolution of interactions between different viruses and parasitic prokaryotes. Amoeba are known to be infected by taxonomically diverse organisms (Greub and Raoult, 2004), and the ability to co-culture viruses and prokaryotic parasites together in a single amoeba provides a potential environment for extensive gene mixture (Boyer *et al.*, 2009). The chimeric nature of the Mimivirus, Mamavirus, and Sputnik genomes have led some researchers to call the amoeba intracellular environment a 'melting pot' of gene transfer where viruses replicate in a soup of foreign DNA (Boyer *et al.*, 2009). The discovery of virophages adds another layer of complexity to these interactions, where they serve as vehicles for horizontal gene transfer between viruses. The presence of Mimivirus genes in the Sputnik genome suggests that horizontal gene transfer between the viruses has already occurred (La Scola *et al.*, 2008).

These kinds of genome–genome interactions may not be unique to a parasitoid–virus symbiosis or amoeba. A replication cycle similar to polydnaviruses has been proposed for the phycodnavirus EsV-1 (Delaroque and Boland, 2008). EsV-1 is a virus of the brown algae *Ectocarpus siliculosus*, infecting algal gametes and spores. The virus is integrated into the algal genome in a proviral state, and spreads through the developing host as a provirus through mitosis. Expression of proviral genes occurs only when triggered by environmental cues. It should be noted that while the EsV-1 genome has a high coding density when compared to polydnaviruses, at 70% it is quite low for a virus in general (Delaroque *et al.*, 2001), which could be indicative of some more intimate link between the genomes of the virus and its host. While the proposed replication cycle of EsV-1 is not as analogous to the polydnavirus replication cycle as Delaroque and Boland claim (2008), as the packaged EsV-1 virus is transmitted by lysis and infection of new algal cells rather than inherited vertically by the alga, it does illustrate how viruses may develop close interactions with the genomes of their hosts. Along similar lines, Bézier *et al.* (2009b) hypothesized based on nudivirus biology that the bracovirus ancestor may have been a sexually transmitted nudivirus which integrated into the wasp chromosome as part of its normal replication cycle.

Transposable elements provide an obvious potential mechanism for movement of genes between the genomes of virus and host. Indeed, extensive remnants of transposable elements have been identified in the genome of CcBV (Drezen *et al.*, 2006; Espagne *et al.*, 2004). Desjardins *et al.* (2008) identified a homologous Maverick-like transposable element in wasp DNA flanking the proviral DNA of GfBV, providing a simple scenario whereby transposable elements could move between flanking and proviral DNA. Furthermore, Desjardins *et al.* identified a p-element-like transposable element present in a small fraction of viral DNA of GiBV, suggesting that the insertion of the transposable element was recent and had not yet gone to fixation. These transposable elements not only provide a mechanism for moving genes between flanking and proviral DNA, but through a polydnavirus vector as a means of horizontal gene transfer between wasp and caterpillar host (Drezen *et al.*, 2006). It has already been shown that DNA from GiBV can be integrated *in vitro* into the genome of the wasp's lepidopteran host *Lymantria dispar* (Gundersen-Rindal and Dougherty, 2000).

In conclusion, while the genomes of Mimivirus and polydnaviruses share few homologous features, in parallel ways their unique elements have challenged our definition of viruses, and greatly expanded how we view the interactions of viruses with their hosts and the environment. The intimate relationship between parasitoid and virus, and amoeba and parasites, may provide ideal environments for close interaction of different genomes over an evolutionary time-scale. However, these systems may only represent a small set of environments predisposed to complex interactions between viruses, other parasites, and their hosts.

REFERENCES

Annaheim, M., Lanzrein, B., 2007. Genome organization of the *Chelonus inanitus* polydnavirus: excision sites, spacers and abundance of proviral and excised segments. J. Gen. Virol. 88, 450–457.

Ashburner, M., Ball, C.A., Blake, J.A., Botstein, D., Butler, H., Cherry, J.M., et al. 2000. Gene ontology: tool for the unification of biology. The Gene Ontology Consortium. Nat. Genet. 25, 25–29.

Bailly-Bechet, M., Vergassola, M., Rocha, E., 2007. Causes for the intriguing presence of tRNAs in phages. Genome Res. 17, 1486–1495.

Bézier, A., Annaheim, M., Herbiniere, J., Wetterwald, C., Gyapay, G., Bernard-Samain, S., et al. 2009a. Polydnaviruses of braconid wasps derive from an ancestral nudivirus. Science 323, 926–930.

Bézier, A., Herbiniere, J., Lanzrein, B., Drezen, J.-M., 2009b. Polydnavirus hidden face: the genes producing virus particles of parasitic wasps. J. Invertebr. Pathol. 101, 194–203.

Bigot, Y., Renault, S., Nicolas, J., Moundras, C., Demattei, M.V., Samain, S., et al. 2009. Symbiotic virus at the evolutionary intersection of three types of large DNA viruses; iridoviruses, ascoviruses, and ichnoviruses. PLoS One 4, e6397.

Bigot, Y., Samain, S., Auge-Gouillou, C., Federici, B.A., 2008. Molecular evidence for the evolution of ichnoviruses from ascoviruses by symbiogenesis. BMC Evol. Biol. 8, 253.

Boyer, M., Yutin, N., Pagnier, I., Barrassi, L., Fournous, G., Espinosa, L., et al. 2009. Giant Marseillevirus highlights the role of amoebae as a melting pot in emergence of chimeric microorganisms. Proc. Natl. Acad. Sci. U.S.A. 106, 21848–21853.

Byrne, D., Grzela, R., Lartigue, A., Audic, S., Chenivesse, S., Encinas, S., et al. 2009. The polyadenylation site of Mimivirus transcripts obeys a stringent 'hairpin rule'. Genome Res. 19, 1233–1242.

Claverie, J.M., 2006. Viruses take center stage in cellular evolution. Genome Biol. 7, 110.

Claverie, J.M., Abergel, C., 2009. Mimivirus and its virophage. Annu. Rev. Genet. 43, 49–66.

Claverie, J.M., Grzela, R., Lartigue, A., Bernadac, A., Nitsche, S., Vacelet, J., et al. 2009. Mimivirus and Mimiviridae: giant viruses with an increasing number of potential hosts, including corals and sponges. J. Invertebr. Pathol. 101, 172–180.

Claverie, J.M., Ogata, H., Audic, S., Abergel, C., Suhre, K., Fournier, P.E., 2006. Mimivirus and the emerging concept of 'giant' virus. Virus Res. 117, 133–144.

Conesa, A., Gotz, S., 2008. Blast2GO: a comprehensive suite for functional analysis in plant genomics. Int. J. Plant Genomics 2008, 619832.

Delaroque, N., Boland, W., 2008. The genome of the brown alga *Ectocarpus siliculosus* contains a series of viral DNA pieces, suggesting an ancient association with large dsDNA viruses. BMC Evol. Biol. 8, 110.

Delaroque, N., Muller, D.G., Bothe, G., Pohl, T., Knippers, R., Boland, W., 2001. The complete DNA sequence of the *Ectocarpus siliculosus* Virus EsV-1 genome. Virology 287, 112–132.

Deng, L., Webb, B.A., 1999. Cloning and expression of a gene encoding a *Campoletis sonorensis* polydnavirus structural protein. Arch. Insect Biochem. Physiol. 40, 30–40.

Desjardins, C., Eisen, J.A., Nene, V., 2005. New evolutionary frontiers from unusual virus genomes. Genome Biol. 6, 212.

Desjardins, C.A., Gundersen-Rindal, D.E., Hostetler, J.B., Tallon, L.J., Fadrosh, D.W., Fuester, R.W., et al. 2008. Comparative genomics of mutualistic viruses of *Glyptapanteles* parasitic wasps. Genome Biol. 9, R183.

Desjardins, C.A., Gundersen-Rindal, D.E., Hostetler, J.B., Tallon, L.J., Fuester, R.W., Schatz, M.C., et al. 2007. Structure and evolution of a proviral locus of *Glyptapanteles indiensis* bracovirus. BMC Microbiol. 7, 61.

Desnues, C., Raoult, D., 2010. Inside the lifestyle of the virophage. Intervirology 53, 293–303.

Drezen, J.-M., Bézier, A., Lesobre, J., Huguet, E., Cattolico, L., Periquet, G., et al. 2006. The few virus-like genes of *Cotesia congregata* bracovirus. Arch. Insect Biochem. Physiol. 61, 110–122.

Espagne, E., Dupuy, C., Huguet, E., Cattolico, L., Provost, B., Martins, N., et al. 2004. Genome sequence of a polydnavirus: insights into symbiotic virus evolution. Science 306, 286–289.

Falabella, P., Varricchio, P., Provost, B., Espagne, E., Ferrarese, R., Grimaldi, A., et al. 2007. Characterization of the IkappaB-like gene family in polydnaviruses associated with wasps belonging to different Braconid subfamilies. J. Gen. Virol. 88, 92–104.

Federici, B.A., Bigot, Y., 2003. Origin and evolution of polydnaviruses by symbiogenesis of insect DNA viruses in endoparasitic wasps. J. Insect Physiol. 49, 419–432.

Forterre, P., 2006. Three RNA cells for ribosomal lineages and three DNA viruses to replicate their genomes: a hypothesis for the origin of cellular domain. Proc. Natl. Acad. Sci. U.S.A. 103, 3669–3674.

Forterre, P., 2010. Giant viruses: conflicts in revisiting the virus concept. Intervirology 53, 362–378.

Ghedin, E., Claverie, J.M., 2005. Mimivirus relatives in the Sargasso sea. Virol. J. 2, 62.

Greub, G., Raoult, D., 2004. Microorganisms resistant to free-living amoebae. Clin. Microbiol. Rev. 17, 413–433.

Gundersen-Rindal, D., Dougherty, E.M., 2000. Evidence for integration of *Glyptapanteles indiensis* polydnavirus DNA into the chromosome of Lymantria dispar in vitro. Virus Res. 66, 27–37.

Iyer, L.M., Aravind, L., Koonin, E.V., 2001. Common origin of four diverse families of large eukaryotic DNA viruses. J. Virol. 75, 11720–11734.

Iyer, L.M., Balaji, S., Koonin, E.V., Aravind, L., 2006. Evolutionary genomics of nucleo-cytoplasmic large DNA viruses. Virus Res. 117, 156–184.

Koonin, E.V., Ilyina, T.V., 1992. Geminivirus replication proteins are related to prokaryotic plasmid rolling circle DNA replication initiator proteins. J. Gen. Virol. 73, 2763–2766.

Kroemer, J.A., Webb, B.A., 2005. Ikappabeta-related vankyrin genes in the *Campoletis sonorensis* ichnovirus: temporal and tissue-specific patterns of expression in parasitized *Heliothis virescens* lepidopteran hosts. J. Virol. 79, 7617–7628.

Krupovic, M., Bamford, D.H., 2008. Virus evolution: how far does the double beta-barrel viral lineage extend? Nat. Rev. Microbiol. 6, 941–948.

Krupovic, M., Ravantti, J.J., Bamford, D.H., 2009. Geminiviruses: a tale of a plasmid becoming a virus. BMC Evol. Biol. 9, 112.

La Scola, B., Audic, S., Robert, C., Jungang, L., de Lamballerie, X., Drancourt, M., et al. 2003. A giant virus in amoebae. Science 299, 2033.

La Scola, B., Campocasso, A., N'Dong, R., Fournous, G., Barrassi, L., Flaudrops, C., et al. 2010. Tentative characterization of new environmental giant viruses by MALDI-TOF mass spectrometry. Intervirology 53, 344–353.

La Scola, B., Desnues, C., Pagnier, I., Robert, C., Barrassi, L., Fournous, G., et al. 2008. The virophage as a unique parasite of the giant mimivirus. Nature 455, 100–104.

Lapointe, R., Tanaka, K., Barney, W.E., Whitfield, J.B., Banks, J.C., Beliveau, C., et al. 2007. Genomic and morphological features of a banchine polydnavirus: comparison with bracoviruses and ichnoviruses. J. Virol. 81, 6491–6501.

Legendre, M., Audic, S., Poirot, O., Hingamp, P., Seltzer, V., Byrne, D., et al. 2010. mRNA deep sequencing reveals 75 new genes and a complex transcriptional landscape in Mimivirus. Genome Res. 20, 664–674.

Monier, A., Claverie, J.M., Ogata, H., 2008. Taxonomic distribution of large DNA viruses in the sea. Genome Biol. 9, R106.

Moreira, D., Brochier-Armanet, C., 2008. Giant viruses, giant chimeras: the multiple evolutionary histories of Mimivirus genes. BMC Evol. Biol. 8, 12.

Moreira, D., Lopez-Garcia, P., 2009. Ten reasons to exclude viruses from the tree of life. Nat. Rev. Microbiol. 7, 306–311.

Raoult, D., Audic, S., Robert, C., Abergel, C., Renesto, P., Ogata, H., et al. 2004. The 1.2-megabase genome sequence of Mimivirus. Science 306, 1344–1350.

Raoult, D., Forterre, P., 2008. Redefining viruses: lessons from Mimivirus. Nat. Rev. Microbiol. 6, 315–319.

Stoltz, D.B., Whitfield, J.B., 1992. Viruses and virus-like entities in the parasitic Hymenoptera. J. Hymenopt. Res. 1, 125–139.

Stoltz, D.B., Whitfield, J.B., 2009. Making nice with viruses. Science 323, 884–885.

Suhre, K., Audic, S., Claverie, J.M., 2005. Mimivirus gene promoters exhibit an unprecedented conservation among all eukaryotes. Proc. Natl. Acad. Sci. U.S.A. 102, 14689–14693.

Suzan-Monti, M., La Scola, B., Barrassi, L., Espinosa, L., Raoult, D., 2007. Ultrastructural characterization of the giant volcano-like virus factory of Acanthamoeba polyphaga Mimivirus. PLoS One 2, e328.

Tanaka, K., Lapointe, R., Barney, W.E., Makkay, A.M., Stoltz, D., Cusson, M., et al. 2007. Shared and species-specific features among ichnovirus genomes. Virology 363, 26–35.

Van Etten, J.L., Lane, L.C., Dunigan, D.D., 2010. DNA viruses: the really big ones (giruses). Annu. Rev. Microbiol. 64, 83–99.

Volkoff, A.N., Jouan, V., Urbach, S., Samain, S., Bergoin, M., Wincker, P., et al. 2010. Analysis of virion structural components reveals vestiges of the ancestral ichnovirus genome. PLoS Pathog. 6, e1000923.

Volkoff, A.N., Ravallec, M., Bossy, J.P., Cerutti, P., Rocher, J., Cerutti, M., et al. 1995. The replication of *Hyposoter didymator* polydnavirus: cytopathology of the calyx cells in the parasitoid. Biol. Cell. 83, 1–13.

Ward, N., Fraser, C.M., 2005. How genomics has affected the concept of microbiology. Curr. Opin. Microbiol. 8, 564–571.

Webb, B.A., Strand, M.R., Dickey, S.E., Beck, M.H., Hilgarth, R.S., Barney, W.E., et al. 2006. Polydnavirus genomes reflect their dual roles as mutualists and pathogens. Virology 347, 160–174.

Whitfield, J.B., Asgari, S., 2003. Virus or not? Phylogenetics of polydnaviruses and their wasp carriers. J. Insect Physiol. 49, 397–405.

Wilson, W.H., Van Etten, J.L., Allen, M.J., 2009. The Phycodnaviridae: the story of how tiny giants rule the world. Curr. Top. Microbiol. Immunol. 328, 1–42.

Wyler, T., Lanzrein, B., 2003. Ovary development and polydnavirus morphogenesis in the parasitic wasp *Chelonus inanitus*. II. Ultrastructural analysis of calyx cell development, virion formation and release. J. Gen. Virol. 84, 1151–1163.

Xiao, C., Chipman, P.R., Battisti, A.J., Bowman, V.D., Renesto, P., Raoult, D., et al. 2005. Cryo-electron microscopy of the giant Mimivirus. J. Mol. Biol. 353, 493–496.

Zauberman, N., Mutsafi, Y., Halevy, D.B., Shimoni, E., Klein, E., Xiao, C., et al. 2008. Distinct DNA exit and packaging portals in the virus *Acanthamoeba polyphaga* mimivirus. PLoS Biol. 6, e114.

Chapter 10

Maintenance of Specialized Parasitoid Populations by Polydnaviruses

Antoine Branca[*], Catherine Gitau[†] and Stéphane Dupas[‡]

[*]*Institute for Evolution and Biodiversity, Hüfferstraße 1, 48149 Münster, Germany*

[†]*EH Graham Centre for Agricultural Innovation, Charles Sturt University, PO Box 883 Orange, NSW 2800, Australia*

[‡]*IRD UR072, CNRS, Laboratoire Evolution Genome et Speciation, 91190 Gif-sur-Yvette, France and Université Paris-Sud, 91400 Orsay, France*

SUMMARY

The parasitoid lifestyle requires an intense specialization at different steps of the life-cycle. As a consequence, most parasitic wasps are specialized and can successfully parasitize very few host species. This host specificity makes them one of the most used biological control agents against insect pests. Unraveling specialization and defining host range within a species or among morphologically identical species might be problematic without the use of molecular markers. In this chapter, we discuss how polydnaviruses are linked to the process of specialization in braconid and ichneumonid wasps and are hence very good markers of wasp specialization. We also review how they are involved in host adaptation and specialization.

ABBREVIATIONS

CrV1	*Cotesia rubecula* virus protein 1
PDV	polydnavirus
SSR	simple sequence repeat

INTRODUCTION

Antagonistic coevolutionary interactions are diversifying forces in host–parasitoid systems (Henry *et al.*, 2008). Indeed, according to the Red Queen Hypothesis (Van Valen, 1973) coevolving species are engaged in an evolutionary arms race of constant innovation prevailing defenses or aggression of antagonist species. In host–parasitoid systems, the host tends to develop defenses that the parasitoids need to overcome in order to successfully develop and reproduce. Larval endoparasitoids are parasites with a free-living adult stage and a parasitic larval stage that develop intimate interactions with their host that ultimately kill the hosts. The basis of the specialization hypothesis is that development

of both virulence (parasitoid) and defense mechanisms (host) is usually costly for the organism, thereby reducing the overall fitness of specialized individuals compared to the ancestral genotype. Expanding the host range of a given parasitoid increases the cost (Henry *et al.*, 2008). From the host's point of view, the cost of developing defense mechanisms could be higher than the cost of parasitism (Sasaki and Godfray, 1999), and more specialized parasitoids take advantage of that situation (Thompson, 2005).

CRYPTIC SPECIALIZATION IS FREQUENTLY OBSERVED IN LARVAL ENDOPARASITOIDS

Cryptic specialization is defined in this chapter as the existence of different populations of a parasitoid species specialized on different host species, even though the host range of the parasitoid species appears diversified at first glance. For instance, the *Drosophila* parasitoid *Leptopilina boulardi* (Hymenoptera: Figitidae) is comprised of several populations specialized on different *Drosophila* spp. (Diptera: Drosophilidae) (Dupas *et al.*, 2003a). The parasitoid can harbor several host-specific immune suppressive factors and theoretically can develop on all the hosts present in its distribution range (Dupas and Carton, 1999). However, these traits are counter-selected in the absence of the target host [Dupas and Boscaro (1999) in *Leptopilina* parasitoids; Henry *et al.* (2008) in *Aphidius* parasitoids] or involve trade-offs with other fitness-related traits (Kraaijeveld *et al.*, 2001). As a result, no generalist population occurs in nature (Dupas and Carton, 1999; Dupas *et al.*, 2003a). Each population of parasitoid cannot attack every host across the species' geographic range. Such geographic mosaics have also been described in *Asobara tabida* (Hymenoptera: Braconidae) (Kraaijeveld *et al.*, 1995) and *Cotesia sesamiae* (Hymenoptera: Braconidae) (Gitau *et al.*, 2007). In both examples, the parasitoids and hosts vary in geographic location and abundance during different seasons. For the parasitoids, avirulent and virulent populations have been described; the virulent population is able to develop on a host that has developed a resistance to the parasitoid species while the avirulent population is not able to develop on this host. The virulent population distribution matches geographically the distribution of the resistant host species, whereas the avirulent population distribution corresponds geographically to alternative susceptible host species (Kraaijeveld and Godfray, 1997; Mochiah *et al.*, 2002; Gitau *et al.*, 2007). More concretely, the *A. tabida* host community is composed of immune-resistant hosts (*D. melanogaster*) and immune susceptible hosts (*D. subobscura*). The *C. sesamiae* host community is composed of immune resistant *Busseola fusca* (Lepidoptera: Noctuidae) and immune susceptible hosts *Sesamia calamistis* (Lepidoptera: Noctuidae) and *Chilo partellus* (Lepidoptera: Crambidae). Thus the patterns of host/parasite distribution represent geographic mosaics of interactions where the parasitoid species attacks different host species over its distribution range, but each population is specialized on local hosts (Kraaijeveld and Godfray, 1997; Dupas *et al.*, 2003a; Gitau *et al.*, 2010).

WHY ARE LARVAL ENDOPARASITOIDS SELECTED FOR CRYPTIC SPECIALIZATION?

In parasitic wasps, mechanisms of specialization can be diverse and can range from attraction to particular plant chemical components on which the host is feeding (Turlings *et al.*, 1990) to the action of the parasitoid's venom, which is injected into the host at parasitization (Vinson, 1990; Strand and Pech, 1995; Beckage, 1998). As a larva, the parasitoid develops within its host, feeding on nutrients ingested by the host, and as an adult, the parasitoid female exhibits a predator behavior, spending most of its time seeking a suitable host, generally from the same species, for its progeny. Since parasitic wasps do not live for a long time, and thus do not always have multiple trials for parasitism, the behavioral traits for parasitism are strongly selected for. However, the target of selection may be later during the wasp's life-cycle, depending on when the host develops defenses to overcome parasitism attacks and how costly counter-defenses are for the parasitic wasp. Ultimately, throughout their life-cycle, parasitoids need to maintain a set of behavioral and physiological traits specific to a particular host or hosts through several steps of interaction to reach maximum developmental success.

As adults, parasitoids need to:

- Recognize chemical cues of suitable hosts to induce parasitoid oviposition.

- Sometimes fight with the host and avoid being killed during oviposition.
- Determine host suitability preparatory to parasitization to optimize developmental success.

As larvae, they need to:

- Avoid or counteract host humoral and cellular immune defenses.
- Adjust to the host's developmental stage and nutritional milieu.
- Coordinate their development with that of the host and finally successfully kill it.

Each step requires specific mechanisms of virulence that may represent energetic costs and create trade-offs with other fitness related traits (Dupas *et al.*, 1999; Kraaijeveld *et al.*, 2001). In this respect, specializing on a few hosts should be less costly than developing virulence factors against a large range of hosts especially when they have different resistance strategies. By natural selection, such diversification of resistance strategies in the host is expected in order to counteract specific parasitoid virulence traits. Virulence traits may be counteracted by any kind of resistance from the host and may coevolve with the resistance. Dupas *et al.* (2003a), described such specific gene-for-gene interactions between a parasitoid and its hosts in the *D. melanogaster/L. boulardi* system (Fig. 1). Polymorphisms in the immune resistance capacity against parasitoids have been observed between targeted host species populations in *D. melanogaster* (Kraaijeveld and Van Alphen, 1995; Dupas *et al.*, 2003a), in pea aphids *Acyrthosiphon pisum* (Henter and Via, 1995) and in the lepidopteran stem borer *Busseola fusca* (Gitau *et al.*, 2010). In Lepidoptera, affinities of binding molecules eliciting immune encapsulation process of parasitoid eggs vary among resistant hosts (Lavine and Strand, 2002).

At the initial behavioral steps, interaction polymorphisms have been described. *Drosophila melanogaster* larvae vary in their locomotion activity between rover-type larvae, which move while eating in the substrate, and the sitter morph, which eats staying still. Reciprocally, parasitoids of *Drosophila* vary in their searching strategies between species detecting host vibrations that are adapted to rover hosts, and species detecting host chemicals that are adapted to sitter hosts (Carton and Sokolowski, 1992). At each step of the interaction, only a part of the available host population will allow the parasitoid success. If the part is different at the different steps, the parasitoid will not develop. The ultimate host specificity of a parasite is the proportion of hosts that combine all the specificities occurring at each step (Combes, 1995; Gitau *et al.*, 2010). The energy required for the parasitoid to enlarge its host range will be expected to increase with the number of steps limiting the host range. Since parasitoids develop complex multi-step interactions with their hosts at

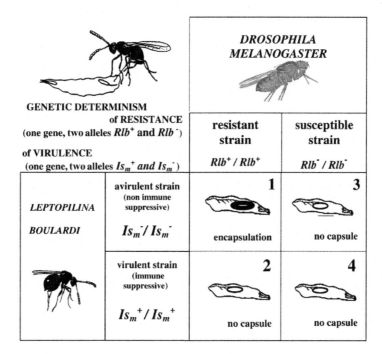

FIGURE 1   Leptopilina determinism of virulence (Dupas *et al.*, 2003). Matching interactions in the *D. melanogaster–L. boulardi* reference system. The interactions between two reference strains of the host (*D. melanogaster*) and two reference strains of the parasitoid (*L. boulardi*) are illustrated. Resistance in the host and virulence in the parasitoid are both conferred by one major gene. The host alleles Rlb+ and Rlb− correspond to the resistant and susceptible phenotypes, respectively, and the alleles Ism+ and Ism− are responsible for virulence and avirulence of the parasitoid. If a resistance allele and an avirulence allele occur in the host and the parasitoid, respectively (case 1), the parasitoid egg is encapsulated. In the three other situations (cases 2, 3 and 4), the parasitoid develops successfully. Thus, the outcome of parasitism will differ depending on whether the host or parasitoid strains differ by a single allele at the resistance or virulence loci.

different stages of their development, they are expected to be more specific than other parasites. Within parasitoids, koinobionts that develop on a living host are observed to be more host-specific than idiobionts that develop on a killed or paralyzed host (Althoff, 2003). The hypothesis is that the specificity of the interaction increases with the number of traits involved and with time duration of the interaction during lifespan. In addition, parasitoids can choose their hosts but the host cannot choose its parasitoid, allowing them to invest more specifically in finely tuned mechanisms of virulence by attacking only the suitable hosts (Lapchin and Guillemaud, 2005). Therefore, we would expect mechanisms of parasitoid virulence to be more specific than mechanisms of host resistance, and the parasitoid to be selected for high level of specialization on their hosts (Lapchin and Guillemaud, 2005). Although most groups of parasitoids are specialists, others appear to be generalists, with the egg parasitoid Trichogramma serving as an example. This may represent a paradox, but the advantage of being a generalist is the ability of parasitoids with wide host ranges to survive in changing host communities. The evolutionary advantage for the parasitoid is to specialize on different hosts in different geographic regions and to maintain the ability to adapt to changing host communities. The importance of geographic mosaics

of interactions, such as those described in the previous paragraph, may explain some of the reasons why parasitoid species may appear to be generalists while in fact they are not. They are able to develop within a wide range of host species, but each geographically isolated population may be specialized on one or a few hosts. One expectation from this theory is that cryptic specialization may be a common feature in apparent generalist parasitoid species. But how is cryptic specialization unraveled?

WHY ARE POLYDNAVIRUSES GOOD MARKERS FOR STUDYING CRYPTIC SPECIALIZATION?

We may predict that the opportunity to generate specialized local lineages should increase with the number of steps determining the interaction and may be higher in koinobiont than idiobiont parasitoids. Polydnaviruses (PDVs) are particularly well studied agents of durable and intimate successful koinobiont interaction in braconid and ichneumonid subclades (Stoltz and Vinson, 1979; Edson *et al.*, 1981; Federici and Bigot, 2003). They mostly encode for virulence factors, notably in the early steps of parasitism, and specifically target the host immune system causing immunosuppression (Strand, Chapter 12 of this volume; Espagne *et al.*, 2004). It has been recently demonstrated that bracoviruses originated from the integration of an ancestral nudivirus (Bézier *et al.*, 2009), a known entomopathogen, and that ichnoviruses originated from a different virus family although the genes encoding structural components of the particles have no similarities with known viruses (Volkoff *et al.*, 2010). Diversification of both ichneumonid and braconid wasps have been suggested to result from the integration of the ancestral virus inside the parasitoid genome (Webb, 1998; Whitfield, 2002). Although not all braconids and ichneumonids harbor PDVs, Braconidae and Ichneumonidae are among the most species-rich families of Hymenoptera and exhibit processes of positive diversification as compared to their sister groups (Davis *et al.*, 2010). We can assume that virus–parasitoid coevolution has facilitated the mechanisms of speciation by allowing the parasitic wasps to specialize on a broader taxonomic range of hosts. Hence, PDV genes should be important targets of selection because they code for major components of the virulence factors of parasitoids with the ovarian and venom proteins.

Specialization of braconid and ichneumonid wasps have been described through studies of parasitoid behavior (Van Alphen and Janssen, 1981; Janssen, 1989; Kraaijeveld *et al.*, 1995; Potting *et al.*, 1997), or through the characterization of neutral genetic differentiation with mitochondrial markers (Baer *et al.*, 2004; Smith *et al.*, 2006, 2008), or in microsatellite markers (Lozier *et al.*, 2009), or by the use of a combination of selected traits and molecular markers (Kankare *et al.*, 2005a, b). Most of the studies fail to distinguish between species and populations and several so-called 'species complexes' may represent different locally adapted populations of the same species. However parasitic wasps are critical biological control agents in agriculture and one caveat is finding markers of host specialization on the target pest which needs to be controlled. Behavioral and physiological studies provide a good clue about specialization but generally necessitate extensive laboratory work to rear the different populations to be tested. Analyses of molecular markers are faster and facilitate studying more lineages at the same time. However, suitable molecular markers should not be neutral and be related to behavioral or physiological traits. For instance, mitochondrial markers can reflect more vicariance, i.e., the effect of abiotic factors on the genome, and old history, and thereby can fail to detect any specialization. *Wolbachia* or other maternally inherited reproductive parasites, frequently associated with parasitic wasps can also blur the specialization signal on such markers (Hurst and Jiggins, 2005). More variable markers such as simple sequence repeats (SSRs) are more precise and constitute the markers of choice when one attempts to identify populations and quantify gene flow between them. This approach can be powerful if gene flow between locally specialized populations of parasitoids is much reduced either in allopatry

or sympatry. Problems can arise when specialized lineages do not reflect actual gene flow. The mismatch between gene flow and selection for specialization can occur when divergent host species occur in sympatry and more related host species in allopatry. In this case, parasitoids specialized on divergent hosts can have more extensive gene flow than parasitoids specialized on related hosts. That is why a candidate gene approach directly linked to virulence seems a more cost-effective and precise approach when studying specialization. Genes coding for different components of the parasitoid venom and PDVs provide good candidates since they are subjected to strong selection by host defense mechanisms in early steps of parasitism. In *L. boulardi*, a wasp protected by its venom (Rizki and Rizki, 1984; Dupas *et al.*, 1996), a relationship exists between geographic distribution and immune suppression controlled by one segregating unit with two alleles, *ISm* and *ISy*, respectively, adapted to a different host species, *D. melanogaster* and *D. yakuba*, and the relative abundance of these hosts (Dupas *et al.*, 1999). We expect PDV genes to display such patterns. Indeed, at the interspecific level, positive Darwinian selection (nonsynonymous changes in DNA occurring more frequently than synonymous changes) such as found for many genes involved in host–pathogen interactions has been described in several PDV genes (Dupas *et al.*, 2003; Serbielle *et al.*, 2008) suggesting their involvement in the host–parasite arms race.

THE ASSOCIATION BETWEEN POLYDNAVIRUS VARIANTS AND THE EVOLUTION OF HOST RACES

In wasps of the genus *Cotesia*, the CrV1 PDV gene encodes for a glycoprotein implicated in the disruption of the actin cytoskeleton of host hemocytes, therefore leading to immunosuppression and hemocyte inactivation (Asgari and Schmidt, 2001). As for its evolution, the gene was successfully used to show the co-diversification of *Cotesia* wasps and their associated bracovirus at the genus level (Whitfield, 2000). CrV1 expression is required for parasitism success and is a good candidate for studying specialization.

In the generalist parasitoid *Cotesia sesamiae*, CrV1 was shown to vary in its expression level depending on the virulence of the wasp population and the resistance of its host (Gitau *et al.*, 2007). *Cotesia sesamiae* is a parasitoid of lepidopteran stem borers in sub-Saharan Africa. In Kenya, the host community on maize and sorghum differs between western highlands, where the endemic noctuid *B. fusca* dominates, and the eastern lowlands where the invasive crambid *C. partellus* and the endemic noctuid *S. calamistis* dominate. Different populations of *C. sesamiae* parasitoids have been identified in lowlands and highlands (Gitau *et al.*, 2007). The lowland variant is able to develop only in the lowland host species

C. partellus and *S. calamistis*, whereas the highland variant is able to develop only in the highland host species *B. fusca*. The avirulent population of *C. sesamiae* occurs in lowland coastal localities, where *B. fusca* is rare or absent (Dupas *et al.*, 2008; Gitau *et al.*, 2010). Further studies showed that virulent and avirulent populations differ in the nucleotides comprising the CrV1 variant sequence (Gitau *et al.*, 2007; Dupas *et al.*, 2008; Branca *et al.*, 2011). This variation between parasitoid populations is correlated with the variation in host community structure (Dupas *et al.*, 2008).

More recently, many more variants of this same gene have been discovered that are correlated with host species specialization by using 28 different species (Branca *et al.*, 2011), suggesting that the variants target specific host genera or species (Fig. 2). In this latter study, the host species associated with a given CrV1-like sequence was the main determinant among several environmental factors explaining the CrV1 genotypic polymorphism observed in *C. sesamiae*. To the best of our knowledge, this is the only example of cryptic specialization revealed by the polymorphism of a PDV gene. The correlation observed is not perfect and some CrV1 allelic forms might be associated with successful development of some *C. sesamiae* parasitoids having a broader host range than others. However, one has access only to the progeny when sampling in wild habitats and not the ovipositing female. Thus, the individuals sequenced are the progeny, not the parents that produced

the virus allowing the development of parasitoids on the hosts, which potentially harbors a different genotype from her daughters. Superparasitism may also blur the signal by allowing some genotypes to develop on unsuitable hosts. Finally, the target of selection might not be CsV1, but another PDV gene linked to CsV1 within the major locus of PDV genes in the wasp genome, comprising approx 70% of the segments (Desjardins *et al.*, 2008).

HOW TO MAINTAIN HOST RACES WITHOUT PRODUCING NEW SPECIES?

How can cryptic specialization be stable? Indeed, on the one hand, gene flow between parasitoid populations specialized on different hosts can disrupt genetic linkage between co-specialized genes, notably those affecting parasitoid behavior and virulence. On the other hand, if gene flow is completely stopped, genetic diversity and the ability of some genotypes to survive in changing host communities might be lost.

Among the mechanisms by which parasitoids maintain the ability to specialize on local hosts, reproductive preferences can maintain locally adapted races without achieving speciation. Assortative mating or assortative pairing within host races usually result from behavioral mating preferences. However, such pre-mating isolation is frequently a

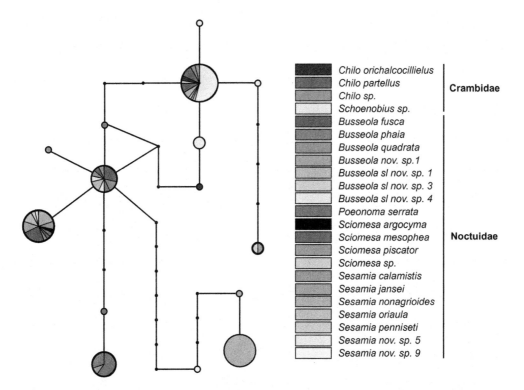

Chilo orichalcocillielus
Chilo partellus
Chilo sp.   **Crambidae**
Schoenobius sp.
Busseola fusca
Busseola phaia
Busseola quadrata
Busseola nov. sp.1
Busseola sl nov. sp. 1
Busseola sl nov. sp. 3
Busseola sl nov. sp. 4
Poeonoma serrata
Sciomesa argocyma
Sciomesa mesophea   **Noctuidae**
Sciomesa piscator
Sciomesa sp.
Sesamia calamistis
Sesamia jansei
Sesamia nonagrioides
Sesamia oriaula
Sesamia penniseti
Sesamia nov. sp. 5
Sesamia nov. sp. 9

FIGURE 2   Parsimonious network of association between CrV1 variants and host species (from Branca *et al.*, in press). Please see color plate section at the back of the book.

result of an active speciation process and long-term evolution. Post-mating isolation is therefore more likely to primarily occur in order to maintain the diversity of mechanisms and genotype. Bacteria commonly associated with insects might be involved in post-mating isolation. Telschow *et al.* (2005) and Branca *et al.* (2009) showed by theoretical modeling the potential importance of bidirectional cytoplasmic incompatibility and sibmating in the maintenance of differentiation between populations of specialized parasitoids existing in sympatry (Fig. 3). These two mechanisms do not need any active adaptation process from the parasitoid. First, bidirectional cytoplasmic incompatibility is a common phenomenon resulting from the infection of reproductive parasitic bacteria, such as *Wolbachia*, the most important members (Tregenza and Wedell, 2000). It is a result of a poison–antidote system where infected males bear the poison and infected females the antidote to this poison (for a review, see Werren, 1997). Different strains of bacteria correspond generally to different poisons. Hence, mutual cytoplasmic incompatibilities can arise between differentially infected populations. In this model, the differentiation between host races was maintained by *Wolbachia* in reciprocally incompatible strains for incompatibility rates above

50%, selection coefficients above 0.5, and migration rates below 0.1. Sibmating could also maintain host races in the absence of bidirectional incompatibilities and interacted synergistically with them when present (Fig. 3). Two main reasons make this mechanism a good candidate for partial isolation of specialized populations: incompatibilities may be only partial and infection may be transient and acquired by horizontal transfer.

The second mechanism, inbreeding, is naturally occurring in parasitic wasp populations, especially gregarious haplodiploid species where sibmating is common (Elias *et al.*, 2010). Inbreeding among specialized parasitoids can actively maintain the specialization by limiting gene flow. In addition, sibmating and cytoplasmic incompatibility can act synergistically in isolated parasitoid populations (Branca *et al.*, 2009).

THE POLYDNAVIRUS ASSOCIATION IN THE SPECIALIZATION PROCESS

Another hypothetical non-exclusive strategy is the role of PDV machinery in maintaining at low-cost cryptic specialization. The PDV genomes are composed of relatively few

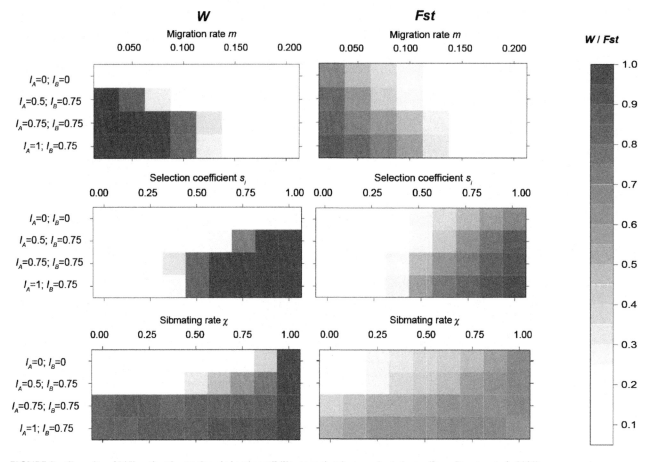

FIGURE 3 Capacity of bidirectional cytoplasmic incompatibility to maintain on a single locus (from Branca *et al.*, 2008).
W, Wolbachia density; *Fst*, genetic differentiation on selected locus; *IA*, Incompatibility strain a; *IB*, incompatiblite strain b.

but very diversified gene families, themselves composed of many slightly diverging duplicate genes. The first complete genome sequence of a PDV was determined for the braconid wasp *Cotesia congregata* (Espagne *et al.*, 2004) and showed this importance of gene duplication in the structure of the genome. This is illustrated by the Protein Tyrosine Phosphatase (PTP) gene family, which is composed of 27 members, the gene products of which are potentially involved in hemocyte cytoskeleton regulation. Francino (2005) suggested that gene duplications may have been selected for during PDV evolution since the beginning in order to increase gene expression. In the case of PDV genes, enhanced expression levels and increased protein abundance may expand their range of action toward new hosts, and provide a new adaptive niche to exploit for one of the copies against a new host, or a new molecular target in the same host, which may be achieved by positive selection.

Positive selection was found to be very important among members of the cystatin gene family, which represents another putative virulence factor gene family (Serbielle *et al.*, 2008). The specific role of each expression product is not known. One hypothesis may be that the different gene family members are selected to target specific hosts, and that only the battery of genes that are adapted to a specific host is expressed. Differential expression in compatible and incompatible hosts has been observed in the ichneumonid wasp *Campoletis sonoresis* for the WHW1.4 gene (in later stages of parasitism) (Cui *et al.*, 2000) and in the braconid wasp *Cotesia congregata* for Early Expressed PDV gene 1 (EP1) (Harwood *et al.*, 1998). In both systems, PDV EP1 gene mRNA or protein expression levels were correlated with host compatibility. In *Campoletis sonorensis* the WHV4.1 PDV gene was expressed in permissive host *Heliothis virescens* but inhibited after five days in the non-permissive host *Manduca sexta*. The gene product was unable to attach to host hemocytes in the nonpermissive hosts. In *C. congregata*,

a correlation between EP genes expression and compatibility was observed among four sphingid hosts, two suitable, one partially suitable and one unsuitable, although the linkage between quantitative levels of EP production and the extent of encapsulation was variable. This suggests an evolution toward a fine-tuning of PDV expression products during host–parasitoid antagonistic coevolution. Gitau *et al.* (2008) provide a more convincing argument for the hypothesis of coevolution of gene expression and gene efficiency according to host suitability (Fig. 4).

This study is the first that compares expression of PDV genes between host races of population from the same parasitoid species. The CrV1 homolog and hemocyte inactivation caused by gene expression was studied in four populations, two from lowland, avirulent against *B. fusca* and two from highland, virulent against *B. fusca*. The *Cotesia sesaminae* CrV1-like gene was expressed only in the suitable host. Fig. 4 shows that *B. fusca*, the major host of *C. sesamiae* in the inland region of Kenya, expresses CrV1 only when parasitized with the inland parasitoid populations. *S. calamistis*, the major endemic host in the coastal region of Kenya expresses CrV1 only when parasitized with the coastal parasitoid populations.

PDV gene families are assumed to be composed of several genes specific to particular hosts and some 'generalist genes' found in closely related species. Only the necessary genes may be expressed in each host. This would be a unique attribute of parasitoids harboring PDVs to enhance their specialization and regulate host range. As for any parasitoid, it may be able to adapt its behavior and choose to parasitize only the suitable hosts (Gitau *et al.*, 2010), but in addition when parasitizing a host, the PDV may express only the subset of genes adapted to this host. Thus, symbiotic viruses would allow the maintenance of nonexpressed host-specific genes in the population without paying the cost of producing them. The whole battery of coadapted genes packaged in the same PDV segment could be expressed specifically only in the target host. Finally, the integration into a single locus of the PDV genome (Belle *et al.*, 2002) may limit recombination between coadapted gene clusters, particularly for the physically closest linked ones that are located on the same PDV segment.

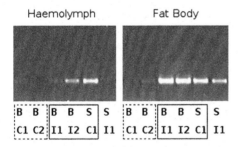

FIGURE 4 Modulation of CrV1 hemolymph inactivation gene expression in hemolymph and fat body of the hosts depending on host species and parasitoid population (*from Gitau et al., 2008*). Solid frame: parasitoid population meets in sympatry the host tested. Absence of solid frame: parasitoid population do not meets in sympatry the host tested. Dashed frame: parasitoid does not develop. B: *Busseola fusca*. S: *Sesamia calamistis*. C1, C2: coastal 1 and 2 *Cotesia sesamiae* populations (respectively Mombasa and Kitui). I1, I2: inland 1 and 2 *C. sesamiae* populations (repectively Kitale and Meru).

CONCLUSION

Cryptic specialization or the presence within parasitoid species of different races adapted to different hosts has been discovered in several host parasitoid systems after detailed population analysis. Cryptic specialization may be problematic when parasitoid wasps are introduced in the field for biological control of pest species. To avoid inappropriate introductions, quality control markers are required. PDV genes have been shown to be one of the best markers that reflect this host race differentiation. This may be explained by the active role played by PDV in the diversification process. Coevolutionary diversification is caused by arms races between the two partners in geographic mosaics of reciprocal selection. In this respect, PDV genes and their expression products reflect a particular dynamic molecular evolution, with frequent duplication events occurring within gene families and important positive Darwinian selection processes (selection for differentiation) on each gene.

REFERENCES

Althoff, D.M., 2003. Does parasitoid attack strategy influence host specificity? A test with New World braconids. Ecol. Entomol. 28, 500–502.

Baer, C.F., Tripp, D.W., Bjorksten, T.A., Antolin, M.F. 2004. Phylogeography of a parasitoid wasp (Diaeretiella rapae): no evidence of host-associated lineages. Mol. Ecol. 13, 1859–1869.

Beckage, N.E., 1998. Modulation of immune responses to parasitoids by polydnaviruses. Parasitology 116, 57–64.

Belle, E., Beckage, N.E., Rousselet, J., et al. 2002. Visualization of polydnavirus sequences in a parasitoid wasp chromosome. J. Virol. 76, 5793–5796.

Bézier, A., Annaheim, M., Herbiniere, J., et al. 2009. Polydnaviruses of braconid wasps derive from an ancestral nudivirus. Science 323, 926–930.

Branca, A., Le Rü, B.P., Vavre, F., Silvain, J.-F., Dupas, S., 2011. Intraspecific specialization of the generalist parasitoid Cotesia sesamiae revealed by polyDNAvirus polymorphism and associated with different Wolbachia infections. Mol. Ecol. 20 (5), 959–971.

Branca, A., Vavre, F., Silvain, J.F., Dupas, S., 2009. Maintenance of adaptive differentiation by Wolbachia induced bidirectional cytoplasmic incompatibility: the importance of sib-mating and genetic systems. BMC Evol. Biol. 9, 185.

Carton, Y., Sokolowski, M.B., 1992. Interactions between searching strategies of Drosophila parasitoids and the polymorphic behavior of their hosts. J. Insect. Behav. 5, 161–175.

Combes, C., 1995. Interactions Durables. Écologie et Evolution du Parasitisme. Masson, Paris.

Cui, L., Soldevila, A.I., Webb, B.A., 2000. Relationships between polydnavirus gene expression and host range of the parasitoid wasp Campoletis sonorensis. J. Insect. Physiol. 46, 1397–1407.

Davis, R., Baldauf, S., Mayhew, P., 2010. The origins of species richness in the Hymenoptera: insights from a family-level supertree. BMC Evol. Biol. 10, 109.

Desjardins, C.A., Gundersen-Rindal, D.E., Hostetler, J.B., et al. 2008. Comparative genomics of mutualistic viruses of Glyptapanteles parasitic wasps. Genome Biol. 9, R183.

Dupas, S., Boscaro, M., 1999. Geographic variation and evolution of immunosuppressive genes in a Drosophila parasitoid. Ecography 22, 284–291.

Dupas, S., Brehelin, M., Frey, F., et al. 1996. Immune suppressive virus-like particles in a Drosophila parasitoid: significance of their intraspecific morphological variations. Parasitology 113, 207–212.

Dupas, S., Carton, Y., 1999. Two non-linked genes for specific virulence of Leptopilina boulardi against Drosophila melanogaster and D. yakuba. Evol. Ecol. 13, 211–220.

Dupas, S., Carton, Y., Poirie, M., 2003. Genetic dimension of the coevolution of virulence–resistance in Drosophila–parasitoid wasp relationships. Heredity 90, 84–89.

Dupas, S., Gitau, C.W., Branca, A., et al. 2008. Evolution of a polydnavirus gene in relation to parasitoid–host species immune resistance. J. Hered. 99, 491.

Dupas, S., Turnbull, M.W., Webb, B.A., 2003. Diversifying selection in a parasitoid's symbiotic virus among genes involved in inhibiting host immunity. Immunogenetics 55, 351–361.

Edson, K.M., Vinson, S.B., Stoltz, D.B., et al. 1981. Virus in a parasitoid wasp: suppression of the cellular immune response in the parasitoid's host. Science 211, 582–583.

Elias, J., Dorn, S., Mazzi, D., 2010. Inbreeding in a natural population of the gregarious parasitoid wasp Cotesia glomerata. Mol. Ecol. 19, 2336–2345.

Espagne, E., Dupuy, C., Huguet, E., et al. 2004. Genome sequence of a polydnavirus: insights into symbiotic virus evolution. Science 306, 286–289.

Federici, B.A., Bigot, Y., 2003. Origin and evolution of polydnaviruses by symbiogenesis of insect DNA viruses in endoparasitic wasps. J. Insect. Physiol. 49, 419–432.

Francino, M.P., 2005. An adaptive radiation model for the origin of new gene functions. Nat. Genet. 37, 573–578.

Gitau, C.W., Gundersen-Rindal, D., Pedroni, M., et al. 2007. Differential expression of the CrV1 haemocyte inactivation-associated polydnavirus gene in the African maize stem borer Busseola fusca (Fuller) parasitized by two biotypes of the endoparasitoid Cotesia sesamiae (Cameron). J. Insect Physiol. 53, 676–684.

Gitau, C.W., Schulthess, F., Dupas, S., 2010. An association between host acceptance and virulence status of different populations of Cotesia sesamiae, a braconid larval parasitoid of lepidopteran cereal stemborers in Kenya. Biol. Control. 54, 100–106.

Harwood, S.H., McElfresh, J.S., Nguyen, A., et al. 1998. Production of early expressed parasitism-specific proteins in alternate sphingid hosts of the braconid wasp Cotesia congregata. J. Invertebr. Pathol. 71, 271–279.

Henry, L.M., Roitberg, B.D., Gillespie, D.R., 2008. Host-range evolution in Aphidius parasitoids: fidelity, virulence and fitness trade-offs on an ancestral host. Evolution 62, 689.

Henter, H.J., Via, S., 1995. The potential for coevolution in a host–parasitoid system. I. Genetic variation within an aphid population in susceptibility to a parasitic wasp. Evolution 49, 427–438.

Hurst, G.D., Jiggins, F.M., 2005. Problems with mitochondrial DNA as a marker in population, phylogeographic and phylogenetic studies: the effects of inherited symbionts. Proc. R. Soc. B: Biol. Sci. 272, 1525.

Janssen, A., 1989. Optimal host selection by Drosophila parasitoids in the field. Funct. Ecol. 3, 469–479.

Kankare, M., Stefanescu, C., et al. 2005. Host specialization by Cotesia wasps (Hymenoptera: Braconidae) parasitizing species-rich Melitaeini (Lepidoptera: Nymphalidae) communities in north-eastern Spain. Biol. J. Linn. Soc. 86, 45–65.

Kankare, M., Van Nouhuys, S., Hanski, I., 2005. Genetic divergence among host-specific cryptic species in Cotesia melitaearum aggregate (Hymenoptera: Braconidae), parasitoids of checkerspot butterflies. Ann. Entomol. Soc. Am. 98, 382–394.

Kraaijeveld, A.R., Godfray, H.C.J., 1997. Trade-off between parasitoid resistance and larval competitive ability in Drosophila melanogaster. Nature 389, 278–280.

Kraaijeveld, A.R., Hutcheson, K.A., Limentani, E.C., et al. 2001. Costs of counterdefenses to host resistance in a parasitoid of Drosophila. Evolution 55, 1815–1821.

Kraaijeveld, A.R., Nowee, B., Najem, R.W., 1995. Adaptive variation in host-selection behaviour of Asobara tabida, a parasitoid of *Drosophila* larvae. Funct. Ecol. 9, 113–118.

Kraaijeveld, A.R., Van Alphen, J.J.M., 1995. Geographical variation in encapsulation ability of *Drosophila melanogaster* larvae and evidence for parasitoid-specific components. Evol. Ecol. 9, 10–17.

Lapchin, L., Guillemaud, T., 2005. Asymmetry in host and parasitoid diffuse coevolution: when the red queen has to keep a finger in more than one pie. Front. Zool., 2, 4.

Lavine, M.D., Strand, M.R., 2002. Insect hemocytes and their role in immunity. Insect Biochem. Mol. Biol. 32, 1295–1309.

Lozier, J.D., Roderick, G.K., Mills, N.J., 2009. Molecular markers reveal strong geographic, but not host associated, genetic differentiation in *Aphidius transcaspicus*, a parasitoid of the aphid genus *Hyalopterus*. Bull. Entomol. Res. 99, 83–96.

Mochiah, M.B., Ngi-Song, A.J., Overholt, W.A., et al. 2002. Variation in encapsulation sensitivity of *Cotesia sesamiae* biotypes to *Busseola fusca*. Entomologia Experimentalis et Applicata 105, 111–118.

Potting, R.P.J., Vet, L.E.M., Overholt, W.A., 1997. Geographic variation in host selection behaviour and reproductive success in the stemborer parasitoid *Cotesia flavipes* (Hymenoptera: Braconidae). Bull. Entomol. Res. 87, 515–524.

Rizki, R.M., Rizki, T.M., 1984. Selective destruction of a host blood cell type by a parasitoid wasp. Proc. Natl. Acad. Sci. U.S.A. 81, 6154.

Sasaki, A., Godfray, H.C.J., 1999. A model for the coevolution of resistance and virulence in coupled host–parasitoid interactions. Proc. R. Soc. B: Biol. Sci. 266, 455.

Serbielle, C., Chowdhury, S., Pichon, S., et al. 2008. Viral cystatin evolution and three-dimensional structure modelling: A case of directional selection acting on a viral protein involved in a host–parasitoid interaction. BMC Biol. 6, 38.

Smith, M.A., Rodriguez, J.J., Whitfield, J.B., et al. 2008. Extreme diversity of tropical parasitoid wasps exposed by iterative integration of natural history, DNA barcoding, morphology, and collections. Proc. Natl. Acad. Sci. U.S.A. 105, 12359.

Smith, M.A., Woodley, N.E., Janzen, D.H., et al. 2006. DNA barcodes reveal cryptic host-specificity within the presumed polyphagous members of a genus of parasitoid flies (Diptera: Tachinidae). Proc. Natl. Acad. Sci. U.S.A. 103, 3657.

Stoltz, D.B. and Vinson, S.B. 1979 Viruses and Parasitism in Insects. Advances in Virus Research, p. 426. Karl Maramorosch.

Strand, M.R., Pech, L.L., 1995. Immunological basis for compatibility in parasitoid–host relationships. Annu. Rev. Entomol. 40, 31–56.

Thompson, J.N., 2005. The Geographic Mosaic of Coevolution. University of Chicago Press, Chicago.

Telschow, A., Hammerstein, P., Werren, J.H., 2005. The effect of Wolbachia versus genetic incompatibilities on reinforcement and speciation. Evolution 22, 1607–1619.

Tregenza, T., Wedell, N., 2000. Genetic compatibility, mate choice and patterns of parentage: invited review. Mol. Ecol. 9, 1013–1027.

Turlings, T.C.J., Tumlinson, J.H., Lewis, W.J., 1990. Exploitation of herbivore-induced plant odors by host-seeking parasitic wasps. Science 250, 1251–1253.

Van Alphen, J.J.M., Janssen, A.R.M., 1981. Host selection by *Asobara tabida* Nees (Braconidae; Alysiinae) a larval parasitoid of fruit inhabiting *Drosophila* species. Neth. J. Zool. 32, 194–214.

Van Valen, L., 1973. A new evolutionary law. Evol. Theory 1, 1–30.

Vinson, S.B., 1990. How parasitoids deal with the immune system of their host: an overview. Arch. Insect Biochem. Physiol., 13.

Volkoff, A.N., Jouan, V., Urbach, S., et al. 2010. Analysis of virion structural components reveals vestiges of the ancestral ichnovirus genome. PLoS Pathog. 6, e1000923.

Webb, B.A., 1998. Polydnavirus biology, genome structure, and evolution. In: Miller, L.K., Ball, A. (Eds.), The Insect Viruses (pp. 105–139). Plenum Press, New York.

Werren, J.H., 1997. Biology of Wolbachia. Annu. Rev. Entomol. 42, 587–609.

Whitfield, J.B., 2000. Phylogeny of microgastroid braconid wasps, and what it tells us about polydnavirus evolution. The Hymenoptera: Evol. Biodivers. Biol. Control, 97–105.

Whitfield, J.B., 2002. Estimating the age of the polydnavirus/braconid wasp symbiosis. Proc. Natl. Acad. Sci. U.S.A. 99, 7508–7513.

Part II

The Biological Roles of Polydnavirus Gene Products

Chapter 11

Polydnavirus Gene Expression Profiling: What We Know Now

Michael R. Strand

Department of Entomology, University of Georgia, Athens, Georgia 30602-7415, U.S.A.

SUMMARY

Polydnaviruses (PDVs) exist in two forms. In wasps, PDVs persist and are transmitted to offspring as stably integrated proviruses, while also replicating in the ovaries of females to produce virions that contain only a portion of the proviral genome. Wasps inject a quantity of these virions into hosts that cause physiological alterations of benefit to the wasp's progeny but do not replicate. A consequence of these contrasting activities is that PDV gene expression differs greatly between wasps and hosts. In this chapter, I summarize what is known about spatio-temporal patterns of PDV gene expression.

ABBREVIATIONS

anks	ankyrin motif genes
BV	bracovirus
Crp	cysteine-rich proteins
Cyst	cysteine protease inhibitors
Cys-motif	cysteine motif containing genes
Egf	epidermal growth factor-like motif
EP	early protein
Glc	glycosylated protein
Inex	innexin genes
IV	ichnovirus
IVSPERs	ichnovirus structural protein encoding regions
LEF	late expression factor
ODV	occlusion-derived virus
ORF	open reading frame
PDV	polydnavirus
PIF	*per os* infectivity factor
pol-res	polar-residue rich proteins
PTP	protein tyrosine phosphatase
Rep	repeat protein
VLF	very late expression factor

INTRODUCTION

Previous chapters summarize the origins, genomics, and replication of polydnaviruses (PDVs). The subject of this chapter is gene expression. As already discussed, PDVs are divided into bracoviruses (BVs) and ichnoviruses (IVs), which are associated with wasps in selected subfamilies of the Braconidae and Ichneumonidae. The hosts for most PDV-carrying wasps are larval-stage Lepidoptera. Each PDV from a given wasp species is genetically distinct and exists in two states: (1) a proviral form that is integrated into the wasp genome and is transmitted to offspring through the germ line; and (2) an encapsidated form, which is produced only in the ovaries of females and which wasps inject into hosts when they oviposit. Importantly, the genes required for replication are not packaged into virions, which instead contain DNAs that encode a mixture of single-copy genes and multimember gene families whose products affect the physiology of hosts (see Huguet *et al.*, Chapter 5 of this volume). One consequence of the differences in gene content between the proviral and encapsidated genomes of PDVs is that viruses injected into hosts cannot replicate. Another is that PDV gene expression markedly differs between the wasp and host environment. Below, I first discuss PDV gene expression patterns in wasps. I then discuss gene expression in hosts with emphasis on the activity patterns of PDVs from different wasp taxa.

THE PROVIRAL AND ENCAPSIDATED GENOMES OF PDVS

While BVs and IVs exhibit similar life-cycles, earlier chapters show that their association with wasps arose independently with BVs evolving from a nudivirus ancestor (Bézier *et al.*, 2009a, b; Drezen *et al.*, Chapter 2 of

Parasitoid Viruses: Symbionts and Pathogens. DOI: 10.1016/B978-0-12-384858-1.00011-4

this volume) and IVs evolving from another viral entity (Volkoff *et al.*, 2010; Volkhoff *et al.*, Chapter 3 of this volume). Prior reviews define the encapsidated genome of BVs and IVs as the complement of circularized dsDNA segments that are packaged into nucleocapsids in non-equimolar amounts (see Fleming, 1992; Stoltz, 1993; Webb and Strand, 2005; Strand, 2010). The encapsidated genomes of four BVs (*Cotesia congregata* bracovirus (CcBV) (Espagne *et al.*, 2004), *Microplitis demolitor* bracovirus (MdBV) (Webb *et al.*, 2006), *Glyptapanteles indiensis* bracovirus (GiBV), and *G. flavicoxis* bracovirus (GfBV) (Desjardins *et al.*, 2008)), three IVs from campoplegine ichneumonids (*Campoletis sonorensis* ichnovirus (CsIV) (Webb *et al.*, 2006), *Hyposoter fugitivus* ichnovirus (HfIV) (Tanaka *et al.*, 2007), and *Tranosema rostrale* ichnovirus (TrIV) (Tanaka *et al.*, 2007)), and one IV from a banchine ichneumonid (*Glypta fumiferanae* ichnovirus, GfIV) (Lapointe *et al.*, 2007) are fully sequenced. Partial or near-complete genomes are also available from a few other BVs including *Chelonus inanitus* bracovirus (CiBV) (Weber *et al.*, 2007) and *Cotesia plutellae* bracovirus (CpBV) (Choi *et al.*, 2009).

In contrast, the proviral genomes of BVs and IVs remain incompletely defined. One component, obviously, is those regions of DNA integrated into the wasp genome that correspond to the episomal DNAs that are packaged into nucleocapids. For BVs, these DNAs cluster in multiple proviral loci, which are excised and amplified during the replication process (Savary *et al.*, 1999; Desjardins *et al.*, 2008). For IVs, DNAs packaged into nucleocapsids are also amplified during replication but appear to be more dispersed in the wasp genome (Fleming and Summers, 1986, 1991; Volkoff *et al.*, 2010). The second component of the proviral genome is the genes that regulate replication such as structural proteins and factors required for DNA packaging. For BVs, several of these predicted proteins have been identified with some sharing similarity with known nudivirus, baculovirus, or eukaryotic proteins, and others sharing no similarity with sequences in current databases (see below). Many of the nudivirus-related genes are clustered in the wasp genome, but they do not excise during replication and they also are not located in the vicinity of the DNAs that are packaged into capsids (Bézier *et al.*, 2009b). For IVs, at least some of the DNA regions encoding structural genes are located in the wasp genome in proximity to regions containing DNAs that are packaged into nucleocapsids (Volkoff *et al.*, 2010). These 'Ichnovirus Structural Protein Encoding Regions' (IVSPERs) amplify during the replication process despite not being packaged. Lastly, none of the genes required for viral DNA replication like polymerases and helicases have been identified in either BVs or IVs, so it remains unclear whether these factors are potential constituents of a proviral genome or are part of the wasp cellular machinery that is required for replication (see Bézier *et al.*, 2009b). For the

purposes of discussing PDV gene expression then, I consider here the proviral genome to be the totality of genes thus far identified that have been implicated in replication plus the regions of DNA and associated genes that are packaged into nucleocapsids.

PDV GENE EXPRESSION IN WASPS

Studies with both BVs and IVs have only detected integrated proviral DNA during the egg and larval stage of wasps (Fleming and Summers, 1986; Gruber *et al.*, 1996; Savary *et al.*, 1999). Replication resulting in DNA packaging and formation of virions is restricted to calyx cells that are located in a domain of the ovary in close proximity to the lateral oviducts (summarized by Fleming, 1992; Stoltz, 1993; Webb and Strand, 2005). Development of the ovary begins when female wasps molt to a pupa but BV and IV replication in calyx cells does not begin until the late pupal stage (Norton and Vinson, 1983; Fleming and Summers, 1986; Webb and Summers, 1992; Volkoff *et al.*, 1995; Wyler and Lanzrein, 2003). Upon emergence, replication continues in the adult female wasp. Several studies have also noted that circularized PDV DNAs are detected in other pupal and adult tissues of females only or wasps of both sexes (Fleming and Summers, 1986; Gruber *et al.*, 1996; Savary *et al.*, 1999). This suggests that at least low levels of proviral DNA excision and/or amplification occur in several tissues although no virions are produced. It is widely assumed that all PDV genes are quiescent during the egg and larval stages of wasps, but to my knowledge no studies have actually examined whether any PDV genes are expressed during these life stages or not. In contrast, a diversity of transcripts encoded by putative proviral genes increase in abundance in the ovaries with the onset of replication in pupae and continuing during the adult stage (Bézier *et al.*, 2009a, b; Wetterwald *et al.*, 2010; Volkoff *et al.*, 2010).

BV Gene Expression in Wasps

For BVs, the identity of genes with roles in replication and their patterns of expression derive from analysis of the microgastrine braconid *C. congregata*, which carries CcBV, and chelonine braconid *Chelonus inanitus*, which carries CiBV. For CcBV, 24 genes share significant similarities with nudiviruses or baculoviruses (Bézier *et al.*, 2009a, b), while for CiBV a total of 44 genes have been identified of which 16 are nudivirus and/or baculovirus related, 14 share similarities with ovarian or other cellular proteins, and 14 share no similarity with known proteins (Wetterwald *et al.*, 2010). Nudivirus/baculovirus-related BV genes include elements of the baculovirus RNA polymerase complex (LEF-4, LEF-8, and p47), which is required for expression of baculovirus 'late genes'. qRT-PCR analysis indicates that p47 expression rises prior to the presence of virions suggesting this factor may be part of a BV polymerase required for transcription of other nudivirus-related genes (Bézier *et al.*, 2009b). Other nudivirus/baculovirus-related genes identified from CcBV and CiBV with predicted roles in regulating gene transcription include LEF-5 and Very Late Factor 1 (VLF-1). The remaining baculovirus and/or nudivirus related genes upregulated during replication encode factors associated with nucleocapsid packaging and assembly (VLF-1, 38K, 19K), occlusion-derived virus (ODV) components of baculovirus envelopes (P74, PIF-1, PIF-2, ODV-E51, ODV-E56, ODV-E66), or other structural constituents (HzNVorf19-like HzNVorf89-like, HzNVorf140-like, PmNV-like) (Bézier *et al.*, 2009a, Wetterwald *et al.*, 2010).

By matching protein sequence data derived from CiBV virions to expressed sequence tags (ESTs) generated from ovary RNA, Wetterwald *et al.* (2010) identified 31 other proteins present in CiBV virions that have no similarity with nudiviruses or baculovirus proteins. Transcript abundance for 28 of these genes increases from a low of 5–10-fold (13c, 95b, 97b) to a high of 4000–100,000-fold (12a, 13a, 13b, 19, and 26) in female ovaries during virion formation. These latter, massively abundant transcripts are most likely components of the protein matrix in calyx fluid that surrounds nucleocapsids because they differ between CiBV and CcBV (Wetterwald *et al.*, 2010). In contrast, we would expect structural components of nucleocapsids to be conserved between CiBV and CcBV owing to their shared nudiviral ancestry (Drezen, personal communication). Sequence analysis indicated that some of these proteins share a TBC1 domain (60a, 60b, 60c, 60d) found in a number of cellular proteins from insects and other organisms, while another (97b) contains a metalloproteinase domain present in some hymenopteran venom proteins. Most, however, share no significant homology with any gene in current databases.

IV Gene Expression in Wasps

The identity of IV genes expressed in wasps is more limited although similar to BVs, current data indicate that the expression of several genes is upregulated in the ovaries with the onset of replication. qRT-PCR studies conducted with *Hyposoter didymator* ichnovirus (HdIV) identified three genes (p12, p53-1, p53-2) encoding transcripts that increased >25-fold in ovaries, suggesting that they encode for structural proteins incorporated into IV virions (Volkoff *et al.*, 2010). Support for this interpretation is further bolstered by earlier studies with CsIV, which also identified p12 and p53 gene homologs (Deng and Webb, 1999; Deng *et al.*, 2000). Interestingly, the p12 gene of CsIV resides in a region of DNA that is packaged into CsIV nucleocapsids, whereas the region of DNA containing the p12 gene of HdIV resides in a non-packaged IVSPER (see above).

Combined with the finding that IVSPERs amplify like the proviral domains which are packaged into virions, these data suggest that both regions possess features recognized by still-unknown components of the viral (or host cellular) machinery that mediate DNA replication. This finding also suggests some variation likely exists among IV isolates in which regions of the proviral genome are packaged into virions and which are not. Sequence analysis of other predicted proteins encoded by genes in HdIV IVSPERs indicate that none share any significant similarity with known sequences except possibly a cyclin domain present in the predicted U12 protein and a domain in U22 with weak similarity to a domain in the P74 envelope protein from a baculovirus (Volkoff *et al.*, 2010). However, homologs of several genes in HdIV IVSPERs (U1, U3, IVSP4-1, IVSP4-2, p12-1, U23, N2) are also expressed during replication of TrIV, supporting a common origin for the IVs associated with campoplegine ichneumonids.

Outside of p12 from CsIV, few genes encoded on DNA segments packaged into nucleocapsids are known to be expressed in wasps. Among IVs, analysis of the *rep* and *inex* gene families from CsIV (see below) indicates that selected members of each are expressed in *C. sonorensis* female pupae and adults (Theilmann and Summers, 1986; Hilgarth and Webb, 2002; Turnbull and Webb, 2002). Expression of some *rep* genes from TrIV are also detected in the ovaries of *T. rostrale* (Rasoolizadeh *et al.*, 2009). Among BVs, studies with MdBV indicate that some members of the *ank* and *ptp* gene families (also see below) are expressed in female pupae and adults albeit at low levels relative to infected hosts (Bitra *et al.*, 2011). However, some single-copy genes encoded by MdBV are expressed at similar levels in wasps and hosts (Bitra *et al.*, 2011). Preliminary observations with CcBV suggest similar patterns to those with MdBV (Provost *et al.*, 2004; Drezen, personal communication).

PDV GENE EXPRESSION IN HOSTS

Unlike the situation in wasps, where conservation exists in the replication-associated nudivirus/baculovirus-like genes of BVs and IVSPER-associated structural genes of IVs, the encapsidated genomes of PDVs differ considerably in gene content (Dupuy *et al.*, Chapter 4 of this volume). This variation reflects a combination of phylogenetic history and the specialized interactions that have evolved between individual wasp species and the hosts they parasitize (Strand, 2009, 2010). Thus, PDV isolates from closely related wasp species share more genes than isolates from distantly related wasps, yet each PDV carried by a given wasp species also exhibits unique features (Dupuy *et al.*, Chapter 4 of this volume). Because of this variation, expression profiles in hosts must be discussed from the perspective of the wasp subfamily and genus that a given PDV isolate belongs to. Recall that BVs are associated exclusively with wasps in the Microgastroid complex, which is a monophyletic assemblage divided into six subfamilies (Microgastrinae, Cardiochilinae, Miracinae, Mendesellinae, Khoikhoiinae, Cheloninae) (Murphy et al., 2008). IVs in contrast are associated with two subfamilies of the Ichneumonidae (Campopleginae, Banchinae). PDV expression data in hosts exist for only a few isolates associated with wasp species in some of these subfamilies. I summarize this literature below.

BV Gene Expression in Hosts: Subfamily Microgastrinae

BVs from Wasps in the Genus Cotesia

As previously noted, only one *Cotesia* BV, CcBV from *C. congregata*, is fully sequenced. Annotation of this encapsidated genome identifies: (1) four gene families with similarity to known proteins (protein tyrosine phosphatases (*ptps*), ankyrin motif genes (*anks*), cysteine-rich proteins (*Crp*), cysteine protease inhibitors (*Cyst*)); (2) five gene families lacking any conserved domains (*EP1-like, hp1, hp2, f1, f2*); and (3) multiple single-copy genes including a homolog of the *CrV1* gene first identified from *C. rubecula* BV (Asgari et al., 1997), an immunolectin, and a histone H4 (Espagne et al., 2004). Other *Cotesia* BVs, like CpBV, encode homologs for most of these genes (Choi et al., 2009). However, expression data in hosts are available for only a portion of the genes *Cotesia* BVs encode.

The largest CcBV gene family is the 27-member *ptp* family which encode predicted functional PTPs as well as non-functional pseudophosphatases. RT-PCR analysis reveals complex patterns of expression for these genes with some family members being transcribed in multiple host tissues (*ptpA, ptpL*) while others are very selectively expressed in, for example, the nervous system (*ptpB*) (Provost et al., 2004). In contrast, Ibrahim et al. (2007) reports that all 14 members of the *ptp* gene family encoded by CpBV are expressed in multiple tissues of *P. xylostella* with transcript abundance for each family member being highest in hemocytes. The second largest gene family in the CcBV genome is the six-member *ank* family (Espagne et al., 2004). Expression patterns in hosts indicate expression of most family members begins within 24h of parasitism and continues several days thereafter (Falabella et al., 2007). Individual family members, however, appear to exhibit differences in the host tissues where they are expressed. Studies of the eight-member *ank* family encoded by CpBV also exhibit variable temporal and spatial expression patterns in parasitized hosts (Bai and Kim, 2009). Immunoblotting data also indicate that at least one member of the CcBV *cyst* family is secreted into

host hemolymph by 24h post parasitism (Serbielle et al., 2009).

While expression data are lacking for the *EP1-like* family from CcBV, studies of the seven-member *EP1-like* family from CpBV indicate that one member (*ELP1*) is expressed in the host *Plutella xylostella* from days 1–8 post parasitism in multiple tissues while others are expressed starting 3 days post parasitism (Kwon and Kim, 2008). In CcBV, EP1 and EP2 proteins are detected by 4h post parasitism in fat body and hemocytes, and continue to be expressed over the course of parasitism (Harwood et al., 1994; Le et al., 2003; Amaya et al., 2005). Two proteins from CrBV, named CrV1, and CrV2 (EP2 homolog) are transiently expressed in parasitized *Pieris rapae* hosts from only 4–8h post parasitism (Asgari et al., 1996, 1997). Transcripts for *CrV1* and *CrV2* are detected in fat body while *CrV1* transcript only is also detected in hemocytes. The CrV1 protein is glycosylated and secreted into hemolymph where it binds to hemocytes (Asgari et al., 1996, 1997). *Cotesia sesamiae* BV (CsBV) also encodes a *CrV1* homolog, which is differentially expressed in biotypes of its host, *Busseola fusca* (Gitau et al., 2007). EP1 gene expression levels also correlate with the host range of *C. congregata* and CcBV. In permissive hosts, EP1 is rapidly and abundantly expressed, whereas nonpermissive hosts exhibit greatly reduced levels of expression (Harwood et al., 1998; Beckage and Tan, 2002).

Other genes encoded by CcBV and other *Cotesia* BVs include lectins, which often function as binding or pattern recognition molecules (Glatz et al., 2003; Teramato and Tanaka, 2003; Lee et al., 2008). CrBV encodes a gene (*CrV3*) related to invertebrate C-type lectins that is expressed within 6h of parasitization in the fat body and hemocytes of *P. rapae* larvae. The CrV3 protein is also secreted into the hemolymph where it forms oligomeric complexes (Glatz et al., 2003). Two lectin genes (*Cky811* and *Crf111*) from *C. kariyai* and *C. ruficrus* BVs (Teramato and Tanaka, 2003) are expressed in host hemocytes and/or fat body by 6h post parasitism. Multiple *Cotesia* BVs also encode a single-copy histone H4 gene, which in hosts infected by CpBV is expressed in multiple tissues during the course of parasitism (Gad and Kim, 2009). CpBV also encodes a unique gene, *CpBV15β*, which is continuously expressed in host larvae and is apparently secreted into the hemolymph (Nalini and Kim, 2009).

BVs From Wasps in the Genus Glyptapanteles

The genus *Glyptapanteles* is the sister taxon to the genus *Cotesia* (Murphy et al., 2008). Associated BVs like GiBV and GfBV encode several of the same single-copy genes and gene families (*ptps, anks, cyst, hp, lectin*) encoded by *Cotesia* BVs as well as some novel genes such as an eight-member sugar transporter gene family (Desjardins et al.,

2008). Expression data, however, in the host *Lymantria dispar* are restricted to analysis of nine members of the *ptp* gene family and one member of the *ank* family encoded on GiBV segment F (Chen *et al.*, 2003; Gundersen-Rindal and Pedroni, 2006), which was renamed segment 25 by Desjardins *et al.* (2008). Transcripts for each GiBV *ptp* family member and the *ank* gene were detected within one day of parasitism and mostly continued for seven days thereafter. Each gene also exhibits tissue-specific patterns of expression consistent with having variable but currently unknown roles in parasitism.

BVs from Wasps in the Genus Microplitis

The genus *Microplitis* resides in the Microgastrinae but is distantly related to *Cotesia* and *Glyptapanteles* sp. (Murphy *et al.*, 2008). Correspondingly, the one sequenced BV from this genus, MdBV, shares only two gene families (*ptps*, *anks*) plus a few single-copy genes with *Cotesia* and *Glyptapanteles* BVs, while encoding several novel genes and gene families (*glcs*, *egfs*) (Webb *et al.*, 2006). Like CcBV and GiBV, detailed expression studies show that the 13-member *ptp* gene family of MdBV exhibits complex patterns of expression in the host *Pseudoplusia* (=*Chrysodeixis*) *includens* (Pruijssers and Strand, 2007; Eum *et al.*, 2010). Expression of most family members begins within 4 h of infection with one family member encoding a functional PTP specifically expressed in hemocytes (*ptpH2*) while other family members encoding predicted functional PTPs or pseudophosphatases are expressed in multiple tissues. Similar complexity is exhibited by the 12-member *ank* family where *ankH4*, which functions as an IκB, is expressed globally in hosts while other family members are selectively expressed in only one or two host tissues like the fat body and gut (Thoetkiattikul *et al.*, 2005; Bitra *et al.*, 2011).

The three members of the *egf* family reside on genomic segment O and encode secreted proteins that are the most abundantly expressed products produced by MdBV (Beck *et al.*, 2007). Each family member (*egf1.0*, *egf1.5*, and *egf0.4*) is preferentially expressed in hemocytes within 2 h of infection with expression of *egf1.0* and *egf1.5* continuing for multiple days thereafter. High-level expression of *egf1.0* and *egf1.5* is fully consistent with their function as competitive inhibitors of specific serine proteases in the host phenoloxidase cascade that are strongly upregulated after immune challenge (Beck and Strand, 2007; Lu *et al.*, 2008, 2010). The second most abundantly expressed genes are the *glc* genes, also located on genomic segment O, which encode heavily glycosylated surface proteins that inhibit adhesion of host hemocytes (Trudeau *et al.*, 2000; Beck and Strand, 2003; Johnson *et al.*, 2010). MdBV encodes 22 single-copy genes of which most share no conserved features with known genes and whose

functions largely remain unknown. Recent expression profiling indicates the majority of these factors are expressed in infected hosts but exhibit highly variable spatio-temporal patterns of activity (Bitra, Zhang, Strand, unpublished observations).

The observation that the highly expressed *egf* and *glc* genes reside on the same high-abundance genomic segment initially suggested that the non-equimolar production of genomic segments evolved as a strategy for adjusting the abundance of different viral gene products produced in parasitized hosts (Webb and Strand, 2005). However, more recent comparative studies by Beck *et al.* (2007) found instead that most multimember gene families are located on high-abundance genomic segments, regardless of expression levels, and that low abundance genomic segments in the MdBV genome tend to encode single copy genes or no identifiable open reading frames (ORFs) at all. Thus, the pattern in MdBV appears to be that most genes expressed in infected hosts are located on high-abundance segments and that low-abundance segments on average encode fewer genes, which are primarily single copy.

BV Gene Expression in Hosts: Subfamily Cardiochilinae

The Cardiochilinae is of intermediate relatedness to the Microgastrinae with BV host expression data restricted to a single isolate, TnBV, from the wasp *Toxoneuron nigriceps*. Partial sequence data of the TnBV encapsided genome identifies two families also present in BVs from microgastrine braconids (*ptps* and *anks*) plus several other genes. Similar to microgastrine BVs, examination of selected TnBV *ptp* and *ank* family members suggest most are transcribed relatively early after parasitism of the host *Heliothis virescens* (Provost *et al.*, 2004; Falabella *et al.*, 2007). By 48 h post parasitism, however, transcript levels appear to decline. Tissue-specific patterns of expression remain largely uncharacterized although one *ptp* family member is expressed in host prothoracic glands (Falabella *et al.*, 2006). Other TnBV expression data are restricted to a predicted aspartyl protease (*TnBV2*) detected 6–48 h post parasitism in host fat body, hemocytes, and prothoracic glands (Falabella *et al.*, 2003), and a gene named *TnBV1* also detected 48 h post parasitism in host prothoracic glands (Varricchio *et al.*, 1999).

BV Gene Expression in Hosts: Subfamily Cheloninae

The Cheloninae is the most distantly related subfamily to the Microgastrinae (Murphy *et al.*, 2008). Unlike most microgastrines and cardiochilines, which are larval endoparasitoids, all members of the Cheloninae are egg–larval parasitoids that oviposit in the egg stage of their hosts and whose offspring complete their immature development in the host's penultimate instar (Lanzrein et al., Chapter 14 of this volume). Genome and host expression data for BVs in this taxon are restricted to CiBV from *C. inanitus* and its host *Spodoptera littoralis*. Importantly, sequence data identify no genes in the encapsidated genome of CiBV with any homology to the genes present in the encapsidated genomes of microgastrine or cardiochiline BVs (Annaheim and Lanzrein, 2007; Bézier et al., 2009a). Also unlike microgastrine and cardiochiline BVs, where most viral transcripts are transcribed within hours of parasitism and continue for days thereafter, CiBV genes appear to exhibit more stage-specific patterns of expression. For example, RT-PCR analysis of eight CiBV genes located on five different genomic segments revealed that some genes (*12g1*, *16.8g1*) are persistently expressed over the course of parasitism (Bonvin et al., 2005). Others, however, are only expressed early in parasitism (*14.5g1*, *21g1*), late in parasitism (*12g2*, *14g1*, *14g2*), or both early and late (*14g3*) (Johner and Lanzrein, 2002; Bonvin et al., 2005). Additional studies further show that CiBV genes exhibiting a given expression pattern (persistent, early, late) tend to cluster on the same genomic segment, and that no viral genes are transcribed at high levels relative to housekeeping genes like host actin (Weber et al., 2007).

IV Gene Expression in Hosts

Sequencing of the CsIV, HfIV, and TrIV encapsidated genomes identified several single-copy genes plus six shared gene families: the cysteine-motif genes (*cys-motif*) which share a cysteine-rich domain related to ω-conotoxins, the repeat element genes (*rep*) distinguished by a shared 540-bp repeat, the ankyrin motif genes (*anks*), the innexin genes (*inex*) that encode predicted proteins involved in forming gap junctions, the N-family genes (*N-family*), and the polar-residue rich genes (*pol-res*) (Webb et al., 2006; Tanaka et al., 2007). The lone banchine-associated IV examined to date (GfIV) shares only one gene family with campoplegine IVs (*ank* genes) but the GfIV *anks* are also structurally distinct from the *anks* encoded by campoplegine IVs and BVs (Lapointe et al., 2007). Other factors identified in GfIV include a 23-member *ptp* family structurally distinct from BV *ptps*, a novel seven-member gene family of NTPase-like proteins containing a pox-D5 domain, genes with similarity to predicted proteins found in other BVs, and genes with no similarity to sequences in existing databases (Lapointe et al., 2007).

IV gene expression patterns in hosts are largely restricted to CsIV from *C. sonorensis* and its *Heliothis virescens*. A few additional data are available from studies of HdIV and selected other campoplegines. In contrast, no published data are yet available for GfIV or other banchine IVs. The 10-member *cys-motif* family is the most studied in the CsIV genome (Blissard et al., 1987; Dib-Hajj et al., 1993; Cui and Webb, 1996). Each gene in this family exhibits a variable cys-motif but shares a very conserved exon–intron structure, hydrophobic signal peptide, and N-terminal glycosylation site (Dib-Hajj et al., 1993; Dupas et al., 2003). Expression patterns of the V and W cys-motif genes (*VHv1.1* and *VHv1.4*, and *WHv1.0* and *WHv1.6*, respectively) have been studied in some detail with each being expressed from 5 h to 8 days post parasitism in host fat body and/or hemocytes (Blissard et al., 1987, 1989). Each product is also secreted into the hemolymph of the host where protein expression data indicate that VHv1.1 binds to hemocytes (Li and Webb, 1994).

The 28-member *rep* gene family is the largest in the CsIV genome (Volkoff et al., 2002; Hilgarth and Webb, 2002; Webb et al., 2006). Segments B, H, and O[1] encode five *rep* genes that are expressed in parasitized *H. virescens* beginning 2 h post parasitism and continuing for up to 10 days (Theilmann and Summers, 1986, 1988). Studies of three other *rep* genes on CsIV genomic segment O show very similar temporal patterns of expression as well as global expression in virtually all host tissues including hemocytes, fat body, and the epidermis (Hilgarth and Webb, 2002). In contrast, studies of 10 *rep* gene homologs from HdIV indicate that family members display differing levels of transcript abundance 24 h post parasitism of *Spodoptera frugiperda* as well as different patterns of expression in host tissues (Galibert et al., 2006). Most but not all family members are also most strongly expressed in the fat body and epidermis, while expression in hemocytes is very low (Galibert et al., 2006). Similar studies with the 17-member *rep* gene family from TrIV also identified tissue-specific patterns of activity (Rasoolizadeh et al., 2009).

The CsIV *ank* family consists of seven members encoded on genomic segments P and I[2] (Webb et al., 2006). Six family members are expressed in parasitized *H. virescens* from 4–8 h through 5 days post parasitism (Kroemer and Webb, 2005). Spatially, these family members divide themselves into two categories with *anks P1, I²2, I²3* detected in host fat body and nervous tissue, and *anks P2-4 and I²1* detected predominantly in hemocytes (Kroemer and Webb, 2005). Immunocytochemical analysis further indicates that Ank1 and Ank4 localize to fat body and hemocyte nuclei, respectively (Kroemer and Webb, 2005). The four members of the CsIV *inex* family are located on genomic segments D, G, and Q[2] (Webb et al., 2006). Three family members, *Csvnx*-d1, *Csvnx*-g1, and *Csvnx*-q2, are expressed in parasitized *H. virescens* with *Csvnx*-q2 protein localizing to membranes of CsIV-infected hemocytes (Turnbull and Webb, 2002; Turnbull et al., 2005).

Other IV genes for which any information exists derive from studies of HdIV in *S. frugiperda* parasitized by *H. didymator*. In addition to encoding the gene families present among other campoplegine IVs, HdIV also encodes a novel family of secreted glycine or proline-rich proteins (M24, M27, and M40) on genome segment M (Volkoff *et al.*, 1999). Transcription of these genes is detected from 4 h through the duration of parasitoid development. Two other novel genes (*s6* and *p30*) encoded on HdIV segments C and G are expressed at high levels by 2 h post-infection of the Sf9 cell line or 24 h post parasitism, primarily in hemocytes (Galibert *et al.*, 2003). The *s6* gene encodes a predicted transmembrane protein while *p30* encodes a predicted secreted and glycosylated protein with mucin-like motifs.

CONCLUDING REMARKS

While BVs and IVs are associated with literally tens of thousands of wasp species, our understanding of the genes they encode and express in wasps and hosts is restricted to only a handful of isolates. Still, the literature collectively suggests that gene expression patterns related to replication is conserved among BVs and IVs, respectively, whereas expression patterns in parasitized hosts are much more variable. However, additional sampling within different wasp taxa is needed to fully gauge how variable gene content of encapsidated genomes actually is among PDVs from closely and more distantly related wasp species. Among BVs, current data suggest the largest and most conserved gene families, like *ptps* and *anks* in microgastrine BVs, exhibit extremely variable spatial patterns of expression suggesting individual family members have specialized to interact with signaling processes associated with particular host cells or tissues (Provost *et al.*, 2004; Pruijssers and Strand, 2007). Sequence analysis of *Cotesia* BV *cyst* and *CrV1* genes as well as IV *cys-motif* genes also provides evidence for individual gene family members specializing to interact with particular proteins in the same host or genetically distinct races of a given host (Dupas *et al.*, 2003; Gitau *et al.*, 2007; Serbielle *et al.*, 2009). In contrast, although the secreted gene products encoded by *Egf1.0* and *Egf1.5* from MdBV differ in primary structure, recent studies indicate they are functional paralogs with no measureable differences in activity as inhibitors of the phenoloxidase cascade (Lu *et al.*, 2010). Likewise, the essentially continuous and global patterns of expression of, for example, CsIV *rep* genes suggest that at least some family members are functionally redundant. In addition, to broadening our understanding of how gene content varies among the encapsidated genomes of PDVs, a second key challenge is to develop approaches for characterizing the function of multimember gene families.

ACKNOWLEDGMENTS

Some of the work discussed was supported by grants from the National Science Foundation (IOS 0749450) and USDA NRI/AFRI (2007-04549 and 2008-04028).

REFERENCES

Amaya, K.E., Asgari, S., Jung, R., Hongskula, M., Beckage, N.E., 2005. Parasitization of *Manduca sexta* by the parastoid *Cotesia congregata* induces an impaired host immune response. J. Insect Physiol. 51, 505–512.

Annaheim, M., Lanzrein, B., 2007. Genome organization of the *Chelonus inanitus* polydnavirus: excision sites, spacers, and abundance of proviral and excised segments. J. Gen. Virol. 8, 450–457.

Asgari, S., Hellers, M., Schmidt, O., 1996. Host hemocyte inactivation by an insect parasitoid: transient expression of a polydnavirus gene. J. Gen. Virol. 77, 2653–2662.

Asgari, S., Schmidt, O., Theopold, U., 1997. A polydnavirus-encoded protein of an endoparasitoid wasp is an immune suppressor. J. Gen. Virol. 78, 3061–3070.

Bai, S., Kim, Y., 2009. IkB genes encoded in *Cotesia plutellae* bracovirus suppress an antiviral response and enhance baculovirus pathogenicity against the diamondback moth, *Plutella xylostella*. J. Invert. Pathol. 102, 79–87.

Beck, M.H., Inman, R.B., Strand, M.R., 2007. *Microplitis demolitor* bracovirus genome segments vary in abundance and are individually packaged in virions. Virology 359, 179–189.

Beck, M., Strand, M.R., 2003. RNA interference silences *Microplitis demolitor* bracovirus genes and implicates glc1.8 in disruption of adhesion in infected host cells. Virology 314, 521–535.

Beck, M.H., Strand, M.R., 2007. A novel polydnavirus protein inhibits the insect prophenoloxidase activation pathway. Proc. Natl. Acad. Sci. U.S.A. 104, 19267–19272.

Beckage, N.E., Tan, F.F., 2002. Development of the braconid wasp *Cotesia congregata* in a semi-permissive noctuid host, *Trichoplusia ni*. J. Invert. Pathol. 81, 49–52.

Bézier, A., Annaheim, M., Herbinière, J., Wetterwald, C., Gyapay, G., Bernard-Samain, S., et al. 2009. Polydnaviruses of braconid wasps derive from an ancestral nudivirus. Science 323, 926–930.

Bézier, A., Herbinière, J., Lanzrein, B., Drezen, J.-M., 2009. Polydnavirus hidden face: the genes producing virus particles of parasitic wasps. J. Invert. Pathol. 101, 194–203.

Bitra, K., Zhang, S., Strand, M.R., 2011. Transcriptomic profiling of *Microplitis demolitor* bracovirus reveals host, tissue, and stage-specific patterns of activity. J. Gen. Virol. (In press.)

Blissard, G.W., Smith, O.P., Summers, M.D., 1987. Two related viral genes are located on a single superhelical DNA segment of the multipartite *Campoletis sonorensis* virus genome. Virology 160, 120–134.

Blissard, G.W., Theilmann, D.A., Summers, M.D., 1989. Segment W of *Campoletis sonorensis* virus: expression, gene products, and organization. Virology 169, 78–89.

Blissard, G.W., Smith, O.P., Summers, M.D., 1987. Two related viral genes are located on a single superhelical DNA segment of the multipartite *Campoletis sonorensis* virus genome. Virology 160, 120–134.

Bonvin, M., Marti, D., Wyder, S., Kojic, D., Annaheim, M., Lanzrein, B., 2005. Cloning, characterization and analysis by RNA interference of various genes of the *Chelonus inanitus* polydnavirus. J. Gen. Virol. 86, 973–983.

Chen, Y.P., Taylor, P.B., Shapiro, M., Gundersen-Rindal, D.E., 2003. Quantitative expression analysis of a *Glyptapanteles indiensis* polydnavirus protein tyrosine phosphatase gene in its natural lepidopteran host, *Lymantria dispar*. Insect Mol. Biol. 12, 271–280.

Choi, J.Y., Kwon, S.J., Roh, J.Y., Yang, T.J., Yoon, S.H., Kim, H., et al. 2009. Sequence and gene organization of 24 circles from the

Cotesia plutellae bracovirus genome. Arch. Virol. 154, 1313–1327.

Cui, L., Webb, B.A., 1996. Isolation and characterization of a member of the cysteine-rich gene family from Campoletis sonorensis polydnavirus. J. Gen. Virol. 77, 797–809.

Deng, L., Stoltz, D.B., Webb, B.A., 2000. A gene encoding a polydnavirus structural polypeptide is not encapsidated. Virology 269, 440–450.

Deng, L., Webb, B.A., 1999. Cloning and expression of a gene encoding a *Campoletis sonorensis* polydnavirus structural protein. Arch. Insect Biochem. 40, 30–40.

Desjardins, C.A., Gundersen-Rindal, D.E., Hostetler, J.B., Tallon, L.J., Fadrosh, D.W., Fuester, R.W., et al. 2008. Comparative genomics of mutualistic viruses of *Glyptapanteles* parasitic wasps. Genome Biol. 9, R183.

Dib-Hajj, S.D., Webb, B.A., Summers, M.D., 1993. Structure and evolutionary implications of a 'cysteine-rich' *Campoletis sonorensis* polydnavirus gene family. Proc. Natl. Acad. Sci. U.S.A. 90, 3765–3769.

Dupas, S., Turnbull, M., Webb, B.A., 2003. Diversifying selection in a parasitoid's symbiotic virus among genes involved in inhibiting host immunity. Immunogenetics 55, 351–361.

Espagne, E., Dupuy, C., Huguet, E., Cattolico, L., Provost, B., Martins, N., et al. 2004. Genome sequence of a polydnavirus: insights into symbiotic virus evolution. Science 306, 286–289.

Eum, J.H., Bottjen, R.C, Clark, K.D., Strand, M.R., 2010. Characterization and kinetic analysis of protein tyrosine phosphatase-H2 from *Microplitis demolitor* bracovirus. Insect Biochem. Mol. Biol. 40, 690–698.

Falabella, P., Cacciaupi, P., Varricchio, P., Malva, C., Pennacchio, P., 2006. Protein tyrosine phosphatases of *Toxoneuron nigriceps* bracovirus as potential disrupters of host prothoracic gland function. Arch. Insect Biochem. Physiol. 61, 157–169.

Falabella, P., Varricchio, P., Gigliotti, S., Tranfaglia, A., Pennacchio, F., Malva, C., 2003. *Toxoneuron nigriceps* polydnavirus encodes a putative aspartyl protease highly expressed in parasitized host larvae. Insect Mol. Biol. 12, 9–17.

Falabella, P., Varricchio, P., Provost, B., Espagne, E., Ferrarese, R., Grimaldi, A., et al. 2007. Characterization of the IkappaB-like gene family in polydnaviruses associated with wasps belonging to different Braconid subfamilies. J. Gen. Virol. 88, 92–104.

Fleming, J.G.W., 1992. Polydnaviruses: multualists and pathogens. Annu. Rev. Entomol. 37, 401–425.

Fleming, J.G.W., Summers, M.D., 1986. *Campoletis sonorensis* endoparasitic wasps contain forms of C. sonorensis virus DNA suggestive of integrated and extrachromosomal polydnavirus DNAs. J. Virol. 57, 552–562.

Fleming, J.G., Summers, M.D., 1991. PDV DNA is integrated in the DNA of its parasitoid wasp host. Proc. Natl. Acad. Sci. U.S.A. 88, 9770–9774.

Gad, W., Kim, Y., 2009. N-terminal tail of a viral histone H4 encoded by *Cotesia plutellae* bracovirus is essential to suppress gene expression of host histone H4. Insect Mol. Biol. 18, 111–118.

Galibert, L., Devauchelle, G., Cousserans, F., Rocher, J., Cerutti, P., Barat-Houari, M., et al. 2006. Members of the *Hyposoter didymator* ichnovirus repeat element gene family are differentially expressed in *Spodoptera frugiperda*. Virol. J. 3, 48.

Galibert, L., Rocher, J., Ravallec, M., Duonor-Cerutti, M., Webb, B.A., Volkoff, A.N., 2003. Two *Hyposoter didymator* ichnovirus genes expressed in the lepidopteran host encode secreted or membrane-associated serine and threonine rich proteins in segments that may be nested. J. Insect Physiol. 49, 441–451.

Gitau, C.W., Gundersen-Rindal, D., Pedroni, M., Mbugi, C., Dupas, S., 2007. Differential expression of the CrV1 hemocyte inactivation-associated polydnavirus gene in the African maize stem borer Busseola fusca (Fuller) parasitized by two biotypes of the endoparasitoid Cotesia sesamiae (Cameron). J. Insect Physiol. 53, 676–684.

Glatz, R., Schmidt, O.., Asgari, S., 2003. Characterization of a novel protein with homology to C-type lectins expressed by the *Cotesia rubecula* bracovirus in larvae of the lepidopteran host, Pieris rapae. J. Biol. Chem. 278, 19743–19750.

Gruber, A., Stettler, P., Heiniger, P., Schumperli, D., Lanzrein, B., 1996. Polydnavirus DNA of the braconid wasp *Chelonus inanitus* is integrated in the wasp's genome and excised only in later pupal and adult stages of the female. J. Gen. Virol. 77, 2873–2879.

Gundersen-Rindal, D.E., Pedroni, M.J., 2006. Characterization and transcriptional analysis of protein tyrosine phosphatase genes and an ankyrin repeat gene of the parasitoid *Glyptapanteles indiensis* polydnavirus in the parasitized host. J. Gen. Virol. 87, 311–322.

Harwood, S.H., Grosovsky, A.J., Cowles, E.A., Davis, J.W., Beckage, N.E., 1994. An abundantly expressed hemolymph glycoprotein isolated from newly parasitized *Manduca sexta* larvae is a PDV gene product. Virology 205, 381–392.

Harwood, S.H., McElfresh, J.S., Nguyen, A., Conlan, C.A., Beckage, N.E., 1998. Production of early expressed parasitism–specific proteins in alternate sphingid hosts of the braconid wasp *Cotesia congregata*. J. Invertebr. Pathol. 71, 271–279.

Hilgarth, R.S., Webb, B.A., 2002. Characterization of *Campoletis sonorensis* ichnovirus segment I genes as members of the repeat element gene family. J. Gen. Virol. 83, 2393–2402.

Ibrahim, A.M.A., Choi, J.Y., Je, Y.H., Kim, Y., 2007. Protein tyrosine phosphatases encoded in *Cotesia plutellae* bracovirus: sequence analysis, expression profile, and a possible role in host immunosuppression. Devp. Comp. Immunol. 31, 978–990.

Johner, A., Lanzrein, B., 2002. Characterization of two genes of the polydnavirus of *Chelonus inanitus* and their stage-specific expression in the host Spodoptera littoralis. J. Gen. Virol. 83, 1075–1085.

Johnson, J.A., Bitra, K., Zhang, S., Wang, L., Lynn, D.E., Strand, M.R., 2010. The UGA-CiE1 cell line from *Chrysodeixis includens* exhibits characteristics of granulocytes and is permissive to infection by two viruses. Insect Biochem. Mol. Biol. 40, 394–404.

Kroemer, J.A., Webb, B.A., 2005. IkB-related vankyrin genes in the *Campoletis sonorensis* ichnovirus: temporal and tissue-specific patterns of expression in parasitized *Heliothis virescens* lepidopteran hosts. J. Virol. 79, 7617–7628.

Kwon, B., Kim, Y., 2008. Transient expression of an EP1-like gene encoded in *Cotesia plutellae* bracovirus suppresses the hemocyte population in the diamondback moth, *Plutella xylostella*. Dev. Comp. Immunol. 32, 932–942.

Lapointe, R., Tanaka, K., Barney, W.E., Whitfield, J.B., Banks, J.C., Béliveau, C., et al. 2007. Genomic and morphological features of a banchine polydnavirus: comparison with bracoviruses and ichnoviruses. J. Virol. 81, 6491–6501.

Le, N.T., Asgari, S., Amaya, K., Tan, F.F., Beckage, N.E., 2003. Persistence and expression of *Cotesia congregata* polydnavirus in host larvae of the tobacco hornworm, *Manduca sexta*. J. Insect Physiol. 49, 533–543.

Lee, S., Nalini, M., Kim, Y., 2008. A viral lectin encoded in *Cotesia plutellae* bracovirus and its immunosuppressive effect on host hemocytes. Comp. Biochem. Physiol. A 149, 351–361.

Li, X., Webb, B.A., 1994. Apparent functional role for a cysteine-rich polydnavirus protein in suppression of the insect cellular immune response. J. Virol. 68, 7482–7489.

Lu, Z., Beck, M.H., Jiang, H., Wang, Y., Strand, M.R., 2008. The viral protein Egf1.0 is a dual activity inhibitor of prophenoloxidase activating proteinases 1 and 3 from Manduca sexta. J. Biol. Chem. 283, 21325–21333.

Lu, Z., Beck, M.H., Strand, M.R., 2010. Egf1.5 is a second phenoloxidase cascade inhibitor encoded by Microplitis demolitor bracovirus. Insect Biochem. Mol. Biol. 40, 497–505.

Murphy, N., Banks, J.C., Whitfield, J.B., Austin, A.D., 2008. Phylogeny of the parasitic microgastroid subfamilies (Hymenoptera: Braconidae) based on sequence data from seven genes, with an improved time estimate of the origin of the lineage. Mol. Phylogenet. Evol. 47, 378–395.

Nalini, M., Kim, Y., 2009. Transient expression of a polydnaviral gene, CpBV15 beta, induces immune and developmental alterations of the diamondback moth, Plutella xylostella. J. Invert. Path. 100, 22–28.

Norton, W.N., Vinson, S.B., 1983. Correlating the initiation of virus replication with a specific pupal developmental phase of an ichneumonid parasitoid. Cell Tissue Res. 231, 387–398.

Provost, B., Varricchio, C., Arana, E., Espagne, E., Falabella, P., Huguet, E., et al. 2004. Bracoviruses contain a large multigene family coding for protein tyrosine phosphatases. J. Virol. 78, 13090–13103.

Pruijssers, A.J., Strand, M.R., 2007. PTP-H2 and PTP-H3 from Microplitis demolitor bracovirus localize to focal adhesions and are antiphagocytic in insect immune cells. J. Virol. 81, 1209–1219.

Rasoolizadeh, A., Beliveau, C., Stewart, D., Cloutier, C., Cusson, M., 2009. Tranosema rostrale ichnovirus repeat element genes display distinct transcriptional patterns in caterpillar and wasp hosts. J. Gen. Virol. 90, 1505–1514.

Savary, S., Drezen, J.-M., Tan, F., Beckage, N.E., Periquet, G., 1999. The excision of PDV sequences from the genome of the wasp Cotesia congregata (Braconidae, microgastrinae) is developmentally regulated but not strictly restricted to the ovaries in the adult. Insect Mol. Biol. 8, 319–327.

Serbielle, C., Moreau, S., Veillard, F., Voldoire, E., Bézier, A., Mannucci, M.A., et al. 2009. Identification of parasite-responsive cysteine proteases in Manduca sexta. Biol. Chem. 390, 493–502.

Stoltz, D.B., 1993. The PDV life-cycle. In: Thompson, S.N., Federici, B.A., Beckage, N.E. (Eds.), Parasites and Pathogens of Insects. Vol. 1: Parasites (pp. 167–187). Academic Press, San Diego.

Strand, M.R., 2009. The interactions between polydnavirus-carrying parasitoids and their lepidopteran hosts. In: Goldsmith, M.R., Marec, F. (Eds.), Molecular Biology and Genetics of the Lepidoptera (pp. 321–336). CRC Press, Boca Raton.

Strand, M.R., 2010. Polydnaviruses. In: Asgari, S., Johnson, K.N. (Eds.), Insect Virology (pp. 171–197). Caister Academic Press, Norwich.

Tanaka, K., Lapointe, R., Barney, W.E., Makkay, A.M., Stoltz, D., Cusson, M., et al. 2007. Shared and species-specific features among ichnovirus genomes. Virology 363, 26–35.

Teramato, T., Tanaka, T., 2003. Similar polydnavirus genes of two parasitoids, Cotesia kariyai and Cotesia rufcrus, of the host Pseudaletia separata. J. Insect Physiol. 49, 463–471.

Theilmann, D.A., Summers, M.D., 1986. Molecular analysis of Campoletis sonorensis virus DNA in the lepidopteran host Heliothis virescens. J. Gen. Virol. 67, 1961–1969.

Theilmann, D.A., Summers, M.D., 1988. Identification and comparison of Campoletis sonorensis virus transcripts expressed from four genomic segments in the insect hosts Campoletis sonorensis and Heliothis virescens. Virology 167, 329–341.

Thoetkiattikul, H., Beck, M.H., Strand, M.R., 2005. Inhibitor kappaB-like proteins from a polydnavirus inhibit NF-kappaB activation and suppress the insect immune response. Proc. Natl. Acad. Sci. U.S.A. 102, 11426–11431.

Trudeau, D., Witherell, R.A., Strand, M.R., 2000. Characterization of two novel Microplitis demolitor PDV mRNAs expressed in Pseudoplusia includens hemocytes. J. Gen. Virol. 81, 3049–3058.

Turnbull, M.W., Volkoff, A.N., Webb, B.A., Phelan, P., 2005. Functional gap junction genes are encoded by insect viruses. Curr. Biol. 15, R491–492.

Turnbull, M.W., Webb, B.A., 2002. Perspectives on polydnavirus origin and evolution. Adv. Virus Res. 58, 203–254.

Varricchio, P., Falabella, P., Sordetti, R., Graziani, F., Malva, C., Pennacchio, F., 1999. Cardiochiles nigriceps polydnavirus: molecular characterization and gene expression in parasitized Heliothis virescens larvae. Insect Biochem. Mol. Biol. 29, 1087–1096.

Volkoff, A-N., Cerutti, P., Rocher, J., Ohresser, M.C.P., Devauchelle, G., Duonor-Cerutti, M., 1999. Related RNAs in lepidopteran cells after in vitro infection with Hyposoter didymator virus define a new polydnavirus gene family. Virology 263, 349–363.

Volkoff, A-N., Jouan, V., Urbach, S., Samain, S., Bergoin, M., Wincker, P., et al. 2010. Analysis of virion structural components reveals vestiges of the ancestral ichnovirus genome. PLoS Pathog. 6, e1000923.

Volkoff, A-N., Ravallec, M., Bossy, J., Cerutti, P., Rocher, J., Cerutti, M., et al. 1995. The replication of Hyposoter didymator PDV: cytopathology of the calyx cells in the parasitoid. Biol. Cell 83, 1–13.

Volkoff, A-N., Rocher, J., Duonor-Cerutti, M., Webb, B.A., Hilgarth, R., Cusson, M., et al. 2002. Evidence for a conserved polydnavirus gene family: homologs of the CsIV repeat element genes from two additional ichnoviruses. Virology 300, 316–331.

Webb, B.A., Strand, M.R., 2005. The biology and genomics of polydnaviruses. In: Gilbert, L.I., Iatrou, K., Gill, S.S. (Eds.), Comprehensive Molecular Insect Science (pp. 323–360). Elsevier Inc., San Diego.

Webb, B.A., Strand, M.R., Deborde, S.E., Beck, M., Hilgarth, R.S., Kadash, K., et al. 2006. Polydnavirus genomes reflect their dual roles as mutualists and pathogens. Virology 347, 160–174.

Webb, B.A., Summers, M.D., 1992. Stimulation of PDV replication by 20-hydroxyecdysone. Experientia 48, 1018–1022.

Weber, B., Annaheim, M., Lanzrein, B., 2007. Transcriptional analysis of polydnavirus gene in the course of parasitization reveals segment-specific patterns. Arch. Insect Biochem. Physiol. 66, 9–22.

Wetterwald, C., Roth, T., Kaeslin, M., Annaheim, M., Wespi, G., Heller, M., et al. 2010. Identification of bracovirus particle proteins and analysis of their transcript levels at the stage of virion formation. J. Gen. Virol. 91, 2610–2619.

Wyler, T., Lanzrein, B., 2003. Ovary development and polydnavirus morphogenesis in the parasitic wasp Chelonus inanitus. II. Ultrastructural analysis of calyx cell development, virion formation and release. J. Gen. Virol. 84, 1151–1163.

Chapter 12

Polydnavirus Gene Products that Interact with the Host Immune System

Michael R. Strand

Department of Entomology, University of Georgia, Athens, Georgia 30602-7415, U.S.A.

SUMMARY

A key function of most polydnaviruses is protection of the wasp's progeny from elimination by the host insect's immune system. The host immune system consists of both cellular and humoral components. Polydnaviruses encode a diversity of genes with some forming multimember families while others are single copy. Functional studies implicate selected members of certain gene families in disabling different host immune defenses. Here I discuss these gene products from the perspective of how they interact with the host immune system and facilitate survival of parasitoid offspring.

ABBREVIATIONS

AMP	antimicrobial peptide
AcMNPV	*Autographa californica* nuclear polyhedrosis virus
BV	bracovirus
GNBP	gram-negative bacteria recognition protein
GRP	glucan recognition protein
HCP	hemocyte chemotactic peptide
IV	ichnovirus
LPS	lipopolysaccharide
PAP	prophenoloxidase activating protease
PDV	polydnavirus
PGRP	peptidoglycan recognition protein
PO	phenoloxidase
PTP	protein tyrosine phosphatase
SPH	serine protease homolog
TEP	thioester-containing proteins
VLPs	virus-like particles

INTRODUCTION

As discussed elsewhere in this volume, the Polydnaviridae is divided into the genus bracovirus (BV), which is associated with wasps in seven subfamilies of the Braconidae (Cardiochilinae, Cheloninae, Dirrhoponae, Mendesellinae, Khoikhoiinae, Miricinae, Microgastrinae; ca. 17,500 species total) and the genus ichnovirus (IV), which is associated with wasps in two subfamilies of the Ichneumonidae (Campopleginae, Banchinae; ca. 13,000 species total). Strict vertical transmission as proviruses results in each wasp species carrying a genetically unique polydnavirus (PDV) isolate. Most PDV-carrying wasps also parasitize only one or a small number of host species, which are primarily larval or egg stage Lepidoptera (Webb and Strand, 2005; Pennacchio and Strand, 2006; Strand, 2010). All PDV-carrying wasps inject the encapsidated form of their virus into host insects when they oviposit. Virions rapidly infect different host tissues and within 1–2 h begin expressing viral gene products. PDVs do not replicate in the wasp's host but a number of the genes they express are implicated in causing physiological alterations of benefit to developing offspring. Among the most important of these alterations is preventing the host's immune system from killing wasp eggs and larvae. Here, I first summarize how the insect immune system responds to parasitoids and other foreign invaders. I then discuss how different PDV isolates interact with the immune system of their wasp's host.

HOST IMMUNE DEFENSES AGAINST PARASITOIDS

Understanding how PDVs interact with the host immune system first requires some background on the insect immune system generally and defense against parasitoids in particular. The innate immune system of insects consists of cellular and humoral components that respond to a variety of potentially pathogenic organisms including viruses, bacteria, fungi, protozoans, and multicellular parasites like parasitoid

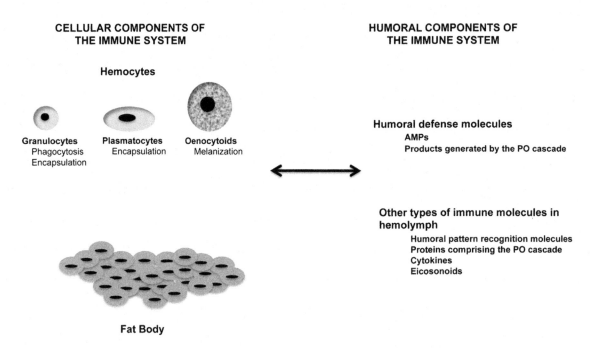

FIGURE 1 Major components of the cellular and humoral arms of the insect immune system. Arrow indicates that the cellular arm of the immune system (hemocytes and fat body) produce most of the immune molecules in hemolymph. Reciprocally, many of these molecules affect the function of hemocytes and the fat body. See text for discussion.

wasps (Fig. 1). Cellular defenses refer to responses directly mediated by hemocytes like encapsulation and phagocytosis (Strand, 2008a) (Fig. 1). Humoral defenses in contrast refer to molecules produced by hemocytes or tissues like the fat body that are secreted into the hemolymph and have the ability to kill foreign intruders (Cerenius and Soderhall, 2004; Imler and Bulet, 2005; Kanost and Gorman, 2008) (Fig. 1). Examples include antimicrobial peptides (AMPs), complement-like proteins, and products generated by the phenoloxidase (PO) cascade. Some of these effector molecules are constitutively produced while others are inducibly expressed following an immune challenge like parasitism. Hemocytes and the fat body also constitutively or inducibly express a number of other immune molecules, which they release into hemolymph (Fig. 1). Several of these factors then regulate the activity of hemocytes and the fat body by modulating signaling processes.

Lepidopteran Hemocyte Types

As previously noted, the majority of PDV-carrying parasitoids parasitize Lepidoptera with most species ovipositing into and progeny completing their juvenile development in the host larval stage. The primary exceptions to this rule are members of the Cheloninae, which oviposit into the egg stage of hosts and whose progeny complete their development in the penultimate instar (see Lanzrein *et al.*, Chapter 14 of this volume). Larval stage Lepidoptera produce hemocytes comprised of four subpopulations named

granulocytes, plasmatocytes, spherule cells, and oenocytoids (Fig. 1). Each subpopulation is distinguished by a combination of morphological, molecular, and functional markers (Lavine and Strand, 2002; Strand, 2008a; Nakahara *et al.*, 2009). Monoclonal antibody markers also suggest that some subpopulations, like plasmatocytes, are themselves specialized into subtypes (see Gardiner and Strand, 1999, for an example). Granulocytes are usually the most abundant subpopulation whose functions include phagocytosis and initiation of an encapsulation response. Plasmatocytes are usually the second most abundant subpopulation and are the main capsule-forming hemocyte. Granulocytes and plasmatocytes circulate in the hemolymph but both rapidly become adhesive and bind to a variety of foreign surfaces following immune challenge. These two cell types also produce some AMPs and other humoral effector molecules. Oenocytoids are non-adhesive cells that express phenoloxidase and other components of the PO cascade, while spherule cells are potential sources of cuticular proteins but have no known role in immunity. Progenitor cells called prohemocytes reside in hematopoietic organs and in some species also occur at low frequency in circulation (Gardiner and Strand, 2000; Nakahara *et al.*, 2003; Nardi *et al.*, 2003, 2006).

Encapsulation of Parasitoids and Related Defense Responses

The most important immune defense mounted by Lepidoptera toward larval endoparasitoids is encapsulation

Host immune factors implicated in regulating capsule formation	Major steps in capsule formation by host insects	PDV gene products implicated in disabling encapsulation

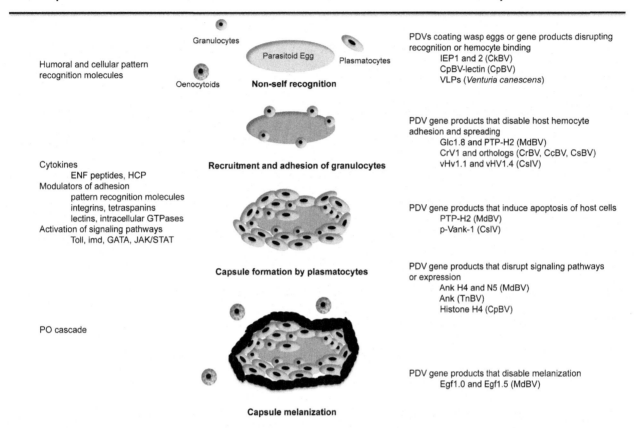

Host immune factors implicated in regulating capsule formation:

Humoral and cellular pattern recognition molecules

Cytokines
ENF peptides, HCP
Modulators of adhesion
pattern recognition molecules
integrins, tetraspanins
lectins, intracellular GTPases
Activation of signaling pathways
Toll, imd, GATA, JAK/STAT

PO cascade

Major steps in capsule formation by host insects:

Granulocytes   Parasitoid Egg   Plasmatocytes
Oenocytoids   **Non-self recognition**

Recruitment and adhesion of granulocytes

Capsule formation by plasmatocytes

Capsule melanization

PDV gene products implicated in disabling encapsulation:

PDVs coating wasp eggs or gene products disrupting recognition or hemocyte binding
IEP1 and 2 (CkBV)
CpBV-lectin (CpBV)
VLPs (*Venturia canescens*)

PDV gene products that disable host hemocyte adhesion and spreading
Glc1.8 and PTP-H2 (MdBV)
CrV1 and orthologs (CrBV, CcBV, CsBV)
vHv1.1 and vHV1.4 (CsIV)

PDV gene products that induce apoptosis of host cells
PTP-H2 (MdBV)
p-Vank-1 (CsIV)

PDV gene products that disrupt signaling pathways or expression
Ank H4 and N5 (MdBV)
Ank (TnBV)
Histone H4 (CpBV)

PDV gene products that disable melanization
Egf1.0 and Egf1.5 (MdBV)

FIGURE 2   Overview of the encapsulation response of lepidopteran insects toward the eggs of PDV-carrying parasitoids. The middle of the figure schematically shows the major steps of capsule formation beginning with recognition of the parasitoid egg as foreign and ending with melanization of a capsule. The column to the left lists some of the host immune molecules that regulate a given step in capsule formation, while the column to the right lists the major effects that PDVs have on capsule formation and the identity of selected PDV-gene products implicated in disrupting a given function. See text for discussion.

(Schmidt *et al.*, 2001; Strand, 2008a, b). This response involves the binding of large numbers of hemocytes to the surface of eggs and/or larvae to form an overlapping sheath of cells (Strand 2008a, 2009) (Fig. 2). Encapsulation is also closely related to phagocytosis, a process where hemocytes internalize small foreign targets like bacteria, and nodulation where small targets are enveloped rather than internalized by hemocytes. Encapsulation, phagocytosis, and nodulation all begin when hemocytes recognize a parasitoid or other intruder as foreign (Fig. 2). This response is mediated by recognition receptors expressed on the surface of hemocytes as well as by humoral pattern recognition receptors, which enhance recognition by binding to the surface of the target (i.e., opsonization). Several cell surface receptors with roles in mediating phagocytosis or encapsulation have been identified from several insects, although most studies focus on *Drosophila*, Lepidoptera, and mosquitoes. Examples include scavenger receptors (Ramet *et al.*, 2001; Kocks *et al.*,

2005; Philips *et al.*, 2005), Down's syndrome cell adhesion protein (Dscam) (Ramet *et al.*, 2002; Moita *et al.*, 2005; Dong *et al.*, 2006), the nimrod superfamily (Somogyi *et al.*, 2008), and integrins (Lavine and Strand, 2003; Irving *et al.*, 2005; Levin *et al.*, 2005; Wertheim *et al.*, 2005; Moita *et al.*, 2005). Humoral molecules that function as pattern recognition receptors include hemolin, lipopolysaccharide (LPS)-binding protein, gram-negative bacteria recognition protein (GNBPs), soluble peptidoglycan recognition proteins (PGRP-SA and PGRP-SD), glucan recognition proteins (GRPs), complement-like thioester-containing proteins (TEPs), and immunolectins (Levashina *et al.*, 2001; Irving *et al.*, 2005; Moita *et al.*, 2005; Dong *et al.*, 2006; Ling and Yu, 2006; Terenius *et al.*, 2007) (Fig. 2).

A few cytokines are also involved in activating adhesion of lepidopteran hemocytes (Fig. 2). The first of these is ENF-peptides, which are secreted as propeptides by the fat body and granulocytes, and which are then processed to their active form upon immune challenge (Strand *et al.*, 2000; Ishii *et al.*, 2010). Evidence that ENF-peptides function as cytokines derives originally from studies by Clark *et al.* (1997) showing that the ENF-peptide plasmatocyte spreading peptide (PSP) rapidly induces adhesion of plasmatocytes to foreign surfaces. Subsequent studies with other family members indicate that ENF-peptides activate adhesion by stimulating tyrosine phosphorylation of integrins (Oda *et al.*, 2010). ENF-peptides also promote phagocytosis and expression of selected AMPs via p38 mitogen-activated protein kinase (MAPK) signaling (Ishii *et al.*, 2010). The second cytokine with an identified role in regulating hemocyte activity is hemocyte chemotactic peptide (HCP), which is expressed by epidermal cells and hemocytes, and which stimulates chemotaxis and aggregation of granulocytes (Nakatogawa *et al.*, 2009). The ligand for the Toll receptor (Spaetzle) is a third cytokine that, like ENF-peptides, is proteolytically processed to its active form in hemolymph following immune challenge in multiple insect taxa including Lepidoptera (Imler and Bulet, 2005; Ao *et al.*, 2008). Recent studies in the beetle *Tenebrio molitor* indicate that proteases involved in activation of Spaetzle are also involved in activation the of the PO cascade (Kan *et al.*, 2008). Binding of Spaetzle to the Toll receptor induces expression of certain AMPs as well as a number of other immune genes (Imler and Bulet, 2005). Microarray studies conducted in larval stage *Drosophila* following parasitoid attack have identified several genes under the control of the Toll and JAK/STAT signaling pathways that are upregulated during the early phases of capsule formation, which have roles in regulating hemocyte proliferation (Wertheim *et al.*, 2005; Irving *et al.*, 2005). Selected AMPs under the control of the Toll and imd pathways along with proteins in the Tep and Tot families involved in pattern recognition, enzyme regulation, and stress responses, are also differentially expressed (Wertheim *et al.*, 2005). A fourth factor that affects hemocyte activity during immune defense responses is eicosonoids, which have diverse functions in both vertebrates and invertebrates (Stanley and Miller, 2008). However, their activities from the perspective of insect immunity have been studied primarily with regard to promoting hemocyte spreading and movement in the clearance of bacteria by nodule formation (Stanley and Miller, 2008). Whether eicosonoids play a role in defense against parasitoids is unclear.

Recognition results in binding of a small number of hemocytes, predominantly granulocytes in Lepidoptera, to the surface of the parasitoid (Pech and Strand, 1996) (Fig. 2). Granulocytes then release cytokines, such as ENF peptides, that recruit additional hemocytes, mainly plasmatocytes, which are most responsible for the formation of a capsule (Fig. 2). As noted above, adhesion to at least some foreign targets involves activation of integrins (Lavine and Strand, 2003; Levin *et al.*, 2005; Moita *et al.*, 2006; Oda

et al., 2010), with tetraspanin proteins serving as ligands for certain integrin subunit combinations (Zhuang *et al.*, 2007). Granulocyte-stimulated activation and binding of plasmatocytes also involves the immunoglobulin superfamily member neuroglian (Nardi *et al.*, 2006). Other factors implicated in hemocyte adhesion to parasitoids in *Drosophila* include products of the *lectin-24A* gene as well as activation of small GTPases which likely regulate formation of actin–myosin filaments that maintain focal adhesions (Wertheim *et al.*, 2005; Williams *et al.*, 2006). In *Pseudoplusia includens* and selected other Lepidoptera, capsule formation terminates when a monolayer of granulocytes is recruited to the surface of the capsule where they apoptose (Pech and Strand, 1996). This terminates binding of additional plasmatocytes to the capsule and prevents the capsule from increasing further in size. Encapsulation of parasitoids in Lepidoptera and other insects usually begins 2–6 h after parasitism and is completed by 48 h (Strand, 2008a).

Melanization of Capsules, Regulation of Phenoloxidase Activity, and Parasitoid Death

Encapsulated targets melanize in some but not all insects due to activation of the PO cascade (Lavine and Strand, 2002) (Fig. 2). A variety of stimuli including wounding, microbial infection, and parasitoid attack activate the PO cascade, which results in the enzyme PO catalyzing the oxidation of diphenolic compounds to quinones (Cerenius *et al.*, 2008; Kanost and Gorman, 2008). These quinones then undergo additional modifications to ultimately form melanin (Sugumaran, 2002). Activation of the PO cascade is thought to depend upon protease zymogens that are capable of autoactivating following immune challenge. Studies with *Manduca sexta* identify one initiation serine protease called HP14 which autoactivates in the presence of microbial elicitors (Wang and Jiang, 2006). Activated HP14 activates pro-HP21, which processes prophenoloxidase activating protease (PAP) 1 and 2, while HP6 activates pro-PAP1 and pro-HP8 (Gorman *et al.*, 2007; Wang and Jiang, 2007; An *et al.*, 2009). All PAPs are clip-domain containing serine proteases that cleave proPO at a conserved Arg–Phe reactive site bond to produce PO. Studies with *B. mori* suggest that its PAP homolog (prophenoloxidase-activating enzyme, PPAE) efficiently processes proPO alone (Satoh *et al.*, 1999) but studies with other insects (*M. sexta*, *H. diomphalia* and *T. molitor*) indicate that proteolysis of proPO by PAPs also requires the presence of serine proteinase homologs (SPHs) as cofactors (Kwon *et al.*, 2000; Yu *et al.*, 2003). SPHs are clip-domain-containing proteins homologous to proteases that lack a serine at the catalytic site (Kwon *et al.*, 2000). SPHs are synthesized as pro-forms, activated themselves by proteolysis, which then interact with PAPs to process proPO. SPHs were first identified in

H. diomphalia and *T. molitor* (Kwon *et al.*, 2000), and then in *M. sexta* (Yu *et al.*, 2003). Once activated, negative regulation of the PO cascade can occur through the action of serpins (serine protease inhibitors), a large family of proteins that irreversibly inhibit their targets by covalent modification at the active nucleophilic residue. Insects encode numerous serpins with some identified as inhibitors of proteases in the PO cascade. In *M. sexta*, these include Serpin4 which inhibits HP21 and Serpins-1J, 3, and 6 which inhibit PAPs 2 and 3 (see Kanost and Gorman, 2008). Several other protein and non-protein factors have also been implicated in inhibition of PO cascade components (summarized by Clark *et al.*, 2010).

Following completion of capsule formation, parasitoids die by either asphyxiation, compounds generated by the PO cascade, or other effector molecules. Experiments conducted by Beck and Strand (2007) show that melanization of eggs from the BV-carrying wasp *Microplitis demolitor* results in high mortality, while inhibition of melanization greatly reduces mortality and promotes successful hatching to larvae. Very recent studies indicate that reactive intermediates generated in association with melanin formation do kill *M. demolitor* (Zhao *et al.*, 2011). However, as noted above, a number of insects produce capsules that do not melanize, yet encapsulated parasitoids still die. Other effector molecules like AMPs are produced during capsule formation, which are implicated in killing organisms other than bacteria and fungi (Imler and Bulet, 2005). However, no studies have yet shown that AMPs kill parasitoids. It is also possible that death within capsules is due to asphyxiation although again experimental data clearly supporting this is, to my knowledge, lacking in the literature. Lastly, while poorly studied, there are a few reports in the literature of PDV-carrying parasitoids dying in the hemocoel of hosts in the absence of any capsule formation. For example, Trudeau and Strand (1998) reported that *M. demolitor* bracovirus (MdBV) fully disables the encapsulation response of the host *Spodoptera frugiperda* (see below), yet *M. demolitor* eggs always die in this species in the early stages of embryogenesis. Whether this failure to develop reflects activity of unknown humoral defense factors or physiological conditions unrelated to the immune system that render *S. frugiperda* an incompatible host for wasp development is unknown.

PDV GENE CONTENT DIFFERS AMONG TAXA

A second consideration in understanding PDV–host immune interactions is recognition that gene content differs among PDVs (see Dupuy *et al.*, Chapter 4 of this volume). As proviruses, these differences provide important evidence that BVs and IVs evolved from different viral ancestors (Bézier *et al.*, 2009; Volkoff *et al.*, 2010; Wetterwald *et al.*, 2010).

In their encapsidated form, differences in gene content are also likely important in interactions with host immune defenses. For example, *Cotesia congregata* is in the subfamily Microgastrinae and carries *C. congregata* bracovirus (CcBV). The encapsidated genome of CcBV encodes five main gene families: (1) the protein tyrosine phosphatase (*ptp*) genes (*ptp*) (27 genes) that encode predicted proteins related to PTPs of the classical subtype; (2) the ankyrin motif (*ank*) genes (6 genes) that are structurally related to inhibitor κB (IκB) proteins; (3) the *crp* gene family (4 genes) that encodes predicted cysteine-rich proteins; (4) the *cyst* gene family (3 genes) that encodes predicted cysteine proteinases; and (5) the *ep1-like* family (6 genes) that encode novel proteins (Espagne *et al.*, 2004). Most other genes in the CcBV genome are single copy. Homologs of these gene families as well as several of the single-copy genes in the CcBV genome are also present in the genomes of BVs carried by other *Cotesia* species (Serbielle *et al.*, 2009; Choi *et al.*, 2009). *Glyptapanteles* is the sister taxon to the genus *Cotesia* (Murphy *et al.*, 2008), and the BVs carried by these wasps encode some of the same gene families (*ptps*, *anks*, *cyst*) found in *Cotesia* BVs as well as some genes absent from known *Cotesia* BVs (Desjardins *et al.*, 2008). The encapsidated genomes of BVs carried by more distantly related taxa exhibit larger differences as illustrated by wasps in the genera *Microplitis* and *Chelonus*. The former also resides in the Microgastrinae but is distantly related to the genus *Cotesia* (Murphy *et al.*, 2008), while the latter is in the subfamily Cheloninae. The encapsidated genome of *Microplitis demolitor* bracovirus (MdBV) shares only two gene families (*ptp*, *ank*) with BVs from *Cotesia* and *Glyptapanteles* sp., while encoding other gene families and several single-copy genes absent in *Cotesia*- and *Glyptapanteles*-associated BVs (Webb *et al.*, 2006). Moreover, the encapsidated genome of *Chelonus inantius* bracovirus (CiBV) appears to share no genes with either CcBV or MdBV (Annaheim and Lanzrein, 2007; Weber *et al.*, 2007).

Among IVs, viruses associated with campoplegine ichneumonids (*Campoletis* and *Hyposoter* sp.) share six gene families: (1) the ankyrin motif genes (*ank*); (2) the innexin genes (*inex*) that encode predicted proteins involved in forming gap junctions; (3) cysteine-motif genes (*cys-motif*) which share a cysteine-rich domain related to ω-conotoxins; (4) the repeat element genes (*rep*) which share a common 540-bp sequence present in one to five copies; (5) the N-family genes (*N-family*); and (6) the polar-residue rich genes (*pol-res*) (Webb *et al.*, 2006; Tanaka *et al.*, 2007). However, the lone banchine-associated IV examined to date (GfIV) shares only one family with campoplegine IVs (*ank* genes) (Lapointe *et al.*, 2007). Taken together, these data suggest that PDV interactions with host insects are likely to be more similar when making comparisons between BVs or IVs from wasps in the same or closely related genera than when making comparisons between more distantly related wasp species.

PDV-MEDIATED SUPPRESSION OF HOST IMMUNE DEFENSES

The first report that PDVs protect wasps from the host immune system involved the IV-carrying wasp *Campoletis sonorensis* (Edson *et al.*, 1981). This study showed that the host, *Heliothis virescens*, always encapsulated *C. sonorensis* eggs in the absence of infection by CsIV but never encapsulated eggs in its presence. This study also showed that prevention of encapsulation depended on CsIV gene expression. Subsequent investigations demonstrated that several other PDVs also protect wasp progeny from encapsulation. However, the specific effects that PDVs have on host immune defenses and the viral genes implicated in causing these alterations differ somewhat among isolates.

BV–Host Immune Interactions

Most studies on BV–immune interactions have been conducted with viruses carried by wasps in the genera *Microplitis* and *Cotesia*. Similar to the first studies conducted with CsIV, inhibition of encapsulation and other immune defenses by BVs from wasps in these genera usually requires infection of the host by viable virus. In the case of *Microplitis demolitor*, MdBV irreversibly suppresses encapsulation and phagocytosis of parasitoid eggs and other foreign targets (Strand and Noda, 1991; Trudeau and Strand, 1998). This occurs because granulocytes and plasmatocytes lose the capacity to bind and spread upon any foreign surface within 4h of infection. These effects also persist for the duration of parasitoid development (Strand and Noda, 1991; Beck and Strand, 2003; Pruijssers and Strand, 2007). Disabled adhesion and spreading is associated with distinct alterations in the distribution of F-actin and other cytoskeletal proteins in the cytoplasm of both hemocyte types. Granulocytes but not plasmatocytes also undergo a high level of apoptosis within the first 36h of infection (Strand and Pech, 1995a; Suderman *et al.*, 2008). These alterations only occur if hemocytes are infected by MdBV, which indicates that suppression of cellular defenses involves viral products expressed within hemocytes rather than gene products that are secreted into the hemolymph (Strand and Noda, 1991; Strand, 1994). Quantitative studies indicate that parasitism results in >99% of circulating host hemocytes being infected by MdBV (Beck *et al.*, 2007). Similar alterations in the adhesive properties of host hemocytes have also been described after infection by BVs from other *Microplitis* species (Tanaka, 1987; Kadash *et al.*, 2003; Luo and Pang, 2006).

More variable effects on host cellular defense responses are reported with BVs from wasps in the genus *Cotesia*. Infection of *Pieris rapae* by *Cotesia rubecula* bracovirus (CrBV) results in transient expression of only a small number of viral genes, which correlates with short-term alterations in the adhesive properties and cytoskeleton of hemocytes (Asgari *et al.*, 1996, 1997; Glatz *et al.*, 2003). Altered adhesion also appears to involve viral gene products that are secreted into the hemolymph which thereafter bind and are internalized by hemocytes. In contrast, alterations in adhesion and the cytoskeleton of host hemocytes after infection by *C. melanoscela* bracovirus (CmBV), *C. kariyai* bracovirus (CkBV), and *C. plutellae* bracovirus (CpBV) persist for longer periods and involve a larger number of viral gene products with some of these proteins being secreted and others acting within infected cells (Guzo and Stoltz, 1987; Tanaka, 1987; Gad and Kim, 2008; Nalini *et al.*, 2009). Some studies report that CcBV causes hemocytes from *Manduca sexta* larvae to more readily aggregate with one another rather than *C. congregata* eggs (Lovallo and Cox-Foster, 1999), while other studies report that CcBV causes host hemocytes to become less adhesive and spread more poorly on foreign surfaces (Amaya *et al.*, 2005). Few studies have been conducted with BVs from wasps in other genera or subfamilies. In the case of BVs associated with wasps in the subfamily Cheloninae, studies with *Chelonus inanitus* bracovirus (CiBV) suggest viral infection protects the wasp larva from encapsulation but does not suppress capsule formation of other foreign targets like latex beads (Stettler *et al.*, 1998).

BVs from *Microplitis* and *Cotesia* spp. also affect host humoral defenses. Several isolates inhibit the melanization of hemolymph (Stoltz and Cook, 1983; Beckage *et al.*, 1990; Strand and Noda, 1991). MdBV and CpBV reduce expression of selected AMPs (Thoetkiattikul *et al.*, 2005; Barandoc *et al.*, 2010), while CpBV also alters expression of a transferrin from *Plutella xylostella*, which in other insects is suggested to sequester iron from pathogens (Yoshiga *et al.*, 1997; Kim and Kim, 2010). Lastly, CcBV affects the abundance of storage proteins in plasma although it is unlikely these alterations impact immune defenses (Beckage and Kanost, 1993).

BV-Encoded Genes Implicated in Altering Host Immune Defenses

As noted above, MdBV broadly immunosuppresses host insects by inhibiting both cellular and humoral defense responses. The encapsidated genome of MdBV encodes four main gene families, *glc* (2 genes), *ptp* (13 genes), *anks* (12 genes), and *egf* (3 genes) (Webb *et al.*, 2006), with selected members of each implicated in immunosuppression. The most important MdBV gene product in inhibition of cell adhesion and phagocytosis is the gene *glc1.8*, which encodes a 518-amino-acid cell surface mucin that localizes to the surface of infected hemocytes and hemocyte-like cell lines (Trudeau *et al.*, 2000; Beck and

Strand, 2003) (Fig. 2). Glc1.8 is characterized by a signal peptide at its N-terminus, five 78-amino-acid repeats arranged in tandem array that is heavily N-glycosylated, and a C-terminal hydrophobic domain encoding an anchor sequence. Expression of Glc1.8 alone prevents hemocytes and hemocyte-derived cell lines from adhering to foreign surfaces including parasitoid eggs, and also greatly reduces phagocytosis (Beck and Strand, 2005; Pruijssers and Strand, 2007). Reciprocally, knockdown of Glc1.8 by RNAi in MdBV infected cells restores normal adhesion and phagocytic activity (Beck and Strand, 2003; Johnson et al., 2010). Phagocytosis and adhesion are also known to involve tyrosine phosphorylation of several intracellular adaptor proteins that link cell surface proteins to the actin cytoskeleton. One member of the MdBV PTP gene family, ptp-H2, is preferentially expressed in host hemocytes and colocalizes to the tyrosine kinase homolog Dfak56, which is a component of focal adhesions associated with integrins that regulate adhesion and phagocytosis (Pruijssers and Strand, 2007). Bioassays indicate that PTP-H2 alone reduces adhesion and phagocytosis of S2 cells while coexpression with Glc1.8 results in near complete inhibition of phagocytosis: a finding that suggests these factors interact to disable hemocyte-mediated defenses (Pruijssers and Strand, 2007). Further studies indicate that PTP-H2 stimulates apoptosis (Suderman et al., 2008) while enzymatic analysis demonstrates that PTP-H2 is a fully functional tyrosine phosphatase (Eum et al., 2010).

Other gene products encoded by MdBV with roles in immunosuppression target melanization and inducible expression of humoral effector molecules. Recall that proPO is activated by PAPs that must also be processed and interact with SPH cofactors to function. Two members of the MdBV Egf gene family, egf1.0 and egf1.5, are secreted proteins that inhibit melanization by blocking the processing of proPAPs, competitively inhibiting activated PAPs, and binding SPHs (Beck and Strand, 2007; Lu et al., 2008, 2010) (Fig. 2). The Toll and imd signaling pathways regulate expression of a diversity of immune genes following parasitoid attack or infection by other organisms. Essential to both pathways are NF-κB transcription factors that are normally negatively regulated by endogenous inhibitor κB (IκB) proteins. Two members of the ank family, Ank-H4 and Ank-N4, disrupt Toll and imd signaling, and AMP expression, by binding insect NF-κBs (Thoetkiattikul et al., 2005) (Fig. 2).

In the case of BVs from Cotesia species, studies with CrBV implicate the CrV1 gene in causing the transient cytoskeletal alterations that occur in hemocytes (Asgari et al., 1997), while more recent studies suggest the protein CrV2 binds heterodimeric G proteins, which suggests altered hemocyte function may involve changes in signaling processes (Cooper et al., 2011). CcBV encodes an ortholog of the CrV1 gene (CcV1) that disables binding

of the pattern recognition protein hemolin to hemocytes (Labropoulou et al., 2008) (Fig. 2). However, it is unclear whether altered binding of hemolin to hemocytes accounts for the effects of CrV1 on hemocyte spreading (Amaya et al., 2005). Differential expression of CrV1 orthologs in BVs from biotypes of Cotesia sesamiae have also been implicated in successful parasitism of hosts: possibly due to their differential efficacy in protecting wasp progeny from encapsulation (Gitau et al., 2007). Sequence analysis combined with activity assays indicate that some members of the CcBV PTP gene family likely encode functional tyrosine phosphatases, and that products of the cystatin gene family function as inhibitors of cathepsin-like cysteine proteases (Provost et al., 2004; Serbielle et al., 2009). Thus, members of both gene families may interact with host immune defenses but exactly what their role might be remains unknown. Although Cotesia spp. inhibit melanization of host hemolymph (Stoltz and Cook, 1983; Beckage et al., 1990), Cotesia BVs do not encode homologs of the MdBV egf genes. However, suppression of melanization by C. rubecula may not involve any CrBV-encoded gene but rather an SPH in wasp venom that interferes with the PO cascade (Zhang et al., 2004).

Recent studies with CpBV report a number of genes including a lectin, histone (H4), an EP1-like gene, and the gene CpBV15 in disrupting adhesion, spreading and proliferation of hemocytes in the host P. xylostella (Gad and Kim, 2008; Kwon and Kim, 2008; Lee et al., 2008; Nalini and Kim, 2009; Kim and Kim, 2010) (Fig. 2). Evidence for these genes altering host hemocyte activity is based on cloning each into a plasmid vector, injecting the plasmid into a host larva together with a cationic lipid transfection agent, assessing its expression and phenotypic effects, and then conducting knockdown studies of the same transiently expressed gene by RNAi. Reported outcomes of these studies, however, are difficult to accept for three reasons. First and most importantly, the plasmids that the authors primarily report using in these studies (pBacPAK8 or 9, and pFast-Bac-Dual) are transfer vectors designed for cloning genes in frame with the polyhedron or p10 promoter of Autographa californica multicapsid nucleopolyhedrosis virus (AcMNPV) (Ibrahim and Kim, 2008; Gad and Kim, 2008; Kwon and Kim, 2008; Nalini and Kim, 2009; Kim and Kim, 2010). Yet the polyhedron and p10 promoters are well known to be nonfunctional in insect cells unless infected by AcMNPV because transcription of these late genes (pol, p10) fully depends on activators and a unique RNA polymerase encoded by the virus itself (see Rohrmann (2008) for a comprehensive summary). Thus, it is unclear how any CpBV or other gene could be expressed by introducing only these plasmid constructs into insect cells. Second, most studies with insects, including my own direct experiences with P. xylostella, indicate that cationic lipid-mediated transfection of plasmids into hemocytes and other tissues *in vivo* is very inefficient; even when using expression vectors with promoters that are functional in Lepidoptera. Third, the phenotype reported in these studies is primarily a loss of hemocyte adhesion, but it is counterintuitive that multiple, different CpBV genes would all have such similar effects.

The only other data on interactions between BV genes and host immune defenses derive from the study of the parasitoid *Toxoneuron nigriceps*, which resides in the subfamily Cardiochilinae. Its associated BV, TnBV, encodes the gene TnBV1, which activates caspases in the absence of apoptosis in selected cell lines (Lapointe *et al.*, 2005). One member of the *ank* gene family encoded by TnBV has also been shown to interact with insect NF-κBs (Falabella *et al.*, 2007) while another alters cytoskeleton organization of some insect cells (Duchi *et al.*, 2010) (Fig. 2).

Immune Interactions Between IVs and Hosts

Although fewer studies have been conducted with IVs, the literature collectively suggests they too protect wasp offspring from encapsulation by altering cellular and humoral components of the host immune system. Infection of hosts with IVs from several species of campoplegine ichneumonids result in altered numbers of hemocytes in circulation and reductions in the ability of granulocytes and plasmatocytes to spread on foreign surfaces (Davies *et al.*, 1987; Guzo and Stoltz, 1987; Doucet and Cusson, 1996; Yin *et al.*, 2003). In the case of CsIV, host hemocytes are unable to encapsulate *C. sonorensis* eggs as well as other foreign targets (Edson *et al.*, 1981; Davies *et al.*, 1987). In contrast, infection of hosts by *Tranosema rostrale* ichnovirus (TrIV) disables encapsulation of *T. rostrale* eggs but has no effect on encapsulation of other targets suggesting suppression of encapsulation is restricted to the parasitoid (Doucet and Cusson, 1996). IV-induced alterations in host humoral defense responses include suppression of hemolymph melanization as well as altered expression of selected AMPs and hemolymph proteins (Stoltz and Cook, 1983; Doucet and Cusson, 1996; Shelby *et al.*, 2000). CsIV infection also reduces transcript levels of the antimicrobial protein lysozyme (Shelby *et al.*, 1998).

IV genes implicated in causing these alterations include the *cys-motif* proteins VHv1.1 and VHv1.4 from CsIV, which are expressed in the host fat body and secreted into the hemolymph. Both proteins bind to the surface of hemocytes (Fig 2). Expression of VHv1.1 also reduces encapsulation of *C. sonorensis* eggs in hosts infected with a VHv1.1-expressing recombinant baculovirus (Li and Webb, 1994; Cui *et al.*, 1997). Other IV genes are implicated in immunosuppression, because of homology to BV genes or because they are expressed in host immune tissues like hemocytes or the fat body. IV-encoded *ank* genes, for example, may function as NF-κB inhibitors, but structural differences with BV *ank* genes as well as recent expression studies suggest they may have other functions (Kroemer and Webb, 2004; Webb *et al.*, 2006; Fath-Goodin *et al.*, 2009) (Fig. 2). Products of the CsIV *inex* genes form functional gap junctions, which have diverse functions in cell–cell communication (Turnbull *et al.*, 2005). Since gap junctions form between hemocytes in capsules, it is possible that IV *inex* genes may play a role in altering hemocyte function. Doucet *et al.* (2008) report that infection of hosts with TrIV transiently reduced expression of host phenoloxidase genes, but this effect does not appear to account for the near-complete inhibition of melanization activity observed in parasitized hosts. Recent transcriptome studies of *Hyposoter didymater* ichnovirus (HdIV) in the host *Spodoptera frugiperda* indicate that infection alters expression of a number of host genes including some with possible immune functions (Provost *et al.*, 2011). Comparison of the alterations in host transcriptome activity between HdIV and MdBV (see above) also reveal numerous differences underscoring that the responses of hosts at the molecular level following infection by IVs and BVs are more generally likely to differ (Provost *et al.*, 2011).

Immunoevasive Gene Products Associated with PDVS

While viral gene expression is usually required for defense against the host immune response, some PDVs together with ovary-produced proteins coat parasitoid eggs, which transiently protects them from encapsulation. Two genes from *Cotesia kariyai* encode immunoevasive ovarial proteins (IEP1 and 2) (Tanaka *et al.*, 2002) (Fig. 2). These genes are not encoded within the CkBV genome but IEP1 and 2 are detected on the surface of CkBV virions. Among ichneumonids, *Venturia canescens* produces virus-like particles (VLPs) morphologically similar to the IVs associated with other members of the Campopleginae but these particles lack any nucleic acid. Hosts parasitized by *V. canescens* remain capable of mounting an encapsulation response against numerous foreign targets but are unable to encapsulate *V. canescens* eggs because of VLPs on their surface (Fedderson *et al.*, 1986) (Fig. 2).

Effects of PDVS on Permissiveness of Insects to Infection by Other Pathogens

A small number of studies indicate that PDVs enhance the susceptibility of hosts to infection by other pathogens. Most of these studies involve experiments with the highly pathogenic baculovirus AcMNPV. As with all baculoviruses, the host range of AcMNPV is limited to only certain species of Lepidoptera that are highly permissive to infection. In contrast, many other species are fully non-permissive or are semipermissive to infection which means that the host must be infected with an initially higher titer of virus in order to establish an infection that leads to death of the host. However, coinfection of the normally non-permissive host *Helicoverpa zea* with CsIV increased permissiveness to AcMNPV infection while coinfection with CcBV increased susceptibility of the semipermissive host *Manduca sexta* (Washburn *et al.*, 1996, 2000). Coinfection with CiBV or expression of the *P-vank-1* gene from CsIV by a recombinant AcMNPV has also been reported to increase the permissiveness of *Spodoptera littoralis* larvae (Rivkin *et al.*, 2006). Similar studies report that PDV infection also increases susceptibility of Lepidoptera to infection by other viruses and bacteria. For example, parasitism by the IV-carrying wasp *Hyposoter exiguae* activated three different but unknown viruses in *Trichoplusia ni* which were thought to have established latent infections (Stoltz and Makkay, 2003), while parasitism of *Pseudaletia separata* by the BV-carrying wasp *Cotesia karyiai* increased susceptibility to infection by a bacterium (Matsumoto *et al.*, 1998). In each of these examples, the effects of parasitism or PDV infection has been attributed to disrupting host immune defenses that otherwise confer resistance to the aforementioned pathogens. However, whether this is actually the case and precisely what the mechanisms are that underlie these effects remains unknown.

CONCLUDING REMARKS

This chapter provides an overview of the interactions between PDVs and host immune defenses. Numerous studies make clear that a key function of most PDVs is suppression of host immune defenses like encapsulation that otherwise would kill the progeny of the parasitoid. However, our understanding of mechanisms underlying PDV-mediated immunosuppression is limited to studies of a much smaller number of PDV isolates and a handful of PDV-encoded genes. Given that the encapsidated genomes of PDVs from different wasp taxa encode quite different types of genes, it is likely that even though many PDVs disable immune defenses like encapsulation, the mechanisms by which they do this are likely to differ: particularly when comparing PDV isolates from phylogenetically disparate taxa of wasps. Thus, a key challenge for the future will be to link expression of specific PDV gene products to particular alterations that occur in hosts at a physiological, cellular, and molecular level. In turn, much more work is needed that rigorously characterizes the function of PDV gene products and how they interact with specific host immune molecules and pathways. Lastly, while this chapter emphasizes the role of PDVs in immunosuppression of hosts, other factors parasitoids introduce into hosts including venom, teratocytes, and products from the developing parasitoid larvae have also been implicated in interactions with host immune defenses. How these factors might interact with the activities of PDVs,

however, is largely unknown. Given that the immuosuppressive effects of PDV infection decline with time in some study systems (see Strand, 2010), it is possible that interactions with other wasp products are important for successful parasitism.

REFERENCES

Amaya, K.E., Asgari, S., Jung, R., Hongskula, M., Beckage, N.E., 2005. Parasitization of *Manduca sexta* by the parastoid *Cotesia congregata* induces an impaired host immune response. J. Insect Physiol. 51, 505–512.

An, C.J., Ishibashi, J., Ragan, E.J., Jiang, H.B., Kanost, M.R., 2009. Functions of *Manduca sexta* hemolymph proteinases HP6 and HP8 in two innate immune pathways. J. Biol. Chem. 284, 19716–19726.

Annaheim, M., Lanzrein, B., 2007. Genome organization of the *Chelonus inanitus* polydnavirus: excision sites, spacers, and abundance of proviral and excised segments. J. Gen. Virol. 8, 450–457.

Ao, J.Q., Ling, E., Yu, X.Q., 2008. A Toll receptor from *Manduca sexta* is activated in response to *Escherichia coli* infection. Mol. Immunol. 45, 543–552.

Asgari, S., Hellers, M., Schmidt, O., 1996. Host hemocyte inactivation by an insect parasitoid: transient expression of a polydnavirus gene. J. Gen. Virol. 77, 2653–2662.

Asgari, S., Schmidt, O., Theopold, U., 1997. A polydnavirus-encoded protein of an endoparasitoid wasp is an immune suppressor. J. Gen. Virol. 78, 3061–3070.

Barandoc, K.P., Kim, J., Kim, Y., 2010. *Cotesia plutella* bracovirus suppresses expression of an antimicrobial peptide, cecropin, in the diamondback moth, *Plutella xylostella*, challenged by bacteria. J. Microbiol. 48, 117–123.

Beck, M.H., Inman, R.B., Strand, M.R., 2007. *Microplitis demolitor* bracovirus genome segments vary in abundance and are individually packaged in virions. Virology 359, 179–189.

Beck, M., Strand, M.R., 2003. RNA interference silences *Microplitis demolitor* bracovirus genes and implicates glc1.8 in disruption of adhesion in infected host cells. Virology 314, 521–535.

Beck, M., Strand, M.R., 2005. Glc1.8 from *Microplitis demolitor* bracovirus induces a loss of adhesion and phagocytosis in insect high five and S2 cells. J. Virol. 79, 1861–1870.

Beck, M.H., Strand, M.R., 2007. A novel polydnavirus protein inhibits the insect prophenoloxidase activation pathway. Proc. Natl. Acad. Sci. U.S.A. 104, 19267–19272.

Beckage, N.E., Kanost, M.R., 1993. Effects of parasitism by the braconid wasp *Cotesia congregata* on host hemolymph proteins of the tobacco hornworm, *Manduca sexta*. Insect Biochem. Mol. Biol. 23, 643–653.

Beckage, N.E., Metcalf, J.S., Nesbit, D.J., Schleifer, K.W., Zetlan, S.R., de Buron, I., 1990. Host hemolymph monophenoloxidase activity in parasitized *Manduca sexta* larvae and evidence for inhibition by wasp polydnavirus. Insect Biochem. 20, 285–294.

Bézier, A., Annaheim, M., Herbinière, J., Wetterwald, C., Gyapay, G., Bernard-Samain, S., et al. 2009. Polydnaviruses of braconid wasps derive from an ancestral nudivirus. Science 323, 926–930.

Cerenius, L., Lee, B.L., Soderhall, K., 2008. The proPO-system: pros and cons for its role in invertebrate immunity. Trends in Immunol. 29, 263–271.

Cerenius, L., Soderhall, K., 2004. The prophenoloxidse-activating system in invertebrates. Immunol. Rev. 198, 116–126.

Choi, J.Y., Kwon, S.J., Roh, J.Y., Yang, T.J., Yoon, S.H., Kim, H., et al. 2009. Sequence and gene organization of 24 circles from the *Cotesia plutellae* bracovirus genome. Arch. Virol. 154, 1313–1327.

Clark, K.D., Lu, Z., Strand, M.R., 2010. Regulation of melanization by glutathione in the moth *Pseudoplusia includens*. Insect Biochem. Mol. Biol. 40, 460–467.

Clark, K.C., Pech, L., Strand, M.R., 1997. Isolation and identification of a plasmatocyte spreading peptide from hemolymph of the lepidopteran insect *Pseudoplusia includens*. J. Biol. Chem. 272, 23440–23447.

Cooper, T.H., Bailey-Hill, K., Leifert, W.R., McMurchie, E.J., Asgari, S., Glatz, R.V., 2011. Identification of an in vitro interaction between an insect immune suppressor protein (CrV2) and G alpha proteins. J. Biol. Chem. 286, 10466–10475.

Cui, L., Soldevila, A.I., Webb, B.A., 1997. Expression and hemocyte-targeting of a *Campoletis sonorensis* polydnavirus cysteine-rich gene in *Heliothis virescens* larvae. Arch. Insect Biochem. Physiol. 36, 251–271.

Davies, D.H., Strand, M.R., Vinson, S.B., 1987. Changes in differential hemocyte count and *in vitro* behaviour of plasmatocytes from host *Heliothis virescens* caused by *Campoletis sonorensis* PDV. J. Insect Physiol. 33, 143–153.

Desjardins, C.A., Gundersen-Rindal, D.E., Hostetler, J.B., Tallon, L.J., Fadrosh, D.W., Fuester, R.W., et al. 2008. Comparative genomics of mutualistic viruses of *Glyptapanteles* parasitic wasps. Genome Biol. 9, R183.

Dong, Y.M., Taylor, H.E., Dimopoulos, G., 2006. AgDscam, a hypervariable immunoglobulin domain-containing receptor of the Anopheles gambiae immune system. PLoS Biol. 4, 1137–1146.

Doucet, D., Béliveau, C., Dowling, A., Simard, J., Feng, Q., Krell, P.J., et al. 2008. Prophenoloxidase 1 and 2 from the spruce budworm, *Choristoneura fumiferana*: molecular cloning and assessment of transcriptional regulation by a polydnavirus. Arch. Insect Biochem. Physiol. 67, 188–201.

Doucet, D., Cusson, M., 1996. Role of calyx fluid in alterations of immunity in *Choristoneura fumiferana* larvae parasitized by *Tranosema rostrale*. Comp. Biochem. Physiol. 114, 311–317.

Duchi, S., Cavaliere, V., Fagnocchi, L., Grimaldi, M.R., Falabella, P., Graziani, F., et al. 2010. The impact on microtubule network of a bracovirus I kappa B-like protein. Cell. Mol. Life Sci. 67, 1699–1712.

Edson, K.M., Vinson, S.B., Stoltz, D.B., Summers, M.D., 1981. Virus in a parasitoid wasp: suppression of the cellular immune response in the parasitoid's host. Science 211, 582–583.

Espagne, E., Dupuy, C., Huguet, E., Cattolico, L., Provost, B., Martins, N., et al. 2004. Genome sequence of a polydnavirus: insights into symbiotic virus evolution. Science 306, 286–289.

Eum, J.H., Bottjen, R.C, Clark, K.D., Strand, M.R., 2010. Characterization and kinetic analysis of protein tyrosine phosphatase-H2 from *Microplitis demolitor* bracovirus. Insect Biochem. Mol. Biol. 40, 690–698.

Falabella, P., Varricchio, P., Provost, B., Espagne, E., Ferrarese, R., Grimaldi, A., et al. 2007. Characterization of the IkappaB-like gene family in polydnaviruses associated with wasps belonging to different Braconid subfamilies. J. Gen. Virol. 88, 92–104.

Fath-Goodin, A., Kroemer, J.A., Webb, B.A., 2009. The *Campoletis sonorensis* ichnovirus vankyrin protein P-vank-1 inhibits apoptosis in insect Sf9 cells. Insect Mol. Biol. 18, 497–506.

Fedderson, I., Sander, K., Schmidt, O., 1986. Virus-like particles with host protein-like antigenic determinants protect an insect parasitoid from encapsulation. Experientia 42, 1278–1281.

Gad, W., Kim, Y., 2008. A viral histone H4 encoded by *Cotesia plutellae* bracovirus inhibits hemocyte-spreading behaviour of the diamondback moth, *Plutellae xylostella*. J. Gen. Virol. 89, 931–938.

Gardiner, E.M.M., Strand, M.R., 1999. Monoclonal antibodies bind distinct classes of hemocytes in the moth Pseudoplusia includens. J. Insect Physiol. 45, 113–126.

Gardiner, E.M.M., Strand, M.R., 2000. Hematopoiesis in larval *Pseudoplusia includens* and *Spodoptera frugiperda*. Arch. Insect Biochem. Physiol. 43, 147–164.

Gitau, C.W., Gundersen-Rindal, D., Pedroni, M., Mbugi, C., Dupas, S., 2007. Differential expression of the CrV1 hemocyte inactivation-associated polydnavirus gene in the African maize stem borer Busseola fusca (Fuller) parasitized by two biotypes of the endoparasitoid Cotesia sesamiae (Cameron). J. Insect Physiol. 53, 676–684.

Glatz, R., Schmidt, O., Asgari, S., 2003. Characterization of a novel protein with homology to C-type lectins expressed by the *Cotesia rubecula* bracovirus in larvae of the lepidopteran host, Pieris rapae. J. Biol. Chem. 278, 19743–19750.

Gorman, M.J., An, C., Kanost, M.R., 2007. Characterization of tyrosine hydroxylase from Manduca sexta. Insect Biochem. Mol. Biol. 37, 1327–1337.

Guzo, D., Stoltz, D.B., 1987. Observations on cellular immunity and parasitism in the tussock moth. J. Insect Physiol. 33, 19–31.

Ibrahim, A.M., Kim, Y., 2008. Transient expression of protein tyrosine phosphatases encoded by *Cotesia plutellae* bracovirus inhibits insect cellular immune responses. Naturwissenschaften 95, 25–32.

Imler, J-L., Bulet, P., 2005. Antimicrobial peptides in *Drosophila*, structures, activities and gene regulation. In: Kabelitz, D., Schroder, J.M. (Eds.), Mechanisms of Epithelial Defense, Vol. 86. Karger, Basel. pp. 1–21.

Irving, P., Ubeda, J., Doucet, D., Troxler, L., Lagueux, M., Zachary, D., et al. 2005. New insights into *Drosophila* larval hemocyte functions through genome-wide analysis. Cell. Microbiol. 7, 335–350.

Ishii, K., Hamamoto, H., Kamimura, M., Nakamura, Y., Noda, H., Imamura, K., et al. 2010. Insect cytokine paralytic peptide (PP) induces cellular and humoral immune responses in the silkworm *Bombyx mori*. J. Biol. Chem. 285, 28635–28642.

Johnson, J.A., Bitra, K., Zhang, S., Wang, L., Lynn, D.E., Strand, M.R., 2010. The UGA-CiE1 cell line from *Chrysodeixis includens* exhibits characteristics of granulocytes and is permissive to infection by two viruses. Insect Biochem. Mol. Biol. 40, 394–404.

Kadash, K., Harvey, J.A., Strand, M.R., 2003. Cross-protection experiments with parasitoids in the genus *Microplitis* (Hymenoptera: Braconidae) suggest a high level of specificity in their associated bracoviruses. J. Insect Physiol. 49, 473–482.

Kan, H., Kim, C.-H., Kwon, H.-M., Park, J.-W., Roh, K.-B., Lee, H., et al. 2008. Molecular control of phenoloxidase-induced melanin synthesis in an insect. J. Biol. Chem. 283, 25316–25323.

Kanost, M.R., Gorman, M.J., 2008. Phenoloxidases in insect immunity. In: Beckage, N.E. (Ed.), Insect Immunity (pp. 69–96). Academic Press, San Diego.

Kim, J., Kim, Y., 2010. A viral histone H4 suppresses expression of a transferrin that plays a role in the immune response of the diamondback moth, *Plutella xylostella*. Insect Mol. Biol. 19, 567–574.

Kocks, C., Cho, J.H., Nehme, N., Ulvila, J., Pearson, A.M., Meister, M., et al. 2005. Eater, a transmembrane protein mediating phagocytosis of bacterial pathogens in *Drosophila*. Cell 123, 335–346.

Kroemer, J.A., Webb, B.A., 2004. Divergences in protein activity and cellular localization within the *Campoletis sonorensis* ichnovirus vankyrin family. J. Virol. 80, 12219–12228.

Kwon, B., Kim, Y., 2008. Transient expression of an EP1-like gene encoded in *Cotesia plutellae* bracovirus suppresses the hemocyte population in the diamondback moth, *Plutella xylostella*. Dev. Comp. Immunol. 32, 932–942.

Kwon, T.H., Kim, M.S., Choi, H.W., Joo, C.H., Cho, M.Y., Lee, B.L., 2000. A masquerade-like serine proteinase homologue is necessary for phenoloxidase activity in the coleopteran insect, *Holotrichia diomphalia* larvae. Eur. J. Biochem. 267, 6188–6196.

Labropoulou, V., Douis, V., Stefanou, D., Magrioti, C., Swevers, L., Iatrou, K., 2008. Endoparasitoid wasp bracovirus-mediated inhibition of hemolin function and lepidopteran host immunosuppression. Cell. Microbiol. 10, 2118–2128.

Lapointe, R., Tanaka, K., Barney, W.E., Whitfield, J.B., Banks, J.C., Béliveau, C., et al. 2007. Genomic and morphological features of a banchine polydnavirus: comparison with bracoviruses and ichnoviruses. J. Virol. 81, 6491–6501.

Lapointe, R., Wilson, R., Vilaplana, L., O'Reilly, D.R., Falabella, P., Douris, V., et al. 2005. Expression of a *Toxoneuron nigriceps* polydnavirus (TnBV) encoded protein, causes apoptosis-like programmed cell death in lepidopteran insect cells. J. Gen. Virol. 86, 963–971.

Lavine, M.D., Strand, M.R., 2002. Insect hemocytes and their role in cellular immune responses. Insect Biochem. Mol. Biol. 32, 1237–1242.

Lavine, M.D., Strand, M.R., 2003. Hemocytes from *Pseudoplusia includens* express multiple alpha and beta integrin subunits. Insect Mol. Biol. 12, 441–452.

Lee, S., Nalini, M., Kim, Y., 2008. A viral lectin encoded in *Cotesia plutellae* bracovirus and its immunosuppressive effect on host hemocytes. Comp. Biochem. Physiol. A 149, 351–361.

Levashina, E.A., Moita, L.F., Blandin, S., Vriend, G., Lagueux, M., Kafatos, F.C., 2001. Conserved role of a complement-like protein in phagocytosis revealed by dsRNA knockout in cultured cells of the mosquito, *Anopheles gambiae*. Cell 104, 709–718.

Levin, D.M., Breuer, L.N., Zhuang, S.F., Anderson, S.A., Nardi, J.B., Kanost, M.R., 2005. A hemocyte-specific integrin requered for hemocytic encapsulation in the tobacco hornworm, *Manduca sexta*. Insect Biochem. Mol. Biol. 35, 369–380.

Li, X., Webb, B.A., 1994. Apparent functional role for a cysteine-rich polydnavirus protein in suppression of the insect cellular immune response. J. Virol. 68, 7482–7489.

Ling, E.J., Yu, X.Q., 2006. Cellular encapsulation and melanization are enhanced by immulectins, pattern recognition receptors from the tobacco hornworm *Manduca sexta*. Dev. Comp. Immunol. 30, 289–299.

Lovallo, N., Cox-Foster, D.L., 1999. Alteration in FAD-glucose dehydrogenase activity and hemocyte behavior contribute to initial disruption of *Manduca sexta* immune response to *Cotesia congregata* parasitoids. J. Insect Physiol. 45, 1037–1048.

Lu, Z., Beck, M.H., Jiang, H., Wang, Y., Strand, M.R., 2008. The viral protein Egf1.0 is a dual activity inhibitor of prophenoloxidase activating proteinases 1 and 3 from *Manduca sexta*. J. Biol. Chem. 283, 21325–21333.

Lu, Z., Beck, M.H., Strand, M.R., 2010. Egf1.5 is a second phenoloxidase cascade inhibitor encoded by *Microplitis demolitor* bracovirus. Insect Biochem. Mol. Biol. 40, 497–505.

Luo, K.J., Pang, Y., 2006. Disruption effect of *Microplitis bicoloratus* polydnavirus EGF-like protein, MbCRP, on actin cytoskeleton in lepidopteran insect hemocytes. Acta Biochem. Biophys. Sinica 38, 577–585.

Matsumoto, H., Noguchi, H., Hayakawa, Y., 1998. Primary cause of mortality in the armyworm larvae simultaneously parasitized by parasitic wasp and infected by bacteria. Eur. J. Biochem. 252, 299–304.

Moita, L.F., Vriend, G., Mahairaki, V., Louis, C., Kafatos, F.C., 2006. Integrins of *Anopheles gambiae* and a putative role of a new beta integrin, BINT2, in phagocytosis of E. coli. Insect Biochem. Mol. Biol. 36, 282–290.

Moita, L.F., Wang-Sattler, R., Michel, K., Zimmermann, T., Blandin, S., Levashina, E.A., et al. 2005. In vivo identification of novel regulators and conserved pathways of phagocytosis in *A. gambiae*. Immunity 23, 65–73.

Murphy, N., Banks, J.C., Whitfield, J.B., Austin, A.D., 2008. Phylogeny of the parasitic microgastroid subfamilies (Hymenoptera: Braconidae) based on sequence data from seven genes, with an improved time estimate of the origin of the lineage. Mol. Phylogenet. Evol. 47, 378–395.

Nakahara, Y., Kanamori, Y., Kiuchi, M., Kamimura, M., 2003. In vitro studies of hematopoiesis in the silkworm, cell proliferation in and hemocyte discharge from the hematopoietic organ. J. Insect Physiol. 49, 907–916.

Nakahara, Y., Shimura, S., Ueno, C., Kanamori, Y., Mita, K., Kiuchi, M., et al. 2009. Purification and characterization of silkworm hemocytes by flow cytometry. Dev. Comp. Immunol. 33, 439–448.

Nakatogawa, S., Oda, Y., Kamiya, M., Kamijima, T., Aizawa, T., Clark, K.D., et al. 2009. A novel peptide mediates aggregation and migration of hemocytes from an insect. Curr. Biol. 19, 779–785.

Nalini, M., Ibrahim, A.M.A., Hwang, I., Kim, Y., 2009. Altered actin polymerization of *Plutella xylostella* (L.) in response to ovarian calyx components of an endoparasitoid *Cotesia plutellae* (Kurdjumov). Physiol. Entomol. 34, 110–118.

Nalini, M., Kim, Y., 2009. Transient expression of a polydnaviral gene, CpBV15 beta, induces immune and developmental alterations of the diamondback moth, *Plutella xylostella*. J Invert. Path. 100, 22–28.

Nardi, J.B., Pilas, B., Bee, C.M., Zhuang, S., Garsha, K., Kanost, M.R., 2006. Neuroglian-positive plasmatocytes of *Manduca sexta* and the initiation of hemocyte attachment to foreign surfaces. Dev. Comp. Immunol. 30, 447–462.

Nardi, J.B., Ujhelyi, E., Pilas, B., Garsha, K., Kanost, M.R., 2003. Hematopoietic organs of *Manduca sexta* and hemocyte lineages. Dev. Genes Evol. 213, 477–491.

Oda, Y., Matsumoto, H., Kurakake, M., Ochiai, M., Ohnishi, A., Hayakawa, Y., 2010. Adaptor protein is essential for insect cytokine signaling in hemocytes. Proc. Natl. Acad. Sci. U.S.A. 107, 15862–15867.

Pech, L.L., Strand, M.R., 1996. Granular cells are required for encapsulation of foreign targets by insect hemocytes. J. Cell Sci. 109, 2053–2060.

Pennacchio, F., Strand, M.R., 2006. Evolution of developmental strategies in parasitic Hymenoptera. Annu. Rev. Entomol. 51, 233–258.

Philips, J.A., Rubin, E.J., Perrimon, N., 2005. *Drosophila* RNAi screen reveals C35 family member requred for mycobacterial infection. Science 309, 1248–1251.

Provost, B., Jouan, V., Hilliou, F., Delobel, P., Bernardo, P., Ravallec, M., et al. 2011. Lepidoptera transcriptome analysis following infection by phylogenetically unrelated polydnaviruses highlights differential and common responses. Insect Biochem. Mol. Biol. (In press.)

Provost, B., Varricchio, C., Arana, E., Espagne, E., Falabella, P., Huguet, E., et al. 2004. Bracoviruses contain a large multigene family coding for protein tyrosine phosphatases. J. Virol. 78, 13090–13103.

Pruijssers, A.J., Strand, M.R., 2007. PTP-H2 and PTP-H3 from *Microplitis demolitor* bracovirus localize to focal adhesions and are antiphagocytic in insect immune cells. J. Virol. 81, 1209–1219.

Ramet, M., Manfruelli, P., Pearson, A., Mathey-Prevot, B., Ezekowitz, R.A.B., 2002. Functional genomic analysis and identification of a *Drosophila* receptor for *E. coli*. Nature 416, 644–648.

Ramet, M., Pearson, A., Manfruelli, P., Li, X., Koziel, H., Gobel, V., et al. 2001. *Drosophila* scavenger receptor CI is a pattern recognition receptor for bacteria. Immunity 15, 1027–1038.

Rivkin, H., Kroemer, J.A., Bronshteinm, A., Belausov, E., Webb, B.A., Chejanovsky, N., 2006. Response of immunocompetent and immunosuppressed *Spodoptera littoralis* larvae to baculovirus infection. J. Gen. Virol. 87, 2217–2225.

Rohrmann, G.F., 2008. Baculovirus Molecular Biology. NCBI Bookshelf E-books. http://hdl.handle.net/1957/9989.

Satoh, D., Horii, A., Ochiai, M., Ashida, M., 1999. Prophenoloxidase-activating enzyme of the silkworm, Bombyx mori. J. Biol. Chem. 274, 7441–7453.

Schmidt, O., Theopold, U., Strand, M., 2001. Innate immunity and its evasion and suppression by hymenopteran endoparasitoids. Bioessays 23, 344–351.

Serbielle, C., Moreau, S., Veillard, F., Voldoire, E., Bézier, A., Mannucci, M.A., et al. 2009. Identification of parasite-responsive cysteine proteases in *Manduca sexta*. Biol. Chem. 390, 493–502.

Shelby, K.S., Adeyeye, O.A., Okot-Kotber, B.M., Webb, B.A., 2000. Parasitism-linked block of host plasma melanization. J. Invert. Pathol. 75, 218–225.

Shelby, K.S., Cui, L., Webb, B.A., 1998. Polydnavirus-mediated inhibition of lysozyme gene expression and the antibacterial response. Insect Mol. Biol. 7, 265–272.

Somogyi, K., Sipos, B., Penzes, Z., Kurucz, E., Zsamboki, J., Hultmark, D., et al. 2008. Evolution of genes and repeats in the nimrod superfamily. Mol. Biol. Evol. 25, 2337–2347.

Stanley, D.W., Miller, J.S., 2008. Insect hemocytes and their role in immunity. In: Beckage, N.E. (Ed.), Insect Immunity (pp. 49–68). Academic Press, San Diego.

Stettler, P., Trenczek, T., Wyler, T., Pfister-Wilhelm, R., Lanzrein, B., 1998. Overview of parasitism associated effects on host hemocytes in larval parasitoids and comparison with effects of the egg–larval parasitoid *Chelonus inanitus* on its host *Spodoptera littoralis*. J. Insect Physiol. 44, 817–831.

Stoltz, D.B., Cook, D.I., 1983. Inhibition of host phenoloxidase activity by parasitoid Hymenoptera. Experientia 39, 1022–1024.

Stoltz, D., Makkay, A., 2003. Overt viral diseases induced from apparent latency following parasitization by the ichneumonid wasp, *Hyposoter exiguae*. J. Insect Physiol. 49, 483–489.

Strand, M.R., 1994. *Microplitis demolitor* polydnavirus infects and expresses in specific morphotypes of *Pseudoplusia includens* hemocytes. J. Gen. Virol. 75, 3007–3020.

Strand, M.R., 2008a. Insect hemocytes and their role in immunity. In: Beckage, N.E. (Ed.), Insect Immunity (pp. 25–47). Academic Press, San Diego.

Strand, M.R., 2008b. Polydnavirus abrogation of the insect immune system. In: Mahy, B.W.J., van Regenmortel, M.H.V. (Eds.), Encyclopedia of Virology (third ed.). Vol. 4 (pp. 250–256). Elsevier, London.

Strand, M.R., 2009. The interactions between polydnavirus-carrying parasitoids and their lepidopteran hosts. In: Goldsmith, M.R., Marec, F. (Eds.), Molecular Biology and Genetics of the Lepidoptera (pp. 321–336). CRC Press, Boca Raton.

Strand, M.R., 2010. Polydnaviruses. In: Asgari, S., Johnson, K. (Eds.), Insect Virology (pp. 171–197). Caister Academic Press, Norfolk.

Strand, M.R., Hayakawa, Y., Clark, K.D., 2000. Plasmatocyte spreading peptide (PSP1) and growth blocking peptide (GBP) are multifunctional homologs. J. Insect Physiol. 46, 817–824.

Strand, M.R., Noda, T., 1991. Alterations in the hemocytes of *Pseudoplusia includens* after parasitism by *Microplitis demolitor*. J. Insect. Physiol. 37, 839–850.

Strand, M.R., Pech, L., 1995a. Encapsulation in the insect Pseudoplusia includens (Lepidoptera: Noctuidae) requires cooperation between granular cells and plasmatocytes. J. Cell Sci. 109, 2053–2060.

Suderman, R.J., Pruijssers, A.J., Strand, M.R., 2008. Protein tyrosine phosphatase-H2 from a polydnavirus induces apoptosis of insect cells. J. Gen. Virol. 89, 1411–1420.

Sugumaran, M., 2002. Comparative biochemistry of eumelanogenesis and the protective roles of phenoloxidase and melanin in insects. Pigment Cell Res. 15, 2–9.

Tanaka, K., Lapointe, R., Barney, W.E., Makkay, A.M., Stoltz, D., Cusson, M., et al. 2007. Shared and species-specific features among ichnovirus genomes. Virology 363, 26–35.

Tanaka, K., Tsuzuki, S., Matsumoto, H., Hayakawa, Y., 2002. Expression of *Cotesia kariyai* polydnavirus genes in lepidopteran hemocytes and Sf9 cells. J. Insect Physiol. 49, 433–440.

Tanaka, T., 1987. Morphological changes in hemocytes of the host *Pseudaletia separata* parasitized by *Microplitis mediator* or *Apanteles kariyae*. Dev. Comp. Immunol. 1, 57–67.

Terenius, O., Bettencourt, R., Lee, S.Y., Li, W., Soderhall, K., Faye, I., 2007. RNA interference of Hemolin causes depletion of phenoloxidase activity in *Hyalophora cecropia*. Dev. Comp. Immunol. 31, 571–575.

Thoetkiattikul, H., Beck, M.H., Strand, M.R., 2005. Inhibitor kappaB-like proteins from a polydnavirus inhibit NF-kappaB activation and suppress the insect immune response. Proc. Natl. Acad. Sci. U.S.A. 102, 11426–11431.

Trudeau, D., Strand, M.R., 1998. The role of Microplitis demolitor polydnavirus in parasitism by its associated wasp, Microplitis demolitor (Hymenotera: Braconidae). J. Insect Physiol. 44, 795–805.

Trudeau, D., Witherell, R.A., Strand, M.R., 2000. Characterization of two novel *Microplitis demolitor* PDV mRNAs expressed in *Pseudoplusia includens* hemocytes. J. Gen. Virol. 81, 3049–3058.

Turnbull, M.W., Volkoff, A.N., Webb, B.A., Phelan, P., 2005. Functional gap junction genes are encoded by insect viruses. Curr. Biol. 15, R491–492.

Volkoff, A-N., Jouan, V., Urbach, S., Samain, S., Bergoin, M., Wincker, P., et al. 2010. Analysis of virion structural components reveals vestiges of the ancestral ichnovirus genome. PLoS Pathog. 6, e1000923.

Wang, Y., Jiang, H., 2006. Interaction of ß-1, 3-Glucan with its recognition protein activates hemolymph proteinase 14, an initiation enzyme of the prophenoloxidase activation system in *Manduca sexta*. J. Biol. Chem. 281, 9271–9278.

Wang, Y., Jiang, H., 2007. Reconstitution of a branch of the *Manduca sexta* prophenoloxidase activation cascade in vitro: snake-like hemolymph proteinase 21 (HP21) cleaved by HP14 activates prophenoloxidase-activating proteinase-2 precursor. Insect Biochem. Mol. Biol. 37, 1015–1025.

Washburn, J.O., Haas-Stapleton, E.J., Tan, F.F., Beckage, N.E., Volkman, L.E., 2000. Coinfection of *Manduca sexta* larvae with polydnavirus from *Cotesia congregata* increases susceptibility to fatal infection by *Autographa californica* M nucleopolyhedrosis virus. J. Insect Physiol. 46, 179–190.

Washburn, J.O., Kirkpatrick, B.A., Volkman, L.E., 1996. Insect protection against viruses. Nature 383, 767.

Webb, B.A., Strand, M.R., 2005. The biology and genomics of polydnaviruses. In: Gilbert, L.I., Iatrou, K., Gill, S.S. (Eds.), Comprehensive Molecular Insect Science (pp. 323–360). Elsevier Inc., San Diego.

Webb, B.A., Strand, M.R., Deborde, S.E., Beck, M., Hilgarth, R.S., Kadash, K., et al. 2006. Polydnavirus genomes reflect their dual roles as mutualists and pathogens. Virology 347, 160–174.

Weber, B., Annaheim, M., Lanzrein, B., 2007. Transcriptional analysis of polydnavirus gene in the course of parasitization reveals segment-specific patterns. Arch. Insect Biochem. Physiol. 66, 9–22.

Wertheim, B., Kraaijeveld, A.R., Schuster, E., Blanc, E., Hopkins, M., Pletcher, S.D., et al. 2005. Genome wide expression in response to parasitoid attack in *Drosophila*. Genome Biol. 6, R94.

Wetterwald, C., Roth, T., Kaeslin, M., Annaheim, M., Wespi, G., Heller, M., et al. 2010. Identification of bracovirus particle proteins and analysis of their transcript levels at the stage of virion formation. J. Gen. Virol. (Epub ahead of print.)

Williams, M.J., Habayeb, M.S, Hultmark, D., 2006. Reciprocal regulation of Rac1 and Rho1 in *Drosophila* circulating immune surveillance cells. J. Cell Sci. 120, 502–511.

Yin, L., Zhang, C., Qin, J., Wang, C., 2003. Polydnavirus of *Campoletis chlorideae*: characterization and temporal effect on host *Helicoverpa armigera* cellular immune response. Arch. Insect Biochem. Physiol. 52, 104–113.

Yoshiga, T., Hernandez, V.P., Fallon, A.M., Law, J.H., 1997. Mosquito transferrin, an acute-phase protein that is up-regulated upon infection. Biochemistry 94, 12337–12342.

Yu, X.-Q., Jiang, H., Wang, Y., Kanost, M.R., 2003. Nonproteolytic serine proteinase homologs involved in the tobacco hornworm, Manduca sexta. Insect Biochem. Mol. Biol. 33, 197–208.

Zhang, G., Lu, Z.-Q., Jiang, H., Asgari, S., 2004. Negative regulation of prophenoloxidase (proPO) activation by a clip-domain serine proteinase homolog (SPH) from endoparasitoid venom. Insect Biochem. Mol. Biol. 34, 477–483.

Zhao, P., Lu, Z., Strand, M.R., Jiang, H., 2011. Antiviral, antiparasitic, and cytotoxic effects of 5, 6-dihyroxyindole (DHI), a reactive compound generated by phenoloxidase during the insect immune response. Insect Biochem. Mol. Biol. (In press.)

Zhuang, S., Kelo, L., Nardi, J.B., Kanost, M.R., 2007. An integrin–tetraspanin interaction required for cellular innate immune responses of an inset, *Manduca sexta*. J. Biol. Chem. 282, 22563–22572.

Chapter 13

Polydnaviruses as Endocrine Regulators

Nancy E. Beckage

Departments of Entomology & Cell Biology and Neuroscience, University of California-Riverside, Riverside, CA 92521, U.S.A.

SUMMARY

Parasitoids are insect parasites that ultimately kill their host prior to its reproductive stage. Killing the host is the essence of being a parasitoid. Hence this explains their exploitation in biological control to reduce host pest populations. Many species of parasitized host lepidopteran larvae are developmentally arrested in their terminal instar when their parasitoids emerge and the normal endocrine signals accompanying pupation are suppressed. Additionally, the host stops feeding prior to parasitoid emergence from the host and this anorexia is irreversible. This chapter summarizes the evidence that parasitoid viruses function as endocrine regulators in the host–parasitoid–polydnavirus relationship and ultimately act as endocrine disruptors leading to death of the host.

ABBREVIATIONS

CcBV	*Cotesia congregata* bracovirus
MdBV	*Microplitis demolitor* bracovirus
JH	juvenile hormone
PDV	polydnavirus
PTTH	prothoracicotropic hormone
PTPs	protein tyrosine phosphatases

INTRODUCTION

Before we summarize the endocrine anomalies seen in parasitized insects, it is worthwhile reviewing the normal endocrine physiology of non parasitized lepidopteran larvae for comparison. I will use the tobacco hornworm, *Manduca sexta*, as a model (see Truman (2009) for a review). Ecdysteroids, juvenile hormone (JH), and neuropeptides are all key players in molting and metamorphosis. Larval–larval molting occurs when JH levels are high so that when prothoracicotropic hormone (PTTH) is released by the brain to stimulate the release of ecdysteroids by the prothoracic glands, a new larval cuticle is produced. In the last instar or stadium, the scenario is different. When a critical weight is attained (which in the tobacco hornworm is 5 g), programming for metamorphosis takes place and JH disappears from the blood midway through the instar due to a rise in hemolymph JH-specific esterase activity and cessation of corpora allata production of JH. Subsequently, a small prewandering peak of ecdysteroid appears in the absence of JH, which is followed a few days later by a large ecdysteroid peak which triggers production of a pupal cuticle. At the wandering peak, the gut is purged and wandering behavior allows the animal to find an appropriate pupation site. If a large amount of JH is still present, then formation of a larval–pupal intermediate occurs in response to the pupal peak in ecdysteroids and the integument produces a cuticle with a mosaic of both larval and pupal characteristics. These intermediates are non viable and die several days after the molt. In normal pupae, adult development begins in response to a rise in ecdysteroids, and eventually the adult moth sheds the pupal cuticle. Ecdyses occur due to release of the peptides eclosion hormone and ecdysis-triggering hormone. Newly ecdysed adult moths initiate mating and egg laying, and the life-cycle begins again with first instar larvae eclosing from the eggs.

Many of the host–parasitoid systems described in this book deal with larval–larval or egg–larval parasitoids. Because the host continues to feed and molt following parasitization, these braconid parasitoids are classified as koinobionts. In contrast, ichneumonid wasps are mainly idiobionts, with hosts that are generally arrested and do not feed or molt following parasitization. With braconid larval–larval parasitoids, the wasps generally emerge from larval stage hosts before the pre wandering peak of ecdysteroids occurs (e.g., *Cotesia* species). With egg–larval parasitoids (wasps in the genera *Chelonus* and *Ascogaster*) the wasps emerge from penultimate instar hosts which undergo a precocious drop in JH due to premature appearance of JH esterase in the hemolymph, which metabolizes JH. The prewandering release of ecdysteroids signaling the onset of metamorphosis occurs an instar earlier than normal, and the parasitoids emerge from arrested host prepupae. Host endocrine programming differs for larval–larval

Parasitoid Viruses: Symbionts and Pathogens. DOI: 10.1016/B978-0-12-384858-1.00013-8

and larval–prepupal parasitoids. The host may live for several days following parasitoid emergence without molting, but ultimately death occurs.

During oviposition, the female wasp injects a cocktail of ovarian fluids containing polydnavirus (PDV) virions along with venom and ovarian proteins into the host hemocoel. The wasp larvae that hatch from the egg are known to secrete both JH and ecdysteroids and also proteins which are detected in the hemolymph of their host. The PDV plays an important role in the endocrine programming of the host, and venom may enhance effects of PDVs on endocrine regulation possibly by increasing membrane permeability in the host (Lanzrein *et al.*, Chapter 14 of this volume).

POLYDNAVIRUSES THAT ACT AS ENDOCRINE REGULATORS

Our appreciation of endocrine-like actions of parasitoid viruses first originated with reports that nonparasitized insects injected with PDVs often experienced (1) a prolongation of the larval stage frequently coupled with (2) disrupted metamorphosis and failure to pupate normally and develop to adulthood. Before the technologies were in place so that JH and ecdysteroid levels could be easily measured (monitoring JH levels is still technically formidable), the analytical data were scarce but a growing number of observations indicated that something was happening in PDV-injected insects to disrupt the onset and progression of the metamorphic transition from larva to pupa in systems where normally the parasitoid egresses from the larval stage of its host.

My laboratory has used the *Manduca sexta–Cotesia congregata* (formerly *Apanteles congregatus*) interaction as a host–parasitoid model system for more than twenty years. Following parasitization of new fourth instar *M. sexta* larvae, the wasps develop inside the host for ca. 10 days then emerge from fifth instar larvae after the host has stopped feeding but before wandering preparatory to pupation begins. In normal parasitized larvae, developmental arrest of the host *M. sexta* larvae is attributed to multiple factors including a high JH level in the terminal stage, caused by the absence of JH esterase and continued synthesis of JH by the corpora allata, coupled with ecdysteroid deficiency and failure of neuropeptides including PTTH to be released in the brain, ventral nerve cord, and gut (see Beckage and Gelman (2001, 2004), for reviews). A low basal level of hemolymph ecdysteroids persists even following emergence, explaining why hosts of this stage are irreversibly arrested and eventually die, without molting or pupating, several days after emergence.

In the *Manduca sexta–Cotesia congregata* system, injection of purified *C. congregata* bracovirus (CcBV) into

a fourth-instar nonparasitized larva causes a developmental arrest in the fifth instar and this arrest can be attributed to disruption of host endocrine signaling pathways in virus-injected larvae (see figs. 1, 2). Such larvae may molt to a supernumerary instar instead of entering the wandering phase preceding pupation, suggestive of an enhancement in JH levels which suppresses the prewandering surge in ecdysteroids. Some virus-injected larvae attempt to pupate but form larval–pupal intermediates instead (Beckage et al., 1994)

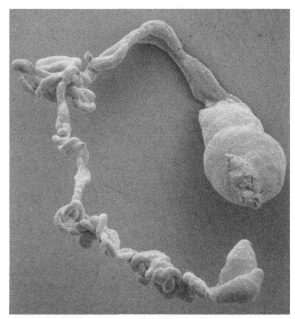
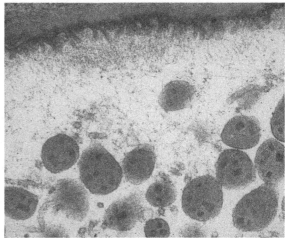

FIGURE 1 (Upper panel) Scanning electron micrograph showing the left ovary with ovarioles of an adult female *Cotesia congregata* parasitoid. The ovary is comprised of the body with the upper calyx region containing the polydnavirus virions comprising the narrower upper portion of the ovary. (Lower panel) Transmission electron micrograph of a cross-section of the ovarian calyx showing the surface of an egg (above) and the virions in the calyx fluid (below). Note the tail structures on the virions and the presence of multiple nucleocapsids within one virion, which is typical of bracoviruses. Scanning and transmission electron micrographs were taken by Dr. Isaure de Buron.

(see Fig. 2). Disruption of both ecdysteroid and JH related regulatory events may induce the developmental arrest caused by PDVs and the infected 'hosts' of the virus die as 'giant' larvae which sometimes attain weights >15 g. Thus, the PDV has virulence factors that act on endocrine regulation as well as many immunity-related targets as described in several chapters of this volume.

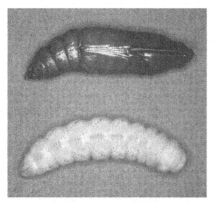

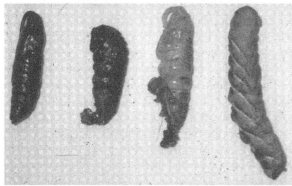

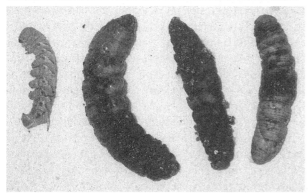

FIGURE 2 (Top panel) Appearance of a normal pupa of an unparasitized *Manduca sexta* (above) and a normal prepupa (below). The cleared dorsal vessel is visible on the dorsal side of the prepupa. (Middle panel) Far left pupa is an unparasitized *M. sexta*. To the right shows appearance of three larvae/pupae that were previously injected with four wasp equivalents of *C. congregata* calyx fluid as new fourth instar larvae. Note the pupae are malformed and the larva (far right) is developmentally arrested. (Bottom panel) Four developmentally arrested *M. sexta* larvae that had previously been injected with four wasp equivalents of calyx fluid as fourth instar larvae. These animals died as larvae prior to pupation.

In several host–parasitoid systems, multiple PDV gene transcripts are expressed within 2 h of parasitization, as occurs in the tobacco hornworm–*C. congregata* system, whereas in other systems viral transcripts do not appear until later, when they may play an important role in causing developmental arrest. In parasitized *M. sexta*, different sets of 'early' and 'late' proteins appear in the hemolymph, and the early proteins appear to be induced by the PDV.

In egg–larval parasitoids systems, the host undergoes precocious metamorphosis in the penultimate instar and the wasp emerges from prepupae. Lanzrein et al. (2001, and Chapter 14 of this volume). The PDV induces JH esterase to be produced so JH disappears during the penultimate instar when precocious metamorphosis is induced.

The PDV of *Microplitis demolitor* inhibits host (*Pseudoplusia includens*) metamorphosis and induces host wasting as shown by Pruijssers et al. (2009). The feeding cessation or 'wasting' of the host that occurs preparatory to wasp emergence may ultimately be found to have a neural or neuroendocrine mechanism that acts on neural networks controlling behavior causing feeding and gut contraction to stop (Cooper and Beckage, 2010). This anorexic state of the host benefits the emerging parasitoids, which might otherwise be consumed by an actively feeding host. Other viruses also cause behavioral changes in host insects including baculoviruses, which induce their host to climb to the top of vegetation. This benefits the virus which is showered on the vegetation below to infect new hosts feeding there. In another example, host larvae actively protect the cocoons of *Glyptapanteles* by fighting with potential predators but how host behavior is regulated is not known (Grosman *et al.*, 2008). In some cases, host behavioral alterations are induced by parasitoid larvae and/or viruses depending on the species involved.

As shown by Dover *et al.* (1988) the PDV of *Campoletis sonorensis* causes the prothoracic glands of last instar *Heliothis virescens* larvae to degenerate, causing host developmental arrest. In the absence of ecdysteroids, the host cannot undergo molting or pupation. Degeneration of the prothoracic glands is stadium-specific (Dover *et al.*, 1995). A similar phenomenon may occur in other species.

As described in Lanzrein *et al.* (Chapter 14 of this volume) the PDV/venom of *Chelonus inanitus* causes developmental arrest and death of the lepidopteran host as shown in X-ray-treated host larvae ('pseudoparasitized' larvae) in which a small dose of radiation killed the parasitoid but did not affect the virus or venom. They found that *Chelonus inanitus* bracovirus effects on host endocrinology and development were not due to alterations in host proteins. However, proteins produced by the parasitoid larvae appear to play major roles in inducing precocious metamorphosis of the host.

In the tobacco hornworm, injection of even a very small dose of CcBV (a small fraction of a 'wasp ovary equivalent') into nonparasitized fourth-instar larvae is sufficient to trigger disrupted development and causes larval death in the fifth instar or molting to a supernumerary sixth instar or production of larval–pupal intermediates (Dushay and Beckage, 1993; Fig. 2). The viral injection affects host growth, developmental fate, pigmentation, and hemolymph proteins (Beckage et al., 1994). Different sets of 'early' versus 'late' hemolymph proteins are produced during parasitism or after injection of CcBV. Larvae injected with PDV become very pale blue or white or even pink, suggesting a loss of insecticyanin from the integument. Whether the virus acts on the corpora allata of the host, causing increased synthesis and release of JH at the critical time when JH needs to be absent to trigger metamorphosis has yet to be elucidated in this system. The peptides allatostatin and allatotropin, which regulate JH release from the corpora allata, may be affected. In addition, in virus-injected larvae, JH degradation could be inhibited due to the absence of JH specific esterase which occurs in naturally parasitized fifth-instar larvae (Beckage and Riddiford, 1982). What is clear is that this is not the only PDV to have this type of juvenilizing effect on host endocrine function as there are several reports of JH-like effects of PDVs on metamorphosis acting alone or in combination with parasitoid venom.

Other insect pathogens have long been implicated with having JH-like disruptive effects on host metamorphosis. For example, the microsporidian Nosema was shown nearly forty years ago to secrete JH-like molecules, and infected lepidopteran hosts experience juvenilizing effects and become nonviable larval–pupal intermediates. Now that PDV genomics has advanced significantly, we can look for genes that likely act on the prothoracic glands or the JH production pathway within the virus genome itself. In the plant world, pathogens can induce hormone-like effects on host plant morphology and development. In insects, it is not yet known whether the viral genome and its gene products are the culprit, or whether virally encoded proteins are acting on host endocrine regulatory pathways indirectly. Host prothoracic gland production of ecdysteroids is also affected by PDVs and/or venom, making the prothoracic glands refractory to the tropic action of PTTH.

The prothoracic glands of Heliothis virescens parasitized by Toxoneuron nigriceps become inactivated during parasitism and a PDV-encoded gene product, protein tyrosine phosphatase (PTP), is thought to be the protein that causes gland inactivity (Pennacchio et al., 1998; Falabella et al., 2006). Reduced ecdysone biosynthesis by the prothoracic glands appears due to underphosphorylation of the PTTH signal transduction pathway which culminates with a translational block of host protein synthesis (Falabella et al., 2006). PTPs are likely the inducing inhibitors of gland function, and would be good candidates to test for

regulating refractory prothoracic glands in parasitized tobacco hornworms.

PDVs of chelonine egg–larval parasitoids and some larval–larval parasitoids have also been implicated in testes degeneration or atrophy of male gonads in host larvae (Bai et al., 2009; Reed and Beckage, 1997; Reed et al., 1997; Reed-Larsen and Brown, 1990). Whether or not virally encoded proteins act on host genes is not clear. Injection of the PDVs mimics the inhibitory effects of parasitization on testis development in M. sexta and other hosts including Plutella xylostella (Bai et al., 2009). In the latter study, female Cotesia vestalis and Diadegma semiclausum wasps were subjected to a dose of gamma irradiation sufficient to kill the parasitoid eggs but which did not affect the virus or venom injected by the female wasp. The 'pseudoparasitized' hosts showed varying degrees of testes degeneration and atrophy. The mechanism(s) by which PDVs or PDV gene products cause parasitic castration remains unknown.

MECHANISTIC EXPLANATIONS FOR PDV-INDUCED HORMONAL INTERVENTIONS

When Microplitis demolitor bracovirus (MdBV) is injected into nonparasitized larvae of the moth Pseudoplusia includens, the virus alone causes feeding cessation, wasting, and anorexia, and inhibits metamorphosis of the infected larva if injected prior to the time the caterpillar reaches the critical weight required for metamorphosis (Pruijssers et al., 2009). If parasitization occurs after attainment of the critical weight for metamorphosis there is no effect on development. MdBV infection of precritical weight larvae induces rapid and persistent hyperglycemia and reduces nutrient stores in the host (Pruijssers et al., 2009). Effects of viral infection on metabolic hormones and feeding behavior are therefore likely to cause the host to fail to achieve the minimal critical weight needed to trigger the hormonal pathways leading to pupation. Illness-induced anorexia is a common effect of infection that occurs in many species throughout the animal kingdom including Lepidoptera (Adamo et al., 2007), and in this case is induced by virus infection with MdBV that induces wasting and prevents host pupation. It will be interesting to identify the metabolic hormone(s) which are involved in this premetamorphic stage prior to the surge of ecdysteroids that triggers pupation. Lack of stimulatory PTTH action on the prothoracic glands may also contribute to arrest and reflect disruption of this signaling pathway. The prothoracic glands can also be refractory to stimulation by PTTH, as occurs when pre critical weight Pseudoplusia includens larvae are parasitized by M. demolitor and are developmentally arrested prior to pupation due to suppression of ecdysteroid release. With respect to inhibition of feeding, polydnaviral DNA and RNA have been detected in the brain and ventral nerve cord in some host species, providing a mechanism to affect the nervous system and behavior.

Virally mediated endocrine disruption can represent a 'fail-safe' mechanism whereby parasitism ensures that the host lingers in the stage most beneficial for parasitoid development. However, some mechanisms of host developmental arrest remain unexplained. Transplantation of a single *Cotesia congregata* second instar parasitoid from a naturally parasitized host which is arrested in the fourth larval stage, to a naïve unparasitized 'surrogate host', is sufficient to trigger arrest in the surrogate host while the parasitoid larva successfully emerges from the 'new' arrested host in the absence of PDV or venom (Lavine and Beckage, unpublished observations). The arrested surrogate host eventually dies similar to naturally parasitized hosts following wasp emergence. So, a single parasitoid larva, even in the presumed absence of any viral intervention whatsoever, can induce irreversible host arrest, and the larva successfully emerges. Thus, factors of parasitoid origin may contribute to the induction of host developmental arrest. The parasitoid *Chelonus inanitus* secretes proteins that play critical roles in endocrine modulation of the host by inducing host developmental arrest (Hochuli *et al.*, 1999). Juvenile hormone (Cole *et al.*, 2002) as well as several (10–20) proteins (Beckage, 1993) are secreted *in vitro* by *C. congregata* and these molecules may similarly contribute to developmental arrest of host *M. sexta* larvae. Ecdysteroids are also released by parasitoids (Brown *et al.*, 1993; Gelman *et al.*, 1998, 1999) and may contribute to endocrine regulation of the host.

Why, then, would the parasitoid need the virus? The parasitoid eggs are encapsulated without the virus. Perhaps a redundancy of 'arrest-inducing factors' is a fail-safe-proof mechanism evolved by the parasitoid to ensure success of wasp progeny. Perhaps this represents a combined approach for assurance of wasp success in an uncertain host environment. Similarly, the parasitoid and its PDV and venom in the egg–larval parasitoid *Chelonus inanitus* manipulate host development and host ecdysteroids (Grossniklaus-Burgin, *et al.*, 1998; Pfister-Wilhelm and Lanzrein, 2009; see Lanzrein *et al.*, Chapter 14 of this volume). Jones (1985, 1986, 1987) studied *Trichoplusia ni* larvae pseudoparasitized by *Chelonus insularis* and similarly demonstrated a critical PDV role in causing host developmental arrest in the prepupal stage.

Le *et al.* (2003) showed that CcBV DNAs persist in the host tobacco hornworm larvae for the duration of parasitism and viral DNA can be detected in host fat body genomic DNA. Polydnavirus DNA integration into host caterpillar genomic DNA (Gundersen-Rindal, Chapter 8 of this volume) can have a major influence on host gene expression. Integration of PDV DNA sequences into host genomic DNA opens avenues for host gene regulation by transposable elements in the viral genome that may affect host gene expression. In newly parasitized larvae, several viral transcripts are produced in fat body and hemocytes but the roles of the transcripts are not yet known (Le *et al.*, 2003). The host's nervous system is also infected with polydnaviral DNAs leaving open the possibility that viral genes, or host genes regulated by viral sequences, are expressed to cause downregulation of feeding behavior. Viral genes may also possibly affect release of neuropeptides by neurosecretory cells in the brain, gut, and ventral nerve cord, which accumulate to high levels in the nervous system in the absence of release. There appears to be a global shutdown of neuropeptide release in last-instar parasitized larvae (Zitnan *et al.*, 1995a, b). PTTH may not be released in post-emergence hosts, or the prothoracic glands may be refractory to the peptide, so that the hosts eventually die as arrested pre metamorphic larvae.

The function of several early-expressed viral transcripts could be as regulators of host endocrine physiology as well as immunity. While PDV genomics has generated a wealth of new information about the role of different transcripts and their protein products in suppressing host immunity, PDV functional genomics is still a field in its infancy. How viral transcripts impact endocrine pathways remains to be elucidated, but some interesting candidate genes have been proposed including protein tyrosine phosphatases as discussed above. Polydnavirus-induced impacts on food consumption and growth have endocrine consequences contributing to the induction of developmental arrest, linking behavior and development via endocrine mechanisms. Identification of the genes whose products include feeding inhibitors and those causing developmental arrest are likely to have application potential for expression in plants and biopesticides as discussed by Pennacchio *et al.* (Chapter 22 of this volume). We have made progress in identifying the viral gene products (PTPs and others) that cause arrest, but not those causing host anorexia. Thus, agriculture and pest control will benefit from PDV research in the short as well as longterm.

REFERENCES

Adamo, S.A., Fidler, T.L., Forestell, C.A., 2007. Illness-induced anorexia and its possible function in the caterpillar, *Manduca sexta*. Brain Behav. Immun. 21, 292–300.

Bai, S.F., Cal, D.Z., Li, X., Chen, Z.Z., 2009. Parasitic castration of *Plutella xylostella* larvae induced by polydnaviruses and venom of *Cotesia vestalis* and *Diadegma semiclausum*. Arch. Insect Biochem. Physiol. 70, 30–43.

Beckage, N.E., 1993. Games parasites play: The dynamic roles of proteins and peptides in the relationship between parasite and host. In: Beckage, N.E., Thompson, S.N., Federici, B.A. (Eds.), Parasites and Pathogens of Insects. Vol. 1: Parasites (pp. 25–57). Academic Press, San Diego.

Beckage, N.E., Gelman, D.B., 2001. Parasitism of *Manduca sexta* by *Cotesia congregata*: a multitude of disruptive endocrine effects. In: Edwards, J.P., Weaver, R.J. (Eds.), Endocrine Interactions of Insects Parasites and Pathogens (pp. 59–81). BIOS, Oxford.

Beckage, N.E., Gelman, D.B., 2004. Wasp parasitoid disruption of host development: implications for new biologically based strategies for insect control. Annu. Rev. Entomol. 49, 299–330.

Beckage, N.E., Riddiford, L.M., 1982. Effects of parasitism by *Apanteles congregatus* on the endocrine physiology of the tobacco hornworm *Manduca sexta*. Gen. Comp. Endocrin. 47, 308–322.

Beckage, N.E., Tan, F.F., Schleifer, K.W., Lane, R.D., Cherubin, L.L., 1994. Characterization and biological effects of *Cotesia congregata* polydnavirus on host larvae of the tobacco hornworm. Arch. Insect Biochem. Physiol. 26, 165–195.

Brown, J.J., Kiuchi, M., Kainoh, Y., Takeda, S., 1993. In vitro release of ecdysteroids by an endoparasitoid, *Ascogaster reticulates* Watanabe. J. Insect Physiol. 39, 229–234.

Cole, T.J., Beckage, N.E., Srinivasan, A., Ramaswamy, S.B., 2002. Parasitoid–host endocrine relations: self-reliance or co-optation? Insect Biochem. Mol. Biol. 2, 1673–1679.

Cooper, P.D., Beckage, N.E., 2010. Effects of starvation and parasitism on foregut contraction in larval *Manduca sexta*. J. Insect Physiol. 56, 1958–1965.

Dover, B.A., Davies, D.H., Vinson, S.B., 1988. Degeneration of last instar *Heliothis virescens* prothoracic glands by *Campoletis sonorensis* polydnavirus. J. Invertebr. Pathol. 51, 80–91.

Dover, B.A., Tanaka, T., Vinson, S.B., 1995. Stadium-specific degeneration of host prothoracic glands by *Campoletis sonorensis* calyx fluid and its association with host ecdysteroid titers. J. Insect Physiol. 41, 947–955.

Dushay, M.S., Beckage, N.E., 1993. Dose-dependent separation of *Cotesia congregata*-associated effects of polydnavirus on *Manduca sexta* larval development and immunity. J. Insect Physiol. 39, 1029–1040.

Falabella, P., Caccialupi, P., Varricchio, P., Malva, C., Pennacchio, F., 2006. Protein tyrosine phosphatases of *Toxoneuron nigriceps* bracoviruses as potential disruptors of host prothoracic gland function. Arch. Insect Biochem. Physiol. 61, 157–169.

Gelman, D.B., Kelly, T.J., Reed, D.A., Beckage, N.E., 1999. Synthesis and release of ecdysteroids by *Cotesia congregata*, a parasitoid wasp of the tobacco hornworm. Arch. Insect Biochem. Physiol. 40, 17–29.

Gelman, D.B., Reed, D.A., Beckage, N.E., 1998. Manipulation of fifth-instar host (*Manduca sexta*) ecdysteroid levels by the parasitoid wasp *Cotesia congregata*. J. Insect Physiol. 44, 833–843.

Grosman, A.H., Janssen, A., de Brito, E.F., Cordeiro, E.G., Colares, F., et al. 2008. Parasitoid increases survival of its pupae by inducing hosts to fight predators. PLoS ONE 3 (6), e2276.

Grossniklaus-Burgin, C., Pfister-Wilhelm, R, Meyer, V., Treiblmayr, K., Lanzrein, B., 1998. Physiological and endocrine changes associated with polydnavirus/venom in the parasitoid–host system *Chelonus inanitus–Spodoptera littoralis*. J. Insect Physiol. 44, 305–321.

Hochuli, A, Pfister-Wilhelm, R, Lanzrein, B., 1999. Analysis of endopar-asitoid-released proteins and their effects on host development in

the system Chelonus inanitus (Braconidae)–Spodoptera littoralis (Noctuidae). J. Insect Physiol. 45, 823–833.

Jones, D., 1985. The endocrine basis for developmentally stationary pre-pupae in larvae of *Trichoplusia ni* pseudoparasitized by *Chelonus insularis*. J. Comp. Physiol. 155B, 235–240.

Jones, D., 1986. *Chelonus* sp. suppression of host ecdysteroids and devel-opmentally stationary pseudoparasitized prepupae. Exp. Parasitol. 61, 471–475.

Jones, D., 1987. Material from adult female *Chelonus* sp. directs expres-sion of altered developmental programme of host Lepidoptera. J. Insect Physiol. 33, 129–134.

Lanzrein, B., Pfister-Wilhelm, R., von Niederhausen, F., 2001. Effects of an egg–larval parasitoid and its polydnavirus on development and the endocrine system of the host. In: Edwards, J.P., Weaver, R.J. (Eds.), Endocrine Interactions of Insect Parasites and Pathogens (pp. 95–109). BIOS, Oxford.

Le, N.T., Asgari, S., Amaya, K., Tan, F.F., Beckage, N.E., 2003. Persistence and expression of Cotesia congregata polydnavirus in host larvae of the tobacco hornworm, *Manduca sexta*. J. Insect Physiol. 49, 533–543.

Pennacchio, F., Falabella, P., Vinson, S.B., 1998. Regulation of *Heliothis virescens* prothoracic glands by *Cardiochiles nigriceps* polydnavirus (CnPDV). Arch. Insect Biochem. Physiol. 38, 1–10.

Pfister-Wilhelm, R., Lanzrein, B., 2009. Stage dependent influences of polydnaviruses and the parasitoid larva on host ecdysteroids. J. Insect Physiol. 55, 707–715.

Pruijssers, A.J., Falabella, P., Eum, J.H., Pennacchio, F., Brown, M.R., Strand, M.R., 2009. Infection by a symbiotic polydnavirus induces wasting and inhibits metamorphosis of the moth *Pseudoplusia inclu-dens*. J. Exp. Biol. 212, 2998–3006.

Reed, D.A., Beckage, N.E., 1997. Inhibition of testicular growth and development in *Manduca sexta* larvae parasitized by the braconid wasp *Cotesia congregata*. J. Insect Physiol. 43, 29–38.

Reed, D.A., Loeb, M.J., Beckage, N.E., 1997. Inhibitory effects of parasitism by the gregarious endoparasitoid *Cotesia congregata* on host testicular development. Arch. Insect Biochem. Physiol. 36, 95–114.

Reed-Larsen, D.A., Brown, J.J., 1990. Embryonic castration of the codling moth *Cydia pomonella* by an endoparasitoid *Ascogaster qua-dridentata*. J. Insect Physiol. 36, 111–118.

Truman, J.W., 2009. Hormonal control of the form and function of the nervous system. In: Gilbert, L.I. (Ed.), Insect Development: Morphogenesis, Molting, and Metamorphosis (pp. 133–161). Academic Press, San Diego.

Zitnan, D., Kingan, T.G., Beckage, N.E., 1995. Parasitism-induced accu-mulation of FMRFamide-like peptides in the gut innervation and endocrine cells of *Manduca sexta*. Insect Biochem. Mol. Biol. 25, 669–678.

Zitnan, D., Kingan, T.G., Kramer, S.J., Beckage, N.E., 1995. Accumulation of neuropeptides in the cerebral neurosecretory system of *Manduca sexta* larvae parasitized by the braconid wasp *Cotesia congregata*. J. Comp. Neurol. 356, 83–100.

The Orchestrated Manipulation of the Host by *Chelonus Inanitus* and its Polydnavirus

Beatrice Lanzrein, Rita Pfister-Wilhelm, Martha Kaeslin, Gabriela Wespi and Thomas Roth

Institute of Cell Biology, University of Bern, Baltzerstrasse 4, CH-3012 Bern, Switzerland

SUMMARY

The interactions of the *Chelonus inanitus* bracovirus (CiBV) and the parasitoid larva with the host are complex and variable and venom is essential only in the initial phase of parasitization. *C. inanitus* is an egg–larval parasitoid and in the beginning it is a conformer and hardly influences the host. But towards the end of its first larval instar, when the host is a fourth-instar larva, CiBV and the parasitoid larva begin to manipulate the host at various levels in an orchestrated manner. CiBV plays *piano pianissimo* as only low amounts of viral transcripts and only very few virus-related proteins of low abundance are found in the host. Expression of viral genes varies along with host and parasitoid development. Nevertheless, the effects of CiBV are strong: the virus prevents encapsulation of the parasitoid larva, alters the host's nutritional physiology, and induces a developmental arrest in the prepupal stage by acting on the host's ecdysone production. The late first instar parasitoid larva plays *mezzo forte* and releases some proteins into the host; along with CiBV, it induces a precocious onset of metamorphosis by acting on host juvenile hormones. It also increases host lipids. The second-instar parasitoid larva consumes host hemolymph and releases ecdysteroids before egression from the host. Thus, the effects of CiBV and the parasitoid larva are finely tuned and involve many physiological targets in the host.

ABBREVIATIONS

BV	bracovirus
JH	juvenile hormone
L1 to L6	first to sixth larval instar
PDV	polydnavirus
PTTH	prothoracicotropic hormone

INTRODUCTION

We have devoted the past twenty years of research to investigating the various types of interactions of the parasitic wasp *Chelonus inanitus* with its natural host *Spodoptera littoralis*. We have used physiological, biochemical, ultrastructural, and molecular methods, and this broad approach clearly shows that the interactions of the *Chelonus inanitus* bracovirus (CiBV) venom and the parasitoid larva are multifaceted and change in the course of parasitoid and host development. Let us first consider the life-cycles of the host and the parasitoid, which are displayed in Fig. 1. Nonparasitized *S. littoralis* pass through six larval instars, then dig into the ground, pupate, and emerge as adults. When parasitized by the solitary egg–larval parasitoid *C. inanitus,* they only pass through five larval instars, enter metamorphosis precociously, and dig into the ground. They construct a pupal cell and the parasitoid larva then emerges from the precocious prepupa and eats it up. Within the pupal cell of the host, the parasitoid spins a cocoon, pupates, and eventually ecloses.

C. inanitus can successfully parasitize all stages of host eggs and employs three strategies to invade the host, dependent on the time-point of parasitization (Kaeslin *et al.,* 2005c). While the host passes from L1 to L5, the L1 parasitoid goes through dramatic morphological changes (Grossniklaus-Bürgin *et al.,* 1994). It develops an anal vesicle which, according to its ultrastructure, might serve to take up nutrients from host hemolymph (Kaeslin *et al.,* 2006). In the feeding stage of the host's L5, the parasitoid

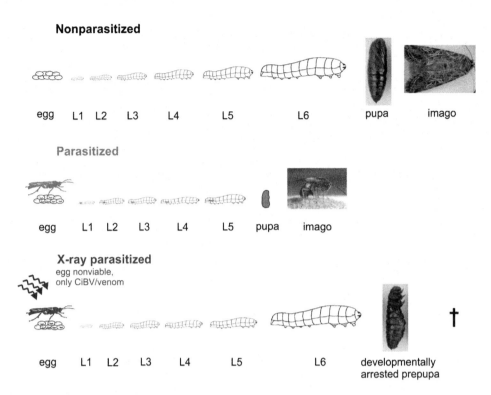

FIGURE 1 **Life-cycles of nonparasitized *S. littoralis* (top) and when parasitized by *C. inanitus* (middle).** On the bottom the method to study CiBV/venom functions in the absence of a developing parasitoid (X-ray treatment of wasps) is shown along with the effect on *S. littoralis* development. Please see color plate section at the back of the book.

molts into L2 and starts to drink host hemolymph; the anal vesicle now disappears. Immediately before emergence from the host, the parasitoid molts into L3 and has sclerotized mandibles. On the bottom of Fig. 1, a method of analyzing the effect of CiBV/venom in the absence of a developing parasitoid (X-ray irradiation of wasps) is shown, along with the impact of CiBV/venom on host development. When we started to investigate the functions of CiBV/venom, we used various approaches, namely injection of calyx fluid and venom, heat-treatment of parasitized eggs to kill the parasitoid embryo, and X-ray irradiation of wasps (Soller and Lanzrein, 1996). The latter technique is the most elegant one as it allows large scale studies. It was thus used to analyze the effects of CiBV/venom alone and to compare them to parasitization with respect to various host parameters (see below). CiBV/venom hardly influence host larval development but cause a developmental arrest in the prepupal stage and eventually death (Fig. 1, bottom). Separate and combined injections of CiBV and venom indicated a synergistic effect of venom (Soller and Lanzrein, 1996; Kaeslin *et al.*, 2010). *In vitro* experiments suggest that venom alters cell membrane permeability (Kaeslin *et al.*, 2010) and this is supported by ultrastructural data (Roth and Lanzrein, unpublished observations). Thus, venom appears to be essential only in the initial phase of parasitism, which is in accordance with the

short persistence of venom proteins in the host (Kaeslin *et al.*, 2010).

CHARACTERIZATION OF THE PLAYERS AND THEIR ANTIGENIC RELATEDNESS

A major player is certainly CiBV, and we have analyzed it in various respects. CiBV particles contain one cylindrical nucleocapsid with a diameter of approx. 34 nm and a length between 8 and 46 nm which is enveloped by a single membrane, and each nucleocapsid contains only one viral segment (Albrecht *et al.*, 1994). Individual packaging of segments was recently also reported for the *Microplitis demolitor* BV (Beck *et al.*, 2007). The relative abundance of segments is variable in all analyzed polydnaviruses (PDVs) and this is also the case for CiBV; in the latter we could show that abundant segments are present in multiple copies in the proviral form (Annaheim and Lanzrein, 2007). Nine of the probably 12–15 segments have been fully sequenced and on all of them genes were identified. Up to now, no or only little similarity to genes of other BVs has been found. This might have to do with the different lifestyles: all other studied BVs are harbored by wasps parasitizing larval stage hosts while *C. inanitus* is an egg–larval parasitoid. The process of calyx cell differentiation and CiBV particle formation and release has been analyzed in detail (Wyler and Lanzrein, 2003). Amplification of viral DNA appears to begin with polyploidization of calyx cells in young pupae followed by selective amplification of integrated proviral segments (Marti *et al.*, 2003). For some of the proviral CiBV segments clustering was shown and spacers could be identified (Wyder *et al.*, 2002; Annaheim and Lanzrein, 2007). Proviral segments have direct repeats at both ends, and it was proposed that excision occurs by juxtaposition of these repeats, formation of a loop, and recombination (Gruber *et al.*, 1996). From the repeats of six CiBV segments, an extended excision site motif was found (Annaheim and Lanzrein, 2007), and interestingly the common GCT in all excision sites was also seen in three other BVs (Desjardins *et al.*, 2007). Similarity with other BVs was also seen with respect to the viral particle proteins and their genes, and relatedness to nudiviruses has recently been demonstrated (Bézier *et al.*, 2009). But these genes reside in the chromosomes and are not excised and encapsidated. Nevertheless, 14 of the 44 identified and sequenced CiBV particle proteins had no similarity to known proteins (Wetterwald *et al.*, 2010), which suggests that there is a set of species- or lineage-specific particle proteins which are essential in the particular parasitoid–host relationship.

A second player is venom, which is coinjected with CiBV and the parasitoid egg. It is a viscous fluid which contains many proteins whereby the majority is glycosylated (Kaeslin *et al.*, 2010). Recently the full protein composition has been resolved by a combined transcriptomic and proteomic approach whereby some conserved venom components as well as many new proteins were found (Vincent *et al.*, 2010).

A third player is the parasitoid larva, which hatches approximately 16 h after parasitization and develops within the host embryo and larva. Parasitoid larvae release several proteins into the host in a stage-specific manner, but only four have been N-terminally sequenced (Hochuli *et al.*, 1999; Hochuli and Lanzrein, 2001; Kaeslin *et al.*, 2005b). It appears that the salivary glands, which are a well-developed organ in L1 and L2 parasitoids, produce and release some of these proteins (Kaeslin *et al.*, 2006).

As some antigenic similarities between PDVs and venom were observed, it was suggested that PDV and venom may share factors with overlapping functions (Webb and Summers, 1990; Asgari, 2006). We thus analyzed all three components (CiBV, venom and parasitoid-released proteins) with respect to antigenic similarities. The protein banding pattern and the reactions and cross reactions with the respective antibodies are shown in Fig. 2. The picture is based on many Western blots for each component (Wespi, unpublished). In general, only a few high-molecular-mass proteins showed cross-reactions with other antibodies. CiBV proteins 250 and 89 reacted with all four antibodies; protein 250 is not sequenced

and protein 89 has similarity to the VP91 capsid protein of the *Gryllus bimaculatus* nudivirus (Wetterwald *et al.*, 2010). Also some venom proteins showed cross reactions, whereby four reacted with all four antibodies. Ve 300b is a metalloprotease (Vincent *et al.*, 2010) and BV 97 has a metalloprotease domain (Wetterwald *et al.*, 2010) which might explain this cross reactivity. Ve 220 is a putative mucin-like pertrophin, Ve 95 has similarity to angiotensin converting enzyme, Ve 14a has similarity to pheromone-binding protein family and 14b has no similarity to known proteins (Vincent *et al.*, 2010). Taken together, CiBV particle proteins and venom proteins share very little similarity and there is no indication of overlapping functions.

Interestingly, many *S. littoralis* egg proteins were seen to react with the CiBV antibody (Wespi *et al.*, unpublished), which might represent some kind of molecular mimicry used to prevent recognition of CiBV particles by the host's immune system. Egg proteins did not react with the venom antibody (Wespi, unpublished). Some CiBV and venom proteins cross-reacted with the antibody against parasitoid-released proteins and some parasitoid-released proteins reacted with the CiBV or venom antibody (Fig. 2) but for those no amino acid sequences are available.

EFFECTS OF CiBV AND THE PARASITOID LARVA ON THE HOST

By using the technique of X-ray treatment of wasps, we have systematically compared the roles of CiBV and parasitization on various host parameters. The major findings are summarized in Table 1. With respect to growth and development, CiBV are essential to prevent encapsulation of the parasitoid egg and larva, but this does not involve persistent effects on host hemocytes (Stettler *et al.*, 1998). Up to the host's L4 instar, nonparasitized, X-ray parasitized, and parasitized larvae grow similarly and *C. inanitus* behaves as a conformer (Lanzrein *et al.*, 2001). Thereafter, *C. inanitus* becomes a regulator which influences nutritional and developmental parameters of the host.

CiBV causes an increase in the concentrations of free sugars in hemolymph and of glycogen in the whole body. The additional presence of a parasitoid larva leads to an accumulation in lipids in the whole body which appears to be important for successful parasitoid development (Kaeslin *et al.*, 2005a, b). In L5, parasitized larvae enter metamorphosis precociously, most likely through release of proteins by the parasitoid larva (Pfister-Wilhem and Lanzrein, 1996; Hochuli and Lanzrein, 2001; Kaeslin *et al.*, 2005b). The first indications of this major developmental effect are seen in the plasma proteome in L4 larvae at the head capsule slippage stage (Kaeslin *et al.*, 2005a). This is followed by slightly altered juvenile hormone (JH) and ecdysteroid levels in freshly molted L5 larvae; at this

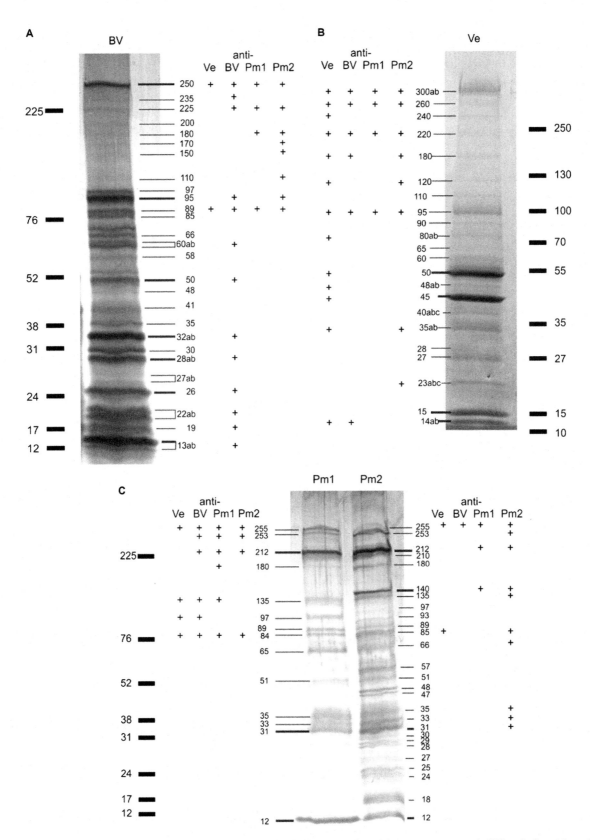

FIGURE 2   Proteins of sucrose gradient-purified CiBV particles (BV) of venom collected from venom reservoir (Ve) and of proteins released by L1 and L2 parasitoid larvae *in vitro* (Pm1 and Pm2). The activities and cross-reactivities of antibodies produced in rats against these four components (anti-Ve, anti-BV, anti-Pm1, anti-Pm2) are indicated with crosses. Gradient SDS gels (4–15%) were used. In (A), 3 eq. of CiBV was loaded and the gel was stained with silver; in (B) 5 eq. of venom was loaded and the gel was stained with Coomassie; in (C) 0.5 eq. of medium in which L1 or L2 larvae had been incubated for 5–7 h was loaded and the gel was stained with silver. Molecular mass markers are shown on the right or left of the gels.

TABLE 1 Summary of Effects of Parasitization and X-Ray Parasitization (=CiBV*) on the Host's Growth, Development, Nutritional Physiology, Endocrinology and Plasma Proteins

Parameter	Major findings	References
Growth and development	Parasitization reduces growth from L4 onwards and causes precocious onset of metamorphosis in L5 and developmental arrest in precocious prepupa	Grossniklaus-Bürgin *et al.*, 1994; Lanzrein *et al.*, 2001
	CiBV protect the parasitoid from encapsulation	Stettler *et al.*, 1998
	CiBV are responsible for the developmental arrest in the prepupal stage	Soller and Lanzrein, 1996
	The parasitoid larva in presence of CiBV is responsible for the precocious onset of metamorphosis	Pfister-Wilhelm and Lanzrein, 1996
Nutritional physiology	CiBV reduces uptake of food in L6	Kaeslin *et al.*, 2005a
	CiBV increases free sugars in the hemolymph and glycogen in the whole body. The additional presence of a parasitoid larva causes an increase in lipids	Kaeslin *et al.*, 2005a, b
Juvenile hormones (JHs)	Parasitization does not alter JH quantities and homolog composition from the egg to L4, but leads to disappearance of JHs in L5	Steiner *et al.*, 1999
	Implantation of a parasitoid larva into X-ray parasitized hosts causes inactivation of C. allata and drop of JHs	Pfister-Wilhelm and Lanzrein, 1996
	CiBV alter hemolymph JH titres only very late, namely in L6 at the pupal cell formation stage	Grossniklaus-Bürgin *et al.*, 1998
Juvenile hormone esterase (JHE)	Parasitization causes a precocious increase in JHE in L5, and CiBV delay the prepupal JHE in X-ray parasitized L6	Steiner *et al.*, 1999; Grossniklaus-Bürgin *et al.*, 1998; Lanzrein *et al.*, 2001
Ecdysteroids	Parasitization causes a reduction of ecdysteroids immediately after the molt to L5; at the late pupal cell formation stage, before egression of the parasitoid, ecdysteroids increase; they appear to be released by the parasitoid larva	Pfister-Wilhelm and Lanzrein, 2009
	CiBV lead to a reduction in hemolymph ecdysteroids in L6, from the early pupal cell formation stage on, by acting on both PTTH and the prothoracic gland. Ecdysone metabolism is only slightly affected	Grossniklaus-Bürgin *et al.*, 1998; Pfister-Wilhem and Lanzrein, 2009
Hemolymph proteins	Hemolymph protein concentrations are reduced in parasitized larvae at digging and cell formation stages; CiBV has no effect. Hemolymph dilution and injection experiments showed that heat-labile factors in the hemolymph play a role in causing the developmental arrest in the prepupal stage. Parasitization or CiBV have only minor effects on the plasma proteome	Kaeslin *et al.*, 2005a, b

*In this table we use the term CiBV when referring to X-ray parasitized larvae as venom is essential only in the initial phase of parasitization.

stage, the precocious onset of metamorphosis is irreversibly determined (Pfister-Wilhelm and Lanzrein, 1996, 2009; Steiner *et al.*, 1999; Lanzrein *et al.*, 2001).

A general look at the effects of parasitization and CiBV on host JHs and ecdysteroids reveals the following. From the egg to L4, JH quantities and homolog composition are not altered by parasitization (Steiner *et al.*, 1999); ecdysteroids were measured from L4 onwards and no differences were seen between nonparasitized, parasitized,

and X-ray parasitized L4 (Pfister-Wilhelm and Lanzrein, 2009). In L5, however, *C. inanitus* and CiBV manipulate the endocrine system of the host at various levels (see also Fig. 3). The parasitoid larva mainly acts on the JH system and causes a drop in JH by inactivation of the corpora allata and an increase in JH esterase (Pfister-Wilhelm and Lanzrein, 1996; Steiner *et al.*, 1999). Parasitoids contain only JH III and highest values are reached in L2, i.e., a time-point when host JHs are undetectable; this shows that

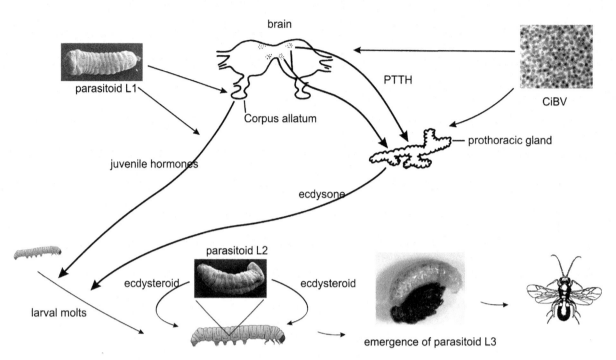

FIGURE 3 The effects of CiBV and the parasitoid L1 larva on the endocrine system of the host, of release of ecdysteroid by L2 parasitoid, and of emergence of L3 parasitoid from host. Please see color plate section at the back of the book.

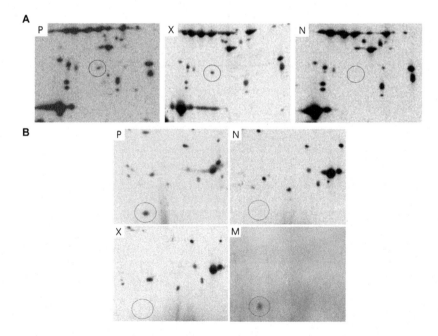

FIGURE 4 Sections of ruthenium-stained 2D gels of plasma from parasitized (P), X-ray parasitized (X), and nonparasitized (N) larvae at the head capsule slippage stage of the penultimate instar. The region showing the virus-related protein V33 (circles) is shown in (A). The region showing the parasitoid-released protein P12 (circles) is shown in (B). Medium in which an L1 parasitoid was incubated for 24 h is shown in (M). Data are from Kaeslin *et al.* (2005b).

JH quality and fluctuations are different in parasitoid and host (Steiner *et al.*, 1999).

CiBV, on the other hand, targets the ecdysteroid system (Grossniklaus-Bürgin *et al.*, 1998; Pfister-Wilhelm and Lanzrein, 2009). In the pupal cell formation and prepupal stages, release of ecdysone by prothoracic glands is reduced whereby CiBV act on both prothoracicotropic hormone (PTTH) and the prothoracic gland. This leads to reduced hemolymph ecdysteroid titres which cause a developmental arrest and eventually death of X-ray parasitized larvae.

Interestingly, in parasitized larvae hemolymph ecdysteroids increase at the late pupal cell formation stage, i.e., shortly before the parasitoid egresses from the host. They appear to be released by the parasitoid larva and we speculate that they might prevent the release of eclosion hormone and thus keep the cuticle soft to allow egression of the parasitoid from the host (Pfister-Wilhelm and Lanzrein, 2009).

Dilution and injection experiments with hemolymph of X-ray parasitized larvae indicated that it contains heat-labile factors which play a role in the developmental arrest (Kaeslin *et al.*, 2005b). To identify these and other CiBV related proteins, a comparative analysis of plasma proteins of nonparasitized and X-ray parasitized larvae was carried out (Kaeslin *et al.*, 2005b). Fig. 4 and Table 2 show that only few proteins of low abundance were found in plasma of X-ray parasitized larvae. Similar analyses with fat body, brain, and hemocytes of the same stages revealed only one CiBV-related protein of low abundance in the fat body; in addition, few host proteins of low abundance were seen to be repressed or slightly modulated in X-ray parasitized

larvae (Roth and Lanzrein, unpublished observations). Thus, the effect of CiBV on host endocrinology and development is not associated with strong alterations of host proteins. Also the concentration of hemolymph proteins is not altered by CiBV (Kaeslin *et al.*, 2005b).

In plasma of parasitized larvae, four proteins of parasitoid origin were found between the L4 head capsule slippage stage and the L5 cell formation stage (Fig. 4B, Table 2, Kaeslin *et al.*, 2005b). Proteins P12, P35, P66 and P212 have been N-terminally sequenced (Kaeslin *et al.*, 2005b) and for P212 similarity to a protein released by another *Chelonus* was seen (Soldevila and Jones, 1994; Hochuli and Lanzrein, 2001). Protein P212 is detectable in host hemolymph from the beginning of L4 onwards (Hochuli *et al.*, 1999), and injections of antibodies against this protein suggest that it plays an important role in initiating the precocious onset of metamorphosis (Kaeslin *et al.*, 2005b). For proteins P12, P35 and P66 much less information is available but they might also be involved in this process.

TABLE 2 Overview of Virus-Related (V) and Parasitoid-Related (P) Proteins in Plasma of Various Stages with Molecular Mass (kD) and Isoelectric Point (IP)

kD/IP	X-Ray parasitized					Parasitized			
	L5fm	L5hcs	L6fm	L6f	L6cf	L4hcs	L5fm	L5f	L5cf
V10/5.5		X			X				
V32/7.3		XXX				XX			
V33/5.4		XXX			XXX	X		XXX	XXX
V41/5.8		X				X			
V58/5.5		XX				XX			
V62/5.7		XX				X			
V83 /5.5		X				X			
V93/5.7		X				X			
V112/6.6		XX				XX			
V113/5.5		XX				X			
V123/5.4		X				XX			
P12/5.2*						XXX	XXX	XXX	XX
P35/6.9						XX	XX		
P66/5.6						XXX	XXX		
P71/7.4*							X	XX	XXX
P212/5.2*						XXX	XXX	XXX	XXX

The presence of the proteins in plasma is indicated with crosses whereby the number of crosses reflects the quantity of protein in units as deduced from OD calculations. For each stage, three gels from three independent hemolymph samples (pools from at least three larvae each) were analyzed. Proteins marked with an asterisk were N-terminally sequenced. Data are from Kaeslin et al. (2005b).

CiBV TRANSCRIPTS IN THE COURSE OF PARASITIZATION AND X-RAY PARASITIZATION

The above descriptions show that CiBV and parasitization have only a minor impact on the host's proteins as deduced from the extensive analyses of 2D gels from various host tissues and stages. CiBV particles are seen within the host embryo 5 h after parasitization (Roth and Lanzrein, unpublished observations) and CiBV DNA is seen all over the host embryo 15 h after parasitization, and in all analyzed tissues in L5 hosts (Wyder *et al.*, 2003). We also investigated relative amounts of CiBV gene transcripts in parasitized and X-ray parasitized *S. littoralis* from the egg stage until emergence of the parasitoid in the former or until developmental arrest in the latter (Fig. 5). Here we show genes which have different patterns of expression. For gene 16.8g1, transcripts are found from the beginning of parasitization and quantities reach much higher levels in

parasitized than X-ray parasitized larvae in the last instar. Thus, the parasitoid larva positively influences transcript quantities of this gene at this stage. Gene 12g1 is mainly early and late expressed and transcript quantities are rather negatively affected by the parasitoid larva in the embryonic stage and the last instar. Gene 12g2 is a late expressed gene and its transcripts reach much higher levels in X-ray parasitized larvae mainly at the late cell formation stage; in parasitized larvae this is shortly before emergence of the parasitoid from the host. Gene 14.5g1 is early expressed and the presence of a parasitoid embryo and larva only weakly influences relative transcript levels.

Taken together, transcript quantities vary in a gene-specific manner in the course of host development and the presence of a parasitoid can influence relative amounts of transcripts in both directions, depending on the gene and the stage. To get an idea about the quantities of CiBV gene transcripts in relation to transcripts of regular host genes, we also measured actin transcripts and roughly compared

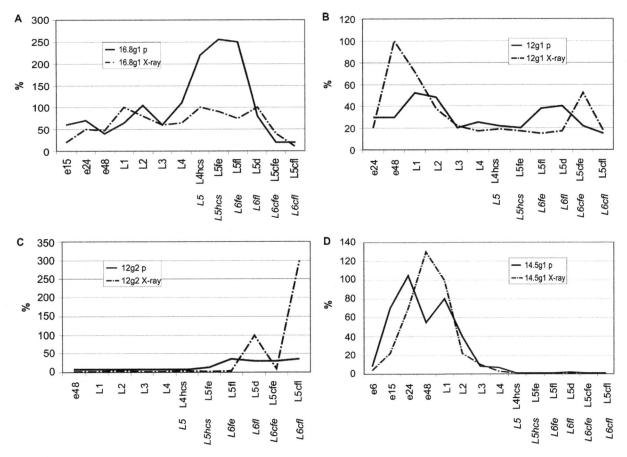

FIGURE 5   Relative amounts of transcripts of four CiBV genes in parasitized (straight lane) or X-ray parasitized (dotted lane) hosts from the egg stage until the late pupal cell formation stage. After L4, when developmental effects of parasitization become manifest, the stages are indicated separately for parasitized (regular fonts) or X-ray parasitized (italic fonts) hosts. 100% values were set as follows: 16.8g1 at X-ray L1; 12g1 at X-ray eggs 48 h; 12g2 at X-ray L5 digging; 14.5g1 at X-ray L1. Abbreviations: e, eggs with age in h; L1–L6, first to sixth larval instar; hcs, head capsule slippage stage; fe, feeding; fl, feeding late; d, digging; cfe, cell formation early; cfl, cell formation late. The drawing is based largely on data obtained by Bonvin *et al.* (2004).

maximal values between actin and CiBV gene transcripts (Bonvin *et al.,* 2004). Of the genes shown in Fig. 5, 12g1 and 12g2 are among the CiBV genes which reach the highest values, but their maximal values are around 250-fold lower than those of actin. Maximal values for 14.5g1 and 16.8g1 are 10-times lower than those of 12g1 and 12g2 and are thus approximately 2500-fold lower than values of actin. In total we have analyzed transcripts of over twenty CiBV genes (Johner and Lanzrein, 2002; Bonvin *et al.,* 2004; Weber *et al.,* 2007; Bieri, unpublished observations) and maximal values never exceeded those of 12g1 and 12g2. This indicates that a multitude of CiBV genes are expressed in a stage-dependent manner but that transcript quantities are never high. The tissue distribution has been analyzed for transcripts of several CiBV genes and hemolymph was found to harbor a large portion (Bonvin *et al.,* 2004). However, depending on the gene, also fat body, midgut, and integument were seen to contain important portions of transcripts (Roth and Lanzrein, unpublished observations). All information available thus indicates that CiBV influences the host by a concerted and subtle expression of various genes. None of the identified genes has high similarity to known PDV genes but 12g1 and 12g2 have similarity to 14g1 and 14g2 respectively (Bonvin *et al.,* 2005). Almost nothing is known about their functions with the exception of 14g1 and 14g2, which appear to be involved in causing the developmental arrest in the prepupal stage (Bonvin *et al.,* 2005). Interestingly, the transcriptional analysis of CiBV genes in the course of parasitization indicated segment-specific patterns; this suggests that one function of the segmentation of the CiBV genome might be the grouping together of genes which are regulated in a similar manner (Weber *et al.,* 2007).

CONCLUSION

The orchestrated manipulation of *S. littoralis* by the egg–larval parasitoid *C. inanitus* can be characterized as follows. In the early phase of parasitism, *C. inanitus* affects the host in a subtle manner and CiBV prevents encapsulation of the parasitoid. Various CiBV genes are expressed, although at a low level, they play *piano pianissimo*. When the host becomes an L4, the parasitoid, which is still an L1, begins to release detectable quantities of protein; some of them are presumably involved in inducing the precocious onset of metamorphosis and the increase in host lipids. At the end of the host's L4, various CiBV-related and four parasitoid-released proteins are seen in the hemolymph, and the first signs of the precocious onset of metamorphosis are detectable. Glycogen levels are elevated as well. After the molt into L5, JH production is suppressed and JH esterase begins to increase, which leads to undetectable JH levels; ecdysteroids are at very low levels. Three of the parasitoid-released proteins are still present in host hemolymph and

transcripts of several CiBV genes are detectable. The players now play *mezzo forte*. In the feeding stage of L5, the parasitoid molts into L2 and massively consumes host hemolymph. Transcripts of several CiBV genes increase and parasitoid-released proteins are still present. Towards pupal cell formation, some CiBV transcripts reach their highest level and ecdysone production is reduced by effects of CiBV on both PTTH and the prothoracic gland; this prevents pupation. Before egression from the prepupal host, the late L2 parasitoid releases ecdysteroids, possibly to keep the host's cuticle soft. In this phase the players play *fortissimo*, and this final crescendo marks the end of the host, which is then consumed by the L3 parasitoid.

REFERENCES

Albrecht, U., Wyler, T., Pfister-Wilhelm, R., Gruber, A., Stettler, P., Heiniger, P., et al. 1994. Polydnavirus of the parasitic wasp *Chelonus inanitus* (Braconidae): characterization, genome organization and time point of replication. J. Gen. Virol. 75, 3353–3363.

Annaheim, M., Lanzrein, B., 2007. Genome organization of the *Chelonus inanitus* polydnavirus: excision sites, spacers and abundance of proviral and excised segments. J. Gen. Virol. 88, 450–457.

Asgari, S., 2006. Venom proteins from polydnavirus-producing endoparasitoids: their role in host–parasite interactions. Arch. Insect Biochem. Physiol. 61, 146–156.

Beck, M.H., Inman, R.B., Strand, M.R., 2007. *Microplitis demolitor* bracovirus genome segments vary in abundance and are individually packaged in virions. Virology 359, 179–189.

Bézier, A., Annaheim, M., Herbinière, J., Wetterwald, C., Gyapay, G., Bernard-Samain, S., et al. 2009. Polydnaviruses of braconid wasps derive from an ancestral nudivirus. Science 323, 926–930.

Bonvin, M., Kojic, D., Blank, F., Annaheim, M., Wehrle, I., Wyder, S., et al. 2004. Stage-dependent expression of *Chelonus inanitus* polydnavirus genes in the host and the parasitoid. J. Insect Physiol. 50, 1015–1026.

Bonvin, M., Marti, D., Wyder, S., Kojic, D., Annaheim, M., Lanzrein, B., 2005. Cloning, characterisation and analysis by RNAi of various genes of the *Chelonus inanitus* polydnavirus. J. Gen. Virol. 86, 973–983.

Desjardins, C.A., Gundersen-Rindal, D.F., Hostetler, J.B., Tallon, L.J., Fuester, R.W., Schatz, M.C., et al. 2007. Structure and evolution of a proviral locus of *Glyptapanteles indiensis* bracovirus. BMC. Microbiol. 7, 61.

Grossniklaus-Bürgin, C., Pfister-Wilhelm, R., Meyer, V., Treiblmayr, K., Lanzrein, B., 1998. Physiological and endocrine changes associated with polydnavirus/venom in the parasitoid–host system *Chelonus inanitus–Spodoptera littoralis*. J. Insect Physiol. 44, 305–321.

Grossniklaus-Bürgin, C., Wyler, T., Pfister-Wilhelm, R., Lanzrein, B., 1994. Biology and morphology of the parasitoid *Chelonus inanitus* (Braconidae, Hymenoptera) and effects on the development of its host *Spodoptera littoralis* (Noctuidae, Lepidoptera). Invertebr. Reprod. Dev. 25, 143–158.

Gruber, A., Stettler, P., Heiniger, P., Schümperli, D., Lanzrein, B., 1996. Polydnavirus DNA of the braconid wasp *Chelonus inanitus* is integrated in the wasp's genome and excised only in later pupal and adult stages of the female. J. Gen. Virol. 77, 2873–2879.

Hochuli, A., Lanzrein, B., 2001. Characterization of a 212 kD protein released into the host by the larva of the endoparasitoid *Chelonus inanitus* (Hymenoptera, Braconidae). J. Insect Physiol. 47, 1313–1319.

Hochuli, A., Pfister-Wilhelm, R., Lanzrein, B., 1999. Analysis of endoparasitoid-released proteins and their effects on host development in the system *Chelonus inanitus* (Braconidae)–*Spodoptera littoralis* (Noctuidae). J. Insect Physiol. 45, 823–833.

Johner, A., Lanzrein, B., 2002. Characterization of two genes of the polydnavirus of *Chelonus inanitus* and their stage-specific expression in the host *Spodoptera littoralis*. J. Gen. Virol. 83, 1075–1085.

Kaeslin, M., Pfister-Wilhelm, R., Lanzrein, B., 2005. Influence of the parasitoid *Chelonus inanitus* and its polydnavirus on host nutritional physiology and implications for parasitoid development. J. Insect Physiol. 51, 1330–1339.

Kaeslin, M., Pfister-Wilhelm, R., Molina, D., Lanzrein, B., 2005. Changes in the hemolymph proteome of *Spodoptera littoralis* induced by the parasitoid *Chelonus inanitus* or its polydnavirus and physiological implications. J. Insect Physiol. 51, 975–988.

Kaeslin, M., Reinhard, M., Bühler, D., Roth, T., Pfister-Wilhelm, R., Lanzrein, B., 2010. Venom of the egg–larval parasitoid *Chelonus inanitus* is a complex mixture and has multiple biological effects. J. Insect Physiol. 56, 686–694.

Kaeslin, M., Wehrle, I., Grossniklaus-Bürgin, C., Wyler, T., Guggisberg, U., Schittny, J.C., et al. 2005. Stage-dependent strategies of host invasion in the egg–larval parasitoid *Chelonus inanitus*. J. Insect Physiol. 51, 287–296.

Kaeslin, M., Wyler, T., Grossniklaus-Bürgin, C., Lanzrein, B., 2006. Development of the anal vesicle, salivary glands and gut in the egg–larval parasitoid *Chelonus inanitus*: tools to take up nutrients and to manipulate the host? J. Insect Physiol. 52, 269–281.

Lanzrein, B., Pfister-Wilhelm, R., von Niederhäusern, F., 2001. Effects of an egg–larval parasitoid and its polydnavirus on development and the endocrine system of the host. In: Edwards, J.P., Weaver, R. (Eds.), Endocrine Interactions of Insect Parasites and Pathogens (pp. 95–109). BIOS, Oxford.

Marti, D., Grossniklaus-Bürgin, C., Wyder, S., Wyler, T., Lanzrein, B., 2003. Ovary development and polydnavirus morphogenesis in the parasitic wasp *Chelonus inanitus*. I. Ovary morphogenesis, amplification of viral DNA and ecdysteroid titres. J. Gen. Virol. 84, 1141–1150.

Pfister-Wilhelm, R., Lanzrein, B., 1996. Precocious induction of metamorphosis in *Spodoptera littoralis* (Noctuidae) by the parasitic wasp *Chelonus inanitus* (Braconidae): identification of the parasitoid larva as the key element and the host corpora allata as a main target. Arch. Insect Biochem. Physiol. 32, 511–525.

Pfister-Wilhelm, R., Lanzrein, B., 2009. Stage dependent influences of polydnaviruses and the parasitoid larva on host ecdysteroids. J. Insect Physiol. 55, 707–715.

Soldevila, A.I., Jones, D., 1994. Characterization of a novel protein associated with the parasitization of lepidopteran hosts by an endoparasitic wasp. Insect Biochem. Mol. Biol. 24, 29–38.

Soller, M., Lanzrein, B., 1996. Polydnavirus and venom of the egg–larval parasitoid *Chelonus inanitus* (Braconidae) induce developmental arrest in the prepupa of its host *Spodoptera littoralis* (Noctuidae). J. Insect Physiol. 42, 471–481.

Steiner, B., Pfister-Wilhelm, R., Grossniklaus-Bürgin, C., Rembold, H., Treiblmayr, K., Lanzrein, B., 1999. Titres of juvenile hormone I, II and III in *Spodoptera littoralis* (Noctuidae) from the egg to the pupal molt and their modification by the egg–larval parasitoid *Chelonus inanitus* (Braconidae). J. Insect Physiol. 45, 401–413.

Stettler, P., Trenczek, T., Wyler, T., Pfister-Wilhelm, R., Lanzrein, B., 1998. Overview of parasitism associated effects on host hemocytes in larval parasitoids and comparison with effects of the egg–larval parasitoid *Chelonus inanitus* on its host *Spodoptera littoralis*. J. Insect Physiol. 44, 817–831.

Vincent, B., Kaeslin, M., Roth, T., Heller, M., Poulain, J., Cousserans, F., et al. 2010. The venom composition of the parasitic wasp *Chelonus inanitus* resolved by combined expressed sequence tag analysis and proteomic approach. BMC Genomics 11, 693.

Webb, B.A., Summers, M.D., 1990. Venom and viral expression products of the endoparasitic wasp *Campoletis sonorensis* share epitopes and related sequences. Proc. Natl. Acad. Sci. U.S.A. 87, 4961–4965.

Weber, B., Annaheim, M., Lanzrein, B., 2007. Transcriptional analysis of polydnaviral genes in the course of parasitization reveals segment-specific patterns. Arch. Insect Biochem. Physiol. 66, 9–22.

Wetterwald, C., Roth, R., Kaeslin, M., Annaheim, M., Wespi, G., Heller, M., et al. 2010. Identification of bracovirus particle proteins and analysis of their transcript levels at the stage of virion formation. J. Gen. Virol. 91, 2610–2619.

Wyder, S., Blank, F., Lanzrein, B., 2003. Fate of polydnavirus DNA of the egg–larval parasitoid *Chelonus inanitus* in the host *Spodoptera littoralis*. J. Insect Physiol. 49, 491–500.

Wyder, S., Tschannen, A., Hochuli, A., Gruber, A., Saladin, V., Zumbach, S., et al. 2002. Characterization of *Chelonus inanitus* polydnavirus segments: sequences and analysis, excision site and demonstration of clustering. J. Gen. Virol. 83, 247–256.

Wyler, T., Lanzrein, B., 2003. Ovary development and polydnavirus morphogenesis in the parasitic wasp *Chelonus inanitus*, part II: ultrastructural analysis of calyx cell development, virion formation and release. J. Gen. Virol. 84, 1151–1163.

Unique Attributes of Viruses and Virus-Like Particles Associated with Parasitoids

Diversity of Virus-Like Particles in Parasitoids' Venom: Viral or Cellular Origin?

Jean-luc Gatti, Antonin Schmitz, Dominique Colinet and Marylène Poirié

UMR Interaction Biotiques et Santé Végétale, INRA (1301), CNRS (6243), Université Nice Sophia Antipolis; Sophia Agrobiotech, 400 Route des Chappes, 06903 Sophia Antipolis, France

SUMMARY

During oviposition in their insect hosts, most endoparasitic wasps inject venom and/or ovarian products that will either prevent the egg from being recognized by the host immune system or protect it from the immune response. This response, the encapsulation process, involves both hematopoietic cells that form successive layers around the parasitoid egg, and induction of the humoral melanization process, due to activation of the phenoloxidase cascade. Among components injected by the wasp with its eggs, several immune suppressive proteins have been described that suppress either the cellular or the humoral defenses. In species from specific braconid and ichneumonid groups, viral capsids containing multiple DNA circles (polydnaviruses, PDVs) produced in the ovary fluid are injected into the host and enter host cells. Expression of viral genes then leads to the production of specific proteins that interfere with the host response. In other species from a large range of hymenopterans' families, vesicles, named virus-like particles (VLPs), have been observed in the ovarian tissue or in the venom apparatus. These VLPs, demonstrated or assumed to be devoid of DNA or RNA, are injected with the egg and participate in its protection from host immune defenses. Comparison of published electron microscopy pictures of VLPs reveals a large diversity in form and size between wasp species with more or less resemblance to viruses, but also an important heterogeneity within a given species. Besides, the few VLPs-associated proteins characterized to date show no homology with viral proteins, and their function is unknown. In this paper, we review data available on parasitoid wasps' VLPs, mainly on the basis of their structure and composition. This leads us to discuss how to discriminate between virus-resembling VLPs and viruses, and to suggest that some of the VLPs might be cellular secreted vesicles, such as microvesicles and exosomes described in many epithelium and cells, aimed to transfer toxins inside host cells. Finally, we propose to avoid the misleading name of VLPs for vesicles produced in venom and rather use the term of 'venosomes'.

ABBREVIATIONS

LC-MS/MS	liquid chromatography coupled to mass spectrometry
PDVs	polydnaviruses
TEM	transmission electron microscopy
VLPs	virus-like particles

INTRODUCTION

Endoparasitoid wasps develop inside the body of their arthropod host, which will die as a result of the interaction (Eggleton and Belshaw, 1992; Godfray, 1994; Quicke, 1997). Not surprisingly, they have evolved original and specific strategies to face the physiological challenges associated with this life-cycle, and notably the host immune response and encapsulation of the parasitoid egg (Smilanich *et al.*, 2009; Cerenius *et al.*, 2010; Nappi, 2010). The mechanisms used by parasitoids to protect their eggs from encapsulation and subsequent death have been investigated in several models and some of the molecular virulence factors have been the subjects of intensive investigation (Poirié *et al.*, 2009). Parasitoids can avoid activation of host immune defenses by laying eggs in a host tissue where they escape recognition (Moreau *et al.*, 2003), or using maternally secreted components mimicking host self-molecules to coat the egg surface (Kinuthia *et al.*, 1999). Teratocytes, specialized cells originating from the extra-embryonic serosal membrane, are known to regulate host development and metabolism and might play a role in hiding the egg and/or altering the immune function (Beckage and de Buron, 1997). However, the most common strategy parasitoids use to deal with host immunity is immune suppression in which venom and/or ovarian components have a pivotal role. They include specific proteins (mostly venom toxins), virus-like particles (VLPs) produced in venom or ovaries, and wasp-specific integrated viruses, the polydnaviruses (PDVs) (Lavine and Beckage, 1995; Pennacchio and Strand, 2006; Dubuffet *et al.*, 2008; Bézier *et al.*, 2009; Asgari and Rivers, 2010).

In this review, we will focus on parasitoid virus-like particles which are known to be associated with parasitism success in different species, but for which only scarce data are available to date. The term 'VLPs' has largely been used to describe particles suggested to be devoid of nucleic acids but that are extremely variable in shape, structure, and organization between different species, even from the same genus, and also between strains of the same species (Poirié *et al.*, 2009, Dubuffet *et al.*, 2009). Through the literature, occurrence of VLPs has been described in phylogenetically distant species (Fig. 1) where they are produced by different tissues (venom apparatus, calyx or ovarian cells) before being finally injected into the host. The nature of VLPs is still largely unknown as are their constituents and origin, and most of them have not been purified or tested for the presence of nucleic acids. We will therefore mainly address the two fundamental questions of the nature and the function of these particles.

A recurrent and confusing problem is that the term 'VLPs' has been used in the literature to describe PDV particles (see a short presentation below), or as yet uncharacterized true viruses. This is not surprising since the morphology of VLPs sometimes strongly resembles that of viruses, and parasitoid reproductive tissues host a large range of viruses (Lawrence and Matos, 2005; Varaldi *et al.*, 2006). These viruses are rarely strongly pathogenic (Whitfield and Asgari, 2003) and can even be mutualistic. Some of them are indeed assumed to increase parasitism success, likely by interfering with the wasp host immunity (Bigot *et al.*, 1997, 2008; Lawrence, 2002).

PDV PARTICLES

Thousands of endoparasitic wasp species are known to bear PDVs integrated into their genome (Beckage and Gelman, 2004; Webb *et al.*, 2006; Bézier *et al.*, 2009). The PDVs' DNA is replicated exclusively in the calyx cells, a specialized part of the wasp ovarian tissue, where it is embedded in virus particles found in the ovarian fluid and delivered to the host along with the eggs. The expression of the virus genes inside the host cells causes immune and developmental changes aimed to favor development of the wasp larva. Two PDV-bearing genera have been described in parasitoid wasps, the ichnoviruses that are associated with species of the *Ichneumonidae* family, and

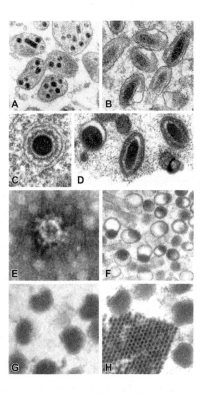

FIGURE 2 Transmission electron microscopy of bracovirus and ichnovirus virions and VLPs from different species. A: Bracovirus virions (dark spots) after their release from the wasp calyx cell. They are included in single-membraned vesicles containing electron dense material. B: Ichnovirus lenticular virions enclosed in a double-membraned vesicle after budding from the calyx cell. C: *Microctonus aethiopoides* VLPs observed in the producing cells. Several membrane layers including a rough reticulum endoplasmic membrane layer surround them. D: The same VLP after its shedding from the cell, appearing as a discoid form surrounded by a double membrane. E: *Opius concolor* VLP observed in the gland cells. F: VLPs released in the lumen of the *Meteorus pulchricornis* venom gland. G and H: *Venturia canescens* VLPs observed in the calyx region in absence (G) or presence of arrays of viruses (H). See the text for the size of the vesicles. Pictures A and B from Frederici and Bigot, 2003; C and D from Barrat *et al.*, 1999; E from Jacas *et al.*, 1997; F from Suziki and Tanaka, 2006; G from Reineke *et al.*, 2006; H from Reineke and Asgari, 2005.

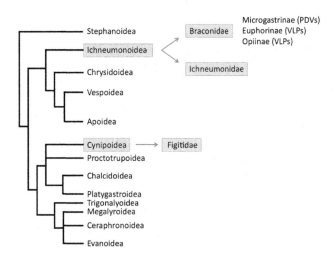

FIGURE 1 Taxonomic position of insects bearing PDVs and the VLPs described in this review. Modified from Whitfield, 1998.

the bracoviruses found in species of the *Braconidae* family that have recently been demonstrated to derive from integration of an ancestral nudivirus (Webb *et al.*, 2006; Bézier *et al.*, 2009). PDV particles are about 150–200 nm and can easily be recognized using electron microscopy since they contain either several rod-shaped virion capsids (bracovirus) or one lenticular capsid (ichnovirus) surrounded by one or two cellular membranes (figs. 2A, 2B).

VLPS

As stated above, the term 'VLPs' should only be used to describe vesicles or particles, usually found in wasp venom or ovaries, and that do not contain viral DNA or RNA. However, these criteria are only met in a few cases. We focus in the following section on parasitoid VLPs for which this information is at least partly available. We first present the two categories of VLPs that have been described, VLPs resembling viruses and VLPs likely of vesicular origin. This last category comprises VLPs from *Drosophila* parasitoids that are clearly of the vesicular type and are probably best described in terms of immune suppressive functions.

VLPs with Virus Size and Shape

VLPs in Microctonus aethiopoides (Hymenoptera: Braconidae: Euphorinae)

Microctonus aethiopoides is an endoparasitoid of the adult stage of *Hypera* and *Sitona* spp. of weevils (Coleoptera: Curculionidae). *M. aethiopoides* virus-like particles (MaVLPs) have been described using transmission electron microscopy (TEM) approaches in the ovarian epithelial cells and the ovarian fluid of females imported from Morocco to fight the alfalfa weevil (*Hypera postica*) in Australia (Barratt *et al.*, 1999; 2006). Similar VLPs have been observed in some other cells (possibly hemocytes) in both sexes, but not in the female venom apparatus (Barratt *et al.*, 1999, 2006).

MaVLPs in ovarian cells are circular (about 100 nm in diameter) with an electron-dense core, and are enclosed by three or more membranes (Fig. 2C). In the ovarian fluid, MaVLPs are ovoid and only surrounded by two membrane envelopes suggesting they undergo a change in structure and shape during the cellular shedding (Fig. 2D). Ovarian fluid MaVLPs are very similar in structure and size to the ichnoviruses and ascoviruses observed in ovarian tissues of other wasp species (Norton *et al.*, 1975; Stoltz and Whitfield, 1992). However, extraction of nucleic acid has been unsuccessful and the protein composition of the particles has not been analyzed. Therefore it is unknown whether they have a viral origin (Murney, 2005, cited in Barrat *et al.*, 2006).

It has been suggested, but not yet demonstrated, that MaVLPs may play a role in suppressing the host immune defenses. Indeed, morphologically similar particles have been observed in cell nuclei of the fat body of the host *Sitona discoideus*, a tissue known to be involved in insect immunity (Barrat *et al.*, 2006). However, MaVLPs are not needed to ensure parasitism success in *M. aethiopoides* since no VLPs were found in ovarian tissues of specimens collected from different locations in Europe. Individuals of the related Australian *M. hyperodae* species or of the *M. zealandicus* New Zealand native species are also free of VLPs (Barratt *et al.*, 2006).

VLPs in Opius concolor (syn. Psyttalia concolor) and Opius caricivorae (Hymenoptera, Braconidae, Opiinae)

The *Opius* genus is one of the largest in the *Braconidae* family, and all opiine braconids are koinobiont endoparasitoids of Cyclorrhapha diptera. Virus-like particles (OcVLPs) have been described within the secretory cells of the venom gland filaments of *O. concolor* (Jacas *et al.*, 1997) and *O. caricivorae* (Wan *et al.*, 2006) adult females. In *O. concolor*, two types of particles have been described. The first one has an icosahedral symmetry (diameter around 70 nm) and presents surface spikes (Fig. 2E). The other one is more pleiomorphic with a diameter ranging from 30 to 60 nm and an envelope with club-shaped projections. Both types of VLP are reminiscent of virus particles and are found embedded in secretory vesicles (Jacas *et al.*, 1997). In *O. caricivorae*, virus-like particles of 50 nm in diameter have been observed in the secretory cells of the venom gland filaments (Wan *et al.*, 2006). Opiinae VLPs have not been purified nor analyzed in detail and whether or not they play a role in parasitism success remains to be determined.

VLPs in Venturia canescens (Hymenoptera: Ichneumonidae)

Venturia canescens (also named *Nemeritis canescens*) is an endoparasitoid wasp that deposits its eggs inside the larvae of pyralid moths such as *Ephestia kuehniella* (Lepidoptera: Phycitidae). The ovarian calyx region produces a viscous fluid containing proteins and virus-like particles that stick to the egg at oviposition (Feddersen *et al.*, 1986; Kinuthia *et al.*, 1999). *V. canescens* VLPs (VcVLPs) are 100–150-nm electron-dense particles surrounded by a membranous envelope, that morphologically resemble ichnoviruses (Feddersen *et al.*, 1986) (Fig. 2G). They are assembled within the nuclei of calyx gland cells (Schmidt and Schuchmann-Feddersen, 1989) but they are devoid of nucleic acids (Bedwin, 1979; Feddersen *et al.*, 1986). Confusingly, the ovaries of *V. canescens* were also reported to produce small RNA-containing virus particles (VcSRV), most likely belonging to the iflavirus family (Reineke and Asgari, 2005). The iflavirus particles form crystalline arrays composed of uniform non-enveloped particles of about 36 nm (Fig. 2E). They are found together with VcVLPs in the lumen of the calyx region, and they are injected with the egg into the host larvae where they enter host hemocytes, suggesting a potential immune suppressive function (Reineke and Asgari, 2005).

VcVLPs have been isolated and analyzed, allowing identification of four major (35-, 40-, 52- and 60-kDa) and two minor proteins (80–94 kDa) (Schmidt and Schuchmann-Feddersen, 1989). The genes encoding the 40-kDa protein, vlp1, the 52-kDa protein, vlp2, immunogically related to the 60-kDa protein, and vlp3 coding for a 80–94 kDa protein, have been cloned and characterized (Asgari et al., 2002; Theopold et al., 1994; Hellers et al., 1996). The gene vlp1 encodes a protein member of the phospholipid hydroperoxide glutathione peroxidases (PHGPx) family, but the active domain lacks an important cysteine and the protein has no enzymatic activity (Hellers et al., 1996, Li et al., 2003). Its function remains to be determined but it has been proposed to play a role in egg-surface protection by masking or removing modified phospholipids from the surface of the parasitoid's eggs. Two alleles of vlp1 exist in wasp populations that differ by the presence of a tandem-repeat sequence of 54 bp, and they produce two structurally distinct proteins (Hellers et al., 1996). Occurrence of pleiotropic effects of vlp1 alleles on the reproductive biology of female wasps has largely been discussed (Beck et al., 1999; Amat et al., 2003; Schmidt et al., 2005). However, V. canescens wasps develop well on E. kuehniella hosts whatever the vlp1 allele they contain.

The gene encoding the 52-kDa protein, vlp2, contains a RhoGAP domain typical of GTPase activating proteins (Reineke et al., 2002). RhoGAPs interact with, and negatively regulate, small GTP-binding proteins (GTPases or G-proteins) that control key cellular processes including actin cytoskeleton assembly/disassembly. The function of this VcVLP protein or of the antigenically related 60-kDa protein is unknown, and it has not been determined whether these two proteins are isoforms or only share antigenic similarities.

The vlp3 (VcNep) gene encodes two proteins of 80 kDa and 94 kDa that correspond, respectively, to non-glycosylated and glycosylated forms. vlp3-encoded proteins have sequence similarity to the mammalian protease neprilysins (NEPs) (Asgari et al., 2002). In mammals, neprilysins are integral membrane zinc metallo-endopeptidases that degrade peptides of up to 40 amino acids, such as peptidic hormones (Turner et al., 2001; Hersh and Rodgers, 2008). This enzyme has also been involved in the metabolic inactivation of neuropeptides in the central nervous system of insects (Isaac et al., 2009) but the role

parasitoid-injected neprilysins can play in the host is still unknown.

The covering of the parasitoid egg by VcVLPs at oviposition—they attach to a fibrous layer of the egg surface—seems to provide a local immune protection to the egg. This was suggested by experiments using antibodies directed against these particles (Feddersen et al., 1986) and by the fact that the cellular defense capacity of the host remains virtually intact after parasitization. How this protection works and what the respective functions of the characterized VcVLP proteins are still remain to be determined.

VLPs with a Vesicular Appearance

VLPs in Meteorus pulchricornis (Hymenoptera: Braconidae: Euphorinae)

Meteorus pulchricornis is a highly polyphagous parasitoid that attacks all larval stages of numerous lepidopteran species (Berry and Walker, 2004). The female ovary lacks the calyx region, and the venom gland is formed of two inter-digited filaments that fuse in a unique long venom reservoir (Suzuki and Tanaka, 2006). Ultrastructural studies using TEM have shown that the lumen of the venom gland is filled with numerous cell-produced VLPs (MpVLPs) that are also observed in the reservoir. MpVLPs are mainly composed of single-membraned vesicles of about 150 nm half-filled or completely filled by an electron-dense material (Fig. 2F). Results of spectroscopy measurements suggested that MpVLPs lack nucleic acids (Suzuki and Tanaka, 2006), but there are no available data on their protein composition that could confirm a nonviral origin.

The role of MpVLPs in suppressing host immunity was assessed by injection of purified particles into larvae of the moth Pseudaletia separata. This resulted in a 10-fold decrease in the ability of host hemocytes to encapsulate fluorescent latex beads. Granulocytes from parasitized or MpVLP-injected P. separata hosts showed a rapid disassembly in the leading edge filopodia and lamellipodia, resulting in reduction of cell diameter and apoptosis, while plasmatocytes remained intact (Suzuki and Tanaka, 2006; Suzuki et al., 2008).

VLPs in Leptopilina heterotoma, L. victoriae and L. boulardi (Hymenoptera, Figitidae)

Cynipid Leptopilina wasps are more or less specialist parasitoids of Drosophila species that ensure development of their eggs by suppressing host immune defenses. They are larval, solitary, koinobiont parasitoids. VLPs have been described in species of the two Leptopilina groups (Allemand et al., 2002), the Heterotoma group (L. heterotoma, L. victoriae) and the Boulardi group (virulent ISm and avirulent ISy lines of L. boulardi; Dubuffet et al.,

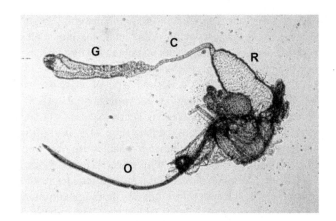

FIGURE 3   Venom apparatus of *Leptopilina boulardi*. The venom apparatus is formed by a unique long gland (G) connected to the reservoir (R) by a canal (C). O, ovipositor. The lumen of the gland, lined by a large epithelium, is visible. The tip or 'nose' region of the gland is half-filled by a darker material. The size of G + C + R is about 1 mm. *Pictures from bright field microscopy.*

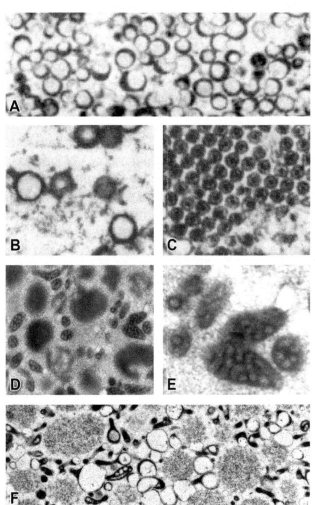

FIGURE 4   Transmission electron microscopy of VLPs from *Leptopilina* species. A–C: VLPs from *L. heterotoma*: reservoir VLPs following purification by density gradient (A), VLPs observed in the reservoir lumen (B) and virus type arrays described in the gland cells (C). **D–F:** VLPs from *L. boulardi*: both large vesicles and VLPs are observed in the reservoir lumen (D); enlarged VLPs (E) are similar to those described in Dupas *et al.* (1996) and Labrosse *et al.* (2003); the reservoir content obtained by centrifugation shows at least three types of vesicles in top of VLPs (F). See text for size of the particles. Pictures A–C from Rizki and Rizki (1990).

2009). VLPs' *sensu stricto*—defined as the structures injected by females at oviposition and retrieved in the host—are only found in the reservoir, but the process leading to their formation begins in the unique venom gland connected with the reservoir (Fig. 3) (Ferrarese *et al.*, 2009; Poirié, unpublished observations). Interestingly, VLPs differ in their shape, size, and structure as well as in the number of particles between *Leptopilina* species, and even between avirulent and virulent strains of *L. boulardi* (Rizki and Rizki, 1984, 1990, 1994; Rizki *et al.*, 1990; Dupas *et al.*, 1996; Labrosse *et al.*, 2003; Chiu *et al.*, 2006; Morales *et al.*, 2005).

L. heterotoma VLPs (LhVLPs) were first studied more than twenty years ago under the name 'lamellolysin' (Rizki and Rizki, 1984). They were so-called based on their ability to induce the lysis of *Drosophila melanogaster* lamellocytes, specialized hemocytes involved in immune encapsulation of parasitoid eggs (Rizki and Rizki, 1984, 1990, 1994; Rizki *et al.*, 1990). This experiment was performed using a light-scattering fraction from the venom reservoir content obtained following separation on density gradients, which contained concentrated VLPs. These particles are about 300 nm in diameter and show an asymmetrical structure (Fig. 4A). Most of them are formed by a clear spherical center surrounded by a variable thickness of electron-dense material, leading to a crescent-shape appearance. A tightly adhering membrane can be observed surrounding the electron-dense material. In some of the particles, long electron-dense surface spikes or knob-like extensions can also be seen. LhVLPs are also present in the reservoir lumen (Fig. 4B). In the study by Rizki and Rizki (1990), occurrence in the gland cells of regular arrays of another type of particle with a smaller diameter

(~30 nm) was reported. These particles are filled with an electron-dense material and resemble viral capsids (Fig. 4A). The relation between these particles and the 300-nm diameter VLPs is unclear (see below).

The biogenesis of VLPs has been described in *L. heterotoma* and *L. victoriae* venom apparatus by Morales *et al.* (2005) and Ferrarese *et al.* (2009). These authors describe at least five different classes of VLP precursors (VLPps) along the venom gland and suggest that they may represent VLPs at different stages of maturation. The first class of VLPps is found in specialized intracellular canals present in the gland cells (Ferrarese *et al.*, 2009). They are spherical, with a diameter of about 90 nm, and contain an electron-dense material resembling the particles reported in Rizki and Rizki (1990) (Fig. 4C). The second class of VLPps may derive from the first type by assembly of several (up to five) VLPps linked to each other by electron-dense connections of 90 nm long. The other classes have more complex structures and are of larger size, maybe because they arise from confluence of the previous VLPps. They form electron-dense particles of various shapes from multi-lobed to stellate, with dimensions up to several hundred nanometers (similar to figs. 4A, 4B). Some empty vesicles with the same shape and in the same range of size can also be observed in the gland lumen. All these vesicles are surrounded by a membrane and are embedded in a fuzzy layer of less dense material. These clustered or merged VLPps are also retrieved in the reservoir.

The VLPs of *L. heterotoma* have been demonstrated to be the main factor acting against host lamellocytes (Rizki and Rizki, 1990, 1994; Rizki *et al.*, 1990). The VLPs' spikes first come in contact with the lamellocyte membrane (maybe *via* a specific receptor), enter the cell by a still unknown mechanism, and are then retrieved free in the cytoplasm. VLPs can be found in *Drosophila* plasmatocytes as well, but they do not alter these cells, likely because plasmatocytes are phagocytic cells and VLPs are always found engulfed in phagolysosomes-type vesicles (Rizki and Rizki, 1994). Although *L. heterotoma* and *L. victoriae* produce similar VLPs, the VLP-containing venom of these species likely differ in their effects against host lamellocytes, *L. heterotoma* venom being more efficient both *in vivo* and *in vitro* (Morales *et al.*, 2005).

In *L. heterotoma*, four major protein bands have been observed from VLPs preparations on SDS-PAGE (at 87.5, 75, 72.5 and 40 kDa) while at least eight bands were observed for *L. victoriae* (Chiu *et al.*, 2006). The 40-kDa protein in *L. heteroma* and the 47.5-kDa protein in *L. victoriae* are antigenically related proteins. The 40-kDa protein has been immunolocalized on the surface and the spike extensions of VLPs in *L. heterotoma*, as well as in lamellocytes of parasitized *Drosophila* hosts (Chiu *et al.*, 2006). However, none of these proteins have been identified at the molecular level nor has their target in the host lamellocyte been identified.

In *L. boulardi*, two lines have been characterized that differ by their virulence properties. The ISm line originates from Tunisia and is representative of *L. boulardi* from Mediterranean areas. It always succeeds in suppressing the immune defenses of *D. melanogaster* but is unsuccessful in *D. yakuba* (Dubuffet *et al.*, 2008). The ISy line comes from tropical West Africa. It can suppress the immune defenses of both *D. melanogaster* and *D. yakuba*, but depending on the susceptible/resistant genotype of the flies. In both parasitoid types, the venom apparatus contains different types of VLPs that mostly resemble vesicles (Dupas

et al., 1996; Labrosse *et al.*, 2003). Filamentous single-membraned vesicles contain dense material punctuated by empty lacunae delimited by membranes (figs. 4D, E). They are 50–100 nm in size and are referred to as 'VLPs' by Dupas *et al.* (1996). A second type of vesicles was observed that corresponded to larger vesicles of about 200–300 nm filled with a homogenous dense material and surrounded by a punctuated membrane (Fig. 4D). In the ISy line, far fewer 50–100-nm diameter vesicles are observed, and they are more elongated with fewer lacunae (Dupas *et al.*, 1996; Labrosse *et al.*, 2003). Hybrid females resulting from crosses between ISm and ISy lines contain more VLPs than ISy females, with an intermediate morphology, being less elongated than in the ISy line (Dupas *et al.*, 1996). Interestingly, these hybrids appear to exhibit half-immunesuppressive ability toward *D. melanogaster* (Dupas *et al.*, 1996), suggesting these VLPs may be involved in variation of virulence between ISm and ISy strains.

In both ISm and ISy lines, the so-called VLPs as well as the large vesicles are retrieved in the reservoir, surrounded by a dense material, and also found in the ovipositor canal. They are also observed attached to the chorion of the deposited egg and within the basement membrane of the host (Dupas *et al.*, 1996). The biogenesis of *L. boulardi* VLPs and the presence of VLP precursors in the lumen of the venom gland (as in *L. heterotoma*) remain to be analyzed.

We recently purified the vesicles from the reservoir of the ISm and ISy lines by centrifugation and we compared the pellet by TEM and SDS-PAGE. The vesicle pellet from ISm and ISy lines contains similar types of vesicles, although in much lower numbers in the ISy sample. We could observe the VLP's vesicles as well as large electron-dense vesicular material and a third type of empty vesicles with asymmetrical membrane thickness (Fig. 4F). Using SDS-PAGE, we showed that the pellets from the two lines contain a large number of proteins leading to very different electrophoretic patterns. They also strongly differ from the patterns of the remaining supernatants, indicating they have specific protein compositions (data not shown). Analysis of the protein components of VLPs will thus likely provide insights into the origin of the particles but also help to explain virulence variation.

VIRAL OR COMMON ORIGIN WITH PDVS?

As exemplified in this review, the definition and description of VLPs remain rather unclear. Whether they have a viral origin independant of that of PDVs, Secondarily evolved from PDVs by losing their ability to encapsulate nucleic acids, or have no link at all to viruses, is still a large subject of debate (Whitfield and Asgari, 2003; Reineke *et al.*, 2006). Among the different types of vesicles referred as VLPs in different species, some are similar in size and shape to

PDVs particles, other resemble true viruses, and some might correspond to wasp secretions. Some VLPs are from ovarian origin, as are PDVs, while others are exclusively produced in the venom gland. Moreover, some described VLPs are assembled in the cell nucleus (resembling viruses or PDVs) while others seems to have a cytoplasmic origin or to be built in an external maturation process during their transit in the venom gland lumen and reservoir.

Currently, the main accepted difference allowing discrimination between VLPs and PDVs or virus particles is the presence of nucleic acids inside the particles. However, this was either not questioned at all or not totally convincingly tested as in *L. boulardi* (Dupas *et al.*, 1996; see also below). This is likely because of the difficulty of purifying enough VLPs to test for the presence of nucleic acids using electrophoretic methods or spectroscopic measurement assays. Overall, the absence of nucleic acids in VLPs was only accurately demonstrated in *V. canescens* and *M. pulchricornis*.

Another important difference between VLPs and viruses (or PDVs) is the fact that the latter replicate their genome or part of it in the wasp, and possibly in the host, thus allowing detection of highly repeated DNA sequences. This may allow their detection, even in absence of viral genome data. For instance, in *L. boulardi*, we extracted genomic DNA from the venom apparatus (or from the total female body as a control), and we used part of it as a probe on a Southern-blot containing the same DNA. If venom VLPs correspond to virions or virus-derived particles containing DNA, the corresponding DNA would be in high copy number and thus discrete bands might have been observed on the hybridization smear which was not the case (Poirié, unpublished data). This favors a nonviral origin for the VLPs in this species. Such genomic probes can also be used on Southern blots from parasitized hosts (which might allow detection of injected viruses)—no signal was found using DNA extracted from *D. melanogaster* hosts parasitized by *L. boulardi* (Poirié, unpublished data)—or on northern blots, allowing the detection of virus gene expression. However, as viruses unrelated to VLPs can infect wasps, as shown for instance for *V. canescens* (Reineke and Asgari, 2005) and *L. boulardi* (Varaldi *et al.*, 2006), a positive signal might correspond to a viral infection not related to VLPs. This can lead to different observations in different laboratories depending on the viral infection status of the parasitoid lines. The same problem occurs in interpreting and comparing data from electron micrographs and care should also be taken in drawing conclusions only on these observations.

Finally, VLPs and PDVs/viruses might be identified at the protein level if particles can be purified in a sufficient amount for proteomic analyses to be performed. Indeed, as described above, even if only few proteins have yet been identified in VLPs, they are wasp cellular proteins and not virus-derived proteins, while PDVs and viral particles may contain viral capsid proteins. This has been recently demonstrated by LC-MS/MS analysis of proteins from purified PDV particles from *Hyposoter didymator* ichnovirus and *Chelonus inanitus* bracovirus (Volkoff *et al.*, 2010; Wetterwald *et al.*, 2010). Among the major proteins found in these PDV particles, most of them were related to viral capsid components, although some cellular and ovarian fluid proteins were also identified. This approach previously helped in identifying the viral ancestor of bracoviruses (Bézier *et al.*, 2009) and also suggested that ichnoviruses derive from an as yet unidentified virus ancestor (Volkoff *et al.*, 2010).

Overall, a large number of the 'VLPs' reported to date were finally proven to be true infectious viruses (see above) and the number of species where occurrence of VLPs is demonstrated or highly supported is still low. Of course, the number of studied parasitoid species is also low and nearly all of them belong to a few groups of the Ichneumonidea super-family (*Braconidae* and *Ichneumonidae*), which mostly contain VLPs morphologically resembling viruses, with the exception of *Meteorus* species. Interestingly, *Meteorus* and *Leptopilina* VLPs, which are more vesicular-like, are both produced in the venom apparatus while other described VLPs are found in the ovary, a difference in their origin that might be related to their different morphological appearance. In the case of *V. canescens* VLPs, which morphologically resemble viruses but do not contain DNA, the debate between the VLPs having a viral origin (VLPs being a type of PDV without any DNA encapsidated in the particles) versus a cellular origin is still ongoing.

CELLULAR SECRETIONS?

Another important point to determine the nature of the VLPs (and other vesicles found in the venom or ovarian fluid) is, of course, their place and mode of production. True viruses replicate either in the nucleus or in the cytoplasm of the infected cells. PDV replication and nucleocapsid assembly occurs within the nuclei of calyx cells. Once formed, ichnovirus particles are released continuously through a 'budding' process without damaging cells (Volkoff *et al.*, 1995). Enveloped viruses bud from cells from either the plasma membrane or intracellular membranes, followed by pinching off or fission. This results in the release of particles coated with a lipid bilayer acquired from the host cell, and then with cell membrane proteins. These particles show a double membrane once released in the calyx fluid. In contrast, bracovirus-producing cells have a greatly enlarged nucleus, occupying most of the cell volume at the end of virus replication, and virus particles are released through cell lysis (de Buron and Beckage,

1992; Wyler and Lanzrein, 2003). Bracoviruses have thus a single membrane envelope.

The production of VLPs or VLP precursors inside cells is not really indicative of their viral or nonviral origin. Indeed, vesicular VLPs might simply originate from cell secretions. Cells usually release different types of membrane vesicles and cytosolic components into the extracellular space. These vesicles are generated within the endosomal system or at the plasma membrane. Among the various kinds of secreted membrane vesicles, exosomes, vesicles of 40–100 nm in diameter, are secreted upon fusion of endosomal multivesicular bodies (MVBs) with the cell surface from almost all organisms from mammals to plants (Théry et al., 2009; Simons and Raposo, 2009; An et al., 2006), and even from bacteria (Silverman et al., 2010). In *Drosophila*, they are named argosomes and are found to be involved in developmental processes (Greco et al., 2001). Interestingly, the shape and size of VLPs from some species, for instance *L. boulardi*, resemble exosomes or slightly larger microvesicles also deriving from MVBs (Théry et al., 2009). These MVB-derived vesicles play an important role in intercellular communication since they can transfer membrane signaling components and proteins but also nucleic acids, particularly mRNAs and miRNAs (Valadi et al., 2007; Miranda et al., 2010; Wang et al., 2010) and even miRNAs from viral origin, in infected cells (Pegtel et al., 2010). An important feature of exosomes is that they target specific cells, thus ensuring specificity of signal transmission. Recently, exosomes have been involved in transmission of several diseases (Février et al., 2005; Anderson et al., 2010) and also as toxin carriers for plant and bacteria *in vitro* (Zhang et al., 2009) as well as for snake venom proteins *in vivo* (Ogawa et al., 2008). Following proteomic analyses, the presence of exosomes in the solitary wasps *Eumenes pomiformis* and *Orancistrocerus drewseni* venom has been suggested in order to explain the presence of proteins without signal sequence-mediated transport processes (Baek and Lee, 2010). Interestingly, VLPs of some parasitoid species target specific cells (hemocytes) and their ability to alter these cells suggest they might contain or carry parasitoid toxins.

However, it is now well described that many viruses are able to 'hijack' cellular secretion pathways and particularly the endosomal multivesicular bodies (MVBs) and the cell membrane lipid rafts. As a consequence, these serve as concentration platforms or delivery vehicles for viral structural elements, facilitating assembly and release of infectious virions (Gould et al., 2003; Wei et al., 2008; Hurley and Hanson, 2010; de Gassart et al., 2009; Izquierdo-Useros et al., 2010; Ono, 2010). Parasitoid VLPs, PDVs or viruses may then act similarly to perform assembly or budding (Thaa et al., 2010) as a 'cheap' way to exit the cell, by using components that mediate endosomal sorting. Then, this might lead to morphological similarities

between virions or viral vesicles and exosomes and microvesicles, adding to an already confused situation (Pelchen-Matthews et al., 2004; Thaa et al., 2010). Exosomes from different sources present a common set of membrane and cytosolic proteins, as well as distinct subsets of proteins that correspond to the cell-type from which they are derived (Simpson et al., 2009, 2010), and, when they contain virus particles, abundant viral proteins (Booth et al., 2006; de Gassart et al., 2009). A more extensive characterization of the proteome of 'exosomes' from parasitoids with vesicular VLPs, such as Leptopilina species, may thus be the simplest way to finally determine the origin of these structures.

FUTURE DIRECTIONS

One of the main efforts that should be carried out in most species is confirmation of the VLP status by purifying VLPs and testing whether or not they contain nucleic acids. Purification techniques (such as density gradients) exist that may avoid contamination by tissues from the venom apparatus or the ovary, and DNA or RNA amplification methods from micro-quantities are now available. These techniques may also allow the separation of the different types of vesicles present in the reservoir/gland or ovaries of the wasps. Purified VLPs will be a valuable source to perform proteomic studies, in parallel with transcriptomic approaches on VLP-producing tissues and setup functional assays.

Biogenesis and function of VLPs produced in the venom apparatus, as in figitid wasps, is by far less understood than that of PDVs produced in the ovary. For instance, nothing is known so far of the role of major VLP proteins in assembly and mode of secretion of the vesicles or in their protective functions in the host. Thus, characterization and identification of a larger number of proteins from VLPs should help in elucidating both their origin and functional properties. For example, *L. boulardi* 'virulent' ISm line produces a RacGAP domain-containing protein, named LbGAP, whose injection in *D. melanogaster* larvae mimics the egg protection provided by parasitism (Labrosse et al., 2003, 2005a, 2005b). LbGAP has a RacGAP activity and induces changes in the morphology of lamellocytes by inactivating Drosophila Rac1 and Rac2 Rho GTPases, (Colinet et al., 2007, 2010), both required for encapsulation of *Leptopilina* eggs (Williams et al., 2005, 2006). These changes more or less resemble those induced by the 'lamellolysine' activity of VLPs from *L. heterotoma*, although no protein with such activity has been described yet in this species (Rizki and Rizki, 1984, 1990, 1994; Rizki et al., 1990; Morales et al., 2005). Interestingly, LbGAP is immunolocalized in *D. melanogaster* lamellocytes as large spots

(Colinet *et al.*, 2007) that might correspond to detection of VLPs. However, the hypothesis that *L. boulardi* VLPs enter *D. melanogaster* lamellocytes and correspond to the detected spots remains to be tested. To do so, it will be important to determine if ISm VLPs contain LbGAP and thus might somehow be used as carriers to transfer the protein inside the lamellocytes. Comparison of the VLPs' contents between ISm and ISy lines of *L. boulardi* and with *L. heterotoma* will also provide information on the role of VLPs and their proteins in the intraspecific and interspecific variation of virulence. Interestingly, a protein with a RhoGAP domain (vlp2) is present on *V. canescens* VLPs. In this species, VLPs are suggested to act as a particle coat on the egg surface. However, a putative cytoplasmic function of VLP2 may be possible if VLPs are taken up by host hemocytes, a hypothesis that is worth testing.

Production of specific antibodies against identified VLPs and venom vesicles proteins may also allow the detailed analysis of their secretion pathway in venom gland cells—answering the question of whether they are exosomal or derived from the MVBs, or if they use classical secretion pathways—and to study the extracellular maturation process of VLPs suggested in *Leptopilina* species.

Another important point is the determination of the way VLPs reach their target cells (usually hemocytes) and enter the cells, which is likely related to specificity of the parasitoid success. Experiments should be performed to search for ligands on the VLPs surface and the presence of corresponding receptors at the surface of host target cells, with the exosome model providing working hypotheses. There is indeed some evidence that ligands present on the mammalian exosome surface are required to dock exosomes on the surface of target cells or to extracellular matrix proteins. MVBs and exosomes are rich in cell adhesion and cell surface molecules like, for instance, tetraspanins. Interestingly, tetraspanin-enriched microdomains such as lipid rafts are specifically adapted to facilitate vesicular fusion and fission (Hemler, 2003; Le Naour *et al.*, 2006), and this type of protein might be searched for at the surface of vesicular VLPs.

We would like to conclude by discussing the following semantic point. Although the 'VLP' terminology has now been used for a long time to refer to the membranous vesicles found in venom or ovarian fluid of wasps, it has some major disadvantages. First, a large diversity of vesicles are present in most VLP-harboring species, notably in the venom. Second, a large part of them are clearly not of viral origin. Moreover, the term 'VLPs' is now commonly used for exosomal types of vesicles containing viral capsid proteins that are produced by *in vitro* manipulations in different cellular systems, in order to obtain vaccines (for a review, see Noad and Roy, 2003: Ludwig and Wagner, 2007). Therefore, to avoid misleading interpretations and

confusion, we suggest using 'venosomes' instead of VLPs as a generic term for wasp venom vesicles devoid of viral nucleic acids, such as *Meteorus* or Leptopilina vesicles.

ACKNOWLEDGMENTS

This work (including personal data) was supported by the ANR PARATOXOSE (ANR-09-BLAN-0243-01) and D. Colinet was financed by the ANR CLIMEVOL (ANR-08-BLAN-0231) from the French Agency for National Research. Antonin Schmitz was supported by a Ph.D. grant from the French National Institute for Agricultural Research (INRA) and the PACA region. Pierre Yves Sizaret (Electron Microscopy Facility of François Rabelais University, Tours, France) and Marc Ravallec (Joint Research Unit for Integrative Biology and Virology of Insects BIVI, INRA, University of Montpellier, France) have been of great help in performing TEM experiments.

REFERENCES

Allemand, R., Lemaitre, C., Frey, F., Boulétreau, M., Vavre, F., Nordlander, G., et al. 2002. Phylogeny of six African *Leptopilina* species (Hymenoptera, Cynipoidea, Figitidae), parasitoids of *Drosophila*, with description of three new species. Annales de la Société Entomologique de France 38, 319–332.

Amat, I., Bernstein, C., van Alphen, J.J., 2003. Does a deletion in a virus-like particle protein have pleiotropic effects on the reproductive biology of a parasitoid wasp? J. Insect Physiol. 49, 1183–1188.

An, Q., Hückelhoven, R., Kogel, K.H., van Bel, A.J, 2006. Multivesicular bodies participate in a cell wall-associated defence response in barley leaves attacked by the pathogenic powdery mildew fungus. Cell. Microbiol. 8, 1009–1019.

Anderson, H.C., Mulhall, D., Garimella, R., 2010. Role of extracellular membrane vesicles in the pathogenesis of various diseases, including cancer, renal diseases, atherosclerosis, and arthritis. Lab. Invest. 90, 1549–1557.

Asgari, S., Reineke, A., Beck, M., Schmidt, O., 2002. Isolation and characterization of a neprilysin-like protein from Venturia canescens virus-like particles. Insect Mol. Biol. 11, 477–485.

Asgari, S., Rivers, D.B., 2010. Venom proteins from endoparasitoid wasps and their role in host–parasite interactions. Annu. Rev. Entomol. 56, 313–335.

Baek, J.H., Lee, S.H., 2010. Identification and characterization of venom proteins of two solitary wasps, *Eumenes pomiformis* and *Orancistrocerus drewseni*. Toxicon 56, 554–562.

Barratt, B.I., Evans, A.A., Stoltz, D.B., Vinson, S.B., Easingwood, R., 1999. Virus-like particles in the ovaries of *Microctonus aethiopoides* loan (Hymenoptera, Braconidae), a parasitoid of adult weevils (Coleoptera, Curculionidae). J. Invertebr. Pathol. 73, 182–188.

Barratt, B.I., Murney, R., Easingwood, R., Ward, V.K., 2006. Virus-like particles in the ovaries of *Microctonus aethiopoides* Loan (Hymenoptera, Braconidae), comparison of biotypes from Morocco and Europe. J. Invertebr. Pathol. 91, 13–18.

Beck, M., Siekmann, G., Li, D., Theopold, U., Schmidt, O., 1999. A maternal gene mutation correlates with an ovary phenotype in a parthenogenetic wasp population. Insect Biochem. Mol. Biol. 29, 453–460.

Beckage, N.E., de Buron, I., 1997. Developmental changes in teratocytes of the braconid wasp *Cotesia congregata* in larvae of the tobacco hornworm, *Manduca sexta*. J. Insect Physiol. 43, 915–930.

Beckage, N.E., Gelman, D., 2004. Wasp parasitoid disruption of host development, implications for new biologically based strategies for insect control. Annu. Rev. Entomol. 49, 299–330.

Bedwin, O., 1979. An insect glycoprotein, a study of the particles responsible for the resistance of a parasitoid's egg to the defense reactions of its insect host. Proc. R. Soc. Lond., B. 205, 271–286.

Berry, J.A., Walker, G.P., 2004. Meteorus pulchricornis (Wesmael) (Hymenoptera, Braconidae, Euphorinae), an exotic polyphagous parasitoid in New Zealand. New Zealand J. Zoo. 31, 33–44.

Bézier, A., Annaheim, M., Herbinière, J., Wetterwald, C., Gyapay, G., Bernard-Samain, S., et al. 2009. Polydnaviruses of braconid wasps derive from an ancestral nudivirus. Science 323, 926–930.

Bigot, Y., Rabouille, A., Doury, G., Sizaret, P.Y., Delbost, F., Hamelin, M.H., et al. 1997. Biological and molecular features of the relationships between *Diadromus pulchellus* ascovirus, a parasitoid hymenopteran wasp (*Diadromus pulchellus*) and its lepidopteran host, *Acrolepiopsis assectella*. J. Gen. Virol. 78, 1149–1163.

Bigot, Y., Samain, S., Augé-Gouillou, C., Federici, B.A., 2008. Molecular evidence for the evolution of ichnoviruses from ascoviruses by symbiogenesis. BMC Evol. Biol. 8, 253.

Booth, A.M., Fang, Y., Fallon, J.K., Yang, J.M., Hildreth, J.E., Gould, S.J., 2006. Exosomes and HIV Gag bud from endosome-like domains of the T cell plasma membrane. J. Cell Biol. 172, 923–935.

Cerenius, L., Kawabata, S., Lee, B.L., Nonaka, M, Söderhäll, K., 2010. Proteolytic cascades and their involvement in invertebrate immunity. Trends Biochem. Sci. 35, 575–583.

Chiu, H., Morales, J., Govind, S., 2006. Identification and immuno-electron microscopy localization of p40, a protein component of immunosuppressive virus-like particles from *Leptopilina heterotoma*, a virulent parasitoid wasp of *Drosophila*. J. Gen. Virol. 87, 461–470.

Colinet, D., Schmitz, A., Cazes, D., Gatti, J.-L., Poirié, M., 2010. The origin of intraspecific variation of virulence in an eukaryotic immune suppressive parasite. PLoS Pathog. 6, e1001206.

Colinet, D., Schmitz, A., Depoix, D., Crochard, D., Poirié, M., 2007. Convergent use of RhoGAP toxins by eukaryotic parasites and bacterial pathogens. PLoS Pathog. 3, e203.

de Buron, I., Beckage, N.E., 1992. Characterization of a polydnavirus and virus-like filaments particle (VLFP) in the braconid wasp Cotesia congregata (Hymenoptera, Braconidae). J. Invertebr. Pathol. 59, 315–327.

de Gassart, A., Trentin, B., Martin, M., Hocquellet, A., Bette-Bobillo, P., Mamoun, R., et al. 2009. Exosomal sorting of the cytoplasmic domain of bovine leukemia virus TM Env protein. Cell Biol. Int. 33, 36–48.

Dubuffet, A, Colinet, D., Anselme, C., Dupas, S., Carton, Y, Poirié, M., 2009. Variation of *Leptopilina boulardi* success in Drosophila hosts, what is inside the black box? Adv. Parasitol. 70, 147–188.

Dubuffet, A., Doury, G., Labrousse, C., Drezen, J.-M., Carton, Y., Poirié, M., 2008. Variation of success of *Leptopilina boulardi* in *Drosophila yakuba*, the mechanisms explored. Dev. Comp. Immunol. 32, 597–602.

Dupas, S., Brehelin, M., Frey, F., Carton, Y., 1996. Immune suppressive virus-like particles in a *Drosophila* parasitoid, significance of their intraspecific morphological variations. Parasitology 113, 207–212.

Eggleton, P., Belshaw, R., 1992. Insect parasitoids, an evolutionary overview. Philos. Trans. R. Soc. B, Biol. Sci. 337, 1–20.

Feddersen, I., Sander, K., Schmidt, O., 1986. Virus-like particles with host protein-like antigenic determinants protect an insect parasitoid from encapsulation. Experientia 42, 1278–1281.

Federici, B.A., Bigot, Y., 2003. Origin and evolution of polydnaviruses by symbiogenesis of insect DNA viruses in endoparasitic wasps. J. Insect Physiol. 49, 419–432.

Ferrarese, R., Morales, J., Fimiarz, D., Webb, B.A., Govind, S., 2009. A supracellular system of actin-lined canals controls biogenesis and release of virulence factors in parasitoid venom glands. J. Exp. Biol. 212, 2261–2268.

Février, B., Vilette, D., Laude, H., Raposo, G., 2005. Exosomes, a bubble ride for prions? Traffic 6, 10–17.

Godfray, H., 1994. Parasitoids: Behavioural and Evolutionary Ecology. Princeton University Press, Princeton.

Gould, S.J., Booth, A.M., Hildreth, J.E., 2003. The Trojan exosome hypothesis. Proc. Natl. Acad. Sci. U.S.A. 100, 10592–10597.

Greco, V., Hannus, M., Eaton, S., 2001. Argosomes, a potential vehicle for the spread of morphogens through epithelia. Cell 106, 633–645.

Hellers, M., Beck, M., Theopold, U., Kamei, M., Schmidt, O., 1996. Multiple alleles encoding a virus-like particle protein in the ichneumonid endoparasitoid *Venturia canescens*. Insect Mol. Biol. 5, 239–249.

Hemler, M.E., 2003. Tetraspanin proteins mediate cellular penetration, invasion, and fusion events and define a novel type of membrane microdomain. Annu. Rev. Cell Dev. Biol. 19, 397–422.

Hersh, L.B., Rodgers, D.W., 2008. Neprilysin and amyloid beta peptide degradation. Curr. Alzheimer Res. 5, 225–231.

Hurley, J.H, Hanson, P.I., 2010. Membrane budding and scission by the ESCRT machinery, it's all in the neck. Nat. Rev. Mol. Cell Biol. 11, 556–566.

Isaac, R.E., Bland, N.D., Shirras, A.D., 2009. Neuropeptidases and the metabolic inactivation of insect neuropeptides. Gen. Comp. Endocrin. 162, 8–17.

Izquierdo-Useros, N., Naranjo-Gomez, M., Erkizia, I., Puertas, M.C., Borras, F.E., Blanco, J., et al. 2010. HIV and mature dendritic cells, Trojan exosomes riding the Trojan horse? PLoS Pathog. 6, e1000740.

Jacas, J.A., Budia, F., Rodriguez-Cerezo, E., Vinuela, E., 1997. Virus-like particles in the poison gland of the parasitic wasp Opius concolor. Ann. Appl. Biol. 130, 587–592.

Kinuthia, W., Li, D., Schmidt, O., Theopold, U., 1999. Is the surface of endoparasitic wasp eggs and larvae covered by a limited coagulation reaction? J. Insect Physiol. 45, 501–506.

Labrosse, C., Carton, Y., Dubuffet, A., Drezen, J.-M., Poirié, M., 2003. Active suppression of *D. melanogaster* immune response by long gland products of the parasitic wasp *Leptopilina boulardi*. J. Insect Physiol. 49, 513–522.

Labrosse, C., Eslin, P., Doury, G., Drezen, J.-M., Poirié, M., 2005. Hemocyte changes in *D. melanogaster* in response to long gland components of the parasitoid wasp *Leptopilina boulardi*, a Rho-GAP protein as an important factor. J. Insect Physiol. 51, 161–170.

Labrosse, C., Stasiak, K., Lesobre, J., Grangeia, A., Huguet, E., Drezen, J.-M., et al. 2005. A RhoGAP protein as a main immune suppressive factor in the *Leptopilina boulardi* (Hymenoptera, Figitidae)–*Drosophila melanogaster interaction*. Insect Biochem. Mol. Biol. 35, 93–103.

Lavine, M.D., Beckage, N.E., 1995. Polydnaviruses, potent mediators of host insect immune dysfunction. Parasitol. Today 11, 368–378.

Lawrence, P.O., 2002. Purification and partial characterization of an entomopoxvirus (DLEPV) from a parasitic wasp of tephritid fruit flies. J. Insect Sci. 2, 1–12.

Lawrence, P.O., Matos, L.F., 2005. Transmission of the *Diachasmimorpha longicaudata* rhabdovirus (DlRhV) to wasp offspring, an ultrastructural analysis. J. Insect Physiol. 51, 235–241.

Le Naour, F., André, M., Boucheix, C., Rubinstein, E., 2006. Membrane microdomains and proteomics, lessons from tetraspanin microdomains and comparison with lipid rafts. Proteomics 6, 6447–6454.

Li, D., Blasevich, F., Theopold, U., Schmidt, O., 2003. Possible function of two insect phospholipid-hydroperoxide glutathione peroxidases. J. Insect Physiol. 49, 1–9.

Ludwig, C., Wagner, R., 2007. Virus-like particles-universal molecular toolboxes. Curr. Opin. Biotechnol. 18, 537–545.

Miranda, K.C., Bond, D.T., McKee, M., Skog, J., Paunescu, T.G., Da Silva, N., et al. 2010. Nucleic acids within urinary exosomes/microvesicles are potential biomarkers for renal disease. Kidney Int. 78, 191–199.

Morales, J., Chiu, H., Oo, T., Plaza, R., Hoskins, S., Govind, S., 2005. Biogenesis, structure, and immunesuppressive effects of virus-like particles of a *Drosophila* parasitoid, *Leptopilina victoriae*. J. Insect Physiol. 51, 181–195.

Moreau, S.J., Eslin, P., Giordanengo, P., Doury, G., 2003. Comparative study of the strategies evolved by two parasitoids of the genus *Asobara* to avoid the immune response of the host, *Drosophila melanogaster*. Dev. Comp. Immunol. 27, 273–282.

Murney, R.J., 2005. Identification of virus-like particles in New Zealand and European ecotypes of *Microctonus aethiopoides* Loan (Hymenoptera: Braconidae). MSc. Thesis, Microbiology and Immunology Department, University of Otago, NZ.

Nappi, A., 2010. Cellular immunity and pathogen strategies in combative interactions involving Drosophila hosts and their endoparasitic wasps. Invertebr. Survival J. 7, 198–210.

Noad, R., Roy, P., 2003. Virus-like particles as immunogens. Trends Microbiol. 11, 438–444.

Norton, W.N., Vinson, S.B., Stoltz, D.B., 1975. Nuclear secretory particles associated with the calyx cells of the ichneumonid parasitoid *Campoletis sonorensis* (Cameron). Cell Tissue Res. 162, 195–208.

Ogawa, Y., Kanai-Azuma, M., Akimoto, Y., Kawakami, H., Yanoshita, R., 2008. Exosome-like vesicles in *Gloydius blomhoffii blomhoffii* venom. Toxicon 51, 984–993.

Ono, A., 2010. Relationships between plasma membrane microdomains and HIV-1 assembly. Biol. Cell 102, 335–350.

Pegtel, D.M., Cosmopoulos, K., Thorley-Lawson, D.A., van Eijndhoven, M.A., Hopmans, E.S., Lindenberg, J.L., et al. 2010. Functional delivery of viral miRNAs via exosomes. Proc. Natl. Acad. Sci. U.S.A. 107, 6328–6333.

Pelchen-Matthews, A., Raposo, G., Marsh, M., 2004. Endosomes, exosomes and Trojan viruses. Trends Microbiol. 12, 310–316.

Pennacchio, F., Strand, M.R., 2006. Evolution of developmental strategies in parasitic hymenoptera. Annu. Rev. Entomol. 51, 233–258.

Poirié, M., Carton, Y., Dubuffet, A., 2009. Virulence strategies in parasitoid Hymenoptera as an example of adaptive diversity. Comptes Rendus de l'Académie des Sciences, Biologie 332, 311–320.

Quicke, D., 1997. Parasitic Wasps. Chapman & Hall, London.

Reineke, A., Asgari, S., 2005. Presence of a novel small RNA-containing virus in a laboratory culture of the endoparasitic wasp *Venturia canescens* (Hymenoptera, Ichneumonidae). J. Insect Physiol. 51, 127–135.

Reineke, A., Asgari, S., Ma, G., Beck, M., Schmidt, O., 2002. Sequence analysis and expression of a virus-like particle protein, VLP2, from the parasitic wasp *Venturia canescens*. Insect Mol. Biol. 11, 233–239.

Reineke, A., Asgari, S., Schmidt, O., 2006. Evolutionary origin of Venturia canescens virus-like particles. Arch. Insect Biochem. Physiol. 61, 123–133.

Rizki, R.M., Rizki, T.M., 1984. Selective destruction of a host blood cell type by a parasitoid wasp. Proc. Natl. Acad. Sci. U.S.A. 81, 6154–6158.

Rizki, R.M., Rizki, T.M., 1990. Parasitoid virus-like particles destroy *Drosophila* cellular immunity. Proc. Natl. Acad. Sci. U.S.A. 87, 8388–8392.

Rizki, T.M., Rizki, R.M., Carton, Y., 1990. *Leptopilina heterotoma* and *L. boulardi*, strategies to avoid cellular defense responses of *Drosophila melanogaster*. Exp. Parasitol. 70, 466–475.

Rizki, T.M., Rizki, R.M., 1994. Parasitoid-induced cellular immune deficiency in *Drosophila*. Ann. N. Y. Acad. Sci. 712, 178–194.

Schmidt, O., Li, D., Beck, M., Kinuthia, W., Bellati, J., Roberts, H.L.S., 2005. Phenoloxidase-like activities and the function of virus-like particles in ovaries of the parthenogenetic parasitoid *Venturia canescens*. J. Insect Physiol. 51, 117–125.

Schmidt, O., Schuchmann-Feddersen, I., 1989. Role of virus-like particles in parasitoid–host interaction of insects. SubCell. Biochem. 15, 91–119.

Silverman, J.M., Clos, J., de'Oliveira, C.C., Shirvani, O., Fang, Y., Wang, C., et al. 2010. An exosome-based secretion pathway is responsible for protein export from *Leishmania* and communication with macrophages. J. Cell Sci. 123, 842–852.

Simons, M., Raposo, G., 2009. Exosomes-vesicular carriers for intercellular communication. Curr. Opin. Cell Biol. 21, 575–581.

Simpson, R.J., Jensen, S.S., Lim, J.W.E., 2008. Proteomic profiling of exosomes: current perspectives. Proteomics 8, 4083–4099.

Simpson, R.J., Lim, J.W., Moritz, R.L., Mathivanan, S., 2009. Exosomes, proteomic insights and diagnostic potential. Expert Rev. Proteomics 6, 267–283.

Smilanich, A.M., Dyer, L.A., Gentry, G.L., 2009. The insect immune response and other putative defenses as effective predictors of parasitism. Ecology 90, 1434–1440.

Stoltz, D.B., Whitfield, J.B., 1992. Viruses and virus-like entities in the parasitic Hymenoptera. J. Hymenoptera Res. 1, 125–139.

Suzuki, M., Miura, K., Tanaka, T., 2008. The virus-like particles of a braconid endoparasitoid wasp, *Meteorus pulchricornis*, inhibit hemocyte spreading in its noctuid host, *Pseudaletia separata*. J. Insect Physiol. 54, 1015–1022.

Suzuki, M., Tanaka, T., 2006. Virus-like particles in venom of *Meteorus pulchricornis* induce host hemocyte apoptosis. J. Insect Physiol. 52, 602–613.

Thaa, B., Hofmann, K.P., Veit, M., 2010. Viruses as vesicular carriers of the viral genome, a functional module perspective. Biochim. Biophys. Acta 1803, 507–519.

Theopold, U., Krause, E., Schmidt, O., 1994. Cloning of a VLP-protein coding gene from a parasitoid wasp *Venturia* canescens. Arch. Insect Biochem. Physiol. 26, 137–145.

Théry, C., Ostrowski, M., Segura, E., 2009. Membrane vesicles as conveyors of immune responses. Nat. Rev. Immunol. 9, 581–593.

Turner, A.J., Isaac, R.E., Coates, D., 2001. The neprilysin (NEP) family of zinc metallo endopeptidases, genomics and function. BioEssays 23, 261–269. 32

Valadi, H., Ekström, K., Bossios, A., Sjöstrand, M., Lee, JJ., Lövall, J.O., 2007. Exosome-mediated transfer of mRNAs and microRNAs is a novel mechanism of genetic exchange between cells. Nat. Cell Biol. 9, 654–659.

Varaldi, J., Ravallec, M., Labrosse, C., Lopez-Ferber, M., Boulétreau, M., Fleury, F., 2006. Artifical transfer and morphological description of virus particles associated with superparasitism behaviour in a parasitoid wasp. J. Insect Physiol. 52, 1202–1212.

Volkoff, A.N., Jouan, V., Urbach, S., Samain, S., Bergoin, M., Wincker, P., et al. 2010. Analysis of virion structural components reveals vestiges of the ancestral ichnovirus genome. PLoS Pathog. 6, e1000923.

Volkoff, A.N., Ravallec, M., Bossy, J., Cerutti, P., Rocher, J., Cerutti, M., et al. 1995. The replication of *Hyposoter didymator* polydnavirus, cytopathology of the calyx cells in the parasitoid. Biol. Cell 83, 1–13.

Wan, Z.W., Wang, H.Y., Chen, X.X., 2006. Venom apparatus of the endoparasitoid wasp *Opius caricivorae* Fischer (Hymenoptera, Braconidae), morphology and ultrastructure. Microsc. Res. Tech. 69, 820–825.

Wang, K., Zhang, S, Weber, J., Baxter, D., Galas, D.J., 2010. Export of microRNAs and microRNA-protective protein by mammalian cells. Nucleic Acids Res. 38, 7248–7259.

Webb, B.A., Strand, M.R., Dickey, S.E., Beck, M.H., Hilgarth, R.S., Barney, W.E., et al. 2006. Polydnavirus genomes reflect their dual roles as mutalists and pathogens. Virology 347, 160–174.

Wei, T., Hibino, H., Omura, T., 2008. Rice dwarf virus is engulfed into and released via vesicular compartments in cultured insect vector cells. J. Gen. Virol. 89, 2915–2920.

Wetterwald, C., Roth, T., Kaeslin, M., Annaheim, M., Wespi, G., Heller, M., et al. 2010. Identification of bracovirus particle proteins and analysis of their transcript levels at the stage of virion formation. J. Gen. Virol. 91, 2610–2619.

Whitfield, J.B., Asgari, S., 2003. Virus or not? Phylogenetics of polydnaviruses and their wasp carriers. J. Insect Physiol. 49, 397–405.

Williams, M.J., Ando, I., Hultmark, D., 2005. Drosophila melanogaster Rac2 is necessary for a proper cellular immune response. Genes to Cells 10, 813–823.

Williams, M.J., Wiklund, M.L., Wikman, S., Hultmark, D., 2006. Rac1 signalling in the *drosophila* larval cellular immune response. J. Cell. Sci. 119, 2015–2024.

Wyler, T., Lanzrein, B., 2003. Ovary development and polydnavirus morphogenesis in the parasitic wasp *Chelonus inanitus*. II. Ultrastructural analysis of calyx cell development, virion formation and release. J. Gen. Virol. 84, 1151–1163.

Zhang, F., Sun, S., Feng, D., Zhao, W.L., Sui, S.F., 2009. A novel strategy for the invasive toxin, hijacking exosome-mediated intercellular trafficking. Traffic 10, 411–424.

Chapter 16

RNA Viruses in Parasitoid Wasps

Sylvaine Renault

Université François Rabelais, UMR CNRS 6239 GICC Génétique, Immunothérapie, Chimie et Cancer, UFR des Sciences et Techniques, Parc Grandmont, 37200 Tours, France

SUMMARY

This chapter describes the different RNA viruses that have been detected at least once in parasitoid wasps. Four different RNA virus families have been reported in parasitoids: corona-like and picorna-like viruses for the positive-sense, single-stranded RNA viruses, rhabdoviruses for negative-sense, single-stranded RNA viruses, and reoviruses for segmented, double-stranded RNA viruses. They have been found in Ichneumonidae, Braconidae and Pteromalidae, in a total of only 10 hymenopteran species. Their morphology, localization in the parasitoid and host, and transmission are described. Their possible involvement in the success of parasitism and pathogenicity are discussed.

ABBREVIATIONS

AFLP	amplified fragment length polymorphism
AGF	accessory gland filaments
DlEPV	Diaschimimorpha longicaudata entomopoxvirus
DpAV-4	*Diadromus pulchellus* ascovirus-4
DpRIV-1	*Diadromus pulchellus* idnoreovirus-1
DpRV-2	*Diadromus pulchellus* reovirus
EST	expressed sequence tag
HeRIV-2	*Hyposoter exiguae* idnoreovirus-2
hpp	hours post-parasitism
ICTV	International Committee on Taxonomy of Viruses
LbFV	*Leptopilina boulardii* filamentous virus
McSRV	*Microplitis croceipes* small RNA containing virus
MlRVLP	*Meteorus leviventris* reovirus-like particle
OpbuCPV19	*Operophtera brumata* cypovirus 19
OpbuRV	*Operophtera brumata* reovirus
ORF	open reading frame
PcColike-V	*Psyttalia concolor* corona-like virus
PcRVLP	*Psyttalia concolor* reovirus like particle
PDV	polydnavirus
PpSRV	*Pteromalus puparum* small RNA containing virus
RdRp	RNA-dependent RNA polymerase
RT-PCR	reverse-transcribed polymerase chain reaction
TEM	transmission electron microscopy
VcSRV	*Venturia canescens* small RNA containing virus
VLP	virus-like particle

INTRODUCTION

A wide variety of viruses have been reported to infect insects. In the 8th International Committee on Taxonomy of Viruses (ICTV) report and ICTV website (http://www.ictvdb.org/Ictv/index.htm), 15 different families are reported: Ascoviridae, Baculoviridae, Birnaviridae, Coronaviridae, Dicistroviridae, Iflaviridae, Metaviridae, Nodaviridae, Parvoviridae, Polydnaviridae, Poxviridae, Pseudoviridae, Reoviridae, Rhabdoviridae, Tetraviridae) (Büchen-Osmond, 2003, Fauquet *et al.*, 2005, Kapoor *et al.*, 2010, Jacas *et al.*, 1997). They included the different types of genomes: double-stranded circular DNA (Ascoviridae, Baculoviridae, Birnaviridae, Polydnaviridae, Poxviridae), single-stranded DNA (Parvoviridae), positive- and negative-sense, single-stranded RNA (Coronaviridae, Dicistroviridae, Iflaviridae, Metaviridae, Nodaviridae, Picornaviridae, Rhabdoviridae, Tetraviridae), and segmented, double-stranded RNA (Reoviridae).

Surprisingly, only six of these families have been identified in parasitoid wasps: Ascoviridae, Coronaviridae, Iflaviridae, Polydnaviridae, Poxviridae, Reoviridae, and Rhabdoviridae. The absence of other families cannot be definitely concluded, as no systematic search for all types of viruses has been undertaken. It could be noted that the reovirus *Hyposoter exiguae* idnoreovirus-2 (HeRIV-2) was first identified in *Trichoplusia ni* cell cultures infected with calyx fluid from *Hyposoter exiguae*. It was subsequently discovered that the wasp tissues were infected with HeRIV-2. This had not been detected earlier due the abundance of polydnavirus virions in *H. exiguae*, which could have masked the presence of HeRIV-2 (Stoltz and Makkay, 2000). The same phenomenon could occur in the many of the wasps that carry polydnaviruses. Coronaviridae and Polydnaviridae are unusual families of viruses that have not been recovered from any insects other than parasitoid wasps. All the other families are also found in Lepidoptera (Ascoviridae, Poxviridae, Reoviridae), Hymenoptera (Picornaviridae) or Diptera (Rhabdoviridae).

Parasitoid Viruses: Symbionts and Pathogens. DOI: 10.1016/B978-0-12-384858-1.00016-3

The family of viruses most often detected in parasitoids is that of the Polydnaviridae, and it is estimated that tens of thousands of species carry these viruses. This is due to the fact that these viruses have been stably integrated in the genome of ancestral species of whole groups. The other families have been less commonly described, partly because they have not been systematically searched for. They infect very few wasp species. Among those infected by DNA viruses, only one species (*Diadromus pulchellus*) carried an ascovirus (DpAV-4), and one an entomopoxvirus (DlEPV in *Diaschimimorpha (=Biosteres) longicaudata*) (Bigot *et al.*, 1995; Lawrence, 1988). The RNA viruses are very rare in parasitoids. Both the Coronaviridae and the Rhabdoviridae have only been identified in one species, *Psyttalia (=Opius) concolor* (Jacas *et al.*, 1997) and *Diachasmimorpha longicaudata* (Edson *et al.*, 1982; Lawrence, 1988; Lawrence and Akin, 1990; Lawrence and Matos, 2005) respectively. The Picornaviridae have been detected in four species: *Microplitis croceipes* (Hamm *et al.*, 1992), *Venturia canescens* (Reineke and Asgari, 2005) *Nasonia vitripennis* (Oliveira *et al.*, 2010), and *Pteromalus puparum* (Zhu *et al.*, 2008). The Reoviridae are the RNA viruses most commonly reported in parasitoids although only six species have been shown to carry them: *Meteoris leviventris* (Edson, 1981), *P. concolor* (Jacas *et al.*, 1997), *M. croceipes* (Hamm *et al.*, 1994), *H. exiguae* (Stoltz and Makkay, 2000), *D. pulchellus* (Rabouille *et al.*, 1994; Renault *et al.*, 2003) and *Phobocampe tempestiva* (Graham *et al.*, 2006).

Wasps infected with RNA viruses include endoparasitoids: Ichneumonidae (*D. pulchellus, V. canescens, H. exiguae, P. tempestiva*) and Braconidae (*P. concolor, D. longicaudata, M. croceipes, M. leviventris*), and ectoparasitoids: Pteromalidae (*N. vitripennis* and *P. puparum*).

It must be noted that the RNA viruses are usually detected in association with various other families of viruses such as virus-like particles (VLPs), polydnaviruses (PDVs), or ascoviruses, which makes it difficult to investigate the role of RNA viruses in host/parasitoid relationships. With two exceptions, *Diadromus pulchellus* idnoreovirus-1 and *Diadromus pulchellus* reovirus (DpRV-1 and DpRV-2), the physiological impact of the presence of RNA viruses on successful parasitism remains to be elucidated (Bigot *et al.*, 1997, Renault *et al.*, 2003).

In this chapter, I will describe the different RNA viruses according to their genome: positive-sense, single-stranded linear RNA (Coronaviridae and Picornaviridae), negative-sense, single-stranded linear RNA (Rhabdoviridae), and double-stranded, segmented linear RNA viruses (Reoviridae). A new virus, LbFV, was discovered recently in *Leptopilina boulardi*. It does not resemble any conventional virus and so far the DNA or RNA nature of its genome has not been determined (see Varaldi *et al.*, Chapter 17 of this volume; Varaldi *et al.*, 2010).

The localization of the RNA viruses in the wasp and in the lepidopteran host will be described, as well as the type of transmission in wasps, the phylogeny of the viruses, and their role in the host/parasitoid relationships.

POSITIVE-SENSE, SINGLE-STRANDED RNA VIRUSES

Two families of these types of viruses have been detected in parasitoid wasps: coronaviridae-like and picornaviridae-like viruses.

The virions of Coronaviridae are spherical or pleimorphic, and have an envelope. They are 120–160 nm in diameter. The genome consists of a single molecule of linear, positive-sense, single-stranded RNA, and is 25,000–30,000 bases in length. Most of these viruses have been described in mammals, and one of the best known is SARS-CoV, which is responsible for a severe acute respiratory syndrome in humans (Spaan *et al.*, 2005).

As far as I am aware, corona-like viruses have only been reported in a single species of parasitoid, *P. concolor* (PcCo-likeV) (Jacas *et al.*, 1997). They were purified from the venom apparatus of female parasitoids. Transmission electron microscopy (TEM) revealed a virion typical of Coronaviruses, pleomorphic in shape, about 100 nm in diameter, in an envelope with club-shaped projections (15 nm) (Fig. 1A). However, no molecular analysis has been performed to confirm that this virus contains ssRNA or encodes the characteristic genes of the Coronaviridae. At the cellular level, PcCo-likeV was visible as spherical VLPs in the cytoplasmic vesicles of secretory cells (Fig. 2A). However, PcCo-like viruses are always associated with a reovirus, and the VLPs observed could therefore have originated either from the reovirus or the corona-like virus. Unfortunately, nothing is known about the localization of the corona-like virus elsewhere in the parasitoid or in its different tephritid hosts. This makes it impossible to know whether this virus infects the parasitoid or the host, or if it is involved in host–parasitoid interactions. The PcCo-like virus is the only Coronavirus reported in insects and it would be informative to investigate its mechanisms of infection and its evolutionary history.

The virions of Picornavirales consist of an isometric capsid 22–30 nm in diameter. They have no envelope. The 7000–8000-base genome consists of a single molecule of linear, positive-sense, single-stranded RNA. The 5′ extremities are linked to a protein (VpG). They infect a very wide range of hosts, extending from mammals to insects. Two families have been reported in invertebrates, the Dicistroviridae and the Iflaviridae. The type species of the Dicistroviridae is the Cricket paralysis virus. Dicistroviridae are infectious and fatal for many hymenopteran species, such as the honey bee or ants (Christian *et al.,* 2005a). The Iflaviridae have also been described, and they infect the honey bee, and different moths including silkmoths (Christian *et al.,* 2005b).

The picorna-like virus was first identified in *M. croceipes* (Braconidae) using TEM (Hamm *et al.,* 1992). Their presence has more recently been reported in the ichneumonid *V. canescens* (Reineke and Asgari, 2005), and the pteromalid *N. vitripennis* (Oliveira *et al.,* 2010) by identifying sequences homologous to the RNA-dependent-RNA-polymerase (RdRp) of Picornaviridae using cDNA-amplified fragment length polymorphism (AFLP) analysis and the expressed sequence tag (EST) library, respectively. Some colonies of the Pteromalid *P. puparum* are infected with a pathogenic picorna-like virus (Zhu *et al.,* 2008).

The picorna-like virus found in *V. canescens* (VcSRV) was detected during the cDNA-AFLP analysis of two different lines of wasps (RP and RM) (Reineke *et al.,* 2003). This cDNA was expressed differently in the two lines. The analysis of a more complete cDNA has made it possible to identify an open reading frame (ORF) of 515 amino acids displaying homology with the RdRp of other picornaviruses. In *N. vitripennis,* the ESTs showing no homology to the recently sequenced genome of the wasp (The Nasonia working group, 2010) were analyzed, and two sequences of 2789 bp (NvitV-1) and 1523 bp (Nvitv-2), respectively, were identified. The putative ORFs had the eight characteristic domains of the RdRp, and could be used for phylogenetic analysis. Analysis of VcSRV, NvitV-1 and NvitV-2 RdRp has revealed that they are more closely related to the new Iflaviridae family of Picornavirales superfamily than

to the other families (Reineke and Asgari, 2005; Oliveira *et al.,* 2010).

The presence of virions in the parasitoid was confirmed in *V. canescens* calyx region and in the muscle of *M. croceipes.* The 30–36-nm particles are isometric, non-enveloped, arranged in a para-crystalline array typical of picornaviruses (Fig. 1B) (Reineke and Asgari, 2005).

The *M. croceipes* picorna-like virus (McSRV-like) has been detected in deteriorating muscles of larvae, but also in other tissues of pupae and adults. These tissues were too severely damaged to be clearly identified (Hamm *et al.,* 1992). Recently, infections of *P. puparum* with a picorna-like virus (PpSRV) were detected from the abnormal morphology of the venom apparatus in about 5% of the females (Zhu *et al.,* 2008). The same percentage of infected females was observed over a two-year period, and there was no detectable disease phenotype. The molecular analysis of a partial sequence of PpSRV made it possible to carry out the phylogenetic analysis of this virus. PpSRV is similar to the Dicistroviridae, and clearly different from the Iflaviridae detected in *V. canescens* and *N. vitripennis.* Dicistroviridae are pathogens that affect several Hymenoptera, but essentially the Apidae and Formicidae. The honey bee can be infected by 18 different ssRNA viruses, and four have been detected in the same strain. McSRV and PpSRV are picorna-like viruses that have been described as pathogens of parasitoid hymenoptera (Hamm *et al.,* 1994, Zhu *et al.,* 2008).

In contrast, even though reverse-transcribed polymerase chain reaction (RT-PCR) fragments from VcSRVs

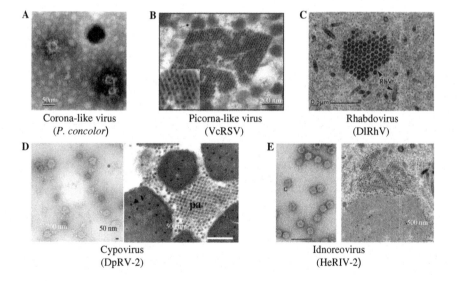

Corona-like virus
(*P. concolor*)

Picorna-like virus
(VcRSV)

Rhabdovirus
(DlRhV)

Cypovirus
(DpRV-2)

Idnoreovirus
(HeRIV-2)

FIGURE 1   Morphology of virions of the various viral genera detected in parasitoids. A: Corona-like virus in *P. concolor* (Jacas *et al.,* 1997) (with permission of *Ann. Appl. Biol.*). **B:** Picorna-like virus (VcSRV) in *V. canescens* (Reineke and Asgari, 2005) (with permission from *J. Ins. Physiology*). **C:** Rhabdovirus in *D. longicaudata* (DlRhV) (Lawrence and Akin, 1990) (with permission of *Can. J. Zool.*). **D:** Cypovirus (DpRV-2) in *D. pulchellus* (Renault *et al.,* 2003) (with permission from *J. Gen. Virol.*) **E:** Idnoreovirus (HeRIV-2) in *H. exiguae* (Stoltz and Makkay, 2000) (with permission from *Virology*).

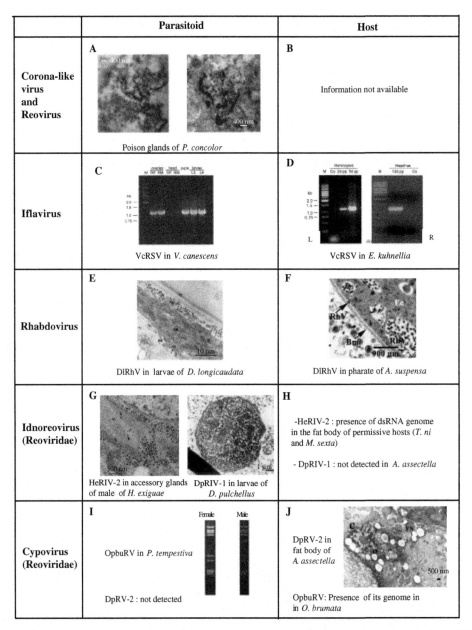

FIGURE 2   Presence and localization of parasitoid RNA viruses in the lepidopteran host and in the parasitoid. A, B: TEM has revealed the corona-like virus in the venom glands of *P. concolor* (A), but provided no information about its lepidopteran host (B) (Jacas *et al.,* 1997) (with permission from *Ann. Appl. Biol.*). Localization of VcRSV in different tissues and different developmental stages of two strains of *V. canescens* (RP and RM) by RT-PCR of RdRp (C) (Reineke and Asgari, 2005) (with permission from *J. Ins. Physiology*). Presence of VcRSV in hemocytes at various days post-parasitization (dpp) (left), and in the head and terminal abdominal segments (as) (right) of the host *E. kuhnellia* 13 dpp. The controls (Co) were unparasitized caterpillars. (D) (Reineke and Asgari, 2005) (with permission from *J. Ins. Physiology*). **E, F:** Localization by TEM of the rhabdovirus DlRhV in the eggs and larvae of the parasitoid *D. longicaudata* (E), and in the pharate of its host, *A. suspensa* (F) (Lawrence and Matos, 2005) (with permission from *J. Ins. Physiology*). **G:** Localization of the idnoreoviruses HeRIV-2 in the accessory glands of *H. exiguae* males, and of DpRIV-1 in the larvae of *D. pulchellus* (Stoltz and Makkay, 2000; Rabouille *et al.,* 1994) **H:** Occurrence of HeRIV-2 in the permissive hosts *T. ni* and *M. sexta* demonstrated by the detection of the genome segments and the absence of DpRIV-1 in *A. assectella* (Stoltz and Makkay, 2000; Rabouille *et al.,* 1994) (with permission from *Virology*). **I:** Detection of the genome segments of the cypovirus OpbuRV in the parasitoid *P. tempestiva,* and the absence of DpRV-2 in *D. pulchellus*. **J:** Localization of DpRV-2 in the fat body of the leek-moth *A. assectella* (Graham *et al.,* 2006; Renault *et al.,* 2005) (with permission from *J. Inv. Pathol.* and *J. Gen. Virol.*).

have been found in the larvae, pupae, and ovaries of *V. canescens*, no deleterious effect was detected on the reproduction of the laboratory colony (Fig. 2C) (Reineke and Asgari, 2005). Similar observations were made for NvitV-1 (Oliveira *et al.,* 2010). VcSRV and NvitV-1 could therefore be considered to be nonpathogenic commensal viruses of the parasitoids. Moreover, VcSRV is not detected in newly collected colonies, and NvitV-1 is not transmitted to sibling species (*N. giraulti* and *N. longicornis*) that have been reared in close proximity for many years (Oliveira *et al.,* 2010). The Iflaviridae, which include both VcSRV and NvitV-1, seem to correspond to nonpathogenic Picornavirales; however, this will have to be confirmed if any other Iflaviridae are discovered in parasitoids. The Picornaviridae are probably the easiest RNA viruses to detect as they can be identified in the cDNA library due their polyadenylation. The extension of genome- and EST-sequencing in Hymenoptera will almost certainly lead to the discovery of new Picornaviridae.

Virions of the Iflaviridae, VcSRV, are detected in the hemocytes, head and terminal abdominal segments of parasitized host caterpillars, *Ephestia kuhnellia* (Fig. 2D). However, it is not known whether the presence of VcSRV promotes the development of the parasitoid larvae in the caterpillar (Reineke and Asgari, 2005). If so, the relationship must be commensal, as it is not deleterious to the parasitoid or to the host. Nothing is known about the presence of virions in the hosts of *M. croceipes* and *N. vitripennis*. As the virus is found in parasitized *E. kuehniella*, the mode of transmission to infect the host and then parasitoid larvae within the host probably resulted from vertical transmission. The same type of transmission probably occurs in NvitV-1 and NvitV-2.

These three different picorna-like viruses are always detected in parasitoids in association with other types of viruses. McSRV is associated with at least one PDV, and in some strains also with a baculovirus, and VcSRV and NvitV-1 with VLPs containing any nucleic acids (Hamm *et al.,* 1992; Reineke and Asgari, 2005; Oliveira *et al.,* 2010). The PDV and the VLPs could play an essential role in inhibiting the immune response and development of the host, suggesting that the picorna-like viruses could be opportunistic viruses. This fits in with the fact that, like NvitV-2 and McSRV, VcSRV is not detected in every parasitoid strain.

It should be noted that a third type of ssRNA virus has been detected in the ESTs of *N. vitripennis* (Oliveira *et al.,* 2010). The cDNA of NvitV-3 was abundantly detected in both adults and pupae. The phylogeny of its RdRp shows that it is close to the Nora virus of *Drosophila melanogaster*, which is no longer classified as a Picornavirus (Oliveira *et al.,* 2010).

NEGATIVE-SENSE, SINGLE-STRANDED RNA VIRUSES

Among the negative-sense, single-stranded RNA viruses found in insects, only one member of the Rhabdoviridae (DlRhV) has been detected in the braconid, *D. longicaudata.*

The virions of Rhabdoviridae consist of an envelope and a nucleocapsid, and they have a characteristic bullet shape. They measure 45–100 nm in diameter, and 100–430 nm in length. The nucleocapsid is elongated with helical symmetry. The complete genome is 11,000–15,000 bases long, and consists of a single, linear molecule of negative-sense, single-stranded RNA. They infect both mammals and plants. Rhabdoviruses are rare in insects, the best described being the Sigma virus that infects *Drosophila melanogaster* and is implicated in the sensitivity to CO_2 (Tordo *et al.,* 2005; Fleuriet, 1999).

The rod-shaped particles observed in the venom apparatus are 250 nm long and 60–70 nm wide; they are rounded at both ends, and are characteristic of the Rhabdoviridae (Fig. 1C) (Edson *et al.,* 1982; Lawrence, 1988; Lawrence and Akin, 1990). However, no molecular analysis has ever been performed to confirm that the genome is composed of a single segment of negative-sense, single-stranded RNA corresponding to that of the Rhabdoviruses. The phylogenic analysis of DlRhV to find out whether it belongs to one of the six genera identified in the Rhabdoviridae remains to be done (Tordo *et al.,* 2005).

The rhabdovirus DlRhV was first reported during larva–pupae apolysis of *Anastrepha suspensa* parasitized by several *D. longicaudata* braconid wasps. DlRhV was not found in nonparasitized larvae. The virus proliferates in the cells of the cuticular epidermis (Fig. 2E) (Lawrence, 1988; Lawrence and Matos, 2005). The presence of the parasite blocks the migration and exocytosis of vesicles that is observed at the apices of cells of the epidermal tissues of the cuticle in nonparasitized pupae. The rhabdovirus is particularly abundant in these vesicles (Lawrence, 1988). During parasitism, DlRhV virions are first observed in the hemolymph 24–36 h post-parasitism (hpp), before the parasitoid eggs hatch, which occurs at about 48 hpp. The DlRhV particles are more abundant in the hemolymph than in the epidermal cells until 48–52 hpp. From 80 hpp, the particles accumulate in the epidermis and appear to replicate (Lawrence and Matos, 2005).

As the DlRhV particles are only detected in superparasitized *A. suspensa*, they were supposed to originate from the parasitoid and to be transmitted during oviposition. Accordingly, DlRhV virions have been detected in the poison apparatus of *D. longicaudata* females. The particles are located in the middle-third of the accessory glands filaments and in a stroma surrounded by vesicles, similar in morphology to that observed in parasitized *A. suspensa* (Fig. 2F) (Lawrence and Akin, 1990). DlRhV particles are not present in the oviduct of *D. longicaudata,* or in the previtellogenic and chorionated vitellogenic ova. They have been detected in the subchorionic space of oviposited eggs. It can therefore be hypothesized that DlRhVs are deposited at the periphery of the ooplasm during its passage through the oviduct below the junction with the venom apparatus (Lawrence and Matos, 2005). The micropyles at the extremity of the chorion could be the entry point for DlRhV. DlRhV particles are also found in midgut lumen of parasitoid first-instar larvae. It is not known if these particles are derived from those observed around the egg inside the chorion (Lawrence and Matos, 2005).

The transmission of DlRhV to the wasp could occur as follows: the virions are produced in the poison glands and enter the egg chorion of *D. longicaudata* during its passage through the lateral oviduct. The egg is oviposited with various accessory materials originating from the venom apparatus. DlRhV particles are probably also directly injected into the host larvae as virions, and are detected in the hemolymph 24–36hpp, i.e., before the wasp hatches. Once the wasp larva has hatched, it can feed on host tissues, thus concomitantly absorbing DlRhV virions that are subsequently found in the midgut lumen. This virus could also be present in other tissues to ensure that it is transmitted into the wasp's venom glands.

DlRhV has always been detected in the wasp in association with an entomopoxvirus (DlEPV). However, these two viruses occupy different tissues, zone II of accessory gland filaments (AGF) for rhabdovirus, and zone III of AGF for the entomopoxvirus. The transmission of these viruses also occurs via different ways: outside the chorion for entomopoxvirus, and inside for rhabdovirus.

The role of DlRhV in the successful development of *D. longicaudata* has not been clearly demonstrated, although viral accumulation in the cell epidermis could block the molting process by inhibiting the migration and apolysis of vesicles (Lawrence, 1988). *A contrario*, the entomopoxvirus is truly beneficial in the development of *D. longicaudata* by inhibiting one of the normal functions of hemocytes: the encapsulation of foreign bodies. The hemocytes show profound alterations after DlEPV infection: blebbing, cytoplasmic fragmentation, and apoptosis (Lawrence, 2005).

SEGMENTED, DOUBLE-STRANDED RNA VIRUSES

The segmented, double-stranded RNA viruses found in the parasitoid wasps all belong to the Reoviridae family. These virions consist of a capsid, a core, and nucleoprotein complex. They are not enveloped. The capsid is isometric and shows icosahedral symmetry. It is 60–80nm in diameter. The capsids are composed of two shells, and sometimes display surface projections. The genome is monomeric, and segmented (10–12 segments), and consists of double-stranded RNA. The complete genome is 18,000–30,000 bases in length. The Reoviridae have been divided into nine genera. Two are found in parasitoids: the Cypoviridae and the idnoreoviridae. Both these reoviruses have an isometric capsid, with icosahedral symmetry, which is

about 55–69nm in diameter for cypovirus and 30nm for idnoreovirus (Fig. 1D, DpRV-2; Fig. 1E, HeRIV-2, respectively) (Renault *et al.*, 2003; Stoltz and Makkay, 2000). The main difference at the morphological level is that cypoviruses are occluded in a polyhedrin matrix, whereas idnoreoviruses are not (figs. 1D, E) (Mertens *et al.*, 2005a, b).

Three types of association between the Reovidae and parasitoids have been described: nonpathogen commensal, mutualist, commensal, and mutualist (Renault *et al.*, 2005).

The nonpathogenic commensal reoviruses are found in the tissues of both female and male wasps which do not show any specific signs of infection, and the populations do not display disease-related collapse. They are always associated with other types of viruses. The reoviruses are detected in *M. leviventris*, associated with VLPs (Edson, 1981), those detected in *P. concolor* with corona-like viruses (Jacas *et al.*, 1997), those in *M. croceipes* with nudiviruses and polydnaviruses (Hamm *et al.*, 1994), and those in *H. exiguae* with polydnaviruses (Stoltz and Makkay, 2000). These four reoviruses are probably all idnoreoviruses, as they are not occluded.

The reoviruses in *M. leviventris* (MlRVLP) and in *P. concolor* (PcRVLP) are found in the venom apparatus of females, and so could be transmitted to the host during ovipositing, as has been observed for DlRhV, but no study has so far been performed to investigate the transmission of these reoviruses. Likewise, it is not known whether these viruses are also present in other tissues of the female or male wasp. MlRVLP is probably commensal as it is associated with VLPs, which are implicated in egg masking during parasitism in other species (Edson *et al.*, 1982). PcRVLPs may play a more active role, as they are associated with a corona-like virus, and nothing is known about these two viruses with regard to host/parasitoid relationships (Jacas *et al.*, 1997). This peculiar association could help to elucidate the use of nonconventional viruses in the deregulation of host immunity and development.

HeRIV-2 has been detected in the calyx of their parasitoid hosts. It is also present in various wasp tissues: ovaries, ovarioles, testes, male accessory glands, midgut, and malphigian tubules. McRVLP has been detected essentially in midgut epithelial cells and oenocytes (Fig. 2G) (Stoltz and Makkay, 2000, Hamm *et al.*, 1994). HeRIV-2 virion morphogenesis has been observed in all tissues except male testes. Moreover, wasp tissue larvae have less RNA than young or older infected adults, indicating that HeRIV-2 replicates in larvae and adults. The transmission of HeRIV-2 to the parasitized lepidopteran host has been explored in three permissive hosts (which allow the wasp to develop) (*Trichoplusia ni*, *Malacosoma disstria*, and *Manduca sexta*) and in two non-permissive hosts (*Orygia leucostigma* and *Lymantria dispar*). Female wasps oviposited on the different hosts, and then the HeRIV-2 RNA genome was looked for. It was found in two permissive hosts and in one non-permissive host, so its replication must depend solely on the host cells, not on the development of parasitoid larvae. No replication was observed in any host when the infection was performed *per os* in first instar larvae, indicating that the transmission in *H. exiguae* more likely involve stinging (Stoltz and Makkay, 2000). HeRIV-2 is probably transmitted to the host during oviposition, and then either HeRIV-2 replicates inside the host and the wasp is infected by feeding on the host tissue, or it directly infects the wasp larvae and then replicates in various different tissues. Transmission in the wasp is probably also vertical and associated with parasitism for McRVLP. A female *M. croceipes* from a noninfected colony was mated with males carrying McRVLP. Adults resulting from this cross were positive for McRVLP (Hamm *et al.*, 1994).

DpRV-2 and *Operophtera brumata* reovirus (OpbuRV) are mutualist viruses and they are usually the only virus present, without any associated viruses, except in 8% of cases where OpbuRV is found associated with *Operophtera brumata* cypovirus 19 (OpbuCPV19) (Renault *et al.*, 2003, Graham *et al.*, 2006). The case of DpRV-2 is quite unusual because its presence could not been detected either by TEM or by nucleic acid extraction in the parasitoid *D. pulchellus*. However, its presence in the genitalia of *D. pulchellus* female wasps has been confirmed by injection into the host of extracts of wasp genitalia that transmit the infection (Renault *et al.*, 2003). In contrast, extraction of the nucleic acids from female and male *P. tempestiva* revealed the presence of OpbuRV, although its precise localization in the wasp tissues is not known (Fig. 2I) (Graham *et al.*, 2006).

After injection into the lepidopteran host, probably during ovipositing, DpRV-2 is mainly detected in the midgut cells where it replicates in a virogenic stroma (Fig. 2J) (Renault *et al.*, 2003). OpbuRV has been detected in the whole extract of the lepidopteran *Operophtera brumata*, but no localization by TEM analysis has been performed (Graham *et al.*, 2006).

DpRV-2 was shown to inhibit the melanization reaction of the host, allowing the wasp larvae to develop. It plays an indispensable role in the host/parasitoid relationship which is equivalent to those demonstrated for the DpAV-4 virus and for polydnaviruses. No mutualist relationship has been demonstrated between OpbuRV and the wasp; however, it is the only virus that has been detected in the parasitoid wasp *Phobocampe tempestiva* and most wasps are infected, and there is a correlation between the presence of infected wasps and the percentage of infected *O. brumata*, suggesting that the parasitoid is at least involved in the dispersion of the virus. It would be very interesting to explore the exact role of OpbuRV in the parasitic success of *P. tempestiva* in *O. brumata*. DpRV-2 resembles a cypovirus in that it is occluded, whereas OpbuRv is not and probably belongs to the new genus of the idnoreoviruses.

The last type of possible relationship between reoviruses and wasps is commensal and mutualist, and has only been described for the reovirus DpRIV-1 (Rabouille *et al.*, 1994). DpRIV-1 is the type species for idnoreoviruses (Mertens *et al.*, 2005b). This reovirus is present in all the natural French populations of *D. pulchellus* examined up to 1999. In *D. pulchellus*, DpRIV-1 is mainly detected in gut epithelial cells, gut lumen, malphigian tubules, and to a lesser extent in the venom glands of the females. The viruses are concentrated in large vesicles present in the epithelial cells, which are released into the gut lumen without cell lysis (Fig. 2G). DpRIV-1 replicates in *D. pulchellus* because more viruses are detected in older wasps than in newly emerged insects. In contrast, DpRIV-1 has not been detected in the lepidopteran host, *A. assectella*, showing that it had not replicated in the lepidopteran host. DpRIV-1 resembles a commensal virus of *D. pulchellus*, as it has no impact on the fitness of the wasp (Rabouille *et al.*, 1994). However, it seems to play a more subtle role in the host/parasitoid relationship because, even if it does not replicate in the host, it does regulate the replication of the associated ascovirus Dp-AV4 (Bigot *et al.*, 1997). Ascovirus DpAV-4 is indispensable for successful parasitism by down-regulating host immunity at the level of melanization (Bigot *et al.*, 1997, Renault *et al.*, 2002). However, when DpAv-4 was injected alone into the lepidopteran host, infection occurred very rapidly and the host died within 72 h (Bigot *et al.*, 1995). In contrast, when co-injected with DpRIV-1 during ovipositing, the replication of DpAV-4 was slower, and allowed *D. pulchellus* larvae to develop (Bigot *et al.*, 1997). In conclusion, DpRIV-1 is commensal but also indispensable for parasitic success, and can therefore be considered as being indirectly mutualist.

Is there any correlation between the genus of reovirus (cypovirus or idnoreovirus) and the type of their relationships with the wasp (commensal, mutualist, commensal and mutualist)? Most studies have only been performed at the morphological level. Only one of the parasitoid reoviruses, DpRV-2, is most probably a cypovirus, because it is the only one to be occluded (Renault *et al.*, 2003) but no genome sequence analysis has been performed to confirm this classification. One other cypovirus, OpbuCPV19, is occasionally detected in the wasp *P. tempestiva* (8% of wasps and always in association with OpbuRV). Most cypoviruses have been detected in Lepidoptera (Mertens *et al.*, 2005a). However it would be very interesting to find out whether some cypoviruses can occur in the wasps that parasitize these Lepidoptera. DpRV-2 is a mutualist reovirus and OpbuRV, which is also suspected of being a mutualist, is an idnoreovirus. Therefore, it looks as though there is no correlation between the type of association with the wasp and phylogenetic classification of the virus.

All the other types of reoviruses seem to belong to the idnoreoviruses, insect viruses which are not occluded (Mertens *et al.*, 2005a, b). Phylogenetic analyses have been performed of only two idnoreoviruses, DpRIV-1 and OpbuRV for which the sequences of the RdRp are available (Bigot *et al.*, 1995, Renault *et al.*, 2005, Graham *et al.*, 2008). They do not show any similarities with other reoviruses. It would be interesting to sequence the RdRp of all the other idnoreoviruses identified so far. The idnoreoviruses include all types of virus: commensal (HeRIV-2, MvRVLP, MlRVLP), commensal and mutualist (DpRIV-1), and suspected mutualist (OpbuRV), so as for cypoviruses there is no correlation between the type of association with the wasp and the virus type. Interestingly, it must be noted that, except for OpbuRV, all the idnoreoviruses are found associated with various other viruses (PDV, corona-like, ascovirus, VLP).

One last remark about idnoreovirus, in the two cases where the genome has been analyzed, a supernumerary segment was found in the females (DpRIV-1 and OpbuRV). This segment is a marker of diploidy rather than being female specific, as it is also detected in the diploid males of *D. pulchellus* (Rabouille *et al.*, 1994) produced in inbred laboratory populations due to the sex determination mechanism (homozygotes at the sexual locus are males). In DpRIV-1, sequencing of the supplemental fragment has revealed that it is composed of a duplication of part of the longest fragment of DpRIV-1 (Bigot *et al.*, 1995). As diploid individuals are most probably mostly females in wild populations, it might play a role in regulating the replication of DpAV-4 (Renault *et al.*, 2005). Unfortunately, no sequence information about the supplemental segment of OpbuRV is available (Graham *et al.*, 2008), and the systematic search for the presence of the supplementary fragment in the other idnoreoviruses could provide information about the exact role of this segment in the life-cycle of these viruses.

CONCLUSION

So far, only four families of RNA viruses have been detected in parasitoids, although eight families have been described in insects. It could be wondered whether the other families of parasitoids are really absent, or if this is due to the fact that inadequate techniques were used to detect viruses. In fact, several different methods have been used: TEM of the venom glands of females, extraction of nucleic acids, followed by DNAse digestion to eliminate the polydnavirus or ascovirus genome, RT-PCR with primers specific of RdRp or data mining in an EST library. To resolve this problem of detecting RNA viruses, a systematic search for viruses in parasitoids should be carried out using a combination of these different methods.

Only a few hymenopteran species (10) are known to be infected by RNA viruses, although thousands of species are known to carry polydnaviruses or VLPs. This could suggest that the presence of polydnaviruses or VLPs may block infections with other viruses. However, some cases of co-infection and coreplication of RNA viruses and polydnaviruses have been described in *H. exiguae*, *M. croceipes*, and *M. leviventris*. The absence of detection of coinfection with polydnaviruses in most cases is perhaps simply due to the fact that the polydnavirus is present in large quantities in parasitoids, and so masks the presence of the other viruses, as in *H. exiguae* (Stoltz and Makkay, 2000).

However, it must be noted that RNA viruses are most often detected in association with various other viruses, such as picorna-like virus, ascoviruses, polydnaviruses, and entomopoxviruses, except in two cases; the reoviruses DpRV-2 and OpbuRV are the only viruses to have been detected as the sole virus in parasitoids. We could suggest that the RNA viruses are opportunistic viruses, which require preliminary infection with other viruses in order to infect the host. This hypothesis is reinforced by the fact that the association with parasitoids is sporadic. Indeed, most of them are detected in some strains but not in others, for example HeRIV-2 or VcRSV. They resemble conventional viruses, where populations or individuals can be either infected or not infected. With the exception of the reovirus DpRV-2, none is directly implicated in regulating host immunity, or promoting successful parasitism.

Information about the RNA viruses is also limited at the molecular level, as very few of them have been sequenced. More data on their genome composition would allow the determination of their phylogeny and facilitate the study of their life-cycle. We hope that this part of our work will expand in the coming years as genome sequencing of parasitoids advances, as exemplified by the recent discovery of the sequences of picorna-like viruses in the EST library of *N. vitripennis* (Oliveira *et al.*, 2010). The limitation in the number of studies of the RNA viruses in parasitoids is probably also due to the fact that infections with these viruses have little impact on the viability of economically important parasitoids used in pest control.

REFERENCES

Bigot, Y., Drezen, J.-M., Sizaret, P.-Y., Rabouille, A., Hamelin, M.-H., Periquet, G., 1995. The genome segments of DpRV, a commensal reovirus of the wasp *Diadromus pulchellus* (Hymenoptera). Virology 210, 109–119.

Bigot, Y., Rabouille, A., Doury, G., Sizaret, P.-Y., Delbost, F., Hamelin, M.-H., et al. 1997. Biological and molecular features of the relationships between *Diadromus pulchellus* ascovirus, a parasitoid hymenopteran wasp (*Diadromus pulchellus*) and its lepidopteran host, *Acrolepiopsis assectella*. J. Gen. Virol. 78, 1149–1163.

Büchen-Osmond, C., 2003. The universal virus database ICTVdB. Comput. Sci. Eng. 5, 16–25.

Christian, P., Carstens, E., Domier, L., Johnson, J., Johnson, K., Nakashima, N., et al. 2005. Dicistroviridae. In: Fauquet, C.M., Mayo, M.A., Maniloff, J., Desselberger, U., Ball, L.A. (Eds.), Virus Taxonomy—Eighth Report of the International Committee on Taxonomy of Viruses (pp. 783–788). Elsevier, Amsterdam.

Christian, P., Carstens, E., Domier, L., Johnson, J., Johnson, K., Nakashima, N., et al. 2005. Iflaviridae. In: Fauquet, C.M., Mayo, M.A., Maniloff, J., Desselberger, U., Ball, L.A. (Eds.), Virus Taxonomy—Eighth Report of the International Committee on Taxonomy of Viruses (pp. 779–782). Elsevier, Amsterdam.

Edson, K.M., 1981. Virus-like and membrane-bound particles in the venom apparatus of a parasitoid wasp (Hymenoptera: Braconidae). In: Bailey, G.W. (Ed.), Proceedings of the Thirty-Nineth Annual Electronic Microscopy Society of America (pp. 610–611). Claitors Publishing Division, Baton Rouge.

Edson, K.M., Barlin, M.R., Vinson, S.B., 1982. Venom apparatus of braconid wasps: comparative ultrastructure of reservoirs and gland filaments. Toxicon 3, 553–562.

Fauquet, C.M., Mayo, M.A., Maniloff, J., Desselberger, U. and Ball, L.A. (Eds.), 2005. Virus Taxonomy—Eighth Report of the International Committee on Taxonomy of viruses. Elsevier, Amsterdam.

Fleuriet, A., 1999. Evolution of the proportions of two sigma viral types in experimental populations of Drosophila melanogaster in the absence of the allele that is restrictive of viral multiplication. Genetics 153, 1799–1808.

Graham, R.I., Rao, S., Sait, R.M., Attoui, H., Mertens, P.P.C., Hails, R.S., et al. 2008. Sequence analysis of a reovirus isolated from the winter moth Operophtera brumata (Lepidoptera: Geometridae) and its parasitoid wasp Phobocambe tempestiva (Hymenoptera: Ichneumonidae). Virus Res. 135, 42–47.

Graham, R.I., Rao, S., Sait, R.M., Possee, R.D., Mertens, P.P.C., Hails, R.S., 2006. Detection and characterization of three novel species of reovirus (Reoviridae), isolated from geographically separate populations of the winter moth Operophtera brumata (Lepidoptera: Geometridae) on Orkney. J. Inverteb. Pathol. 91, 79–87.

Hamm, J.J., Styer, E.L., Lewis, W.J., 1992. Three viruses found in the braconid parasitoid Microplitis croceipes and their implications in biological control programs. Biol. Control 2, 329–336.

Hamm, J.J., Styer, E.L., Steiner, W.M., 1994. Reovirus-like particle in Microplitis croceipes (Hymenoptera: Braconidae). J. Inverteb. Pathol. 63, 304–306.

Jacas, J.A., Budia, F., Rodriguez-Cerezo, E., Vinuela, E., 1997. Virus-like particles in the poison gland of the parasitic wasp Opius concolor. Ann. Appl. Biol. 130, 587–592.

Kapoor, A., Simmonds, P., Lipkin, W.I., Zaidi, S., and Delwart, E., 2010. Use of nucleotide composition analysis to infer hosts for three novel picorna-like viruses. J. Virol. 84, 10322–10328.

Lawrence, P.O., 1988. Ecdysteroid titres and integument changes in superparasitized puparia of Anastrepha suspensa (Diptera: Tephritidae). J. Insect Physiol. 34, 603–608.

Lawrence, P.O., 2005. Morphogenesis and cytopathic effects of the Diachasmimorpha longicaudata entomopoxvirus in host hemocytes. J. Insect Physiol. 51, 221–233.

Lawrence, P.O., Akin, D., 1990. Virus-like particles from the poison glands of the parasitic wasp Biosteres longicaudata (Hymenoptera: Braconidae). Can. J. Zool. 68, 539–546.

Lawrence, P.O., Matos, F.M., 2005. Transmission of the Diachasmimorpha longicaudata rhabdovirus (DlRhV) to wasp offspring: an ultrastructural analysis. J. Insect Physiol. 51, 235–241.

Mertens, P.P.C., Rao, S., Zhou, H., 2005. Cypovirus. In: Fauquet, C.M., Mayo, M.A., Maniloff, J., Desselberger, U., Ball, L.A. (Eds.), Virus Taxonomy—Eighth Report of the International Committee on Taxonomy of Viruses (pp. 522–533). Elsevier, Amsterdam.

Mertens, P.P.C., Rao, S., Zhou, H., 2005. idnoreovirus. In: Fauquet, C.M., Mayo, M.A., Maniloff, J., Desselberger, U., Ball, L.A. (Eds.), Virus Taxonomy—Eighth Report of the International Committee on Taxonomy of Viruses (pp. 517–521). Elsevier, Amsterdam.

Oliveira, D.C.S.G., Hunter, W.B., Ng, J., Desjardins, C.A., Dang, P.M., Werren, J.H., 2010. Data mining cDNAs reveals three new single stranded RNA viruses in Nasonia (Hymenoptera: Pteromalidae). Insect Mol. Biol. 19, 99–107.

Rabouille, A., Bigot, Y., Drezen, J.-M., Sizaret, P.-Y., Hamelin, M.-H., Periquet, G., 1994. A member of the Reoviridae (DpRV) has a ploidy-specific genomic segment in the wasp Diadromus pulchellus (Hymenoptera). Virology 205, 228–237.

Reineke, A., Asgari, S., 2005. Presence of novel small RNA-containing virus in a laboratory culture of the endoparasitic wasp Venturia canescens (Hymenoptera: Ichneumonidae). J. Insect Physiol. 51, 127–135.

Reineke, A., Schmidt, D., Zebitz, C.P., 2003. Differential gene expression in two strains of the endoparasitic wasp Venturia canescens identified by cDNA-amplified fragment length polymorphism analysis. Mol. Ecol. 12, 3485–3492.

Renault, S., Bigot, S., Lemesle, M., Sizaret, P.-Y., Bigot, Y., 2003. The cypovirus Diadromus pulchellus RV-2 is sporadically associated with the endoparasitoid wasp D. pulchellus and modulates the defence mechanisms of pupae of the parasitized leek-moth Acrolepiopsis assectella. J. Gen. Virol. 84, 1799–1807.

Renault, S., Petit, A., Bénédet, F., Bigot, S., Bigot, Y., 2002. Effects of the Diadromus pulchellus ascovirus, DpAV-4, on the hemocytic encapsulation response and capsule melanization of the leek-moth pupa, Acrolepiopsis assectella. J. Insect Physiol. 48, 297–302.

Renault, S., Stasiak, K., Federici, B., Bigot, Y., 2005. Commensal and mutualistic relationships of reoviruses with their parasitoid wasp host. J. Insect Physiol. 51, 137–148.

Spaan, W.J.M., Brian, D., Cavanagh, D., de Groot, R.J., Enjuanes, L., Gorbalenya, A.E., et al. 2005. Coronaviridae. In: Fauquet, C.M., Mayo, M.A., Maniloff, J., Desselberger, U., Ball, L.A. (Eds.), Virus Taxonomy—Eighth Report of the International Committee on Taxonomy of Viruses (pp. 947–964). Elsevier, Amsterdam.

Stoltz, D., Makkay, A., 2000. Co-replication of a reovirus and a polydnavirus in the Ichneumonid Hyposoter exiguae. Virology 278, 266–275.

The Nasonia Working Group, 2010. Functional and evolutionary insights from the genomes of three parasitoid Nasonia species. Science 327, 343–348.

Tordo, N., Benmansour, A., Calisher, C., Dietzen, G.R., Fang, R.-X., Jackson, A.O., et al. 2005. Rhabdoviridae. In: Fauquet, C.M., Mayo, M.A., Maniloff, J., Desselberger, U., Ball, L.A. (Eds.), Virus Taxonomy—Eighth Report of the International Committee on Taxonomy of Viruses (pp. 623–644). Elsevier, Amsterdam.

Varaldi, J., Patot, S., Nardin, M., Gandon, M., 2010. A virus-shaping reproductive strategy in a Drosophila parasitoid. Adv. Parasitol. 70, 333–363.

Zhu, J.-Y., Ye, G.-Y., Fang, Q., Wu, M.-L., Hu, C., 2008. A pathogenic picorna-like virus from the endoparasitoid wasp, Pteromalus puparum: initial discovery and partial genomic characterization. Virus Res. 138, 144–149.

Chapter 17

An Inherited Virus Manipulating the Behavior of its Parasitoid Host: Epidemiology and Evolutionary Consequences

Julien Varaldi[*], Julien Martinez[*], Sabine Patot[*], David Lepetit[*], Frédéric Fleury[*] and Sylvain Gandon[†]

[*]Université de Lyon, F-69000, Lyon; Université Lyon 1; CNRS, UMR5558, Laboratoire de Biométrie et Biologie Evolutive, F-69622, Villeurbanne, France

[†]Centre d'Ecologie Fonctionnelle et Evolutive (CEFE); UMR 5175; 1919 route de Mende; F-34293 Montpellier, France

SUMMARY

Insect endoparasitoids harbor a great diversity of viruses within their reproductive tracts. These viruses are injected within the host together with the parasitoid eggs and are often vertically transmitted from mother to offspring. Some of these viruses have evolved towards a mutualistic relationship with the parasitoid since they protect the parasitoid eggs from the host immune response. However, most parasitoid viruses have no clear phenotypic effects, and the means by which they are maintained in parasitoid populations is unclear. A virus infecting the *Drosophila* parasitoid *Leptopilina boulardi* has been discovered. This virus strongly alters the superparasitism behavior of the parasitoid by increasing its tendency to lay supernumerary eggs in parasitized hosts. The virus is maternally transmitted and also horizontally transmitted in conditions of superparasitism. The virus manipulates the behavior of the parasitoid thus enhancing its own transmission. In this chapter we review the current understanding of this parasitoid/virus system, describing both experimental approaches and theoretical studies. We suggest that virus-induced superparasitism manipulation may be common since the theoritical model used predicts that all parasitoid infectious viruses of the genital tract would be selected for the enhancement of the natural superparasitism tendency of their parasitoid host.

ABBREVIATIONS

ES	evolutionarily stable
ESS	evolutionarily stable strategy
LbFV	*Leptopilina boulardi* filamentous virus
NS	nonsuperparasitizing
PCR	polymerase chain reaction
S	superparasitizing
SSH	suppressive subtractive hybridization
VLP	virus-like particles

INTRODUCTION

Insect endoparasitoids have a very special way of life. They live as parasites inside the body of other arthropods during their larval stages (see Godfray (1994) for a review). During this parasitic phase, the developing parasitoids feed on the host's tissues and/or hemolymph and they ultimately kill the host.

The parasitoid eggs and larvae are potentially exposed to the host immune response. During oviposition, parasitoid females usually inject a complex set of proteins and particles (secreted from the venom gland and/or the ovaries) inside the host. These injected fluids are involved in the protection of the egg from the host immune response (see Poirié *et al.* (2009) for a review). Interestingly, it has been repeatedly found that these fluids may contain viruses or virus-like particles (VLPs). In particular, more than 17,500 braconid species and 15,000 ichneumonid species inject virions of a polydnavirus inside the host. The genes of the polydnavirus are then expressed inside the parasitized host and manipulate the physiology of the host. They are necessary for the development of the parasitoid offspring and all individuals from these species do harbor the virus. Although it is now clear that they have a viral origin (Bézier *et al.*, 2009; Volkoff *et al.*, 2010), they have lost their autonomy and their transmission is ensured by their integration in wasp chromosomes.

Beside polydnaviruses, parasitoid venoms or ovarian fluids may contain a great diversity of viruses and/or VLPs. These viruses were first described by investigators using electron microscopy (Stoltz and Vinson,

Parasitoid Viruses: Symbionts and Pathogens. DOI: 10.1016/B978-0-12-384858-1.00017-5

1977; Barratt *et al.*, 1999; Luo and Zeng, 2010) but their discovery is now facilitated by sequencing technologies (Reineke and Asgari, 2005; Oliveira *et al.*, 2010). They belong to the Ascoviridae (double-stranded DNA virus; Stasiak *et al.*, 2005), Poxviridae (dsDNA virus; Hashimoto and Lawrence, 2005), Reoviridae (dsRNA segmented virus; Renault *et al.*, 2005), Rhabdoviridae (ssRNA negative sense virus; Lawrence and Matos, 2005), picorna-like viruses (single-stranded RNA positive sense virus; Reineke and Asgari, 2005; Oliveira *et al.*, 2010), or are still unclassified (Barratt *et al.*, 2006; Varaldi *et al.*, 2006b; Luo and Zeng, 2010). Contrary to polydnavirus/wasp associations, some populations or individuals of some species are not infected by these viruses, demonstrating that the associations are facultative and likely do not involve integration in the wasp's chromsomes (Renault *et al.*, 2003; Stoltz *et al.*, 1988; Stoltz and Makkay, 2000). However, one of these viruses was shown to be vertically transmitted through wasp generations (Stoltz *et al.*, 1988). In addition, some of them are capable of replicating both within the parasitoid and the host as opposed to polydnaviruses, which replicate exclusively in the wasp ovary (Stoltz *et al.*, 1988; Stoltz and Makkay, 2000). We may speculate that these viruses are on the first steps of an evolution toward a mutualistic relationship with their parasitoid hosts. This step could ultimately lead to a stable mutualistic association similar to polydnavirus/parasitoid associations. This idea is supported by the observation that some of these viruses also facilitate the escape from the host immune system, although the effects are much less dramatic than for polydnaviruses (Renault *et al.*, 2003). From a theoretical point of view, this makes sense since vertically transmitted symbionts are expected to be maintained in populations only if they confer some advantage to their host (here, the parasitoid), or manipulate its reproduction such as some symbiotic bacteria like *Wolbachia* (O'Neill *et al.*, 1997). However, several viruses do not seem to provide any benefit to the parasitoid, raising the question of the nature of their relationship with their parasitoid host and of the mechanism of their persistence in wasp populations.

In this chapter, we describe a virus that manipulates parasitoid behavior to enhance its transmission. The mechanisms allowing the virus to reach high prevalence have been elucidated and will be presented in detail. These mechanisms may apply to other viruses found in parasitoids.

VIRUS DISCOVERY

In solitary parasitoids, one host supports the development of a single parasitoid, whatever the number of parasitoid eggs laid during oviposition. Females are usually able to recognize parasitized from unparasitized hosts (host discrimination) and often avoid laying eggs in the already parasitized hosts. However, when a female chooses to oviposit in a parasitized host, a behavior called superparasitism, parasitoid larval competition ends up in the death of all but one larva. The younger larva is most likely outcompeted and its survival usually depends on the interval between the first and second ovipositions (van Alphen and Visser, 1990). If a parasitoid female parasitizes the same host several times (a behavior called self-superparasitism), the wasp wastes supernumerary eggs since larvae will compete for the possession of the host until all but one dies. Superparasitism is thus expected to be strongly counter-selected in most ecological conditions. However, conspecific-superparasitism (superparasitizing hosts parasitized by other females) may be selected in conditions of strong competition for hosts (van Alphen and Visser, 1990).

In France, the solitary endoparasitoid *Leptopilina boulardi* mainly develops in *D. melanogaster* and *D. simulans*. One remarkable feature of *L. boulardi* is that in some populations, females show a huge tendency to superparasitize, while in others most females lay only one egg per host. In the related parasitoid *L. heterotoma*, however, superparasitism is rarely observed (Varaldi *et al.*, 2005a). We have derived stable 'non-superparasitizing' (NS) and 'superparasitizing' lines (S) from *L. boulardi* populations in order to study the genetic determinism of this behavior. Note that these lines were different with respect to both their self-superparasitism and conspecific superparasitism behavior (Varaldi *et al.*, 2003). Surprisingly, the variations in the superparasitism phenotype were strictly maternally inherited: whatever the nuclear genotype, females adopted the phenotype of their mother (Varaldi *et al.*, 2003). The maternal transmission was further confirmed using two independent lines originating from Antibes (South of France, S phenotype) and Madeira (Portuguese island, NS phenotype) (Fig. 1). Furthermore, when both S and NS lines laid their eggs inside the same host, in the case where NS lines won the within-host competition, the emerging (female) offspring did adopt the 'superparasitizing' phenotype, despite the NS phenotype of its mother (Varaldi *et al.*, 2003)! This was consistent with the hypothesis that some infectious element was causing the 'superparasitizing' phenotype and was transmitted from S infected lines to NS uninfected lines during the short time they coexisted inside the *Drosophila* larva. The newly acquired S phenotype was stably transmitted through generations (Varaldi *et al.*, 2006b). The infectious nature of the S-inducing element was further confirmed by injecting solutions derived from S individuals (ovaries and poison gland) into *Drosophila* larvae parasitized by NS females. Solutions of S females proved able to induce the S phenotype on the emerging parasitoid females (originating from an NS line), whereas NS control injections did not induce

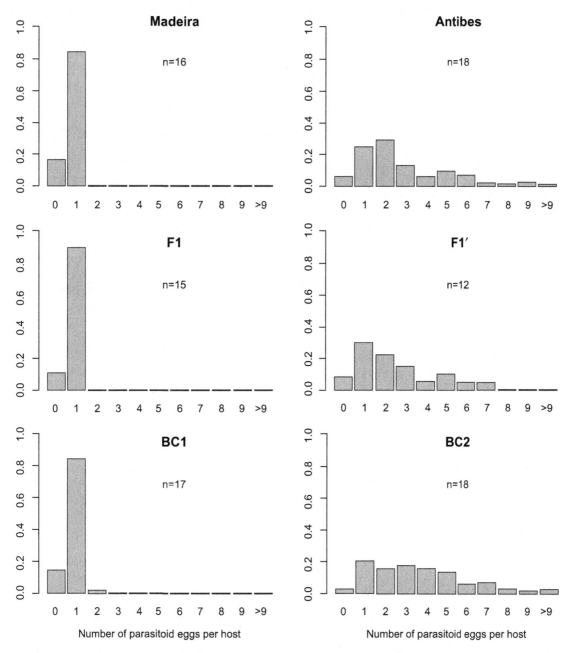

FIGURE 1 Distribution of the number of parasitoid eggs/host larva in Madeira (S phenotype) and Antibes (S phenotype) parental lines, F1 and 2 backcrosses (BC) illustrating two generations of introgression: (mother × father) : F1: M × A ; F19: A × M ; BC1: (M × A) × A; BC2: (A × M) × M. n, number of females tested. The superparasitism phenotype was estimated by providing each female with 10 *Drosophila melanogaster* larvae.

any behavioral change (Fig. 2; Varaldi *et al.*, 2006b). Furthermore, NS females did not become contaminated by sharing food with S females and they did not contaminate their ovipositor by stinging hosts that had been previously attacked by S females (Varaldi *et al.*, 2006a).

The hypothesis that the causative agent was a bacterium was tested and clearly rejected using antibiotic treatments and a polymerase chain reaction (PCR) approach using primers for genes conserved among bacteria (Varaldi

et al., 2006b). The nature of the infectious element was eventually determined by electron microscopical analysis of the ovaries of *L. boulardi* females. It was evident that in S lines, a virus was replicating in cells bordering the lumen of the oviduct (Fig. 3a), contrary to NS females (Varaldi *et al.*, 2003, Varaldi *et al.*, 2006b). Based on its morphology, the superparasitism-inducing virus was called LbFV for *Leptopilina boulardi* filamentous virus (Fig. 3). LbFV is thus vertically transmitted through the female line, and

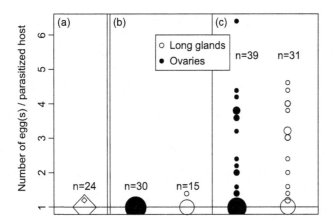

FIGURE 2 Superparasitism behavior of females emerging from hosts parasitized by Sienna0 (NS) females and injected with (a) PBS, (b) long glands or ovaries extracted from Sienna0 (NS) females, or (c) long glands or ovaries extracted from Sienna9 (S) females. Area of symbols is proportional to number of individuals. The superparasitism phenotype was estimated by providing each female with 10 host larvae. After Varaldi *et al.*, 2006b, *Journal of Insect Physiology*.

also horizontally only in conditions of superparasitism. To date, LbFV particles have only been found in *L. boulardi*, but not in sympatric *L. heterotoma* nor in the host *Drosophila melanogaster* (Patot *et al.*, 2009).

MOLECULAR DATA

LbFV was discovered using electron microscopy but we lacked the genomic data required for identifying its phylogenetic position and for developing molecular tools. Since LbFV could be either a DNA or an RNA virus, we focused our attention on the identification of viral mRNAs, because both viral genome types should produce mRNAs. We performed a suppressive subtractive hybridization (SSH) between two lines sharing the same genotypic background but differing in their superparasitism behavior. This work permitted us to identify an 810-bp mRNA that is S-specific. From this mRNA sequence, we derived a simple PCR assay using DNA extracts as templates that showed amplification on all 14 independent S lines whereas no amplification was observed for all 11 independent NS lines (Patot *et al.*, 2009). This perfect correlation between superparasitism phenotype and PCR profiles validates the viral origin of this sequence and has been since confirmed on a larger scale (at least n = 200 for each status). Furthermore, it shows that LbFV has a DNA genome (unless LbFV is a retrovirus, which is very unlikely). Unfortunately, this 810-bp sequence did not display any similarity with public database sequences and did not allow the identification of the phylogenetic position of the virus. However, LbFV share some morphological features with other large double-stranded insect DNA viruses like nudiviruses, or the recently discovered family

Hytrosaviridae, represented by salivary gland hypertrophy viruses found in the tsetse fly and house fly (Lietze *et al.*, 2010), and to a lesser extent Baculoviridae. The GC content of the 810-bp sequence is very low (24%), a feature shared with the gland hypertrophy viruses of the tsetse fly (28% GC content of the genome).

ADAPTIVE SIGNIFICANCE OF THE PARASITOID'S BEHAVIORAL ALTERATION

The induction of superparasitism increases the chance of horizontal transmission, but at the same time leads to egg wastage as all but one progeny dies. Because the virus also benefits from vertical transmission, it remains unclear whether the induction of superparasitism is adaptive for the virus (Varaldi *et al.*, 2003; Reynolds and Hardy, 2004; Gandon, 2005). To demonstrate the adaptive nature of the alteration of the parasitoid's behavior one must show that a virus increasing superparasitism can invade a virus population. In other words, one must demonstrate that the evolutionarily stable (ES) virus strategy of superparasitism is higher than the ES parasitoid superparasitism strategy (in the absence of the virus). To address this question, we developed a model analyzing the evolution of a population of parasitoids (a proovigenic and solitary species) parasitizing a population of hosts (Gandon *et al.*, 2006). This model includes the potential benefit of superparasitism (i.e., the possibility that a parasitoid larva developing in an already parasitized host wins the within-host competition) and the classical costs of superparasitism (i.e., the costs of spending time ovipositing in already parasitized hosts and the cost of producing the eggs). We first used this model in the absence of a virus, to predict the fate of a mutant parasitoid with superparasitism strategy s^* appearing in a parasitoid population dominated by a resident with strategy s (where s indicates the rate of acceptance of parasitized hosts). As expected, the model predicts that the ES strategy (ESS) of superparasitism is zero when the probability to win the within-host competition (c) is low but increases with this probability (gray dotted line in Fig. 4). This further confirmed previous models showing the potential adaptive value of superparasitism under conditions of host scarcity (van Alphen and Visser, 1990).

Then we extended the model to include a virus able to manipulate the superparasitism behavior of the wasp, based on LbFV biology. When females are infected, it is assumed that the parasitoid behavior is strictly under the control of the virus. In other words, the rate of acceptance of parasitized hosts of an infected female is not s (the superparasitism strategy when the female is uninfected), but instead σ, which is a feature of the virus. The virus is vertically transmitted with a rate of tv (<1), and will gain extra routes of transmission via the horizontal transmission that may occur between a larva infected with the virus and an uninfected

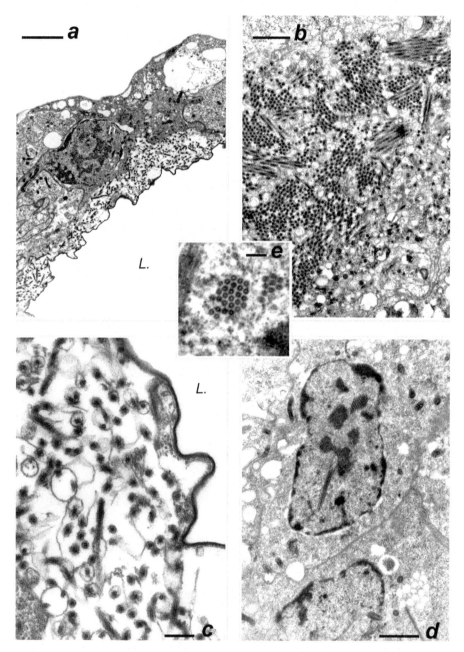

FIGURE 3 Virus replication in the oviduct of superparasitizing _L. boulardi_ (TEM photographs). a: Apparent intranuclear replication in cells bordering the lumen of the oviduct. n, nucleus; L, lumen; Bar: 2.5 mm. **b:** High densities of virions in cell cytoplasm. Bar: 1 mm. **c:** Structure of virions accumulated close to the lumen. L, lumen; bar: 1/4 250 nm. **d:** Structure of viral particles within nucleus. Bar: 1 mm. **e:** Transverse section of nucleocapsids within nucleus. Bar: 200 nm. After Varaldi _et al_., 2006b, _Journal of Insect Physiology_.

larva (with probability τ_h). To allow direct competition between viral strains, it is assumed that a viral strain can replace another one when they compete inside the same _Drosophila_ larva with a probability ϵ. However, no multi-infections at the adult stage are allowed. The model can be used to derive an expression of the fitness of a mutant virus with a strategy σ^* appearing in a population dominated by a resident virus with strategy σ at the epidemiological equilibrium set by the resident virus and the strategy s adopted

by the host. Note that here, only the virus is allowed to evolve, not the parasitoid (s is fixed). In the first part, we fixed $\epsilon = 0$, i.e., a viral strain is not able to replace a resident viral strain in competition within _Drosophila_ larvae. The results indicate that the ES superparasitism is always higher for the virus than that observed for the parasitoid (allowed to evolve to its optimal strategy in the absence of the virus), demonstrating the adaptive value of the behavioral modification from the virus point of view: the virus is

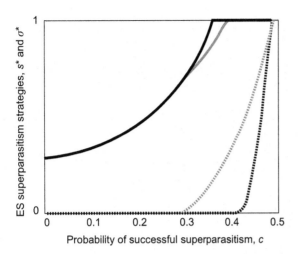

FIGURE 4 Evolutionarily stable superparasitism strategies of the virus (solid lines) and the parasitoid (dotted lines) versus the probability of successful superparasitism. The gray lines indicate a situation where the parasitoid does not coevolve with the virus. The black lines indicate the coevolutionarily stable strategies of the virus and the parasitoid. Parameter values: th $p = 0.75$, $tv = 0.95$, $\epsilon = 0$, $N = 100$, $d = e = 0.2$, $m = 0.1$, $a = 0.01$, $t1 = 0$, $t2 = 0.1$, and $E = 10$. After Gandon et al., 2006, *The American Naturalist*.

always selected to increase the natural superparasitism tendency of the parasitoid (gray full line versus gray dashed line in Fig. 4). The presence of the virus thus induces a conflict of interest between the parasitoid and the virus with respect to superparasitism behavior. The intensity of the evolutionary conflict is even increased if both the virus and the parasitoid are allowed to co evolve: after co evolution, uninfected females (that are produced even in infected populations, due to imperfect vertical transmission) should superparasitize less than uninfected females that did not co evolve with the virus (black lines in Fig. 4). This shows that the presence of the virus in a population should indirectly modify the ESS of a trait for uninfected females. When we allowed direct competition between viral strains within *Drosophila* larvae ($\epsilon > 0$), we found that the virus is even selected for much higher superparasitism strategies, thus strongly increasing the conflict of interest between the parasitoid and the virus (Varaldi et al., 2009). These results clearly show that increasing the superparasitism strategy of the parasitoid is an adaptive strategy from the virus's point of view (whatever ϵ). The intensity of the conflict of interest depends critically on the ability of mutant virus strains to replace resident strains inside *Drosophila* larvae (ϵ) and also on coevolutionary processes.

EFFECTS ON OTHER PHENOTYPIC TRAITS

The effects of LbFV on several phenotypic traits have also been studied. It has been found that LbFV infection has no effect on the survival of parasitoid females

but has a negative impact on male survival, and a slight negative impact on size (female tibia length is reduced by 2%) and on developmental time (increased by 3% for both sexes). Strikingly, the overall locomotor activity of infected females is reduced by 45% while no effect was detected in males. Interestingly, we found that egg-load was increased for infected females (+11%) compared to uninfected females (Varaldi et al., 2005b). Because an infected female is at higher risk of being egg-limited (due to superparasitism), both the virus (since it requires eggs to be transmitted) and the infected parasitoid are selected to increase investment in egg-load. As a consequence, the analysis provided by the theoretical model did not allow us to distinguish between the two following interpretations: the increased eggload observed in infected females corresponds to (1) an adaptive plastic response of the parasitoid to the presence of the virus, or (2) to another component of the manipulation induced by the virus (Gandon et al., 2009).

The influence of LbFV on behavior has also been investigated (Varaldi et al., 2006a, 2009). The behavioral components studied included sexual communication, circadian rhythms, ability of females to detect and discriminate odors of hosts, and trajectometric parameters of foraging females. None of these behavioral repertoires seemed to be perturbed by LbFV infection. Interestingly, it has been found that LbFV does not even impair the ability of the female to discriminate between a 'good host species' (*D. melanogaster*) and a 'bad host species' (*D. subobscura*), although this process probably shares some similarities with the process of discrimination between parasitized and unparasitized hosts. All these results illustrate that the action of the virus is targeted on superparasitism behavior. The influence of the virus on the transcriptome of the parastoid is currently under investigation to unravel the mechanism of action of the virus on the wasp's behavior.

PREVALENCE AND DYNAMICS IN NATURAL POPULATIONS

Leptopilina boulardi is widely distributed under tropical and temperate regions; it has been identified in Africa, Southern Europe, the Middle East, North and South America (California, Brazil), and Australia. It is very abundant in the South of France and its density decreases while going north. In France, *L. boulardi* is completely absent above 45°6 N latitude. During late summer 2004 and 2007, we sampled the *Drosophila* and parasitoid community in the south-east of France including central and northern populations of *L. boulardi* (Patot et al., 2010). In total, 15 locations were sampled along a north to south axis. Two sampling procedures were used for wasp collection. In 11 localities, we placed in each orchard 10–12

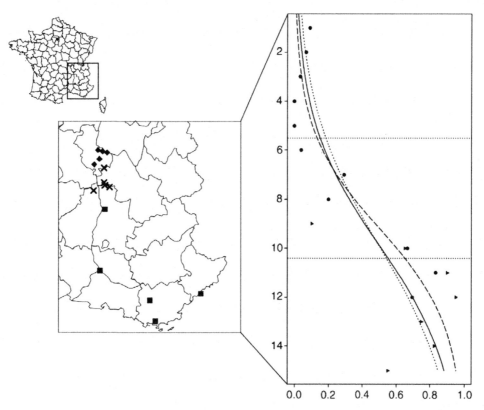

FIGURE 5 **Prevalence of LbFV (horizontal axis) in *L. boulardi* populations in south-eastern France and predictions of the generalized linear model.** The populations were sampled in 2004 (triangles) or 2007 (dots). They are ranked according to their latitude (vertical axis). Continuous line, model prediction for all data (2004 and 2007); dotted line, model prediction for 2004; dashed line, model prediction for 2007. After Patot *et al.*, 2010, *Molecular Ecology.*

closed traps baited with split bananas. They were exposed to natural colonization for 15 days. In the four other sites (7, 8, 10, and 12), we collected 25–30 rotten apples on the ground. Traps and apples were then brought back to the laboratory, and after incubation at approximately 23°C, all emerging *L. boulardi* were collected daily. For 13 populations, individuals were kept in 100% ethanol at −80°C before DNA extraction. For the other two populations (sites 11, 12; samples 2007), a number of isofemale lines (one foundress per line) were established and reared under laboratory conditions. Viral presence was tested by PCR assay, using an insect gene as a control of DNA extraction. The results indicate that the prevalence of LbFV varies greatly from one population to the next with a clear correlation with latitude (Fig. 5). Indeed, infection was much higher in the south with a mean of 77.5% of wasps infected, than in the north (30%), or in the most northern populations (3.3% with two locations out of five without evidence of LbFV presence). Furthermore, this pattern of viral prevalence was observed both in 2004 and 2007, although some slight temporal variations were observed.

Several factors may explain this variable pattern of virus distribution and four hypotheses have been tested: (1) the northern parasitoid genotypes are resistant to infection;

(2) virus strains in the north are not able to induce superparasitism, thus precluding them to invade; (3) virus transmission is temperature-sensitive and variations in mean temperature between northern and southern locations explain the pattern; (4) the expected variations in the densities of *L. boulardi* between north (low densities) and south (high densities) explain the pattern.

By a crossing experiment, we have found that northern genotypes are not refractory to infection (hypothesis (1), Martinez *et al.,* in preparation). We have also found that infected lineages always superparasitize more frequently than uninfected lineages within a given population (Patot *et al.,* 2010), indicating that the virus is able to manipulate the behavior in all populations, thus rejecting hypothesis (2). To test hypothesis (3), we have tested the influence of temperature on the vertical transmission of the virus and on the induction of diapause. Three temperatures were chosen (18°C, 20°C, and 26°C) according to the diapause, which is a physiological state of dormancy allowing the parasitoid to survive under low temperatures. In *L. boulardi*, diapause occurs at the end of the larval development. Claret and Carton (1980) showed that *L. boulardi* exhibits 97.4% diapausing larvae at 17.5°C, 52.7% at 20°C and 0% at 25°C. We confirmed that the temperature had a drastic effect on the probability of a parasitoid larva entering diapause. However, there was no effect of temperature on viral transmission, whether diapause occurred or not. Indeed, the frequency of infection after development was 0.98 (49/50) at 18°C (all wasps diapausing), 1.0 (50/50) at 20°C (both diapausing and nondiapausing females), 0.98 (49/50) at 26°C (only non-diapausing). Thus, the vertical transmission of the virus is excellent and appeared independent of temperature or diapause (all P values >0.99). This led us to reject hypothesis (3) (Patot et al., 2010). Finally, to test hypothesis (4), we adapted the model of Gandon et al. (2006) to study the epidemiology of the virus in populations of L. boulardi. In this model, the virus is both vertically and horizontally transmitted and it manipulates the behavior of the parasitoid by increasing its superparasitism tendency. The parameters used for the model were chosen according to our current knowledge of LbFV characteristics. In particular, vertical transmission is known to be very efficient but imperfect, thus vertical transmission was fixed at 0.98. We used this model to study the prevalence in response to variations in the mortality rate of the parasitoid. Variations in parasitoid mortality rates mimic variations in the density of L. boulardi populations and reflect the conditions near the species range border (L. boulardi has a better survival rate in the South of France). The model was also used to predict the number of parasitoid eggs in parasitized Drosophila larva according to variations in the relative density of L. boulardi (relative density is measured as the proportion of all Drosophila larvae that are parasitized by the wasp). The model predicts a low mean number of eggs per host when the density of L. boulardi is low and an increase in this index with an increase in L. boulardi densities, indicating that opportunities for horizontal transmission may be higher in high-density populations than in low-density populations. Accordingly, the viral prevalence at equilibrium is expected to be low in low-density populations and very high in high-density populations (Fig. 6A). The virus is even unable to persist in the population below a threshold of parasitoid density. Below this threshold, opportunities for horizontal transmission are not sufficient to compensate for the imperfect vertical transmission of the virus and the virus disappears from the population. It is noteworthy that for very low mortality rates (very high parasitoid densities) the model predicts a decrease of the viral prevalence. This is due to the effect of egg limitation. In nature, such low levels of mortality are probably unrealistic. Thus, overall, the model predicts a positive correlation between the proportion of parasitized hosts and the prevalence of the virus. We tested the first prediction of the model by dissecting Drosophila larvae from the field and calculating the relative density of L. boulardi in the sample (proportion of Drosophila larvae that are parasitized) and the mean number of parasitoid eggs found within the

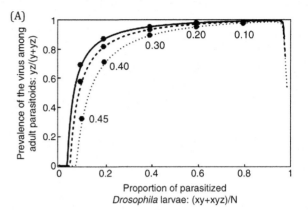

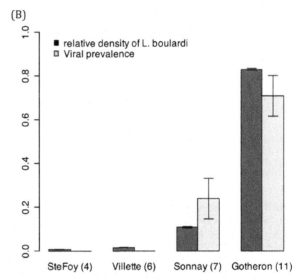

FIGURE 6 **A:** Predictions for the relation between the equilibrium frequency of virus infection in L. boulardi and the rate of parasitism of Drosophila larvae using the epidemiological model presented in Gandon et al. (2006). Variations in the rate of parasitism were obtained by varying the mortality of the parasitoid (m varies between 0.001 and 0.5) and the probability of horizontal transmission: τh takes the values 0.25 (dotted line), 0.5 (dashed line) and 0.75 (solid line). Some values of m are indicated on the figure. Parameter values are: N = 100, d = e = 0.2, Φ = 0.5, a = 0.01, t1 = 0, t2 = 0.1, E = 100, τv = 0.98. **B:** Relative density of L. boulardi and prevalence of LbFV among four populations sampled in 2009. They are ranked according to their latitude from north to south (left to right). After Patot et al., 2010, Molecular Ecology.

parasitized hosts. The plot was qualitatively very similar to the prediction of the model with high occurrence of superparasitim where the relative density of L. boulardi is high (Patot et al., 2010). The second and most important prediction was tested by sampling in 2009 four L. boulardi populations along the north to south axis using a modified version of the previous sampling procedure. In this case, the 12 traps deposited in each location were open traps. This procedure allows the insects to enter the traps, lay their eggs and leave the traps freely, thus providing a reliable estimation of the abundance of each insect species present in the community. In the laboratory, all

Drosophila and parasitoids emerging from traps were collected daily. Based on these counts, we defined an estimation for the density of *L. boulardi* by calculating the ratio between emerging *L. boulardi* and the initial number of potential host flies present in the trap. The number of potential host flies was estimated by summing the number of emerging *Drosophila* and parasitoids since each parasitoid emerged from one *Drosophila* host only. We also studied LbFV prevalence using PCR assays on a sample of *L. boulardi* females. As in 2004 and 2007, we found a much higher prevalence of LbFV in the south compared to the north (Fig. 6B). Furthermore, there was a positive correlation between the density of *L. boulardi* and the prevalence of LbFV as predicted by the model. Taken together, these data suggest that the competition for hosts among *L. boulardi* females is very high in the south with frequent superparasitism and very low in the north with much less superparasitism, which is consistent with previous data (Fleury *et al.*, 2004). Thus, the opportunities for horizontal transmission are much higher in southern than in northern populations. As predicted by the epidemiological model, this pattern of variations in the density of *L. boulardi* results in variations in virus prevalence. This confirms the critical role played by horizontal transmission in the dynamics of the virus within parasitoid populations.

CONSEQUENCES OF THE PRESENCE OF THE VIRUS

As presented in the previous section, LbFV is widespread among *L. boulardi* populations (13 out of 15 were infected) and its prevalence may reach very high levels (up to 95% prevalence). In addition, the prevalence of LbFV appears to be quite stable through time indicating that parasitoids from different populations experience different and quite stable selective pressures. Thus, we can expect that the presence of the virus drives local adaptation in *L. boulardi*.

The presence of the virus induces a dramatic change in the way females manage their egg load and consequently on the distribution of eggs inside hosts. This could have two consequences for the parasitoid. First, because infected females incur a higher risk of being egg-limited, they are selected for increasing investment on this trait, even if there has to be a trade-off with another trait (egg, larval, or adult survival) as expected. Here, two scenarios can be drawn: either the females are able to increase their egg load but only if they are infected (evolution of a plastic response) or they are not able to do so. In this second case, females from heavily infected populations should evolve higher egg loads than females from uninfected populations (evolution of a fixed response). Indeed, we have observed that infected females produce more eggs than uninfected females from the same isofemale line,

which is consistent with the first interpretation (Varaldi *et al.*, 2005b). However, this may also be interpreted as another component of the manipulations induced by the virus because the model shows that the virus would also benefit from an increase in egg load (Gandon *et al.*, 2009). It is known that parasite-induced modification of the host phenotype is often a multidimensional phenomenon. For instance, the presence of the bacterial symbiont Cardinium in the wasp *Encarsia pergandellia* induces two main alterations of the phenotype of the wasp (Zchori-Fein *et al.*, 2001; Kenyon and Hunter, 2007). In this species, females lay fertilized female eggs in hemipteran nymphs (the 'primary hosts'). In contrast, unfertilized eggs are laid in already parasitized hemipteran nymphs (the 'secondary hosts') and develop as male hyperparasitoids (as parasitoids of females). Unmated females, capable of laying only unfertilized male eggs, usually refrain from ovipositing within primary hosts suitable only for female development. Generally, when eggs of the 'wrong' sex are laid in a host, they do not develop. The bacterium *Cardinium* induces thelytoky in some populations of this species (infected females lay all-female offspring without fertilization). As a consequence, these populations are free of males. Together with thelytoky induction, a shift in the egg-laying strategy is observed in infected *E. formosa*: although they are unfertilized, these females choose the appropriate host for female developement and avoid laying eggs in secondary hosts, whereas corresponding controls (unfertilized and uninfected females obtained from antibiotic treatments or from naturally uninfected populations) will choose exclusively this type of host. Here also, because both the symbiont and the parasitoid benefit from this behavioral change, it is not clear whether this is another strategy of the manipulative arsenal of the symbiont or a response of the parasitoid to the presence of the bacteria. In this system, as in the LbFV/wasp system, additional experiments are required to disentangle the two hypotheses.

As mentioned previously, the presence of the virus modifies the distribution of eggs inside *Drosophila* hosts: in a population with no virus, few superparasitisms are expected, whereas in a heavily infected population, superparasitism is expected to be frequent. Note, however, that even within infected populations, both infected and uninfected individuals coexist (this is shown both by theoretical models and by observation). The presence of uninfected individuals is likely due to the imperfect vertical transmission of the virus. Remembering that superparasitism may also have an adaptive value for the parasitoid (see previous sections), we can address the following question: what is the best superparasitism strategy for uninfected females in a population that did not coevolve with the virus (an uninfected population) or in a population that did coevolve with the virus (an infected population)? The model of Gandon *et al.* (2006) has been used to predict the ESS of superparasitism under these two scenarios. As can be seen in Fig. 4, the ESS of superparasitism is lower when the parasitoid has coevolved with the virus (black dotted line versus gray dotted line). This means that uninfected females from an infected population are selected for less superparasitism than in an uninfected population. Indeed, in populations infected with the virus, superparasitism carries an extra cost for the parasitoid: the cost that the progeny could get infected by the virus if the female lays its egg in a host parasitized by an infected larva (we assumed that the female is unable to distinguish between host parasitized by infected or uninfected larvae). This explains why uninfected parasitoids should superparasitize less frequently when they are evolving in an infected population (when the virus is present in the population). As a conclusion, the presence/absence of the virus is expected to drive local adaptation regarding the superparasitism strategy of uninfected females.

Solitary parasitoids have often evolved large mandibles and aggressive behavior toward conspecific parasitoids in their larval stages, as is observed in parasitized *L. boulardi*. The supernumerary larvae are ultimately killed. At the opposite extreme, gregarious parasitoids are typically characterized by nonaggressive larvae that complete development while coexisting peacefully inside the same hosts. Several theoretical investigations have revealed that the conditions for the invasion of a non-aggressive strategy inside an aggressive population are very restricted (Godfray, 1987; Pexton *et al.*, 2003), contrary to the reverse transition. However, phylogeny of parasitoid clutch size suggests that the ancestral state for most parasitoid taxon is solitary, and that frequent transition from solitary to gregarious way of life did occur. Boivin and van Baaren (2000) suggested that the loss of mobility of parasitoid larvae within the host (which reduces the probability of contact with an aggressive larva) may provide a proximal mechanism by which a solitary species may escape the absorbing solitary state and become gregarious (Mayhew *et al.*, 1998; Pexton and Mayhew, 2002; Pexton *et al.*, 2003). Alternatively, we can imagine that the infection by a manipulative parasite or pathogen could facilitate this transition. Indeed, it is adaptive for the virus to allow several infected larvae to emerge because this will enhance horizontal transmission. A virus may be able to colonize all parasitoid larvae within a given host (this is probably true for LbFV) and to manipulate their respective behaviors towards a tolerant behavior (either by inhibiting the development of mandibles, decreasing larval mobility or aggressiveness). This viral mutant would probably invade the parasitoid population very quickly, leading to the development of several parasitoids from the same host due to the manipulation induced by the virus. Subsequently, evolution may lead to an adjustment of the clutch size of females (injection of more than one egg per oviposition), thus leading to the evolution of true gregarious development.

Some intriguing observations suggest a potential implication of microparasites in the control of *Muscidifurax raptorellus* gregariousness (Legner, 1989). Legner reported that the oviposition behavior of females with a low tendency to lay gregarious clutches changed (gregariousness increased) after mating with a male of a strain known to adopt a higher level of gregariousness. The existence of a sexually transmitted virus manipulating the oviposition behavior of infected females could potentially explain these results.

ANY LINK WITH POLYDNAVIRUSES?

Thousands of endoparasitoid species inject polydnavirus particles into their hosts during oviposition. Today, these particles are absolutely necessary for the wasp progeny because they inhibit the immune reaction of the host by causing hemocyte apoptosis or other immunosuppressing mechanisms. Without these particles, parasitoid eggs would be killed due to the formation of a capsule around the egg in the encapsulation and melanization process. The name PDV comes from the fact that the DNA packed within the particles is composed of multiple circular dsDNA molecules. Most of this DNA encodes for virulence genes that do not show similarities with other known viral genomes. Rather, they resemble eukaryotic genes and it is likely that they do not have any viral origin. Particle production is confined to the wasp ovary where viral DNAs are generated from proviral copies maintained within the wasp genome (Espagne *et al.*, 2004). However, the genes involved in the production of the particles are not included within the particles. This led to some difficulties in obtaining their sequences. Nevertheless, these sequences have been obtained for two associations of polydnavirus/wasp (bracovirus and ichnovirus in braconid and ichneumonid wasps, respectively) and a convincing relationship between bracovirus and a sister group of baculoviruses (nudiviruses) has been established (Bézier *et al.*, 2009). Although the genes involved in the production of ichnovirus particles did not reveal any similarities with known viruses (Volkoff *et al.*, 2010), they have most probably an independent viral origin. Possibly, their virus relatives have not yet been sequenced (which is a plausible hypothesis if we consider the poor actual knowledge of the diversity of viruses) or their close relatives became extinct and so only very distant lineages are available. Thus, ichnoviruses and bracoviruses represent two independent events of domestication of a virus to deliver virulence genes into their hosts.

What does the LbFV/*L. boulardi* system tell us regarding these examples of convergent evolution? By definition, when studying ancient symbioses such as the PDV/wasps associations, it is very difficult to understand the first step in the establishment of the association because of the evolution of both the host and the symbiont. Indeed the integration of the ancestral nudivirus that gave rise to the bracoviruses found today in more than 17,000 species occurred around 103 million years ago (Murphy *et al.*, 2008; Bézier *et al.*, 2009).

One question is how was the ancestor virus maintained within wasp populations before the mutualistic association took place? It has been suggested that the ancestor virus might have been a sexually transmitted virus with a tropism for female gonads where bracovirus particles are produced (Bézier *et al.*, 2009). This is the case for the related nudivirus Hz-2V, and LbFV is also found in female reproductive tracts. Furthermore, it is interesting to note that both Hz-2V and LbFV are able to manipulate the behavior of the wasp. In effect, Hz-2V increases the normal pheromone production of females thus increasing their attractivity and facilitating the propagation of virions through male subsequent matings (Burand, 2004; Burand *et al.*, 2005). One can ask whether parasitoid behavioral manipulation has predated the evolution of a mutualistic interaction. To test this idea, it will be exciting to compare the genomic sequences of Hz-2V and LbFV to the viral genes of polydnaviruses when they become available.

CONCLUSION

A fundamental feature of viruses is their ability to colonize new cells or individuals through horizontal transmission. It is highly probable that most viruses found in the reproductive tract of parasitoids have the capacity to be horizontally transmitted. Some virus lineages may have lost this capacity, like polydnaviruses, but horizontal transmission might have played a crucial role during their evolution. The results obtained on the LbFV/*L. boulardi* interaction indicate that horizontal transmission between parasitoid lineages is permitted by the occurrence of superparasitism. Subsequently, natural selection has favored the emergence of a parasitic strategy manipulating the occurrence of superparasitism behavior. Superparasitism is very frequently observed among parasitoid species, even in gregarious parasitoids (Dorn and Beckage, 2007). Theoretically, all infectious viruses could benefit from an increase in the superparasitism tendency of their parasitoid host. Thus we can expect that future studies will document similar phenomena in other parasitoid/virus interactions.

REFERENCES

Barratt, B., Evans, A., Stoltz, D., Vinson, S., Easingwood, R., 1999. Virus-like particles in the ovaries of *Microctonus aethiopoides* loan (Hymenoptera: braconidae), a parasitoid of adult weevils (Coleoptera: curculionidae). J. Invertebr. Pathol. 73, 182–188.

Barratt, B.I.P., Murney, R., Easingwood, R., Ward, V.K., 2006. Virus-like particles in the ovaries of Microctonus aethiopoides Loan (Hymenoptera: Braconidae): comparison of biotypes from Morocco and Europe. J. Invertebr. Pathol. 91, 13–18.

Bézier, A., Annaheim, M., Herbiniere, J., Wetterwald, C., Gyapay, G., Bernard-Samain, S., et al. 2009. Polydnaviruses of braconid wasps derive from an ancestral nudivirus. Science 323, 926–930.

Bézier, A., Herbinière, J., Lanzrein, B., Drezen, J.-M., 2009. Polydnavirus hidden face: the genes producing virus particles of parasitic wasps. J. Invertebr. Pathol. 101, 194–203.

Boivin, G., van Baaren, J., 2000. The role of larval aggression and mobility in the transition between solitary and gregarious development in parasitoid wasps. Ecol. Lett. 3, 469–474.

Burand, J., 2004. Horizontal transmission of Hz-2V by virus infected *Helicoverpa zea* moths. J. Invertebr. Pathol. 85, 128–131.

Burand, J.P., Tan, W., Kim, W., Nojima, S., Roelofs, W., 2005. Infection with the insect virus Hz-2v alters mating behavior and pheromone production in female Helicoverpa zea moths. J. Insect Sci. 5, 6.

Claret, J, Carton, Y., 1980. Diapause in a tropical species, *Cothonaspis boulardi* (parasitic Hymenoptera). Oecologia 45, 32–34.

Dorn, S., Beckage, N., 2007. Superparasitism in gregarious hymenopteran parasitoids: ecological, behavioural and physiological perspectives. Physiol. Entomol. 32, 199–211.

Espagne, E., Dupuy, C., Huguet, E., Cattolico, L., Provost, B., Martins, N., et al. 2004. Genome sequence of a polydnavirus: insights into symbiotic virus evolution. Science 306, 286–289.

Fleury, F., Ris, N., Allemand, R., Fouillet, P., Carton, Y., Boulétreau, M., 2004. Ecological and genetic interactions in *Drosophila*–parasitoids communities: a case study with *D. melanogaster*, *D. simulans* and their common *Leptopilina* parasitoids in south-eastern France. Genetica 120, 181–194.

Gandon, S., 2005. Parasitic manipulation: a theoretical framework may help. Behav. Process. 68, 247–248.

Gandon, S., Rivero, A., Varaldi, J., 2006. Superparasitism evolution: Adaptation or manipulation? Am. Nat. 167, E1–E22.

Gandon, S., Varaldi, J., Fleury, F., Rivero, A., 2009. Evolution and manipulation of parasitoid egg load. Evolution 63, 2974–2984.

Godfray, H.C.J., 1994. Parasitoids: Behavioral and Evolutionary Ecology. Princeton University Press, Princeton.

Godfray, H.C.J., 1987. The evolution of clutch size in parasitic wasps. American Naturalist 129, 221–233.

Hashimoto, Y., Lawrence, P.O., 2005. Comparative analysis of selected genes from *Diachasmimorpha longicaudata* entomopoxvirus and other poxviruses. J. Insect Physiol. 51, 207–220.

Kenyon, S.G., Hunter, M.S., 2007. Manipulation of oviposition choice of the parasitoid wasp, *Encarsia pergandiella*, by the endosymbiotic bacterium *Cardinium*. J. Evol. Biol. 20, 707–716.

Lawrence, P.O., Matos, L.F., 2005. Transmission of the *Diachasmimorpha longicaudata* rhabdovirus (DIRhV) to wasp offspring: an ultrastructural analysis. J. Insect Physiol. 51, 235–241.

Legner, E.F., 1989. Paternal influences in males of *Muscidifurax raptorellus* (Hymenoptera: Pteromalidae). Entomophaga 34, 307–320.

Lietze, V.-U., Abd-Alla, A.M.M., Vreysen, M.J.B., Geden, C.J., Boucias, D.G., 2010. Salivary gland hypertrophy viruses: a novel group of insect pathogenic viruses. Annu. Rev. Entomol. 56, 63–80.

Luo, L., Zeng, L., 2010. A new rod-shaped virus from parasitic wasp *Diachasmimorpha longicaudata* (Hymenoptera: Braconidae). J. Invertebr. Pathol. 103, 165–169.

Mayhew, P.J., Ode, P.J., Hardy, I.C.W., Rosenheim, J.A., 1998. Parasitoid clutch size and irreversible evolution. Ecol. Lett. 1, 139–141.

Murphy, N., Banks, J.C., Whitfield, J.B., Austin, A.D., 2008. Phylogeny of the parasitic microgastroid subfamilies (Hymenoptera: Braconidae) based on sequence data from seven genes, with an improved time estimate of the origin of the lineage. Mol. Phylogenet. Evol. 47, 378–395.

Oliveira, D.C.S.G., Hunter, W.B., Ng, J., Desjardins, C.A., Dang, P.M., Werren, J.H., 2010. Data mining cDNAs reveals three new single stranded RNA viruses in *Nasonia* (Hymenoptera: Pteromalidae). Insect Mol. Biol. 19, 99–107.

O'Neill, S.L., Hoffmann, A.A., Werren, J.H., 1997. Influential Passengers: Inherited Microorganisms and Arthropod Reproduction. Oxford University Press, Oxford.

Patot, S., Lepetit, D., Charif, D., Varaldi, J., Fleury, F., 2009. Molecular detection, penetrance and transmission of an inherited virus responsible for a behavioral manipulation of an insect parasitoid. Appl. Environ. Microbiol. 75, 703–710.

Patot, S., Martinez, J., Allemand, R., Gandon, S., Varaldi, J., Fleury, F., 2010. Prevalence of a virus inducing behavioural manipulation near species range border. Mol. Ecol. 19, 2995–3007.

Pexton, J.J., Mayhew, P.J., 2002. Siblicide and life–history evolution in parasitoids. Behav. Ecol. 13, 690–695.

Pexton, J.J., Rankin, D.J., Dytham, C., Mayhew, P.J., 2003. Asymmetric larval mobility and the evolutionary transition from siblicide to non-siblicidal behavior in parasitoid wasps. Behav. Ecol. 14, 182–193.

Poirié, M., Carton, Y., Dubuffet, A., 2009. Virulence strategies in parasitoid Hymenoptera as an example of adaptive diversity. C. R. Biol. 332, 311–320.

Reineke, A., Asgari, S., 2005. Presence of a novel small RNA-containing virus in a laboratory culture of the endoparasitic wasp *Venturia canescens* (Hymenoptera: Ichneumonidae). J. Insect Physiol. 51, 127–135.

Renault, S., Bigot, S., Lemesle, M., Sizaret, P.-Y., Bigot, Y., 2003. The cypovirus *Diadromus pulchellus* RV-2 is sporadically associated with the endoparasitoid wasp *D. pulchellus* and modulates the defence mechanisms of pupae of the parasitized leek-moth, *Acrolepiopsis assectella*. J. Gen. Virol. 84, 1799–1807.

Renault, S., Stasiak, K., Federici, B., Bigot, Y., 2005. Commensal and mutualistic relationships of reoviruses with their parasitoid wasp hosts. J. Insect Physiol. 51, 137–148.

Reynolds, K.T., Hardy, I.C.W., 2004. Superparasitism: a non-adaptive strategy? Trends Ecol. Evol. (Amst.) 19, 347–348.

Stasiak, K., Renault, S., Federici, B.A., Bigot, Y., 2005. Characteristics of pathogenic and mutualistic relationships of ascoviruses in field populations of parasitoid wasps. J. Insect Physiol. 51, 103–115.

Stoltz, D.B., Krell, P., Cook, D., MacKinnon, E.A., Lucarotti, C.J., 1988. An unusual virus from the parasitic wasp *Cotesia melanoscela*. Virology 162, 311–320.

Stoltz, D., Makkay, A., 2000. Co-replication of a reovirus and a polydnavirus in the ichneumonid parasitoid *Hyposoter exiguae*. Virology 278, 266–275.

Stoltz, D., Vinson, S.B., 1977. Baculovirus-like particles in the reproductive tracts of female parasitoid wasps II: the genus Apanteles. Can. J. Microbiol. 23, 28–37.

van Alphen, J.J., Visser, M.E., 1990. Superparasitism as an adaptive strategy for insect parasitoids. Annu. Rev. Entomol. 35, 59–79.

Varaldi, J., Boulétreau, M., Fleury, F., 2005. Cost induced by viral particles manipulating superparasitism behaviour in the parasitoid *Leptopilina boulardi*. Parasitology 131, 161–168.

Varaldi, J., Fouillet, P., Boulétreau, M., Fleury, F., 2005. Superparasitism acceptance and patch-leaving mechanisms in parasitoids: a comparison between two sympatric wasps. Anim. Behav. 69, 1227–1234.

Varaldi, J., Fouillet, P., Ravallec, M., Lopez-Ferber, M., Boulétreau, M., Fleury, F., 2003. Infectious behavior in a parasitoid. Science 302, 1930.

Varaldi, J., Patot, S., Nardin, M., Gandon, S., 2009. A virus-shaping reproductive strategy in a *Drosophila* parasitoid. Adv. Parasitol. 70, 333–363.

Varaldi, J., Petit, S., Boulétreau, M., Fleury, F., 2006. The virus infecting the parasitoid *Leptopilina boulardi* exerts a specific action on superparasitism behaviour. Parasitology 132, 747–756.

Varaldi, J., Ravallec, M., Labrosse, C., Lopez-Ferber, M., Boulétreau, M., Fleury, F., 2006. Artifical transfer and morphological description of virus particles associated with superparasitism behaviour in a parasitoid wasp. J. Insect Physiol. 52, 1202–1212.

Volkoff, A.-N., Jouan, V., Urbach, S., Samain, S., Bergoin, M., Wincker, P., et al. 2010. Analysis of virion structural components reveals vestiges of the ancestral ichnovirus genome. PLoS Pathog. 6, e1000923.

Zchori-Fein, E., Gottlieb, Y., Kelly, S.E., Brown, J.K., Wilson, J.M., Karr, T.L., et al. 2001. A newly discovered bacterium associated with parthenogenesis and a change in host selection behavior in parasitoid wasps. Proc. Natl. Acad. Sci. U.S.A. 98, 12555–12560.

Venoms of Parasitoids

Chapter 18

Venoms from Endoparasitoids

Sassan Asgari

School of Biological Sciences, The University of Queensland, St Lucia, QLD 4072, Australia

SUMMARY

Successful development of endoparasitoids inside the host is dependent on a variety of maternal factors that accompany the egg upon parasitization. These include venom, ovarian proteins, and viruses/virus-like particles that are involved in the regulation of host physiology. Endoparasitoid venoms are rich sources of biomolecules that consist mainly of peptides and proteins, which play key roles in venom metabolism and host–parasitoid interactions. Suppression of the host immune system, either independently or in association with mutualistic viruses, and developmental alterations are the main functions of venom from endoparasitoids. In this chapter, the functional role of venom in host regulation will be discussed with a focus on endoparasitoids that produce polydnaviruses.

ABBREVIATIONS

EPV	entomopoxvirus
γ-GT	γ-glutamyl transpeptidase
PAP	proPO activating proteinase
PDV	polydnavirus
PO	phenoloxidase
proPO	prophenoloxidase
PTG	prothoracic gland
SPH	serine proteinase homolog
THC	total hemocyte count
VLPs	virus-like particles

INTRODUCTION

Endoparasitism commonly occurs in Hymenoptera, although instances are found in other insect orders, such as Strepsiptera and Diptera (Godfray, 1994; Kathirithamby *et al.*, 2003). However, venom has only been reported from hymenopteran endoparasitoids. Transition from ectoparasitism to endoparasitism, which generally overlaps with idiobiontic versus koinobiontic life strategies, respectively, required modification of venom components from being mainly paralytic to host regulatory factors. In endoparasitoids that are phylogenetically basal and closer to ectoparasitoids, venom seems to have more potent paralytic effects on the host. Therefore, induction of paralysis by venom is considered to be an ancestral state, whereas venom serving diverse host regulatory functions is regarded as more derived (Whitfield, 1998).

In many host–parasitoid systems, venom seems to be essential for successful parasitism. For example, when venom glands were removed from *Cardiochiles nigriceps* (Hymenoptera: Braconidae), no parasitoids emerged from host larvae, *Heliothis virescens* (Lepidoptera: Noctuidae) (Tanaka and Vinson, 1991). Dissections of larvae parasitized by glandectomized wasps showed that emerged parasitoid larvae were all encapsulated by the host's hemocytes in the absence of venom. In *Cotesia glomerata* (Hymenoptera: Braconidae) (Kitano, 1986; Wago and Tanaka, 1989) and *C. kariyai* (Tanaka, 1987), both venom and PDVs are required for successful parasitism. Endoparasitoid venom components may have direct effects on alterations observed in the host physiology (e.g., immune suppression) and development, or may simply synergize the effects of associated virus-like particles, such as polydnaviruses (PDVs), found in some braconid and ichneumonid wasps. In endoparasitoids devoid of any virus-like particles, venom seems to play a bigger role in host regulation. In PDV-associated endoparasitoids, except in a few reports in which specific effects of venom have been studied, analysis of the direct role of venom on the effects seen following parasitization is unfeasible.

Research on the hymenopteran venoms has primarily focused on social wasps and bees due to their clinical relevance to human and animal health (Monteiro *et al.*, 2009). Nevertheless, the limited available information on the endoparasitoid venom compositions provides evidence for the complex nature of venom components. Most of the peptides and proteins characterized, especially those from PDV-associated endoparasitoids, exhibit similarities to insect proteins providing evidence for their recruitment and modifications throughout evolution into noxious components used for host manipulation. In this chapter, known venom constituents from endoparasitoids and their function in host–parasitoid interactions will be discussed.

Parasitoid Viruses: Symbionts and Pathogens. DOI: 10.1016/B978-0-12-384858-1.00018-7

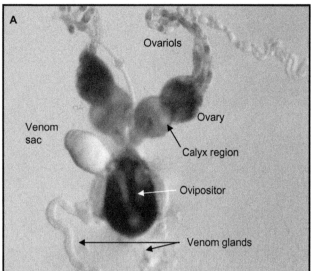 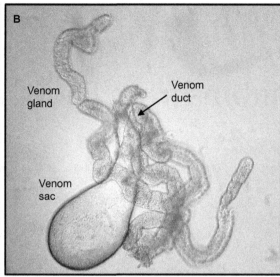

FIGURE 1   A representative image of the type 2 venom apparatus from the endoparasitoid, *Cotesia rubecula* (Hymenoptera: Braconidae).
A: Female reproductive organ including venom glands and venom sac. B: Venom apparatus. Please see color plate section at the back of the book.

VENOM APPARATUS

The anatomies of venom glands are very diverse among parasitoids; however, there seems to be some degree of correlation between the life strategy of parasitoids and the overall anatomy of their glands (Edson et al., 1982; Edson and Vinson, 1979). Examining the venom apparatus from 160 braconid species, Edson and Vinson categorized them into two main types (Edson and Vinson, 1979). The type 1 venom apparatus is mainly found in ectoparasitoids or endoparasitoids that pupate inside the host. In this type, the venom sac is cone-shaped with a relatively thick wall lined by longitudinal muscles. The type 2 venom apparatus is only found in endoparasitoids, in which pupation takes place outside the host (Fig. 1). In addition, the type 2 venom sac is translucent and has a small number of muscles. All the braconid PDV-associated wasps seem to have a type 2 venom apparatus. Such information for ichneumonid wasps is unavailable.

The secretory units in venom glands in Hymenoptera, which appear to be conserved regardless of their life strategies (Edson et al., 1982), constitute a supracellular system of canals in the gland (Ferrarese et al., 2009). Venom secretory units are of ectodermal origin and belong to class 3 cells according to the classification of insect gland cells proposed by Noirot and Quennedey (1974). Each secretory unit consists of a secretory cell, with the plasma membrane organized into brush border invaginations to increase cellular exchange, a canal which leads to the gland's lumen, and a cell that lines the lumen in the distal part of the canal forming the intimal layer (Ferrarese et al., 2009).

Regarding the development of the venom apparatus, there is not much information available. The only reports

are from wasps of the genus *Chelonus* (Hymenoptera: Braconidae), in which the venom apparatus appears about half way through the pupal stage. Venom proteins can be detected at the very late stages of pupation, almost close to adult eclosion (Kaeslin et al., 2010). In earlier observations, Jones and Wozniak (1991) reported that within 24 h after eclosion, venom proteins had reached their final and maximal titer in the venom reservoir.

WHAT IS THE COMPOSITION OF VENOM IN ENDOPARASITOIDS?

To date, only proteinaceous components have been reported from endoparasitoid venoms, including peptides, proteins, and amines. Low-molecular-weight proteins are less common in endoparasitoid wasps, but not totally absent as in studies in which more sensitive proteomics approaches have been used, low-molecular-weight peptides have been detected (Asgari et al., 2003a; Uçkan et al., 2004; Zhang et al., 2004b). Venom proteins range from basic to acidic and most of them are glycosylated (Kaeslin et al., 2010; Leluk et al., 1989). Persistence of venom proteins after injection into the host is variable ranging from a few hours to days. For example, *Microplitis demolitor* (Hymenoptera: Braconidae) venom was shown to persist from 6–12 h after parasitization (Strand et al., 1994). In *Cotesia rubecula* (Hymenoptera: Braconidae), a venom protein (Vn50) was detected up to 72 h post parasitization using western blotting (Zhang et al., 2004a). Persistence of *Chelonus inanitus* venom proteins in parasitized *Spodoptera littoralis* (Lepidoptera: Noctuidae) eggs is egg-stage dependent (one to three days). Accordingly, the earlier the host eggs are parasitized, the longer the venom proteins persist (Kaeslin et al., 2010). Compared to braconid wasps, the constituents and the role of venom from ichneumonid wasps associated with PDVs are less well characterized.

Commonly, the amount of venom used in experimentations is expressed as venom gland equivalents. The amount of venom proteins in the venom reservoir varies among parasitoids. For example, in *Pimpla turionellae* (Hymenoptera: Ichneumonidae), the total amount of protein was determined to be 0.04 μg per venom sac (Uçkan et al., 2004), whereas in a closely related species, *P. hypochondriaca*, it was about 180 μg (Parkinson and Weaver, 1999). In a study that focused on *P. turionellae*, it was shown that the concentration of venom proteins increases as the parasitoid ages and significantly declines following each parasitization incidence (Uçkan et al., 2006). It was also shown that the titer of the main venom protein of a *Chelonus* sp. wasp progressively decreased during *ad libitum* stinging of the host (Jones and Wozniak, 1991). Peptides and proteins characterized in endoparasitoid venoms include enzyme inhibitors, enzymes, and others with variable functions.

ENZYME INHIBITORS

Protease inhibitors in the venom sac may serve two functions: (1) while stored in the reservoir, they inhibit activation of enzymes present in the venom fluid; and (2) when injected into the host, they inhibit host enzymes that play roles in immune responses. Both types of enzyme inhibitor have been found in endoparasitoid wasps.

In *P. hypochondriaca* venom, two cystein-rich proteins that each contain one trypsin inhibitory-like domain, cvp1 and cvp6, were found. The protein cvp1 physically binds to venom phenoloxidases (POs) present in the venom fluid, while in the storage sac, to avoid the enzymes' detrimental effects on the wasp tissues (Parkinson and Weaver, 1999; Parkinson et al., 2004). When venom is injected into the host, dilution of venom in the host hemolymph apparently neutralizes the effect of the inhibitor (Parkinson and Weaver, 1999). Two other venom proteins, cvp2 and cvp4, with similarities to different protease inhibitor families were also identified (Parkinson et al., 2004).

In the ISy strain of *Leptopilina boulardi* (Hymenoptera: Figitidae), a serine protease inhibitor (LbSPN$_y$) in the venom, which belongs to the serpin superfamily, inhibits the PO cascade in the hemolymph of the host, *Drosophila yakuba* (Colinet et al., 2009). However, the host enzyme targeted by this serpin has not been identified. A 50-kDa protein from *C. rubecula* venom with similarity to non-catalytic serine protease homologs also inhibits melanization by preventing activation of prophenoloxidase (proPO) and its conversion to active PO (see 'Venom as an Immune Regulatory Factor' below) (Asgari et al., 2003b).

ENZYMES

Enzymes constitute the most abundant components of venoms from endoparasitoids. Antigenic similarity of braconid venom proteins with those of higher hymenopterans, including non-parasitic wasps, has been recorded (Leluk et al., 1989), which could be due to similarities among the enzymes present in the venoms. The enzymes reported thus far belong to oxidoreductases (e.g., POs), hydrolases and transferases (e.g., γ-glutamyl transpeptidase (γ-GT)).

In endoparasitic wasps, POs have only been reported from *P. hypochondriaca* venom (Parkinson and Weaver, 1999; Parkinson et al., 2001). The enzymes seem to exist in an active form in the venom reservoir; however, their activity is suppressed by inhibitors present in the venom sac to avoid detrimental effects of the enzyme while stored in the reservoir. The functional role of *P. hypochondriaca* POs is not well-understood but it seems they cause apoptotic-type cell death in host cells (Rivers et al., 2009). PO is a key enzyme in a proteolytic cascade activated upon immune induction leading to melanization (Cerenius et al., 2008; Nappi and Christensen, 2005). Melanization is involved in wound healing, immune recognition, and cellular and humoral encapsulation. Therefore, the injection of POs into the host's hemolymph by the parasitoid seems paradoxical. Two proteins with PO activity have also been identified in the venom from an ectoparasitoid, *Nasonia vitripennis* (Hymenoptera: Pteromalidae) (Abt and Rivers, 2007). In this species, the venom enzyme could be critical in the intoxication pathway leading to the death of host cells. In addition to POs, a laccase cDNA was also found in *P. hypochondriaca* venom library (Parkinson et al., 2003). In insects, these oxidoreductases are involved in cuticle hardening and tanning.

The only venoms that have chitinase and reprolysin (zinc-dependent metalloproteinases) are from the egg–parasitoid *Chelonus* near *curvimaculatus* (Hymenoptera: Braconidae) (Krishnan et al., 1994) and *P. hypochondriaca* (Parkinson et al., 2002a), but their functional role in parasitism is not known. A 52-kDa chitinase is also secreted from teratocytes of the larval endoparasitoid *Toxoneuron nigriceps* (Viereck) (Hymenoptera: Braconidae) (Cônsoli et al., 2007).

Teratocytes are extraembryonic cells derived from serosal membrane that are released into the host hemolymph following hatching of endoparasitoid eggs. They enlarge several-fold in size and have been shown to play various roles in host–parasitoid interactions, including parasitoid nutrition (Dahlman and Vinson, 1993), developmental arrest (Dahlman et al., 2003), changes in hemolymph protein content (Zhang et al., 1997), as well as immunosuppressive properties (Hoy and Dahlman, 2002). This protein shows the highest structural similarity to the *C. curvimaculatus* chitinase (54% identity, 70% similarity) (Cônsoli *et al.*, 2007). Similarly, the exact role of this protein in parasitism is not understood, but it was speculated that it might be involved in the suppression of the host immune system or facilitate emergence of the parasitoid larva from its host larval cuticle by digesting the chitin present in the exocuticle.

In a comprehensive biochemical analysis, six hydrolases were detected in the venom of *P. hypochondriaca* (Dani *et al.*, 2005). These include acid phosphatase, β-glucosidase, esterase, β-galactosidase, esterase lipase, and lipase. The role of these enzymes in parasitism or venom metabolism is not known. In addition, a trehalase of 61 kDa was also found secreted by the venom glands of *P. hypochondriaca* (Parkinson *et al.*, 2003). Expression of acid (Zhu *et al.*, 2008) and alkaline (Zhu *et al.*, 2010b) phosphatases and their storage in the venom reservoir with measurable enzyme activities has been reported from *Pteromalus puparum* L. (Hymenoptera: Pteromalidae). Likewise, their role in host–parasitoid interactions or venom metabolism has not been explored. Acid phosphatase has also been isolated from honey bee venom that is an allergen (Hoffman *et al.*, 2005).

A 94-kDa heterotetrameric protein from *Asobara tabida* (Hymenoptera: Braconidae) venom shows sequence homology to several aspartylglucosaminidases (Moreau *et al.*, 2004; Vinchon *et al.*, 2010). The enzyme activity was optimal in acidic conditions (Vinchon *et al.*, 2010). A paralytic effect on the host, *Drosophila melanogaster*, was suggested for the protein since aspartylglucosaminidase activity, in general, leads to the production of aspartate, which was previously identified as an excitory neurotransmitter in *Drosophila*. Therefore, the 94-kDa protein may be responsible for the transient paralytic effect observed in host larvae (Moreau *et al.*, 2002).

A γ-GT is present in *Aphidius ervi* (Hymenoptera: Braconidae) venom (Falabella *et al.*, 2007). γ-GTs are enzymes that play key roles in amino acid trafficking, homeostasis, and protecting cells from oxidative damage (H. Zhang *et al.*, 2005). *Aphidius ervi* γ-GT induces apoptosis in the pea aphid ovary resulting in an early castration. This effect was initially observed after injections of crude venom or chromatographic fractions enriched with this abundant protein (Digilio *et al.*, 2000). The mechanism of induction of apoptosis is not clearly understood but the mode of action is possibly by interfering with the balance of glutathione which causes oxidative stress in ovarian cells triggering apoptosis (Falabella *et al.*, 2007) (also see Chapter 20 of this volume).

OTHER VENOM PROTEINS/PEPTIDES

Apart from enzymes and enzyme inhibitors, other peptides/proteins are present in the venom that may be involved in immune suppression or facilitating the effects of PDVs. For example, a small peptide consisting of 14 amino acids (Vn1.5) from *C. rubecula* venom is essential for expression of *C. rubecula* bracovirus (CrBV) genes (Zhang *et al.*, 2004b). In the absence of venom or the purified peptide, CrBV particles enter host cells, but PDV genes are not expressed. It is not yet known how the peptide facilitates PDV gene expression, but it may assist in the uncoating of the particles in the cytoplasm or act at the nuclear membrane.

Other proteins reported so far include, a serine proteinase homolog (Vn50) from *C. rubecula* which inhibits melanization (Asgari *et al.*, 2003b), calreticulins from *C. rubecula* (Zhang *et al.*, 2006), *Microctonus* spp. (Crawford *et al.*, 2008) and *P. puparum* (Zhu *et al.*, 2010a), a RhoGAP protein from *Leptopilina boulardi* affecting cell adhesion and motility (Labrosse *et al.*, 2005b), and tetraspanin and ferritin from *Microctonus* spp. (Crawford *et al.*, 2008). The function of these proteins will be discussed in the following sections.

WHAT ROLES DOES VENOM PLAY IN HOST–ENDOPARASITOID INTERACTIONS?

Transient Paralysis and Cytotoxicity

In contrast to the majority of ectoparasitoids, which permanently paralyze their hosts (with some exceptions; e.g., Coudron *et al.*, 1990), endoparasitoids usually do not paralyze their host, and if they do, paralysis is only transient. As a result, paralytic factors (e.g., neurotoxins) have been lost or their function changed in most endoparasitoids, especially in those associated with PDVs. For example, a 4.6-kDa peptide (Vn4.6) from *C. rubecula* venom shows significant similarities to a very potent ω-atracotoxin from the Australian funnel-web spider (*Hadronyche versuta*), but a paralytic effect on the host is not evident (Asgari *et al.*, 2003a). This could be reminiscent of neurotoxins present in ancestral idiobionts. Transient paralysis of the host, however, is often found in parasitism by endoparasitoids devoid of PDVs (e.g. *A. tabida* (Moreau *et al.*, 2002), *P. hypochondriaca* (Parkinson and Weaver, 1999), *P. turionellae* (Ergin *et al.*, 2006), *Binodoxys communis* and *B. koreanus* (Desneux *et al.*, 2009); also see Chapter 21 of this volume)). In *B. communis*, it was shown that the temporary paralysis may be used by the parasitoid as a means to avoid self-superparasitism (Desneux *et al.*, 2009). Contrasted results were obtained upon injections of venom secretions from *Asobara japonica* into larvae of host and nonhost *Drosophila* species. Mabiala-Moundoungou *et al.* (2009) found that venom induced permanent paralysis followed by death of *D. melanogaster* larvae. Remarkably, these effects could be reversed by injection of ovarian extracts from female wasps. Similar experiments performed by Furihata and Kimura (2009) on other host species (*D. simulans*, *D. auraria* and *D. lutescens*) confirmed the toxicity of venom injected alone and the ability of additional factors provided by the parasitoid females to rescue the deleterious effects. However, the venom-induced paralysis was only found transiently in these species, most of the larvae recovering usually within 1 h. In addition, the venom of *A. japonica* had little or even no effect on two non-host *Drosophila* species, which suggests that the venom and/or the ovarian compensatory factor(s) could be adapted to a restricted host range. Venom from *Chelonus inanitus*, which is an egg–larval parasitoid associated with PDVs, causes transient paralysis in the host, *Spodoptera littoralis* (Kaeslin *et al.*, 2010). This effect is dose dependent.

Cell death and degradation of host tissues and/or hemocytes is often observed following parasitization, which could be targeted specifically towards a particular cell type or tissue resulting in developmental alteration, castration or suppression of cellular immunity. For example, degeneration of prothoracic gland (PTG) in hosts parasitized by *Cotesia kariyai* (Tanaka *et al.*, 1987), *C. congregata* (Kelly *et al.*, 1998) and *Cardiochiles nigriceps* (Pennacchio *et al.*, 1998a) has been reported. However, the role of venom, in particular, in these degradations has not been investigated since in all these cases venom and PDVs were injected together. The outcome of PTG degeneration is developmental arrest which could be due to the inability of the gland to produce prothoracicotropic hormone and/or insensitivity of target tissues to the hormone (Kelly *et al.*, 1998).

Another target of PDV/venom is the host reproductive tissues. For instance, *Cotesia plutellae* (Hymenoptera: Braconidae) and *Diadegma semiclausum* (Hymenoptera: Ichneumonidae) cause degeneration of larval testes in *Plutella xylostella* (Bai *et al.*, 2009), but it has not been shown whether it can be caused by venom alone. Also, venom from *A. ervi*, an endoparasitoid without PDVs, causes castration of the host *Acyrthosiphon pisum* (Homoptera: Aphididae) by specifically inducing apoptosis in the ovaries (Digilio *et al.*, 2000; Falabella *et al.*, 2007) (also see 'Host Castration' below). Cytotoxicity has also been reported from another nonPDV wasp venom, *P. hypochondriaca*. Venom induces apoptosis both in *Lacanobia oleracea* (Lepidoptera: Noctuidae) host hemocytes (Richards and Parkinson, 2000), and in *Spodoptera frugiperda* Sf9 cells (Parkinson and Weaver, 1999) and in *Trichoplusia ni* (Rivers *et al.*, 2009) cell lines. The exact molecule(s) that triggers apoptosis has not been identified, but it appears to be independent of active venom PO (Richards and Parkinson, 2000; Parkinson *et al.*, 2001) and could be caused by the calreticulin present in the venom (Rivers *et al.*, 2009). Cell death also occurs in hemocytes from *Galleria mellonella* following natural parasitization or injection of *P. turionellae* venom (Er *et al.*, 2010). Venom from this parasitoid is implicated in

paralysis and cytotoxicity in dipteran and lepidopteran hosts (Ergin *et al.*, 2006).

Venom as Polydnavirus Partner

The degree of dependency of PDVs on venom is variable in different host–parasitoid systems. In the ichneumonid systems studied, venom does not seem to be required for PDV functions (Davies *et al.*, 1987; Dover *et al.*, 1987; Guzo and Stoltz, 1987; Webb and Luckhart, 1994). However, in braconid systems, we find a range of interactions from total dependency of PDV on venom to complete independency. For example, in several studies, expression of *Microplitis demolitor* bracovirus genes in the absence of venom has been reported (e.g., PTP (Pruijssers and Strand, 2007), glc1.8 and egf1.0 genes (Beck and Strand, 2003)), whereas in others, this may not be the case (see below).

Considering that venom in many braconid host–parasitoid systems synergizes the effects of PDVs, this effect might mainly be due to facilitating the entry of PDVs or uncoating of the particles in the host cells. Alternatively, venom proteins may interact with PDV-expressed proteins and enhance their activity, although there is no study that supports this. In *Cotesia melanoscela*, venom is required for persistence of PDV DNA in the host (Stoltz *et al.*, 1988). In addition, in the absence of venom, uptake and uncoating of PDVs does not occur in host cells, suggesting that venom is essential for those events. Venom from *C. inanitus* increased the cell membrane permeability of *S. littoralis* hemocytes, measured by uptake of Dapi stain, but calyx fluid had no effect (Kaeslin *et al.*, 2010). By increasing the permeability of host cells, uptake of PDVs may be facilitated. In *C. rubecula*, venom is not essential for the entry of PDVs into host cells, but in its absence viral genes are not expressed (Zhang *et al.*, 2004b). A 1546-Da peptide, either purified from venom or chemically synthesized, was found to facilitate expression of *Cotesia rubecula* bracoviruses (CrBVs). The mechanism by which the peptide facilitates expression of CrBV genes is not known, but it could be involved in uncoating of the particles at the nuclear membrane to allow insertion of DNA into the nucleus and subsequently expression of CrBV genes.

Venom as an Immune Regulatory Factor

Since endoparasitoids live in the host hemolymph and consequently are exposed to the host immune responses, they require a suppression of those responses in order to successfully complete their parasitic stage. Evolutionary pressures have led to the evolution of a variety of strategies to avoid/suppress various aspects of the host immunity, both at the cellular and non-cellular (humoral) levels. These strategies can be broadly grouped into passive and active mechanisms. Passive mechanisms include attacking

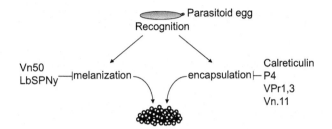

FIGURE 2 The main immune responses against parasitoid eggs are encapsulation and melanization. In the majority of host–parasitoid systems, maternal products (such as polydnaviruses, venom or calyx proteins) inhibit melanization and interfere with the encapsulation response either by affecting cell spreading and aggregation behavior or by affecting the cell numbers. Known venom proteins that block melanization or encapsulation are shown in the figure.

eggs or young stages of the host that have less developed immune systems, deposition of the egg inside host tissues to avoid recognition, and molecular disguise in which parasitoid eggs are covered by components that do not elicit an immune response. On the other hand, active mechanisms lead to the suppression of the host humoral and/ or cellular immune responses. In these instances, venom, viruses, or virus-like particles introduced by the female wasp at oviposition interact with host factors leading to inactivation of host responses. However, often, combinations of passive and active mechanisms are employed by individual parasitoids. Venom may act alone or in conjunction with other factors, such as PDVs, to enhance their effects. In parasitoids that do not produce PDVs or VLPs, venom has a central role in the regulation of host immunity. Below are some examples of direct or synergistic effects of venom on host immunity.

The major immune response against endoparasitoid eggs and larvae is the encapsulation response carried out by immunocytes; granulocytes and plasmatocytes or lamellocytes (Fig. 2). The response requires recognition of the egg surface as foreign, initial attachment and lysis of granulocytes leading to recruitment of circulating hemocytes to build up multiple cellular layers around the egg. This response is accompanied generally by melanization.

In *M. demolitor*, venom alone does not seem to have any effect on the host hemocyte activities, such as spreading or encapsulation (Strand and Noda, 1991). However, venom has a synergistic effect in a dose-dependent manner, since addition of venom enhances the effects of MdBVs or calyx fluid in inhibiting cell spreading and prolongs their effects compared to when they are applied alone (Strand and Noda, 1991). In *Choristoneura fumiferana* (Lepidoptera: Tortricidae) parasitized by *Tranosema rostrale* (Hymenoptera: Ichneumonidae), significant reductions in total hemocyte counts (THC) and PO activity were observed three days after parasitization (Doucet and

Cusson, 1996b). Although injection of calyx fluid alone mimicked the above-mentioned effects, venom alone had no effect on THC and PO activity. In addition, a synergistic effect was not observed when both fluids were injected together into the host, indicating that TrIV effects are independent from venom.

In *P. hypochondriaca*, which does not harbor PDVs, venom consists of several proteins and peptides involved in host regulation, including the inhibition of encapsulation (Richards and Parkinson, 2000). Injection of venom into pupae of the tomato moth, *L. oleracea*, leads to failure of hemocytes to form capsules around injected Sephadex beads. This is due to changes in spreading and aggregation behavior of host hemocytes caused by two homologous venom proteins: vpr3 (Richards and Dani, 2008; Dani and Richards, 2009) and vpr1 (Dani and Richards, 2010).

In *Leptopilina* spp., there is intra-specific genetic variation which influences their success in parasitism directly correlated with their differential ability to suppress the encapsulation response in various *Drosophila* species (Dubuffet *et al.*, 2007; Schlenke *et al.*, 2007). Analysis of venom proteins from virulent (suppress encapsulation; IS_m strain) and avirulent species (IS_y) has shown markedly different profiles in the strains which may contribute to the different effects on the host (Labrosse *et al.*, 2005b). In addition, venom components from the IS_m strain are able to cross-protect IS_y eggs in *D. melanogaster* by significantly reducing the encapsulation rate of eggs deposited by IS_y strain when IS_m venom fluid is injected prior to parasitization by IS_y or when superparasitization by both strains occurs (Labrosse *et al.*, 2003). A major immunosuppressive protein in the IS_m line is a RacGAP (Rac GTPase activating protein) which affects lamellocytes' behaviors and consequently suppresses their encapsulation capacity (Labrosse *et al.*, 2005a). Apparently, the venom protein enters *D. melanogaster* hemocytes and targets Rac1 and Rac2 (Colinet *et al.*, 2007), which are key players in several cellular pathways including the regulation of the actin cytoskeleton, vesicular transport and proliferation (Bourne *et al.*, 1991). Colinet *et al.* (2010) demonstrated that the RacGAP protein differs only quantitatively between virulent and avirulent strains, which suggests the existence of a threshold effect of this molecule on parasitoid virulence.

A 24.1-kDa venom protein (Vn.11) was characterized from *Pteromalus puparum* that mimics the same effects of the total venom causing namely a reduction of THC, spreading of hemocytes, and encapsulation of Sephadex beads (Cai *et al.*, 2004; Wu *et al.*, 2008). Inhibition of hemocyte spreading by *P. puparum* venom seems to affect hosts beyond its habitual hosts, *P. rapae* and *Papilio xuthus*. In comparison, venom from the ectoparasitoid pteromalid wasp *N. vitripennis* only affects hemocyte spreading in its natural hosts, *M. domestica* and *Sarcophaga peregrina* (Zhang *et al.*, 2005).

Calreticulin, a molecular chaperone, has been reported in various parasitoid tissues, including parasitoid venoms. Growing evidence suggests that the protein may mediate a broad array of cellular functions and play key roles in host–parasite interactions (Ferreira *et al.*, 2004; Nakhasi *et al.*, 1998; Valck *et al.*, 2010). In insects, the role of calreticulin in encapsulation and phagocytosis has been demonstrated (Asgari and Schmidt, 2003; Choi *et al.*, 2002). In addition to the endoplasmic reticulum, the protein is found in a variety of nonendoplasmic reticulum locations, such as the nucleus, cytotoxic granules in T cells, tick saliva, blood serum, sperm acrosomes, and the cell surface (Ferreira *et al.*, 2004; Spiro *et al.*, 1996). The protein has Ca^{2+} and lectin binding properties with chaperone functions. A calreticulin is expressed in *C. rubecula* venom glands and injected into the host, *P. rapae* (Zhang *et al.*, 2006). *In vitro* assays showed that the protein inhibits hemocyte spreading behavior and, as a result, hemocytes failed to encapsulate Sephadex beads. The mechanism of inhibition is not known, but the protein may potentially interfere with a host hemocyte calreticulin function that plays a role in early-encapsulation reactions (Choi *et al.*, 2002). In another instance, when venom from *P. hypochondriaca* was pretreated with an anti-*G. mellonella* calreticulin antibody and applied to *T. ni* cells grown *in vitro*, cell death was significantly reduced compared to control. This suggested that the venom may contain a calreticulin-like protein that affects the intracellular calcium balance leading to apoptotic events (Rivers *et al.*, 2009). In addition to endoparasitoids, a calreticulin-like protein has also been reported from an ectoparasitoid venom, *N. vitripennis*, which is involved in cell death in the host, *Sarcophaga bullata* (Diptera: Sarcophagidae) (Rivers and Brogan, 2008).

Melanization is one of the major insect immune responses launched against foreign intruders, including parasitoid eggs. In most instances, the encapsulation response is accompanied by melanization of the parasitoid egg. This involves the activation of circulating and attached proPO on the surface of granulocytes and spherulocytes (Ling and Yu, 2005). During this process, the toxic molecules produced may kill the developing egg, even before completion of the capsule. The activated PO enzyme oxidizes tyrosine into L-3,4-dihydroxyphenylalanine (DOPA) leading to the production of toxic monophenols, diphenols and catechols that eventually form the insoluble melanin (Nappi and Christensen, 2005). Serine protease inhibitors (serpins) play central roles in regulating the cascade (e.g. (Scherfer *et al.*, 2008; Zhu *et al.*, 2003)).

There is no experimental evidence to demonstrate whether melanization is essential for the demise of the encapsulated egg or not. In the wild-type *D. melanogaster* larvae, *L. boulardi* eggs are encapsulated and subsequently melanized; however, parasitoid eggs were also encapsulated in mutant strains of the fly deficient for PO activity without being melanized (Rizki and Rizki, 1990a). Therefore, melanization may not be required for recognition or formation of capsule, but it could be required for strengthening the capsule and killing the developing parasitoid egg prior to completion of the capsule by production of toxic intermediate phenolic compounds during the melanization cascade (Nappi and Christensen, 2005). Consequently, this may prevent further development and potential escape of the parasitoid larva from incomplete capsules. The fact that in most host–endoparasitoid systems studied the parasitoid-injected factors inhibit melanization suggests that suppression of melanization is advantageous or essential for successful parasitism. Venom from several parasitoids has been shown to inhibit melanization by interfering with the proteolytic cascade that leads to the activation of proPO to PO. In addition to venom, PDV expressed genes (Beck and Strand, 2007) or calyx fluid (Beck *et al.*, 2000) may also interfere with melanization.

A 50-kDa protein from *Cotesia rubecula* venom (Vn50) reduces melanization by inhibiting activation of PO (Asgari *et al.*, 2003b). The protein has significant similarities to insect serine proteinase homologs (SPHs), which are catalytically inactive but mediate activation of PO. Based on the proposed model (Jiang and Kanost, 2000; Yu *et al.*, 2003), SPHs bind to proPO facilitating its cleavage to PO by proPO activating proteinases (PAPs). However, addition of Vn50 to reactions containing proPO, PAP and an SPH, inhibits activation of proPO into PO suggesting that due to its structural similarity to SPHs, Vn50 may compete with SPHs and therefore block activation of proPO (Zhang *et al.*, 2004a). In contrast to SPHs that are cleaved into a clip and a protease-like domain (Jiang and Kanost, 2000), Vn50 remains stable and uncleaved in the host hemolymph after parasitization. A 50-kDa serpin from *L. boulardi* (ISy type) also inhibits melanization in *D. yakuba* by an unknown mechanism (Colinet *et al.*, 2009). The protein belongs to the serine protease inhibitor superfamily and is different from Vn50.

Developmental Alterations

Koinobiont parasitoids, which mostly include endoparasitoids, allow further development of their host after parasitization (Askew and Shaw, 1986). Nevertheless, in many cases, development of the host is regulated; e.g., larval development is prolonged and pupation is inhibited (Beckage and Gelman, 2004). The regulatory pathway may involve disruptions of the host hormonal titers by causing damage to the endocrine glands (Dover and Vinson, 1990; Pennacchio *et al.*, 1998b; Tanaka *et al.*, 1987) or by suppressing proteins that are involved in the regulation of development (Hayakawa, 1995).

In endoparasitoids associated with PDVs, venoms either are not required for suppressing metamorphic changes (Beckage *et al.*, 1994; Dushay and Beckage, 1993) or are only required as a cofactor injected with PDVs to trigger developmental alterations. For example, venom in combination with calyx fluid from *Cardiochiles nigriceps* prolongs larval development and inhibits pupation; calyx fluid alone is not able to induce the same effects in the host (Tanaka and Vinson, 1991). However, venom from two other parasitoids of *H. virescens*, *Microplitis croceipes* and *Campoletis sonorensis*, mixed with *C. nigriceps* calyx fluid, was not able to compensate for venom from *C. nigriceps* to inhibit pupation, suggesting host-specificity of the venom proteins. The active protein component from *C. nigriceps* venom responsible for delaying larval development and inhibiting pupation in combination with calyx fluid was narrowed down to a venom protein of about 66 kDa in size (Tanaka and Vinson, 1991). However, there are no molecular details available on the protein.

In *M. demolitor* and *C. karyai*, venom alone does not seem to have any effect on the host development; although it synergizes the effects of PDVs (Strand and Dover, 1991; Tanaka, 1987). These effects included delays in larval period, induction of supernumerary instars, and formation of larval–pupal intermediates. In *C. nigriceps*, parasitization or co-injection of venom and calyx fluid leads to cessation of growth in *Heliothis virescens* larvae. This seems to be due to degradation of prothoracicotropic glands (PTGs) in the host resulting in their inactivation (Pennacchio *et al.*, 1998a, b). Nevertheless, the outcome was not reproduced when they were injected separately into the host (Edson and Vinson, 1979). In addition, surgical removal of the venom apparatus prolonged or inhibited emergence of the parasite, suggesting that venom is required for successful parasitism. In *Tranosema rostrale*, venom alone has no effect on host development (Doucet and Cusson, 1996a).

Parasitization of *Helicoverpa zea* by *M. croceipes* leads to prolongation of larval development. Injection experiments showed that calyx fluid alone is able to delay larval development but venom is not (Gupta and Ferkovich, 1998). However, venom had a synergistic effect by speeding up reduction in growth and extending the inhibitory effects compared to when calyx fluid alone was injected. In contrast, venom alone was able to inhibit larval growth when it was injected to an atypical host, *Galleria mellonella*, and had no synergistic effect on calyx fluid which equally affected larval growth. Injection of calyx fluid or venom or a combination did not have significant effects on larval growth in another nonhabitual host, *Spodoptera exigua* (Gupta and Ferkovich, 1998). Again, this shows host-specific function of venom proteins.

In the third larval stage of *Spodoptera littoralis*, injection of *C. inanitus* venom alone had no effect on host

development, and injection of calyx fluid had a moderate effect on development leading to formation of deformed moths (30%). However, venom had a synergistic effect as injection of the fluids in combination increased the percentage of deformed moths to 70% at 0.5 wasp equivalent and caused mortality in the host at larval or pupal stage when 1 wasp equivalent of calyx fluid plus venom was injected (Kaeslin *et al.*, 2010).

In parasitoids devoid of PDVs, developmental arrest has also been shown. In these systems, venom acts as the main factor. For example, in the pea aphid, *Acyrthosiphon pisum*, parasitized by *Aphidius ervi* or injected with venom alone, aphids undergo developmental arrest (Digilio *et al.*, 1998). In *Pteromalus puparum*, parasitization or artificial injection of venom into *P. rapae* pupae disrupts the host's endocrine system causing significant elevations in juvenile hormone titers (Zhu *et al.*, 2009).

Determination of Host Range and Host Stage

As we find out more about the properties of venoms from endoparasitoids, it becomes apparent that they have evolved specific functions tailored to their host which may affect the behavior of parasitic wasps in their preference for specific hosts or host stages to increase their survival chance. Physiological compatibility of parasitoids with the host environment is crucial for their success. As a consequence, effectiveness of their maternal factors injected along with the egg may influence their host selection behavior. For example, *Chelonus* near *curvimaculatus* prefers to parasitize host eggs in their early stages rather than later (Jones, 1987). Investigations showed that venom is largely degraded if injected into older eggs, possibly due to an increase in serine protease activity in older eggs (Leluk and Jones, 1989). In addition, several reports have shown species-specific activity of venom (see above for examples) which may also influence the host range of endoparasitoids.

Host Castration

Effects of parasitism on host reproductive organs are fairly common. However, there is only one example of a direct effect of venom on host castration, which is from a wasp without PDVs. Venom from *A. ervi*, an aphid endoparasitoid, causes castration in the host *A. pisum* by inducing apoptosis in ovarial tissues (Digilio *et al.*, 2000, 1998). The active protein is a γ-glutamyl transpeptidase (γ-GT) (Falabella *et al.*, 2007). Although the enzymatic activity of *A. ervi* γ-GT has been documented, the molecular mechanism by which the enzyme induces apoptosis has not been determined. As speculated by the investigators, it is highly likely that the enzyme interferes with the delicate balance of glutathione leading to oxidative stress in ovarian cells triggering apoptosis (Falabella *et al.*, 2007) (for details see Chapter 20 of this volume). Other examples of host castration induced by endoparasitoids are from PDV-carrying endoparasitoids, in which the role of venom alone, in most cases, has not been tested. In the systems tested, either venom has no role in host castration or it synergizes the effects of PDVs.

Cotesia vestalis (=*plutellae*) and *Diadegma semiclausum* cause degeneration of testis in *Plutella xylustella*. However, the effect of venom alone has not been tested (Bai *et al.*, 2009). Similarly, inhibition of spermatogenesis and damage to testicular cells of *Pseudaletia separata* parasitized or injected with a mixture of PDV–venom from *C. kariyai* was demonstrated (Tanaka *et al.*, 1994), which was found to be host stage dependent (Tagashira and Tanaka, 1998). In contrast, in *P. separata* larvae parasitized by *C. congregata*, only natural parasitism led to reduction in host testicular volume (Reed and Beckage, 1997). Injection of a PDV–venom mixture did not produce the same effect in the host. Similarly, in *Manduca sexta*, injection of PDV–venom from *C. congregata* did not have the same effect as natural parasitization. Although malformed testes in host larvae were observed (Reed *et al.*, 1997), it was proposed that the castration effect is not due to the PDV–venom but perhaps due to nutrient deprivation of gonads in competition with the developing parasitoid (Reed and Beckage, 1997).

Ascogaster reticulatus (Braconidae: Cheloninae), an egg parasitoid, also castrates its host *Adoxophyes* sp. Pseudoparasitism, in which venom and ovarian proteins are injected, led to complete castration or reduction in gonadal growth in both males and females (Brown and Kainoh, 1992). However, venom alone did not produce the same effects suggesting that, if involved, venom on its own is not responsible for castration.

Antimicrobial Activity

The presence of antimicrobial peptides/activity has been reported from several non parasitoid social hymenopterans (Čeřovský *et al.*, 2008; Krishnakumari and Nagaraj, 1997; Viljakainen and Pamilo, 2008; Xu *et al.*, 2009), but there are only a couple of examples from endoparasitoid wasps (Table 1). Venom from *P. hypochondriaca* was shown to have activity against gram-negative bacteria (*Escherichia coli* and *Xanthamonas campestris*) but not against gram-positive bacteria tested (*Bacillus cereus* and *B. subtilis*) (Dani *et al.*, 2003). The peptides responsible for the antibacterial activities have not been identified. In a more recent report, using an expression library made from *P. puparum*, investigators screened for antimicrobial peptides in the library by inducing the library and screening for colonies whose growth was inhibited. Three peptides (PP13, PP102, and PP113) were identified with activities against

both gram-negative and -positive bacteria. However, no activity was found against human red blood cells or fungi (Shen *et al.*, 2009). Based on the partial sequences available from these peptides, they are all cationic with high isoelectric points (IP) (~10–12), and have high ratio of hydrophobic residues.

Viruses and Virus-Like Particles Produced in the Venom

One of the most fascinating evolutionary novelties in hymenopteran endoparasitoids is their symbiotic association with viruses and virus-like particles which are injected into the host at oviposition interfering with host immunity and development. In most cases, these occur in the ovaries, including PDVs, picorna-like viruses, filamentous viruses, ascoviruses, and virus-like particles (VLPs) devoid of nucleic acids. However, there are endoparasitoids in which viruses or VLPs are produced in their venom glands and are delivered during oviposition into the host, facilitating parasitization. With regard to biogenesis, a distinction between PDVs/VLPs produced in the calyx cells and VLPs produced in venom is that PDVs are assembled in the nuclei of calyx cells whereas VLPs associated with venom glands are assembled in the lumen of the gland from proteins produced in glandular cells and secreted into the lumen (Chiu *et al.*, 2006; Ferrarese *et al.*, 2009).

The only conventional virus that has been identified and shown in association with venom glands from an endoparasitoid wasp is an entomopoxvirus (EPV). The virus replicates in the venom apparatus of *Diachasmimorpha longicaudata* (Hymenoptera: Braconidae) which parasitizes the larvae of the Caribbean fruit fly, *Anastrepha suspense* (Diptera: Tephritidae). However, replication of DlEPV does not seem to be venom-specific because the particles replicate in the male Hagen's glands as well (Khoo and Lawrence, 2002; Lawrence, 2002). Similar to other EPVs, DlEPV replicates in the host cytoplasm. Considering that this virus, in contrast to other EPVs, does not produce spheroidin that forms occlusion bodies, and that it replicates in hosts from two different insect orders, it was proposed that it may represent a new group of EPVs or belong to group C viruses (Lawrence, 2002). DlEPV replicates both in the wasp and in the host-infected hemocytes (Lawrence, 2005). Infected hemocytes exhibit typical apoptotic symptoms including blebbing and DNA concatenation (Lawrence, 2005). As a consequence, they lose their adhesive properties and fail to perform the encapsulation response. These pathological effects do not seem to occur within the carrying wasps and therefore it is considered as a mutualistic virus that facilitates the parasite's survival in the host.

VLPs that are produced exclusively in the venom glands have been reported from several parasitoids. These

TABLE 1 A List of Endoparasitoid Venom Proteins/Peptides Reported with Known Functions

Protein/peptide	Function	Parasitoid species	Reference
Hemocyte inactivation			
Calreticulin	Inhibits hemocyte spreading	*Cotesia rubecula*	Zhang *et al.*, 2006
		Microctonus aethiopoides	Crawford *et al.*, 2008
		Microctonus hyperodae	Crawford *et al.*, 2008
Virulence protein, P4	Reduces cell adhesion	*Leptopilina boulardi*	Labrosse *et al.*, 2005b
VPr1	Inhibits hemocyte aggregation	*Pimpla hypochondriaca*	Dani and Richards, 2010
VPr3	Inhibits hemocyte aggregation	*P. hypochondriaca*	Richards and Dani, 2008
Vn.11	Inhibits encapsulation	*Pteromalus puparum*	Wu *et al.*, 2008
Inhibiting melanization			
LbSPNy	Inhibits melanization	*L. boulardi*	Colinet *et al.*, 2009
Vn50	Inhibits melanization	*C. rubecula*	Asgari *et al.*, 2003b
Transient paralysis			
Pimplin	Paralytic factor	*P. hypochondriaca*	Parkinson *et al.*, 2002b
Enhancing PDV			
Vn1.5	Facilitates PDV gene expression	*C. rubecula*	Zhang *et al.*, 2004b
Castration			
γ-Glutamyl transpeptidase	Induces apoptosis in ovaries	*Aphidius ervi*	Falabella *et al.*, 2007
Antimicrobial			
PP13, PP102, and PP113	Activity against gram + and − bacteria	*P. puparum*	Shen *et al.*, 2009

particles lack nucleic acids and the constituents characterized so far provide evidence for their insect rather than viral origin. VLPs produced in *Meteorus pulchricornis* (Hymenoptera: Braconidae) venom glands (MpVLPs) affect spreading behavior of *Pseudaletia separata* hemocytes as early as 30 min following their exposure to the particles, providing immediate protection for the parasitoid egg (Suzuki *et al.*, 2008). In addition, MpVLPs induce apoptosis in hemocytes, providing a longer term protection from encapsulation (Suzuki *et al.*, 2009; Suzuki and Tanaka, 2006). The two phenomena seem to be independent of each other since inhibition of apoptosis using a caspase inhibitor does not prevent loss of spreading and adhesion in hemocytes after exposure to the particles (Suzuki *et al.*, 2008). The composition of MpVLPs has not been determined.

VLPs are also produced in the venom glands (so-called long glands) of *Leptopilina* spp. (Hymenoptera: Figitidae). Morphology and composition of VLPs in virulent and avirulent strains of *L. boulardi* are different, the former being able to evade the host's immune system,

Drosophila melanogaster (Dupas *et al.*, 1996; Labrosse *et al.*, 2003). *Leptopilina heterotoma* VLPs specifically cause cell death in *Drosophila* lamellocytes (Rizki and Rizki, 1990b). Similarly, VLPs from *Leptopilina victoriae* have immunosuppressive properties and cause lysis of lamellocytes. LvVLPs show antigenic and morphological similarities to LhVLPs (Morales *et al.*, 2005). Two proteins, p40 and p47.5, from *L. heterotoma* and *L. victoriae* VLPs respectively show antigenic similarity and antibodies to p40 completely block lysis of lamellocytes from *D. melanogaster* induced by LhVLPs. This demonstrates that both proteins are involved in VLP–lamellocyte interactions, either used for recognition of lamellocytes or induction of the observed changes in the lamellocytes.

Filamentous viruses that replicate in the ovaries as well as venom glands of *L. boulardi* (LbFV) females induce superparasitism behavior in the wasp (Varaldi *et al.*, 2006). In addition, the behavior could be horizontally transferred to daughters of a non-superparasitizing strain of the wasp by injecting venom gland or ovary extracts into host larvae in which they develop. This effect also remains stable in the next generations. Ovary or venom gland extracts that do not contain LbFV do not transfer the superparasitism behavior, which strongly suggests that the viruses are responsible for this behavior. These viruses have not been classified into any virus family as yet.

CONCLUSIONS

Despite the important role of venoms in host–parasitoid interactions, very few of these components have been isolated and characterized at the molecular level in endoparasitoid wasps, and we have information from a very limited number of hymenopteran families. Therefore, a very large pool of unexplored molecules exists in nature which we can expect to have very interesting and valuable properties. Among those identified so far are enzymes, protease inhibitors, paralytic factors, and several others with unknown functions. Interestingly, these proteinaceous venom products are very similar to cellular proteins, which suggests that they have been derived from cellular components with their function(s) modified throughout the evolutionary processes. Certainly, more research is required to elucidate the roles and constituents of venom from more endoparasitic wasps. These molecules can have critical roles in insect immunity, development, taxonomy, evolutionary biology, protein chemistry and potentially practical biological control.

REFERENCES

Abt, M., Rivers, D.B., 2007. Characterization of phenoloxidase activity in venom from the ectoparasitoid *Nasonia vitripennis* (Walker) (Hymenoptera: Pteromalidae). J. Invertebr. Pathol. 94, 108–118.

Asgari, S., Schmidt, O., 2003. Is cell surface calreticulin involved in phagocytosis by insect hemocytes? J. Insect Physiol. 49, 545–550.

Asgari, S., Zareie, R., Zhang, G., Schmidt, O., 2003. Isolation and characterization of a novel venom protein from an endoparasitoid, *Cotesia rubecula* (Hym: Braconidae). Arch. Insect Biochem. Physiol. 53, 92–100.

Asgari, S., Zhang, G., Zareie, R., Schmidt, O., 2003. A serine proteinase homolog venom protein from an endoparasitoid wasp inhibits melanization of the host hemolymph. Insect Biochem. Mol. Biol. 33, 1017–1024.

Askew, R.R., Shaw, M.R., 1986. Parasitoid communities: their size, structure and development. In: Waage, J., Greathead, D. (Eds.), Insect Parasitoids (pp. 225–263). Academic Press, London.

Bai, S., Cai, D., Li, X., 2009. Parasitic castration of *Plutella xylostella* larvae induced by polydnaviruses and venom of *Cotesia vestalis* and *Diadegma semiclausum* polydnaviruses and venom of *Cotesia vestalis* and *Diadegma semiclausum*. Arch. Insect Biochem. Physiol. 70, 30–43.

Beck, M., Strand, M.R., 2003. RNA interference silences *Microplitis demolitor* bracovirus genes and implicates glc1.8 in disruption of adhesion in infected host cells. Virology 314, 521–535.

Beck, M., Theopold, U., Schmidt, O., 2000. Evidence for serine protease inhibitor activity in the ovarian calyx fluid of the endoparasitoid *Venturia cancescens*. J. Insect Physiol. 46, 1275–1283.

Beck, M.H., Strand, M.R., 2007. A novel polydnavirus protein inhibits the insect prophenoloxidase activation pathway. Proc. Natl. Acad. Sci. U.S.A. 104, 19267–19272.

Beckage, N.E., Gelman, D.B., 2004. Wasp parasitoid disruption of host development: implications for new biologically based strategies for insect control. Annu. Rev. Entomol. 49, 299–330.

Beckage, N.E., Tan, F.F., Schleifer, K.W., Lane, R.D., Cherubin, L.L., 1994. Characterization and biological effects of *Cotesia congregata* polydnavirus on host larvae of the tobacco hornworm, *Manduca sexta*. Arch. Insect Biochem. Physiol. 26, 165–195.

Bourne, H.R., Sanders, D.A., McCormick, F., 1991. The GTPase superfamily: conserved structure and molecular mechanisms. Nature 349, 117–127.

Brown, J.J., Kainoh, Y., 1992. Host castration by *Ascogaster* spp. (Hymenoptera: Braconidae). Ann. Entomol. Soc. Am. 85, 67–71.

Cai, J., Ye, G.-y., Hu, C., 2004. Parasitism of *Pieris rapae* (Lepidoptera: Pieridae) by a pupal endoparasitoid, *Pteromalus puparum* (Hymenoptera: Pteromalidae): effects of parasitization and venom on host hemocytes. J. Insect Physiol. 50, 315–322.

Cerenius, L., Lee, B.L., Söderhäll, K., 2008. The proPO-system: pros and cons for its role in invertebrate immunity. Trends Immunol. 29, 263–271.

Čeřovský, V., Slaninová, J., Fučír, V., Hulačová, H., Borovičková, L., Ježek, R., et al. 2008. New potent antimicrobial peptides from the venom of Polistinae wasps and their analogs. Peptides 99, 992–1003.

Chiu, H., Morales, J., Govind, S., 2006. Identification and immuno-electron microscopy localization of p40, a protein component of immunosuppressive virus-like particles from *Leptopilina heterotoma*, a virulent parasitoid wasp of *Drosophila*. J. Gen. Virol. 87, 461–470.

Choi, J.Y., Whitten, M.M.A., Cho, M.Y., Lee, K.Y., Kim, M.S., Ratcliffe, N.A., et al. 2002. Calreticulin enriched as an early-stage encapsulation protein in wax moth *Galleria mellonella* larvae. Dev. Comp. Immunol. 26, 335–343.

Colinet, D., Dubuffet, A., Cazes, D., Moreau, S., Drezen, J.-M., Poirié, M., 2009. A serpin from the parasitoid wasp *Leptopilina boulardi* targets the *Drosophila* phenoloxidase cascade. Dev. Comp. Immunol. 33, 681–689.

Colinet, D., Schmitz, A., Cazes, D., Gatti, J.L., Poirié, M., 2010. The origin of intraspecific variation of virulence in an eukaryotic immune suppressive parasite. PLoS Pathog. 6 (11), e1001206.

Colinet, D., Schmitz, A., Depoix, D., Crochard, D., Poirié, M., 2007. Convergent use of RhoGAP toxins by eukaryotic parasites and bacterial pathogens. PLoS Pathog. 3 (12), e203.

Cônsoli, F.L., Lewis, D., Keeley, L., Vinson, S.B., 2007. Characterization of a cDNA encoding a putative chitinase from teratocytes of the endoparasitoid *Toxoneuron nigriceps*. Ent. Exp. Appl. 122, 271–278.

Coudron, T.A., Kelly, T.J., Puttler, B., 1990. Developmental responses of *Trichoplusia ni* (Lepidoptera: Noctuidae) to parasitism by the ectoparasite *Euplectrus plathypenae* (Hymenoptera: Eulophidae). Arch. Insect Biochem. Physiol. 13, 83–94.

Crawford, A.M., Brauning, R., Smolenski, G., Ferguson, C., Barton, D., Wheeler, T.T., et al. 2008. The constituents of *Microctonus* sp. parasitoid venoms. Insect Mol. Biol. 17, 313–324.

Dahlman, D.L., Rana, R.L., Schepers, E.J., Schepers, T., DiLuna, F.A., Webb, B.A., 2003. A teratocyte gene from a parasitic wasp that is associated with inhibition of insect growth and development inhibits host protein synthesis. Insect Mol. Biol. 12, 527–534.

Dahlman, D.L., Vinson, S.B., 1993. Teratocytes: developmental and biochemical characteristics. In: Beckage, N.E., Thompson, S.N., Federici, B.A. (Eds.), Parasites and Pathogens of Insects. Vol. 1 Parasites (pp. 145–166). Academic Press, San Diego.

Dani, M.P., Edwards, J.P., Richards, E.H., 2005. Hydrolase activity in the venom of the pupal endoparasitic wasp, *Pimpla hypochondriaca*. Comp. Biochem. Physiol. B: Biochem. Mol. Biol. 141, 373–381.

Dani, M.P., Richards, E.H., 2009. Cloning and expression of the gene for an insect haemocyte anti-aggregation protein (VPr3), from the venom of the endoparasitic wasp, *Pimpla hypochondriaca*. Arch. Insect Biochem. Physiol. 71, 191–204.

Dani, M.P., Richards, E.H., 2010. Identification, cloning and expression of a second gene (vpr1) from the venom of the endoparasitic wasp, *Pimpla hypochondriaca* that displays immunosuppressive activity. J. Insect Physiol. 56, 195–203.

Dani, M.P., Richards, E.H., Isaac, R.E., Edwards, J.P., 2003. Antibacterial and proteolytic activity in venom from the endoparasitic wasp *Pimpla hypochondriaca* (Hymenoptera: Ichneumonidae). J. Insect Physiol. 49, 945–954.

Davies, D.H., Strand, M.R., Vinson, S.B., 1987. Changes in differential haemocyte count and *in vitro* behaviour of plasmatocytes from host *Heliothis virescens* caused by *Campolethis sonorensis* polydnavirus. J. Insect Physiol. 33, 143–153.

Desneux, N., Barta, R.J., Delebecque, C.J., Heimpel, G.E., 2009. Transient host paralysis as a means of reducing self-superparasitism in koinobiont endoparasitoids. J. Insect Physiol. 55, 321–327.

Digilio, M.C., Isidoro, N., Tremblay, E., Pennacchio, F., 2000. Host castration by *Aphidius ervi* venom proteins. J. Insect Physiol. 46, 1041–1050.

Digilio, M.C., Pennacchio, F., Tremblay, E., 1998. Host regulation effects of ovary fluid and venom of *Aphidius ervi* (Hymenoptera: Braconidae). J. Insect Physiol. 44, 779–784.

Doucet, D., Cusson, M., 1996. Alteration of developmental rate and growth of *Choristoneura fumiferana* parasitized by *Tranosema rostrale* – role of the calyx fluid. Entomol. Exp. Appl. 81, 21–30.

Doucet, D., Cusson, M., 1996. Role of calyx fluid in alterations of immunity in *Choristoneura fumiferana* larvae parasitized by *Tranosema rostrale*. Comp. Biochem. Physiol. 114, 311–317.

Dover, B.A., Davies, D.H., Strand, M.R., Gray, R.S., Keeley, L.L., Vinson, S.B., 1987. Ecdysteroid-titre reduction and developmental arrest of last instar *Heliothis viriscens* larvae by calyx fluid from the parasitoid *Campoletis sonorensis*. J. Insect Physiol. 33, 333–338.

Dover, B.A., Vinson, S.B., 1990. Stage–specific effects of *Campoletis sonorensis* parasitism on *Heliothis virescens* development and prothoracic glands. Physiol. Entomol. 15, 405–414.

Dubuffet, A., Dupas, S., Frey, F., Drezen, J.-M., Poirié, M., Carton, Y., 2007. Genetic interactions between the parasitoid wasp *Leptopilina boulardi* and its *Drosophila* hosts. Heredity 98, 21–27.

Dupas, S., Brehelin, M., Frey, F., Carton, Y., 1996. Immune suppressive virus-like particles in a *Drosophila* parasitoid: significance of their intraspecific morphological variations. Parasitology 113, 207–212.

Dushay, M.S., Beckage, N.E., 1993. Dose-dependent separation of *Cotesia congregata*-associated polydnavirus effects on *Manduca sexta* larval development and immunity. J. Insect Physiol. 39, 1029–1040.

Edson, K.M., Barlin, M.R., Vinson, S.B., 1982. Venom apparatus of braconid wasps: comparative ultrastructure of reservoirs and gland filaments. Toxicon 20, 553–562.

Edson, K.M., Vinson, S.B., 1979. A comparative morphology of the venom apparatus of female braconids (Hymenoptera; Braconidae). Can. Entomol. 111, 1013–1024.

Er, A., Uçkan, F., Rivers, D.B., Ergin, E., Sak, O., 2010. Effects of parasitization and envenomation by the endoparasitic wasp *Pimpla turionellae* (Hymenoptera: Ichneumonidae) on hemocyte numbers, morphology, and viability of its host *Galleria mellonella* (Lepidoptera: Pyralidae) and toward an embryonic cell line from *Trichoplusia ni* (Lepidoptera: Noctuidae). Ann. Entomol. Soc. Am. 103, 273–282.

Ergin, E., Uçkan, F., Rivers, D.B., Sak, O., 2006. In vivo and in vitro activity of venom from the endoparasitic wasp *Pimpla turionellae* (L.) (Hymenoptera: Ichneumonidae). Arch. Insect Biochem. Physiol. 61, 87–97.

Falabella, P., Riviello, L., Caccialupi, P., Rossodivita, T., Valente, M.T., De Stradis, M.L., et al. 2007. A γ-glutamyl transpeptidase of *Aphidius ervi* venom induces apoptosis in the ovaries of host aphids. Insect Biochem. Mol. Biol. 37, 453–465.

Ferrarese, R., Morales, J., Fimiarz, D., Webb, B.A., Govind, S., 2009. A supracellular system of actin-lined canals controls biogenesis and release of virulence factors in parasitoid venom glands. J. Exp. Biol. 212, 2261–2268.

Ferreira, V., Molina, M.C., Valck, C., Rojas, Á., Aguilar, L., Ramírez, G., et al. 2004. Role of calreticulin from parasites in its interaction with vertebrate hosts. Mol. Immunol. 40, 1279–1291.

Furihata, S.X., Kimura, M.T., 2009. Effects of *Asobara japonica* venom on larval survival of host and nonhost *Drosophila* species. Physiol. Entomol. 34, 292–295.

Godfray, H.J.C., 1994. Parasitoids: Behavioural and Evolutionary Ecology. Princeton University Press, Princeton.

Gupta, P., Ferkovich, S.M., 1998. Interaction of calyx fluid and venom from *Microplitis croceipes* (Braconidae) on developmental disruption of the natural host, *Helicoverpa zea*, and two atypical hosts, *Galleria mellonella* and *Spodoptera exigua*. J. Insect Physiol. 44, 713–719.

Guzo, D., Stoltz, D.B., 1987. Observation on cellular immunity and parasitism in the tussock moth. J. Insect Physiol. 33, 19–31.

Hayakawa, Y., 1995. Growth-blocking peptide: an insect biogenic peptide that prevents the onset of metamorphosis. J. Insect Physiol. 41, 1–6.

Hoffman, D.R., Weimer, E.T., Sakell, R.H., Schmidt, M., 2005. Sequence and characterization of honeybee venom acid phosphatase. J. Allergy Clin. Immunol. 115, S107.

Hoy, H.L., Dahlman, D.L., 2002. Extended in vitro culture of *Microplitis croceipes* teratocytes and secretion of TSP14 protein. J. Insect Physiol. 48, 401–409.

Jiang, H., Kanost, M.R., 2000. The clip-domain family of serine proteinases in arthropods. Insect Biochem. Mol. Biol. 30, 95–105.

Jones, D., 1987. Material from adult female *Chelonus* sp. directs expression of altered developmental programme of host lepidoptera. J. Insect Physiol. 33, 129–134.

Jones, D, Wozniak, M., 1991. Regulatory mediators in the venom of *Chelonus* sp.: their biosynthesis and subsequent processing in homologous and heterologous systems. Biochem. Biophysi. Res. Commun. 178, 213–220.

Kaeslin, M., Reinhard, M., Bühler, D., Roth, T., Pfister-Wilhelm, R., Lanzrein, B., 2010. Venom of the egg–larval parasitoid *Chelonus inanitus* is a complex mixture and has multiple biological effects. J. Insect Physiol. 56, 686–694.

Kathirithamby, J., Ross, L.D., Johnston, J.S., 2003. Masquerading as self? Endoparasitic Strepsiptera (Insecta) enclose themselves in host-derived epidermal bag. Proc. Natl. Acad. Sci. U.S.A. 100, 7655–7659.

Kelly, T.J., Gelman, D.B., Reed, D.A., Beckage, N.E., 1998. Effects of parasitization by *Cotesia congregata* on the brain-prothoracic gland axis of its host, *Manduca sexta*. J. Insect Physiol. 44, 232–332.

Khoo, C.C.H., Lawrence, P.O., 2002. Hagen's glands of the parasitic wasp *Diachasmimorpha longicaudata* (Hymenoptera: Braconidae): ultrastructure and the detection of entomopoxvirus and parasitism-specific proteins. Arthropod Struct. Dev. 31, 121–130.

Kitano, H., 1986. The role of *Apanteles glomeratus* venom in the defensive response of its host, *Pieris rapae crucivora*. J. Insect Physiol. 32, 369–375.

Krishnakumari, V., Nagaraj, R., 1997. Antimicrobial and hemolytic activities of crabrolin, a 13-residue peptide from the venom of the European hornet, *Vespa crabro*, and its analogs. J. Peptide Res. 50, 88–93.

Krishnan, A., Nair, P.N., Jones, D., 1994. Isolation, cloning and characterization of new chitinase stored in active form in chitin-lined venom reservoir. J. Biol. Chem. 269, 20971–20976.

Labrosse, C., Carton, Y., Dubuffet, A., Drezen, J.-M., Poirie, M., 2003. Active suppression of *D. melanogaster* immune response by long gland products of the parasitic wasp *Leptopilina boulardi*. J. Insect Physiol. 49, 513–522.

Labrosse, C., Eslin, P., Doury, G., Drezen, J.-M., Poirie, M., 2005. Haemocyte changes in *D. melanogaster* in response to long gland components of the parasitoid wasp *Leptopilina boulardi*: a Rho-GAP protein as an important factor. J. Insect Physiol. 51, 161–170.

Labrosse, C., Stasiak, K., Lesobre, J., Grangeia, A., Huguet, E., Drezen, J.-M., et al. 2005. A RhoGAP protein as a main immune suppressive factor in the *Leptopilina boulardi* (Hymenoptera, Figitidae)–*Drosophila melanogaster* interaction. Insect Biochem. Mol. Biol. 35, 93–103.

Lawrence, P.O., 2002. Purification and partial characterization of an entomopoxvirus (DlEPV) from a parasitic wasp of tephritid fruit flies. J. Insect Sci. 2, 10.

Lawrence, P.O., 2005. Morphogenesis and cytopathic effects of the *Diachasmimorpha longicaudata* entomopoxvirus in host haemocytes. J. Insect Physiol. 51, 221–233.

Leluk, J., Jones, D., 1989. *Chelonus* sp. near *curvimaculatus* venom proteins: analysis of their potential role and processing during development of host *Trichoplusia ni*. Arch. Insect Biochem. Physiol. 10, 1–12.

Leluk, J., Schmidt, J., Jones, D., 1989. Comparative studies on the protein composition of hymenopteran venom reservoirs. Toxicon 27, 105–114.

Ling, E., Yu, X.-Q., 2005. Prophenoloxidase binds to the surface of hemocytes and is involved in hemocyte melanization in *Manduca sexta*. Insect Biochem. Mol. Biol. 35, 1356–1366.

Mabiala-Moundoungou, A.D.N., Doury, G., Eslin, P., Cherqui, A, Prévost, G., 2009. Deadly venom of *Asobara japonica* parasitoid needs ovarian antidote to regulate host physiology. J. Insect Physiol. 56, 35–41.

Monteiro, M.C., Romão, P.R.T., Soares, A.M., 2009. Pharmacological perspectives of wasp venom. Protein Peptide Lett. 16, 944–952.

Morales, J., Chiu, H., Oo, T., Plaza, R., Hoskins, S., Govind, S., 2005. Biogenesis, structure, and immune-suppressive effects of virus-like particles of a *Drosophila* parasitoid, *Leptopilina victoriae*. J. Insect Physiol. 51, 181–195.

Moreau, S.J.M., Cherqui, A., Doury, G., Dubois, F., Fourdrain, Y., Sabatier, L., et al. 2004. Identification of an aspartylglucosaminidase-like protein in the venom of the parasitic wasp *Asobara tabida* (Hymenoptera: Braconidae). Insect Biochem. Mol. Biol. 34, 485–492.

Moreau, S.J.M., Dingremont, A., Doury, G., Giordanengo, P., 2002. Effects of parasitism by *Asobara tabida* (Hymenoptera: Braconidae) on the development, survival and activity of *Drosophila melanogaster* larvae. J. Insect Physiol. 48, 337–347.

Nakhasi, H.L., Pogue, G.P., Duncan, R.C., Joshi, M., Atreya, C.D., Lee, N.S., et al. 1998. Implication of calreticulin function in parasite biology. Parasitol. Today 14, 157–160.

Nappi, A.J., Christensen, B.M., 2005. Melanogenesis and associated cytotoxic reactions: Applications to insect innate immunity. Insect Biochem. Mol. Biol. 35, 443–459.

Noirot, N., Quennedey, A., 1974. Fine structure of insect epidermal glands. Annu. Rev. Entomol. 19, 61–80.

Parkinson, N., Conyers, C., Smith, I., 2002. A venom protein from the endoparasitoid wasp *Pimpla hypochondriaca* is similar to snake venom reprolysin-type metalloproteases. J. Invertebr. Pathol. 79, 129–131.

Parkinson, N., Smith, I., Audsley, N., Edwards, J.P., 2002. Purification of pimplin, a paralytic heterodimeric polypeptide from venom of the parasitoid wasp *Pimpla hypochondriaca*, and cloning of the cDNA encoding one of the subunits. Insect Biochem. Mol. Biol. 32, 1769–1773.

Parkinson, N., Smith, I., Weaver, R., Edwards, J.P., 2001. A new form of arthropod phenoloxidase is abundant in venom of the parasitoid wasp pimpla hypochondriaca. Insect Biochem. Mol. Biol. 31, 57–63.

Parkinson, N.M., Conyers, C., Keen, J., MacNicoll, A., Smith, I., Audsley, N., et al. 2004. Towards a comprehensive view of the primary structure of venom proteins from the parasitoid wasp *Pimpla hypochondriaca*. Insect Biochem. Mol. Biol. 34, 565–571.

Parkinson, N.M., Conyers, C., Keen, J.N., MacNicoll, A.D., Weaver, I.S.R., 2003. cDNAs encoding large venom proteins from the parasitoid wasp *Pimpla hypochondriaca* identified by random sequence analysis. Comp. Biochem. Physiol. C 134, 513–520.

Parkinson, N.M., Weaver, R.J., 1999. Noxious components of venom from the pupa-specific parasitoid *Pimpla hypochondriaca*. J. Invertebr. Pathol. 73, 74–83.

Pennacchio, F., Falbella, P., Sordetti, R., Varricchio, P., Malva, C., Vinson, S.B., 1998. Prothoracic gland inactivation in *Heliothis virescens* (F.) (Lepidoptera: Noctuidae) larvae parasitized by *Cardiochiles nigriceps* Viereck (Hymenoptera: Braconidae). J. Insect Physiol. 44, 845–857.

Pennacchio, F., Falabella, P., Vinson, S.B., 1998. Regulation of *Heliothis virescens* prothoracic glands by *Cardiochiles nigriceps* polydnavirus. Arch. Insect Biochem. Physiol. 38, 1–10.

Pruijssers, A.J., Strand, M.R., 2007. PTP-H2 and PTP-H3 from Microplitis demolitor bracovirus localize to focal adhesions and are antiphagocytic in insect immune cells. J. Virol. 81, 1209–1219.

Reed, D.A., Beckage, N.E., 1997. Inhibition of testicular growth and development in *Manduca sexta* larvae parasitized by the braconid Wasp *Cotesia Congregata*. J. Insect Physiol. 43, 29–38.

Reed, D.A., Loeb, M.J., Beckage, N.E., 1997. Inhibitory effects of parasitism by the gregarious endoparasitoid *Cotesia congregata* on host testicular development. Arch. Insect Biochem. Physiol. 36, 95–114.

Richards, E.H., Dani, M.P., 2008. Biochemical isolation of an insect haemocyte anti-aggregation protein from the venom of the endoparasitic wasp, *Pimpla hypochondriaca*, and identification of its gene. J. Insect Physiol. 54, 1041–1049.

Richards, E.H., Parkinson, N.M., 2000. Venom from the endoparasitic wasp *Pimpla hypochondriaca* adversely affects the morphology, viability, and immune function of hemocytes from larvae of the tomato moth, *Lacanobia oleracea*. J. Invertebr. Pathol. 76, 33–42.

Rivers, D.B., Brogan, A., 2008. Venom glands from the ectoparasitoid *Nasonia vitripennis* (Walker) (Hymenoptera: Pteromalidae) produce a calreticulin-like protein that functions in developmental arrest and cell death in the flesh fly host, *Sarcophaga bullata* Parker (Diptera: Sarcophagidae). In: Maes, R.P. (Ed.), Insect Physiology: New Research. Nova Science Publishers, New York.

Rivers, D.B., Dani, M.P., Richards, E.H., 2009. The mode of action of venom from the endoparasitic wasp *Pimpla hypochondriaca* (Hymenoptera: Ichneumonidae) involves Ca^{+2}-dependent cell death pathways. Arch. Insect Biochem. Physiol. 71, 173–190.

Rizki, R.M., Rizki, T.M., 1990. Encapsulation of parasitoid eggs in phenoloxidase-deficient mutants of *Drosophila melanogaster*. J. Insect Physiol. 36, 523–529.

Rizki, R.M., Rizki, T.M., 1990. Parasitoid virus-like particles destroy *Drosophila* cellular immunity. Proc. Natl. Acad. Sci. U.S.A. 87, 8388–8392.

Scherfer, C., Tang, H., Kambris, Z., Lhocine, N., Hashimoto, C., Lemaitre, B., 2008. *Drosophila* Serpin-28D regulates hemolymph phenoloxidase activity and adult pigmentation. Dev. Biol. 323, 189–196.

Schlenke, T.A., Morales, J., Govind, S., Clark, A.G., 2007. Contrasting infection strategies in generalist and specialist wasp parasitoids of *Drosophila melanogaster*. PLoS Pathog. 3, e158.

Shen, X., Ye, G., Cheng, X., Yu, C., Yao, H., Hu, C., 2009. Novel antimicrobial peptides identified from an endoparasitic wasp cDNA library. J. Peptide Sci. 16, 58–64.

Spiro, R.G., Zhu, Q., Bhoyroo, V., Söling, H.-D., 1996. Definition of the lectin-like properties of the molecular chaperone, calreticulin, and demonstration of its copurification with endomannosidase from rat liver Golgi. J. Biol. Chem. 271, 11588–11594.

Stoltz, D.B., Guzo, D., Belland, E.R., Lucarotti, C.J., MacKinnon, E.A., 1988. Venom promotes uncoating in vitro and persistence in vivo of DNA from a braconid polydnavirus. J. Gen. Virol. 69, 903–907.

Strand, M.R., Dover, B.A., 1991. Developmental disruption of *Pseudaplusia includens* and *Heliothis virescens* larvae by calyx fluid and venom of *Microplitis demolitor*. Arch. Insect Biochem. Physiol. 18, 131–145.

Strand, M.R., Johnson, J.A., Noda, T., Dover, B.A., 1994. Development and partial characterization of monoclonal antibodies to venom of the parasitoid *Microplitis demolitor*. Arch. Insect Biochem. Physiol. 26, 123–136.

Strand, M.R., Noda, T., 1991. Alterations in the haemocytes of *Pseudoplusia includens* after parasitism by *Microplitis demolitor*. J. Insect Physiol. 37, 839–850.

Suzuki, M., Miura, K., Tanaka, T., 2008. The virus-like particles of a braconid endoparasitoid wasp, *Meteorus pulchricornis*, inhibit hemocyte spreading in its noctuid host, *Pseudaletia separata*. J. Insect Physiol. 54, 1015–1022.

Suzuki, M., Miura, K., Tanaka, T., 2009. Effects of the virus-like particles of a braconid endoparasitoid, *Meteorus pulchricornis*, on hemocytes and hematopoietic organs of its noctuid host, *Pseudaletia separata*. Appl. Entomol. Zool. 44, 115–125.

Suzuki, M., Tanaka, T., 2006. Virus-like particles in venom of *Meteorus pulchricornis* induce host hemocyte apoptosis. J. Insect Physiol. 52, 602–613.

Tagashira, E., Tanaka, T., 1998. Parasitic castration of *Pseudaletia separata* by *Cotesia kariyai* and its association with polydnavirus gene expression. J. Insect Physiol. 44, 733–744.

Tanaka, T., 1987. Calyx and venom fluids of *Apanteles kariyai* (Hymenoptera: Braconidae) as factors that prolong larval period of the host, *Pseudaletia separata* (Lepidoptera: Noctuidae). Ann. Entomol. Soc. Am. 80, 530–533.

Tanaka, T., Agui, N., Hiruma, K., 1987. The parasitoid *Apanteles kariyai* inhibits pupation of its host, *Pseudoletia separata*, via disruption of prothoracicotropic hormone release. Gen. Comp. Endocrinol. 67, 364–374.

Tanaka, T., Tagashira, E., Sakurai, S., 1994. Reduction of testis growth of *Pseudaletia separata* larvae after parasitization by *Cotesia kariyai*. Arch. Insect Biochem. Physiol. 26, 111–122.

Tanaka, T., Vinson, S.B., 1991. Interaction between venom and calyx fluids of three parasitoids, *Cardiochiles nigriceps*, *Microplitis croceipes* (Hymenoptera: Braconidae), and *Campoletis sonorensis* (Hymenoptera: Ichneumonidae) in affecting a delay in the pupation of *Heliothis virescens* (Lepidoptera: Noctuidae). Ann. Entomol. Soc. Am. 84, 87–92.

Uçkan, F., Ergin, E., Rivers, D.B., Gençer, N., 2006. Age and diet influence the composition of venom from the endoparasitic wasp *Pimpla turionellae* L. (Hymenoptera: Ichneumonidae). Arch. Insect Biochem. Physiol. 63, 177–187.

Uçkan, F., Sinan, S., Savasci, S., Ergin, E., 2004. Determination of venom components from the endoparasitoid wasp *Pimpla turionellae* L. (Hymenoptera: Ichneumonidae). Ann. Entomol. Soc. Am. 97, 775–780.

Valck, C., Ramirez, G., Lopez, N., Ribeiro, C.H., Maldonado, I., Sanchez, G., et al. 2010. Molecular mechanisms involved in the inactivation of the first component of human complement by *Trypanosoma cruzi* calreticulin. Mol. Immunol. 47, 1516–1521.

Varaldi, J., Ravalle, M., Labross, C., Lopez-Ferber, M., Boulétreau, M., Fleury, F., 2006. Artifical transfer and morphological description of virus particles associated with superparasitism behaviour in a parasitoid wasp. J. Insect Physiol. 52, 1202–1212.

Viljakainen, L., Pamilo, P., 2008. Selection on an antimicrobial peptide defensin in ants. J. Mol. Evol. 67, 643–652.

Vinchon, S., Moreau, S.J.M., Cherqui, A., Drezen, J.-M., Prévost, G., 2010. Molecular and biochemical analysis of an aspartylglucosaminidase from the venom of the parasitoid wasp *Asobara tabida* (Hymenoptera: Braconidae). Insect Biochem. Mol. Biol. 40, 38–48.

Wago, H., Tanaka, T., 1989. Synergistic effects of calyx fluid and venom of *Apanteles kariyai* Watanabe (Hymenoptera: Braconidae) on the granular cells of *Pseudaletia separata* Walker (Lepidoptera: Noctuidae). Zool. Sci. 6, 691–696.

Webb, B.A., Luckhart, S., 1994. Evidence for an early immunosuppressive role for related *Campoletis sonorensis* venom and ovarian proteins in *Heliothis virescens*. Arch. Insect Biochem. Physiol. 26, 147–163.

Whitfield, J.B., 1998. Phylogeny and evolution of host–parasitoid interactions in hymenoptera. Annu. Rev. Entomol. 43, 129–151.

Wu, M.-L., Ye, G.-Y., Zhu, J.Y., Chen, X.-X., Hu, C., 2008. Isolation and characterization of an immunosuppressive protein from venom of the pupa-specific endoparasitoid *Pteromalus puparum*. J. Invertebr. Pathol. 99, 186–191.

Xu, P., Shi, M., Chen, X.X., 2009. Antimicrobial peptide evolution in the Asiatic honey bee *Apis cerana*. PLoS ONE 4, e4239.

Yu, X.-Q., Jiang, H., Wang, Y., Kanost, M.R., 2003. Nonproteolytic serine proteinase homologs are involved in prophenoloxidase activation in the tobacco hornworm, *Manduca sexta*. Insect Biochem. Mol. Biol. 33, 197–208.

Zhang, D., Dahlman, D.L., Jarlfors, U.E., 1997. Effect of *Microplitis croceipes* teratocytes on host haemolymph protein content and fat body proliferation. J. Insect Physiol. 43, 577–585.

Zhang, G., Lu, Z.-Q., Jiang, H., Asgari, S., 2004. Negative regulation of prophenoloxidase (proPO) activation by a clip-domain serine proteinase homolog (SPH) from endoparasitoid venom. Insect Biochem. Mol. Biol. 34, 477–483.

Zhang, G., Schmidt, O., Asgari, S., 2004. A novel venom peptide from an endoparasitoid wasp is required for expression of polydnavirus genes in host hemocytes. J. Biol. Chem. 279, 41580–41585.

Zhang, G., Schmidt, O., Asgari, S., 2006. A calreticulin-like protein from endoparasitoid venom fluid is involved in host hemocyte inactivation. Dev. Comp. Immunol. 30, 756–764.

Zhang, H., Forman, H.J., Choi, J., 2005. Gamma-glutamyl transpeptidase in glutathione biosyntesis. Method. Enzymol. 401, 468–483.

Zhang, Z., Ye, G.-Y., Cai, J., Hu, C., 2005. Comparative venom toxicity between *Pteromalus puparum* and *Nasonia vitripennis* (Hymenoptera: Pteromalidae) toward the hemocytes of their natural hosts, non-target insects and cultured insect cells. Toxicon 46, 337–349.

Zhu, J.-Y., Fang, Q., Wang, L., Hu, C., Ye, G.-Y., 2010. Proteomic analysis of the venom from the endoparasitoid Pteromalus puparum (Hymenoptera: Pteromalidae). Arch. Insect Biochem. Physiol. 75, 28–44.

Zhu, J.-Y., Ye, G.-Y., Dong, S.-Z., Fang, Q., Hu, C., 2009. Venom of *Pteromalus puparum* (Hymenopera: Pteromalidae) induced endocrine changes in the hemolymph of its host, *Pieris rapae* (Lepidoptera: Pieridae). Arch. Insect Biochem. Physiol. 71, 45–53.

Zhu, J.-Y., Ye, G.-Y., Fang, Q., Hu, C., 2010. Alkaline phosphatase from venom of the endoparasitoid wasp, *Pteromalus puparum*. J. Insect Sci. 10, 14.

Zhu, J.-Y., Ye, G.-Y., Hu, C., 2008. Molecular cloning and characterization of acid phosphatase in venom of the endoparasitoid wasp *Pteromalus puparum* (Hymenoptera: Pteromalidae). Toxicon 51, 1391–1399.

Zhu, Y., Wang, Y., Gorman, M.J., Jiang, H., Kanost, M.R., 2003. *Manduca sexta* serpin-3 regulates prophenoloxidase activation in response to infection by inhibiting prophenoloxidase-activating proteinases. J. Biol. Chem. 278, 46556–46564.

Chapter 19

Proteomics of the Venom of the Parasitoid *Nasonia vitripennis*

Ellen M. Formesyn, Ellen L. Danneels and Dirk C. de Graaf

Laboratory of Zoophysiology, Ghent University, B-9000 Ghent, Belgium

SUMMARY

Nasonia vitripennis is an ectoparasitoid that uses host flies as a food source for its progeny, whereby the venom of adult females is used to subdue the host. The venom of this tiny wasp is able to elicit several host responses ranging from an altered immunity to developmental changes. Over the last few years, the proteomic research of venomous substances has gained more and more interest and several powerful tools have become available. With the completion of the genome sequences of *N. vitripennis*, it became possible to perform an in depth investigation of the venom composition using bioinformatic tools and mass spectrometry. This combined approach enabled the discovery, for the first time, of a full set of venom proteins from a single parasitoid.

ABBREVIATIONS

CID	collision-induced dissociation
ESI-MS	electrospray ionization mass spectrometry
ESI-FT-ICR-MS	electrospray ionization-Fourier transform-ion cyclotron resonance-mass spectrometry
ESI-Q-TOF-MS	electrospray ionization-quadrupole-time of flight-mass spectrometry
FT-ICR-MS	Fourier transform-ion cyclotron resonance-mass spectrometry
GC	gas chromatography
HPLC	high-performance liquid chromatography
Hsc	heat shock cognate protein
Hsp	heat shock protein
LC	liquid chromatography
L-DOPA	L-3,4-di-hydroxyl-phenylalanine
MALDI	matrix assisted laser desorption ionization
MS	mass spectrometry
MudPIT	multidimensional protein identification technology
pI	isoelectric point
PO	phenoloxidase
RP	reversed phase
SCX	strong cation exchange
SDS-PAGE	sodium dodecyl sulfate-poly acrylamide gel electroforesis
TOF	time of flight
2D-PAGE	two dimensional-poly acrylamide gel electroforese

INTRODUCTION

Animal venoms contain a complex mixture of proteins, enzymes, peptides, and small organic compounds. Venomics, the analysis of venom proteomes, has gained a lot of interest because venoms are used for predation or defense and contain toxic compounds that are associated with specific pathologies.

In the past, identification of venom components in parasitoid wasps was restricted due to the technological limitations and lack of genome sequences. Intensive studies of the parasitoid *Pimpla hypochondriaca* could be launched once a cDNA library from its venom gland was constructed (Parkinson *et al.*, 2001). Thereafter, cDNA libraries became available from venom glands of several other parasitoid wasps: *Cotesia rubecula* (Asgari *et al.*, 2003a), *Aphidium ervi* (Falabella *et al.*, 2007), *Microctonus hyperodae* (Crawford *et al.*, 2008), *Eulophus pennicornis* (Price *et al.*, 2009), *Pteromalus puparum* (Zhu *et al.*, 2010), *Eumenes pomiformis* (Baek and Lee, 2010a), and *Chelonus inanitus* (Vincent *et al.*, 2010). Nevertheless, most studies on parasitoid venoms were directed towards a limited number of components, like for instance a calreticulin-like protein (Zhang *et al.*, 2006), a serine proteinase homolog venom protein (Asgari *et al.*, 2003b) and a phenoloxidase (PO; Parkinson *et al.*, 2001), and until 2009 none of the parasitoids' venoms were fully identified.

Nasonia vitripennis (Pteromalidae) is a small ectoparasitoid wasp. Females parasitize pupae and pharate adults belonging to the families Sarcophagidae and Calliphoridae. Other species can also be parasitized, although they are less susceptible to the venom and their sensitivity differs among several life stages (Rivers *et al.*, 1993). Recently, the complete genome of *N. vitripennis* was sequenced (The *Nasonia* Genome Working Group, 2010), which made it possible to accomplish a complete screening of its venom. This unique attempt to unravel the complete venom composition of an ectoparasitoid wasp using both bioinformatic and proteomic approaches resulted in the identification of 79 venom proteins (de Graaf *et al.*, 2010). In this chapter, we will try to place this research outcome in its broader context. The first part will deal with the venom system and venom collection, and will give an overview of the effects of the venom as a whole. The second part will give an overview of the proteomic tools that became available for venomics' studies and will deal with the individual venom components of *N. vitripennis*.

THE VENOM APPARATUS

The ovipositor and venom glands of *N. vitripennis* are much smaller than these in honeybees and their structure is slightly different. For instance, the ovipositor of *N. vitripennis* is folded back in the abdomen. The structure that builds up the venom system is an elongated acid gland, also known as the venom gland, and the venom reservoir that serves as storage room for venom coming from the acid gland (Fig. 1).

Both the acid gland and the reservoir produce the toxic proteins that result in diverse modifications to the immune system, physiology, and the development of the host (Rivers *et al.*, 2006). The venom gland consists of a long, tubular, and folded structure, composed of large columnar cells, which surround the central cuticle-lined lumen. These cells contain secretory granules and invaginations of the apical cell membrane, which are lined with the cuticle of the lumen and end in a vesicular organelle bearing microvilli.

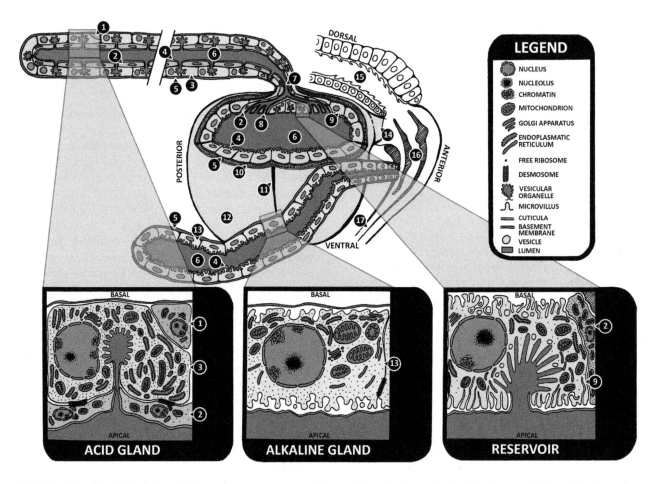

FIGURE 1   **Overview of the *Nasonia vitripennis* venom apparatus.** (1) Interstitial cell (acid gland), (2) chitogenous cell (acid gland), (3) columnar cell (acid gland), (4) cuticula, (5) basement membrane, (6) lumen, (7) duct, (8) ductile, (9) secretory cell (reservoir), (10) striated muscle (reservoir), (11) medial constriction (reservoir), (12) squamous region (reservoir), (13) secretory cell (alkaline gland), (14) efferent duct, (15) vagina, (16) funnel, (17) ovipositor shaft. Please see color plate section at the back of the book.

This vesicular organelle serves as the secretory site of the gland. Between the columnar cells and the lumen, chitogenous or interstitial cells form a thin layer that most probably produces the lining of the lumen and the basement membrane. The contents of the acid gland are stored in a two-lobed venom reservoir and consist of a two-layered cell wall composed of squamous cells and a muscle sheet on the outside. The squamous cells are on the inside covered with a cuticular lining. The mid-dorsal part of the reservoir forms the region where the acid gland has its entry and is composed of secretory epithelium with numerous vesicles and vacuoles. The apical plasma membranes of these cells possess a cuticular involution, which gives rise to long, apical microvilli that are arranged to form a modified vesicular organelle. Therefore, it is likely that the secretory epithelium can also produce several venom proteins, some of which are not produced elsewhere and serve to activate the toxins of the acid glands. At the position where the acid gland is connected to the venom reservoir, the ductus divides into several branches through which the secretions flow into the reservoir (Ratcliffe and King, 1969; Whiting, 1967).

The alkaline or Dufour gland, also derived from the female accessory reproduction glands in female Hymenoptera, is much shorter than the acid gland. This gland has no vesicular organelles, but the secretory product is delivered into the lumen by accumulation of granular material within the apices of the microvilli. The alkaline gland ends in the discharge duct that seems to be connected to a chitinous funnel of the vagina and is not connected to the venom reservoir. The viscous contents of this gland are injected independently of the products of the acid gland and do not contain toxic substances. Probably, these secretions have a lubricatory function and are involved in greasing the ovipositor components or used to smear the eggs (Ratcliffe and King, 1967; King and Ratcliffe, 1969).

VENOM COLLECTION

Because of the small size of the wasps (1.0–3.5 mm for females), it is impossible to employ the manual or electric milking technique performed to collect the venom of, for instance, honeybees, that delivers substantial quantities of venom (Deyrup and Matthews, 2003; de Graaf *et al.*, 2009). Therefore, dissection of the female wasps is inevitable (Fig. 2).

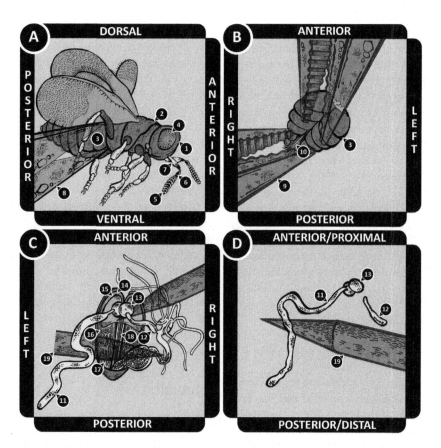

FIGURE 2   Dissection of the venom apparatus in *Nasonia vitripennis*. (1) Cephalon, (2) thorax, (3) abdomen, (4) compound eye, (5) flagellum (antenna), (6) pedicel (antenna), (7) scapus (antenna), (8) scissors, (9) forceps, (10) ovipositor, (11) acid gland, (12) alkaline gland, (13) reservoir, (14) stylet, (15) ovipositor ramus, (16) inner ovipositor plate, (17) outer ovipositor plate, (18) sheath and stylets, (19) needle. Please see color plate section at the back of the book.

The dissection starts by cutting the abdomen from the thorax with a fine pair of scissors. Subsequently, the abdomen is placed dorsally and held with tweezers at the side where the thorax was attached. At the posterior side of the abdomen, the ovipositor, which is folded back in the abdomen and partly covered by the sterna, is gripped firmly. The venom structures are separated from the surrounding tissue by slowly removing the ovipositor from the abdomen and subsequently the whole structure is placed in a droplet of insect saline buffer. The acid gland has a characteristic long, white, tubular structure and is often loosely attached to the venom gland. After removing the ovaries and other tissues, the colorless, two-lobed venom reservoir and a short alkaline gland become visible. The alkaline gland has a bluish-white and shiny appearance and is firmly attached to the base of the ovipositor. By means of a fine needle placed under the venom reservoir, the efferent duct from the reservoir is disconnected from the ovipositor. After removal of the acid and alkaline gland, the content of the venom reservoir is liberated by centrifugation in a microtube.

MODE OF ACTION OF THE VENOM AS A WHOLE

Many parasitoids, and in particular endoparasitoids, coinject a mix of venom, calyx fluid, polydnaviruses and virus-like particles to assure the successful development of offspring. There are several records demonstrating that the injected substances alone are not enough to subdue the host. Indeed, teratocytes, which are derived from the serosal membrane of endoparasitoid eggs, seem to play a role in modifying the host physiology. These large cells start to circulate in the host hemolymph when the eggs hatch and are involved in bypassing immune responses, secretion of proteins and larval nutrient uptake (Andrew *et al.*, 2006; Nakamatsu *et al.*, 2002; Hotta *et al.*, 2001). In addition, the feeding larvae secrete factors that affect metamorphosis, development, hemocytes and other immune reactions (Richards and Edwards, 2001, 2002a, 2002b; Edwards *et al.*, 2006; Price *et al.*, 2009). The main insights into the functionality of the *N. vitripennis* venom were obtained using bioassays and suggest that, in contrast to some endoparasitoid wasps, the venom on its own is able to modify the host physiology, immune responses and biochemical profile.

Metabolism

In order to create the optimal food conditions for the developing larvae, the host metabolism is modified and synchronized with the development of the wasps. Several metabolic changes are observed during parasitism and seem to depend on the life stages involved. Soon after venom injection, both trehalose and glycogen levels will drop significantly. Envenomated pharate adults also display a decrease in the rate of their oxygen consumption following the first 12h; consumption remains suppressed afterwards. This response is not affected by the wound made during envenomation but is probably caused by the venom. In the long term, the levels of pyruvate will increase, whereas whole body lipid levels will rise during the first four days, followed by a decrease eight days after envenomation. These high lipid levels are the most striking changes observed and seem to be synchronized with the development of the parasitoid larvae (Rivers and Denlinger, 1994, 1995). This response is also observed in other parasitoid–host interactions of the ectoparasitoid *Euplectrus separatae* and its lepidopteran host *Pseudaletia separata* (Nurullahoglu *et al.*, 2004; Nakamatsu and Tanaka, 2003, 2004). Moreover, the adult wasps seem to have lost their ability to synthesize lipids *de novo* and become completely dependent on the lipids of the host (Rivero and West, 2002).

Development

Parasitoid venoms are known to induce host paralysis and developmental arrest in order to provide the best conditions for parasitoid progeny. These responses will eventually elicit the death of the host. Both responses are regularly seen in ectoparasitoids, in which paralysis is used to immobilize the host and the induction of a developmental arrest is necessary to prevent elimination of the larvae during molting. The developmental arrest induced by endoparasitoids is more subtle and can be provoked by the venom, polydnaviruses or teratocytes (Rivers *et al.*, 2009; Reed *et al.*, 1997; Dahlman *et al.*, 2003). However, unlike other ectoparasitoids, the venom of *N. vitripennis* seems to be nonparalytic in nontarget flies and its preferred hosts, but young pupae of *Trichoplusia ni* and *Tenebrio molitor* (Lepidoptera) display a dose-dependent loss of abdominal mobility (Rivers *et al.*, 1993). These differences in functionality are not well understood.

The venom of *N. vitripennis* is known to induce a developmental arrest of the immobile host. Previous research demonstrated that this phenomenon could not be reversed by applying exogenous ecdysteroids, but it is possible that metabolic changes are involved in this arrest (Rivers and Denlinger, 1994). Recent studies revealed that under this arrest, the venom causes incomplete development of bristles and eye pigment deposition, which is probably under the influence of venom calreticulin and PO (Rivers and Brogan, 2008).

Immunity

When an appropriate host encounters a parasitoid, it will mount an innate immune response. This immune response can be subdivided into the humoral and cellular immune responses. The former include the synthesis of antimicrobial proteins, melanization, and clotting induced by the phenoloxidase cascade. The cellular immune responses involve phagocytosis, nodule formation and encapsulation. These responses must be evaded by the parasitoid in order to secure the development of the progeny. This is achieved by injecting fluids that alter the immune response. Melanization, an important process of the defense system, is controlled by the PO cascade. This cascade is stimulated after injury and detection of pathogen associated molecular patterns (PAMPS) and is mainly mediated by a subset of the hemocyte population, the crystal cells. Furthermore, the PO cascade plays a role in wound healing, encapsulation, and the production of toxic intermediates. Melanization occurs when prophenoloxidase is activated by a proteolytic serine protease, which results in the formation of melanin and toxic phenolic compounds. Coagulation, another immune process involved in wound healing, is supported by plasmatocytes that carry out a permanent surveillance in circulation. Encapsulation is used to prevent the invasion of parasites, which starts with the detection and attachment of plasmatocytes around the particle. Over the course of a few hours, the lymph gland will increase the proliferation and differentiation of hemocytes into lamellocytes, which are subsequently released to form a multilayered capsule around the particle (Lemaitre and Hoffmann, 2007). Encapsulation is frequently inhibited in endoparasitoids that lay their eggs in the hemocoel of the host (Lu *et al.*, 2006; Parkinson *et al.*, 2002a, b). However, the ectoparasitic wasp *Eulophus pennicornis* is also able to suppress this encapsulation in its lepidopteran host *Lacanobia oleracea*. In this way, wound healing cannot occur and the larvae are able to continue their feeding (Richards and Edwards, 2002a). Many parasitoids are known to inhibit these processes when their venom intermingles with host hemolymph (Richards and Edwards, 2000; Zhang *et al.*, 2004a; Ergin *et al.*, 2006).

Although the venom of *N. vitripennis* is able to suppress these immune responses, which makes feeding possible for female wasps and larvae, melanization is often observed at the wound site. However, *in vitro* assays with hemocytes showed that the venom is able to inhibit host hemolymph PO activity and to suppress the adhesion and spreading behavior of granular cells and plasmatocytes. These plasmatocytes and other susceptible cells are found to retract their pseudopods, have a spherical shape, a granular cytosol, and display plasma membrane blebbing. The two cell types involved in wound healing are plasmatocytes and granular cells. In fact, plasmatocytes recruit granular cells at the wound site using chemo-attractant molecules. However, only plasmatocytes undergo venom-induced cell death by lysis (Rivers *et al.*, 2002). This is also seen with the venoms of *P. hypochondriaca* and *Eulophus pennicornis*, which affect hemocyte morphology, behavior and viability (Rivers *et al.*, 2009; Richards and Edwards, 2002a). Furthermore, previous studies have demonstrated

that the venom-induced cell death is a result of Ca^{2+} mobilization from intracellular stores (e.g., mitochondria and endoplasmic reticulum) into the cytosol, through G-protein dependent signaling. This Ca^{2+} release is stimulated by venom calreticulin in *N. vitripennis* and can lead to irreversible cell damage. This calreticulin was detected by immunoblotting (Rivers and Brogan, 2008). A calreticulin-like protein in *C. rubecula* has the ability to suppress the spreading behavior of host hemocytes, thereby preventing encapsulation. This protein might be an antagonist of host calreticulin and possibly competes for the same binding sites (Zhang *et al.*, 2006). The observed PO activity of the venom can cause the same cell damage as seen with calreticulin (Rivers *et al.*, 2005). Probably this Ca^{2+} mobilization is controlled by phospholipase C and inositol trisphosphate, which subsequently activate phospholipase A_2. The latter is perhaps involved in the regulation of fatty acid synthesis and release, which can lead to toxic accumulation, thus inducing death in the long term (Rivers *et al.*, 2002).

Stress Response

Parasitized hosts also display a molecular response that results in an altered gene expression of proteins, including parasitism-specific proteins and stress proteins. These stress proteins are part of the stress response and can be up- or downregulated in envenomated hosts. Observations of the expression profiles of stress proteins in *Sarcophaga crassipalpis*, using northern blot hybridizations 13 h after envenomation, revealed the upregulation of both heat shock protein 23 (*hsp23*) and *hsp70*. Additionally, *hsp90* and heat shock cognate protein 70 (*hsc70*) transcripts are downregulated. These transcriptional changes are possibly involved in the immune response or developmental arrest following envenomation (Rinehart *et al.*, 2002). A similar response is observed after envenomation of the ectoparasitoid *Bracon hebetor* and its lepidopteran host *Plodia interpunctella*. In this case, *hsc70* and small heat shock protein transcripts are upregulated while *hsp90* is downregulated (Shim *et al.*, 2008).

VENOM COMPOSITION IN THE PRE-GENOME AREA

Venoms of parasitoid wasps contain a complex mixture of peptides and proteins, which are mostly acidic in the Ichneumonoidea superfamily (Moreau and Guillot, 2005). Recent analysis with infrared spectroscopy revealed the proteinaceous and acidic nature of *N. vitripennis* venom. The absence of the absorption bands at position 3600 (OH peak), 2900 (C-H stretching) and 1700 (C=O stretching), indicates that the venom is unglycosylated, a quite uncommon feature for parasitoid venoms. This property was also found with a periodic acid Schiff staining performed in our laboratory. Failure to stain venom spots with this technique demonstrates the absence of some carbohydrates including polysaccharides, mucopolysaccharides, glycoproteins and glycolipids. Furthermore, the venom is composed of mid- to high-molecular-weight proteins, ranging from 13 to 200 kDa, and it is most likely that multiple proteins are involved in the modification of host development and physiology (Rivers et al., 2006).

High-performance liquid chromatography (HPLC) analysis of crude venom revealed the presence of two major components of crude venom, namely apamin and histamine. An enzymatic property of crude venom concerns the PO activity. Two venom proteins of 68 and 160 kDa are found to perform this PO activity, through sodium dodecyl sulfate-poly acrylamide gel electrophoresis (SDS-PAGE) and subsequent oxidation of L-3,4-dihydroxyl-phenylalanine (L-DOPA) as substrate (Abt and Rivers, 2007). Recently, two laccase proteins were found to be present in the venom using a bioinformatic approach. Probably one of these proteins, with a molecular weight of 68 kDa, is performing this L-DOPA oxidizing activity (de Graaf et al., 2010). Venom of the endoparasitoid wasp *P. hypochondriaca* also possesses this PO activity, which is performed by three genes (Parkinson et al., 2001). Rivers and colleagues detected also a calreticulin-like protein with an approximate molecular weight of 68 kDa in the venom (Rivers and Brogan, 2008). The possible effects of these compounds on the host have already been discussed above (see 'Metabolism' and 'Immunity') and Labrosse and colleagues discovered a Ras homologous GTPase activating protein (RhoGAP) as a main immune suppressive factor in the *Leptopilina boulardi–Drosophila* interaction (Labrosse et al., 2005; Colinet et al., 2007).

TOOLS FOR VENOM CHARACTERIZATION

For many years, research towards snake and arthropod venoms has been performed with several proteomic tools, to reveal the structure and biological role of these venomous components. Single purified venom components can be characterized using Edman degradation whereby peptides or proteins are sequenced by stepwise chemical degradation from the N-terminus, with subsequent identification of the released amino acid derivatives by UV-absorbance spectroscopy. This sequencing method can be carried out on peptides consisting of 70 residues, while mass spectrometric (MS) methods can only sequence peptides up to 25 residues. However, Edman degradation has the limitation that sequencing is impossible with a blocked amino-terminus due to modifications, and large quantities of purified peptides are needed in comparison with MS (Favreau et al., 2006). Although the former method is still being used, mass spectrometric analyses have gained more and more interest because of their superior sensitivity and high sample throughput. Substantial amounts of

hymenopteran venoms have already been analyzed using these MS strategies, sometimes in combination with Edman sequencing, in order to characterize a few venom components. This methodology was used for the study of the venoms from the endoparasitoid wasp *C. rubecula* (Asgari et al., 2003a), the solitary bee *Osmia rufa* (Stöcklin et al., 2010), and the jack jumper ant *Myrmecia pilosula* (Davies et al., 2004). The combination of Edman degradation and matrix assisted laser desorption ionization-time of flight-MS (MALDI-TOF-MS) was also used to characterize a defensin-like antimicrobial peptide in the venom of *N. vitripennis* (Ye et al., 2010).

Before complex protein mixtures can enter the mass spectrometer, they are separated using SDS-PAGE, liquid chromatography (LC) or two dimensional-poly acrylamide gel electrophoresis (2D-PAGE) in order to reduce the sample complexity (Steen and Mann, 2004). For several years, 2D-PAGE was the main tool in proteomics to separate proteins, but it has many drawbacks. 2D-PAGE can detect isoforms, but favors only abundant proteins, shows difficulties in resolving and visualizing proteins with extreme properties (molecular weight and isoelectric point (pI)), has a limited dynamic range, and decreases the detection sensitivity in the mass spectrometer (Aerts, 2010; Delahunty and Yates, 2005). Subsequently, the proteins undergo a digest and the resulting peptide samples are offered to the ion source, in which MALDI and electrospray ionization (ESI) are the most common methods used (discussed in 'MALDI-TOF/TOF-MS' and 'ESI-FT-ICR-MS', below). However, complete protein mixtures can also undergo a digest before they are separated. The resulting peptides are then fractionated using the same separation methods including 1D, 2D (strong cation exchange (SCX)/reversed phase (RP)) or 3D (SCX/avidin/RP) LC and this in an online or offline setup. Gas chromatography (GC) is also applied for sample separation, although less frequently. According to previous observations, the off-line approach increases the number of detected peptides with MALDI-TOF-MS and electrospray ionization mass spectrometry (ESI-MS), unlike the on-line analysis, which reduces sample manipulation (Gilar et al., 2005). The former approach is also more compatible with higher acetonitrile concentrations and optimization of the separations between dimensions is possible. Furthermore, a continuous linear gradient in SCX-LC makes it possible to achieve a high resolution and peak capacity (Aerts, 2010).

There are currently five types of mass analyzers used for proteomics research, these are the ion trap, TOF, quadrupole, orbitrap, and Fourier transform ion cyclotron (FT-ICR-MS) analyzers (Aebersold and Mann, 2003). Together with several separation options and ion source possibilities, various combinations of high-power hybrid instruments are available to analyze the proteome of venomous species (Table 1).

Venom fingerprinting is performed with single-stage MS, which analyzes complete unfractionated venom

TABLE 1 Mass Spectrometry on Hymenoptera Venoms

Species[*]	Extract	Proteomic technique	Result	Reference
Anoplius samariensis and *Batozonellus maculifrons* (solitary spider wasp)	Venom of 60 Anoplius females, venom of 29 Batozonellus females	RP-HPLC, MALDI-TOF-MS, MALDI-CID	Isolation and structure of 2 novel peptide neurotoxins	Konno *et al.*, 1998
Polybia occidentalis and *P. sericea* (polistine wasps)	3 pooled venom sacs	Silica chromatography, enantioselective GC-MS	Diverse volatile constituents in both species	Dani *et al.*, 2000
Anterhynchium flavomarginatum micado (eumenine solitary wasp)	Venom of 136 female wasps	RP-HPLC, MALDI-TOF-MS, FAB-MS/MS (fast-atom-bombardment MS/MS), automated Edman degradation	Structure of a new mast cell degranulating peptide	Konno *et al.*, 2000
Cyphononyx dorsalis (solitary spider wasp)	Venom of 2 female wasps	MALDI-TOF MS-guided fractionation, RP-HPLC and MALDI-TOF-MS, Edman degradation	Isolation and sequence determination of 2 novel peptides and 1 known peptide	Konno *et al.*, 2001
Anoplius samariensis and *Batozonellus maculifrons* (pompilid wasps)	Venom	HPLC, MALDI-TOF-MS, CID/PSD (collision-induced dissociation/post-source decay)	Sequence determination of 2 peptides	Hisada *et al.*, 2002
Megacampsomeris prismatica, *Campsomeriella annulata annulata* and *Carinoscolia melanosoma fascinata* (solitary scoliid wasps)	Venom of 24 female wasps	RP-HPLC, Edman degradation, MALDI-TOF-MS	Identification of bradykinins in scoliid wasp venoms	Konno *et al.*, 2002
Pimpla hypochondriaca (Ichneumonidae, endoparasitoid wasp)	Venom	RP-HPLC, SDS-PAGE, MALDI-TOF-MS	Identification of a paralytic peptide (Pimplin)	Parkinson *et al.*, 2002a, b
Agelaia pallipes pallipes (social wasp)	Venom	RP-HPLC/ESI-MS, CID-LC/ESI-MS/MS	Structural characterization of 2 novel mastoparan peptides	Mendes *et al.*, 2004
Polybia paulista (social wasp)	Venom of 500 worker wasps	RP-HPLC, Edman degradation, CID-ESI-MS/MS	Structural characterization of 2 novel peptides	Ribeiro *et al.*, 2004
Cotesia rubecula (Braconidae, endoparasitoid wasp)	Venom of 50 females	RP-HPLC/ESI-MS, automated Edman degradation	Identification and characterization of Vn1.5	Zhang *et al.*, 2004b
Vespula vulgaris (major yellow jacket wasp)	Venom	Reducing SDS-PAGE, MALDI-TOF-MS, RP-HPLC, MALDI-MS or ESI-MS	Glycan structure determination of venom hyaluronidases	Kolarich *et al.*, 2005
Protopolybia exigua (neotropical social wasp)	Venom	RP-HPLC, ESI-Triple Quadrupole MS, Edman degradation	Structural characterization of 3 novel peptides	Mendes *et al.*, 2005
Polybia paulista (social wasp)	Venom of 3000 worker wasps	RP-HPLC, ESI-MS, automated Edman degradation	Structural characterization of 2 novel inflammatory peptides	Souza *et al.*, 2005
Polistes dominulus, P. gallicus, P. nimphus, P. sulcifer and *P. olivaceus* (social paper wasps)	Venom	GS-MS	Differences in mass spectra between species	Bruschini *et al.*, 2006

(Continued)

TABLE 1 (Continued)

Species*	Extract	Proteomic technique	Result	Reference
Ampulex compressa (Ampulicidae, endoparasitoid wasp)	Milked venom	Dabsylation, HPLC, MALDI-TOF-MS	Identification of GABA, taurine and β-alanine	Moore et al., 2006
Polistes dominulus (social paper wasp)	Venom	HPLC-UV, RP-HPLC, ESI-TOF-MS, ESI-IT-MS, MALDI-TOF-MS	Identification of 2 new antibacterial peptides	Turillazzi et al., 2006
4 Polistes species (social wasps) and Vespa crabro	Venom from single specimens	MALDI-TOF/TOF-MS	Different venom composition between species in medium MW fraction	Turillazzi et al., 2007
Polistes major major (neotropical social wasp)	Venom from 5 specimens	RP-HPLC, MALDI-TOF-MS, ESI-Q-TOF-MS, Edman degradation	Characterization of 3 novel peptides	Čeřovský et al., 2007
Aphidius ervi (Braconidae, endoparasitoid wasp)	Venom from 100 females	HPLC, MALDI-TOF-MS, Edman degradation	γ-glutamyl transpeptidase (Ae-γ-GT)	Falabella et al., 2007
6 Vespula species (yellow jacket wasps)	Venom	SDS-PAGE, MALDI-MS, LC-ESI-Q-TOF-MS/MS	Differences in major allergens from yellow jacket wasps	Kolarich et al., 2007
Polybia occidentalis (social wasp)	Venom	RP-HPLC, ESI-MS/MS	Isolation and structural determination of an anti-nociceptive peptide	Mortari et al., 2007
Polistes major major, Polistes dorsalis dorsalis and Mischocyttarus phthisicus (neotropical social wasps)	Venom	RP-HPLC, MALDI-TOF-MS, ESI-TOF-MS	Characterization of 4 novel antimicrobial peptides	Čeřovský et al., 2008
Vespa bicolor Fabricius (hornet wasp)	Venom	RP-HPLC, Edman degradation, FAB-MS	Characterization of antimicrobial peptides	Chen et al., 2008
Microctonus hyperodae (Braconidae, endoparasitoid wasp)	15 venom reservoirs	PAGE, trypsin digest, RP-HPLC, LTQ ion trap tandem MS	Identification of 3 most abundant venom proteins	Crawford et al., 2008
Leptopilina boulardi (Figitidae, endoparasitoid wasp)	Venom of 130 glands	Native-PAGE, SDS-PAGE, MALDI-TOF-MS	Characterization of a serpin	Colinet et al., 2009
Orancistrocerus drewseni drewseni (solitary wasp)	Venom of 9 female wasps	RP-HPLC, MALDI-TOF-MS, CID/PSD, ESI-TOF-MS/MS, automated Edman degradation	Isolation of novel mastoparan and protonectin analogs	Murata et al., 2009
Vespula vulgaris (major yellow jacket wasp)	Venom (200mg)	RP-HPLC, MALDI-TOF-MS, Q-TOF LC-MS/MS	Characterization of N-glycans from allergen Ves v 2	Seppälä et al., 2009
Eumenes pomiformis (Eumenidae, solitary hunting wasp)	Venom of 4 reservoirs	LC, ESI-Q-TOF-MS/MS	Identification of 2 most abundant peptides	Baek and Lee, 2010a
Orancistrocerus drewseni (solitary wasp)	Venom of 5 female wasps	MALDI-TOF-MS, LC-ESI-Q-TOF-MS/MS	Isolation of 3 mastoparan-like venom peptides	Baek and Lee, 2010b
Nasonia vitripennis (Pteromalidae, ectoparasitoid wasp)	Venom of 10 specimens	2D-LC-MALDI-TOF/TOF, 2D-LC-ESI-FT-ICR	14 proteins by 2D-LC-MALDI-TOF MS, 61 proteins by 2D-LC-ESI-FT-ICR	de Graaf et al., 2010

(Continued)

TABLE 1 (Continued)

Species[*]	Extract	Proteomic technique	Result	Reference
Nasonia vitripennis (Pteromalidae, ectoparasitoid wasp)	Venom of 2000 reservoirs	Edman degradation, MALDI-TOF-MS	Characterization and purification of defensin-NV	Ye *et al.*, 2010
Chelonus inantius (Braconidae, endoparasitoid wasp)	For each protein separation, venom of four wasps was used	Nano-LC-MS/MS	Identification of 29 venom proteins	Vincent *et al.*, 2010
Apis mellifera (honeybee)	Venom of 35 worker bees	2D gel electrophoresis, MALDI-TOF/TOF-MS, nano-HPLC Q-TRAP LC-MS/MS	6 known proteins + 3 new proteins	Peiren *et al.*, 2005
Apis mellifera (honeybee)	Venom	ICP-MS (inductively coupled plasma MS)	Low levels of metal contamination in honeybee venom	Kokot and Matsysiak, 2008
Apis mellifera (honeybee)	Venom of 10 specimens, Cuticular compounds of 10 specimens	MALDI-TOF/TOF-MS, LC-ESI high resolution MS	Differences of venom composition between bee castes, presence of venom peptides in cuticle of females	Baracchi and Turillazzi, 2010
Myrmecia pilosula (jack jumper ant)	Venom of 3000 ants	HPLC-MS, MS/MS, Edman sequencing	>50 peptides in 4–9-kDa range	Davies *et al.*, 2004
Solenopsis invicta and *S. geminate* (ants)	Metapleural gland secretions (10 venom glands)	GC-MS	Small differences in chemical composition between 2 species	Cabrera *et al.*, 2004
Bombus lapidarius (bumblebee)	Venom from 1 specimen	Nano-ESI-MS, ESI-QqTOF, MALDI-LIFT-TOF-TOF	24 compounds	Favreau *et al.*, 2006
Myrmicaria melanogaster (Brunei ant)	Venom	GC-MS	5 new alkaloids	Jones *et al.*, 2007
Pachycondyla chinensis (ant)	Venom	2-DE, western blots, N-terminal amino acid sequencing, ESI-MS/MS, RT-PCR	2 major allergens of 23 and 25 kDa	Lee *et al.*, 2009
Solenopsis invicta (red imported fire ant)	Venom	GC-MS	6 major venom alkaloids in monogyne and polygyne colonies	Lai *et al.*, 2008
Solenopsis richteri (black fire ant)	Venom	GC-MS	7 novel piperideines	Chen and Fadamiro, 2009
Dinoponera australis (giant Neotropical hunting ant)	Venom of multiple members of a colony	HPLC, Q-TOF-MS	>75 proteinaceous components	Johnson *et al.*, 2010
Osmia rufa (solitary bee)	Venom of 30 bees	On-line LC-ESI-MS, MS/MS *de novo* sequencing, automated Edman degradation	>50 compounds in 0.4–4-kDa range	Stöcklin *et al.*, 2010
Solenopsis invicta Buren, *Solenopsis richteri* Forel and hybrid imported fire ant (*S. invicta* × *S. richteri*)	Venom	GC-MS	Abundance of 6 $\Delta^{1,6}$-piperideines in 2 species and their hybrid	Chen *et al.*, 2010

FAB, fast atom bombardment; ICP, inductively coupled plasma; LC-ESI-MS liquid chromatography electrospray ionization mass spectrometry; PSD, post-source decay.
*Parasitoids in bold.

samples to reveal the molecular mass of peptides. Mostly, MALDI-TOF-MS or liquid chromatography coupled to ESI-MS are applied for the study of proteinaceous venom components, and have become useful taxonomic tools (Escoubas *et al.*, 2008; Souza *et al.*, 2008). Tandem MS (MS/MS) on the other hand, is performed to unravel the exact amino acid sequence of peptides and usually only a few venom fractions are used (Domon and Aebersold, 2006). Tandem MS analysis starts with the ionization and detection of primary ions and subsequent fragmentation of these ions with collision-induced dissociation (CID), surface-induced dissociation (SID), ultra-violet photodissociation (UVPD) or other dissociation methods. The resulting secondary ions are then analyzed in another round of MS (Delahunty and Yates, 2005).

The most widely used and powerful method in tandem MS is shotgun proteomics, also known as bottom-up proteomics. It allows the identification of proteins unbiased towards their molecular weight, pI and hydrophobicity although this method suffers from limited dynamic range and high mass spectra redundancy (Domon and Aebersold, 2006). However, shotgun proteomics are facilitated by the use of multidimensional protein identification technology (MudPIT). The MudPIT technique incorporates strong cationic exchange and reversed-phase HPLC, tandem mass spectrometry, and database screening to reveal the identity of the protein (Wu and MacCoss, 2002).

VENOM COMPOSITION IN THE GENOME AREA

Venom Analysis by Means of Bioinformatics

In order to identify potential venom proteins of *N. vitripennis*, the genome was screened for homologs using the BLASTP program, with a query file that contained 383 known hymenopteran venom proteins. The hits found were further narrowed down to a set of 59 protein sequences by an in-depth study of their conserved-domain architecture and presence of a signal peptide. Twenty one of them could be confirmed by reverse-transcription PCR (de Graaf *et al.*, 2010).

Proteomic Approach

A proteomic study was done on the content of 10 *N. vitripennis* venom reservoirs using two different ionization methods and tandem MS instruments; i.e., MALDI-TOF/TOF-MS and electrospray ionization-Fourier transform-ion cyclotron resonance-MS (ESI-FT-ICR-MS). Prior to MS analysis, samples underwent a tryptic digest and subsequent 2D-LC separation (de Graaf *et al.*, 2010). Two-dimensional LC is a commonly used separation method to detect the lowest abundant peptides and uses

a biphasic nanocolumn (nano-LC format) (Malmström *et al.*, 2007). In the first dimension, an SCX column was placed upstream from the RP segment (second dimension) in the nanocolumn. The former column has high loading capacities and acts as a peptide reservoir. The RP part includes two C_{18} columns, the first one is a trapping column followed by a second analytical column. These result in a reduced solvent flow rate, an increased peak capacity, higher sensitivity of downstream tandem MS applications, and higher efficiency of ESI (Delahunty and Yates, 2005).

MALDI-TOF/TOF-MS

MALDI-TOF is commonly used for the rapid screening of simple peptide mixtures or particular target compounds in a number of venom extracts with minute amounts and allows the analysis of mass ranges up to 500,000 Da (Favreau *et al.*, 2006; Hisada *et al.*, 2002). MALDI, known as a soft ionization method, ionizes peptides using a laser beam and produces mass spectra with few or no fragment-ions. This approach generates singly charged peptides under low pressure and fixed voltages, regardless of the molecular mass. In TOF instruments, the ions are analyzed according to their flight time. MALDI-TOF/TOF-MS allows peptide fingerprinting of compounds in a tryptic digest mixture and is known for the high throughput, sensitivity, resolution (above 10,000), and good mass accuracy (approx. 10 ppm). This type of mass analyzer has a broader mass to charge (*m/z*) range compared to FT-ICR instruments (Schrader and Klein, 2004; Domon and Aebersold, 2006).

In the case of *N. vitripennis*, nine SCX venom fractions were analyzed using the 4700 Proteomic analyzer, Applied Biosystems. The first TOF mass analyzer allows the transmission of all sample ions. After an automatic selection of precursor peaks per MALDI spot, ions with a certain *m/z* value are fragmented in a collision cell by CID with N_2 (Aebersold and Mann, 2003). Subsequently, the second TOF analyzer monitors the fragment ions. In this way, the MALDI-TOF/TOF-MS analysis contributed to the identification of 29 peptides, representing 14 proteins (de Graaf *et al.*, 2010).

ESI-FT-ICR-MS

FT-ICR-MS is one of the most powerful tools in proteomics and the LTQ-FTUltra mass spectrometer (Thermo Fisher Scientific) was used in parallel to analyze the venom composition of *N. vitripennis* (de Graaf *et al.*, 2010). This was the first record whereby FT-ICR-MS was used to analyze the venom composition of a parasitoid wasp. Before the samples entered the mass analyzer, they were ionized using ESI. This soft ionization method adds or removes a proton under high pressure and renders multiply charged ions, which reduce the *m/z* ratio. In addition, large molecules acquire many charges, which is favorable for FT-ICR detection (Russell and Edmondson, 1997). ESI is often coupled to ion traps such as FT-ICR and triple quadrupole instruments, which generate fragment ion spectra (CID) of selected precursor ions (see 'Tools for Venom Characterization'). ESI is used for complex mixtures, containing thermally labile and high-molecular-weight compounds, which are observed at lower m/z values. Furthermore, using ESI leads to a more precise mass determination than MALDI, but both techniques are complementary to each other. However, ESI-MS seems to be more accurate in mass determination compared to MALDI-TOF (Favreau *et al.*, 2006).

FT-ICR instruments are comprised of a cyclotron device that traps ions in a magnetic field, which travel in a circular path. The ions are detected according to their cyclotron frequency while rotating in a magnetic field. This signal is converted to the mass spectrum using the Fourier transformation. FT-ICR-MS has an extremely high mass resolution (approx. 150,000), which is higher than the resolution levels of other MS instruments (Schrader and Klein, 2004; Quinton *et al.*, 2005). As a result, two ions with similar *m/z* values are detected as two different ions. Furthermore, FT-ICR-MS generates a better signal-to-noise ratio and the masses are determined with excellent accuracy (approx. 2 ppm), because the superconducting magnet is more stable than radio frequency voltage (Domon and Aebersold, 2006).

The LC-ESI-FT-MS analysis of *N. vitripennis* venom used nine selected SCX fractions, which resulted in the identification of 258 unique peptides. In this way, the analysis contributed to the identification of 61 venom proteins including the 14 proteins previously identified with MALDI-TOF/TOF analysis. However, this resulted also in the identification of 15 additional nonsecretory proteins (de Graaf *et al.*, 2010).

When comparing both mass spectrometric approaches, it seems that LC-ESI-FT-MS is superior in determining the venom composition of *N. vitripennis* with 61 of the 79 components identified using this technique. MALDI-TOF/TOF-MS, on the other hand, could reveal only 14 proteins, which had already been found with LC-ESI-FT-MS. Both MS analyses yielded 52 proteins that were not found with the bioinformatic study, while nine proteins found with bioinformatics could be confirmed using MS. The difference in coverage between the two MS techniques is probably caused by both technical and biological limitations (de Graaf *et al.*, 2010).

CONCLUSIONS AND FUTURE DIRECTIONS

The venom of *N. vitripennis* is able to elicit a wide range of host responses, which makes it an interesting research objective to unravel its pharmacopoeia. The availability of the complete genome sequence in combination with powerful proteomic techniques has lifted venom proteomics to

another level, as demonstrated by the recent discovery of 79 *N. vitripennis* venom proteins (de Graaf *et al.*, 2010). The possible functions and interactions of these 79 venom proteins with host pathways were discussed previously, but remain largely speculative (Danneels *et al.*, 2010). Moreover, the main challenge now is to discover the functions of a subset of 23 venom compounds that do not display similarities to any known protein. Knowing the fact that the venom as a whole influences host immunity and development, it seems reasonable to believe that at least some of these 23 proteins are involved. However, genome sequencing of wasps is not absolutely required, as was demonstrated recently by Vincent and colleagues, who discovered successfully the venom composition of *Chelonus inanitus* by random expressed sequence tags sequencing in combination with nano-LC-MS/MS (Vincent *et al.*, 2010).

Thus, in the next decades, applications of the venom in medicine, biotechnology, and agriculture can be expected. With the new sequencing techniques that have become available, an important threshold to genome sequencing related parasitoids has been eliminated, which will eventually boost research in the field of venomics.

ACKNOWLEDGMENT

We thank Dieter De Koker for the artwork.

REFERENCES

Abt, M., Rivers, D.B., 2007. Characterization of phenoloxidase activity in venom from the ectoparasitoid *Nasonia vitripennis* (Walker) (Hymenoptera: Pteromalidae). J. Invertebr. Pathol. 94, 108–118.

Aebersold, R., Mann, M., 2003. Mass spectrometry-based proteomics. Nature 422, 198–207.

Aerts, M., 2010. The impact of high-resolution mass spectrometry in shotgun proteomics: two case studies. PhD-dissertation, Ghent University.

Andrew, N., Basio, M., Kim, Y., 2006. Additive effect of teratocyte and calyx fluid from *Cotesia plutellae* on immunosuppression of *Plutella xylostella*. Physiol. Entomol. 31, 341–347.

Asgari, S., Zareie, R., Zhang, G.M., Schmidt, O., 2003. Isolation and characterization of a novel venom protein from an endoparasitoid, *Cotesia rubecula* (Hym: Braconidae). Arch. Insect. Biochem. Physiol. 53, 92–100.

Asgari, S., Zhang, G.M., Zareie, R., Schmidt, O., 2003. A serine proteinase homolog venom protein from an endoparasitoid wasp inhibits melanization of the host hemolymph. Insect. Biochem. Mol. Biol. 33, 1017–1024.

Baek, J.H., Lee, S.H., 2010a. Differential gene expression profiles in the venom gland/sac of *Eumenes pomiformis* (Hymenoptera: Eumenidae). Toxicon 55, 1147–1156.

Baek, J.H., Lee, S.H., 2010b. Isolation and molecular cloning of venom peptides from Orancistrocerus drewseni (Hymenoptera: Eumenidae). Toxicon 55, 711–718.

Baracchi, D., Turillazzi, S., 2010. Differences in venom and cuticular peptides in individuals of Apis mellifera (Hymenoptera: Apidae) determined by MALDI-TOF MS. J. Insect Physiol. 56, 366–375.

Bruschini, C., Dani, F.R., Pieraccini, G., Guarna, F., Turillazzi, S., 2006. Volatiles from the venom of five species of paper wasps (Polistes dominulus, P. gallicus, P. nimphus, P. sulcifer and P. olivaceus). Toxicon 47, 812–825.

Cabrera, A., Williams, D., Hernandez, J.V., Caetano, F.H., Jaffe, K., 2004. Metapleural- and postpharyngeal-gland secretions from workers of the ants Solenopsis invicta and S. geminata. Chem. Biodivers. 1, 303–311.

Čeřovský, V., Pohl, J., Yang, Z.H., Alam, N., Attygalle, A.B., 2007. Identification of three novel peptides isolated from the venom of the neotropical social wasp Polistes major major.. J. Pept. Sci. 13, 445–450.

Čeřovský, V., Slaninova, J., Fucik, V., Hulacova, H., Borovickova, L., Jezek, R., et al. 2008. New potent antimicrobial peptides from the venom of Polistinae wasps and their analogs. Peptides 29, 992–1003.

Chen, J., Shang, H.W., Jin, X.X., 2010. Interspecific variation of Delta(1,6)-piperideines in imported fire ants. Toxicon 55, 1181–1187.

Chen, L., Fadamiro, H.Y., 2009. Re-investigation of venom chemistry of Solenopsis fire ants. I. Identification of novel alkaloids in S. richteri. Toxicon 53, 469–478.

Chen, W.H., Yang, X.B., Yang, X.L., Zhai, L., Lu, Z.K., Liu, J.Z., et al. 2008. Antimicrobial peptides from the venoms of Vespa bicolor Fabricius. Peptides 29, 1887–1892.

Colinet, D., Dubuffet, A., Cazes, D., Moreau, S., Drezen, J.-M., Poirie, M., 2009. A serpin from the parasitoid wasp Leptopilina boulardi targets the Drosophila phenoloxidase cascade. Dev. Comp. Immunol. 33, 681–689.

Colinet, D., Schmitz, A., Depoix, D., Crochard, D., Poirié, M., 2007. Convergent use of RhoGAP toxins by eukaryotic parasites and bacterial pathogens. Plos Pathog. 3, 2029–2037.

Crawford, A.M., Brauning, R., Smolenski, G., Ferguson, C., Barton, D., Wheeler, T.T., et al. 2008. The constituents of Microctonus sp. parasitoid venoms. Insect. Mol. Biol. 17, 313–324.

Dahlman, D.L., Rana, R.L., Schepers, E.J., Schepers, T., DiLuna, F.A., Webb, B.A., 2003. A teratocyte gene from a parasitic wasp that is associated with inhibition of insect growth and development inhibits host protein synthesis. Insect. Mol. Biol. 12, 527–534.

Dani, F.R., Jeanne, R.L., Clarke, S.R., Jones, G.R., Morgan, E.D., Francke, W., et al. 2000. Chemical characterization of the alarm pheromone in the venom of Polybia occidentalis and of volatiles from the venom of P. sericea. Physiol. Entomol. 25, 363–369.

Danneels, E.L., Rivers, D.B., de Graaf, D.C., 2010. Venom proteins of the parasitoid wasp Nasonia vitripennis: recent discovery of an untapped pharmacopee. Toxins 2, 491–519.

Davies, N.W., Wiese, M.D., Browne, S.G.A., 2004. Characterisation of major peptides in 'jack jumper' ant venom by mass spectrometry. Toxicon 43, 173–183.

de Graaf, D.C., Aerts, M., Brunain, M., Desjardins, C.A., Jacobs, F.J., Werren, J.H., et al. 2010. Insights into the venom composition of the ectoparasitoid wasp Nasonia vitripennis from bioinformatic and proteomic studies. Insect. Mol. Biol. 19, 11–26.

de Graaf, D.C., Aerts, M., Danneels, E., Devreese, B., 2009. Bee, wasp and ant venomics pave the way for component-resolved diagnosis of sting allergy. J. Proteomic. 2, 145–154.

Delahunty, C., Yates, J.R., 2005. Protein identification using 2D-LC-MS/MS. Methods 35, 248–255.

Deyrup, L.D., Matthews, R.W., 2003. A simple technique for milking the venom of a small parasitic wasp, Melittobia digitata (Hymenoptera: Eulophidae). Toxicon 42, 217–218.

Domon, B., Aebersold, R., 2006. Review—Mass spectrometry and protein analysis. Science 312, 212–217.

Edwards, J.P., Bell, H.A., Audsley, N., Marris, G.C., Kirkbride-Smith, A., Bryning, G., et al. 2006. The ectoparasitic wasp Eulophus pennicornis (Hymenoptera: Eulophidae) uses instar-specific endocrine disruption strategies to suppress the development of its host Lacanobia oleracea (Lepidoptera: Noctuidae). J. Insect. Physiol. 52, 1153–1162.

Ergin, E., Uckan, F., Rivers, D.B., Sak, O., 2006. In vivo and in vitro activity of venom from the endoparasitic wasp Pimpla turionellae (L.) (Hymenoptera: Ichneumonidae). Arch. Insect. Biochem. Physiol. 61, 87–97.

Escoubas, P., Quinton, L., Nicholson, G.M., 2008. Venomics: unravelling the complexity of animal venoms with mass spectrometry. J. Mass Spectrom. 43, 279–295.

Falabella, P., Riviello, L., Caccialupi, P., Rossodivita, T., Valente, M.T., De Stradis, M.L., et al. 2007. A gamma-glutamyl transpeptidase of Aphidius ervi venom induces apoptosis in the ovaries of host aphids. Insect. Biochem. Mol. Biol. 37, 453–465.

Favreau, P., Menin, L., Michalet, S., Perret, F., Cheyneval, O., Stocklin, M., et al. 2006. Mass spectrometry strategies for venom mapping and peptide sequencing from crude venoms: Case applications with single arthropod specimen. Toxicon 47, 676–687.

Gilar, M., Olivova, P., Daly, A.E., Gebler, J.C., 2005. Two-dimensional separation of peptides using RP-RP-HPLC system with different pH in first and second separation dimensions. J. Sep. Sci. 28, 1694–1703.

Hisada, M., Konno, K., Itagaki, Y., Naoki, H., Nakajima, T., 2002. Sequencing wasp venom peptides by endopeptidase digestion and nested collision-induced dissociation/post-source decay methods. Rapid Commun. Mass Spectrom. 16, 1040–1048.

Hotta, M., Okuda, T., Tanaka, T., 2001. Cotesia kariyai teratocytes: growth and development. J. Insect. Physiol. 47, 31–41.

Johnson, S.R., Copello, J.A., Evans, M.S., Suarez, A.V., 2010. A biochemical characterization of the major peptides from the venom of the giant Neotropical hunting ant Dinoponera australis. Toxicon 55, 702–710.

Jones, T.H., Voegtle, H.L., Miras, H.M., Weatherford, R.G., Spande, T.F., Garraffo, H.M., et al. 2007. Venom chemistry of the ant Myrmicaria melanogaster from Brunei. J. Nat. Prod. 70, 160–168.

King, P.E., Ratcliffe, N.A., 1969. Structure and possible mode of functioning of female reproductive system in Nasonia vitripennis (Hymenopters-Pteromalidae). J. Zool. 157, 319–344.

Kokot, Z.J., Matysiak, J., 2008. Inductively coupled plasma mass spectrometry determination of metals in honeybee venom. J. Pharm. Biomed. Anal. 48, 955–959.

Kolarich, D., Leonard, R., Hemmer, W., Altmann, F., 2005. The N-glycans of yellow jacket venom hyaluronidases and the protein sequence of its major isoform in Vespula vulgaris. Febs J. 272, 5182–5190.

Kolarich, D., Loos, A., Leonard, R., Mach, L., Marzban, G., Hemmer, W., et al. 2007. A proteomic study of the major allergens from yellow jacket venoms. Proteomics 7, 1615–1623.

Konno, K., Hisada, M., Itagaki, Y., Naoki, H., Kawai, N., Miwa, A., et al. 1998. Isolation and structure of pompilidotoxins, novel peptide neurotoxins in solitary wasp venoms. Biochem. Biophys. Res. Commun. 250, 612–616.

Konno, K., Hisada, M., Naoki, H., Itagaki, Y., Kawai, N., Miwa, A., et al. 2000. Structure and biological activities of eumenine mastoparan-AF (EMP-AF), a new mast cell degranulating peptide in the venom of the solitary wasp (Anterhynchium flavomarginatum micado). Toxicon 38, 1505–1515.

Konno, K., Hisada, M., Naoki, H., Itagaki, Y., Yasuhara, T., Juliano, M.A., et al. 2001. Isolation and sequence determination of peptides in the venom of the spider wasp (Cyphononyx dorsalis) guided by matrix-assisted laser desorption/ionization time of flight (MALDI-TOF) mass spectrometry. Toxicon 39, 1257–1260.

Konno, K., Palma, M.S., Hitara, I.Y., Juliano, M.A., Juliano, L., Yasuhara, T., 2002. Identification of bradykinins in solitary wasp venoms. Toxicon 40, 309–312.

Labrosse, C., Staslak, K., Lesobre, J., Grangeia, A., Huguet, E., Drezen, J.-M., et al. 2005. A RhoGAP protein as a main immune suppressive factor in the Leptopilina boulardi (Hymenoptera, Figitidae)–Drosophila melanogaster interaction. Insect. Biochem. Mol. Biol. 35, 93–103.

Lai, L.C., Huang, R.N., Wu, W.J., 2008. Venom alkaloids of monogyne and polygyne forms of the red imported fire ant, Solenopsis invicta, in Taiwan. Insect. Sociaux 55, 443–449.

Lee, E.K., Jeong, K.Y., Lyu, D.P., Lee, Y.W., Sohn, J.H., Lim, K.J., et al. 2009. Characterization of the major allergens of Pachycondyla chinensis in ant sting anaphylaxis patients. Clin. Exp. Allergy 39, 602–607.

Lemaitre, B., Hoffmann, J., 2007. The host defense of *Drosophila melanogaster*. Annu. Rev. Immunol. 25, 697–743.

Lu, J.F., Hu, J., Fu, W.J., 2006. Levels of encapsulation and melanization in two larval instars of *Ostrinia furnacalis* Guenee (Lep., Pyralidae) during simulation of parasitization by *Macrocentrus cingulum* Brischke (Hym., Braconidae). J. Appl. Entomol. 130, 290–296.

Malmström, J., Lee, H., Aebersold, R., 2007. Advances in proteomic workflows for systems biology. Curr. Opin. Biotechnol. 18, 378–384.

Mendes, M.A., de Souza, B.M., dos Santos, L.D., Palma, M.S., 2004. Structural characterization of novel chemotactic and mastoparan peptides from the venom of the social wasp Agelaia pallipes pallipes by high-performance liquid chromatography electrospray ionization tandem mass spectrometry. Rapid Commun. Mass Spectrom. 18, 636–642.

Mendes, M.A., de Souza, B.M., Palma, M.S., 2005. Structural and biological characterization of three novel mastoparan peptides from the venom of the neotropical social wasp Protopolybia exigua (Saussure). Toxicon 45, 101–106.

Moore, E.L., Haspel, G., Libersat, F., Adams, M.E., 2006. Parasitoid wasp sting: a cocktail of GABA, taurine, and beta-alanine opens chloride channels for central synaptic block and transient paralysis of a cockroach host. J. Neurobiol. 66, 811–820.

Moreau, S.J.M., Guillot, S., 2005. Advances and prospects on biosynthesis, structures and functions of venom proteins from parasitic wasps. Insect. Biochem. Mol. Biol. 35, 1209–1223.

Mortari, M.R., Cunha, A.O.S., Carolino, R.O.G., Coutinho-Netto, J., Tomaz, J.C., Lopes, N.P., et al. 2007. Inhibition of acute nociceptive responses in rats after i.c.v. injection of Thr(6)-bradykinin, isolated from the venom of the social wasp, Polybia occidentalis. Br. J. Pharmacol. 151, 860–869.

Murata, K., Shinada, T., Ohfune, Y., Hisada, M., Yasuda, A., Naoki, H., et al. 2009. Novel mastoparan and protonectin analogs isolated from a solitary wasp, Orancistrocerus drewseni drewseni. Amino Acids 37, 389–394.

Nakamatsu, Y., Fujii, S., Tanaka, T., 2002. Larvae of an endoparasitoid, Cotesia kariyai (Hymenoptera: Braconidae), feed on the host fat body directly in the second stadium with the help of teratocytes. J. Insect. Physiol. 48, 1041–1052.

Nakamatsu, Y., Tanaka, T., 2003. Venom of ectoparasitoid, Euplectrus sp. near plathypenae (Hymenoptera: Eulophidae) regulates the physiological state of Pseudaletia separata (Lepidoptera : Noctuidae) host as a food resource. J. Insect. Physiol. 49, 149–159.

Nakamatsu, Y., Tanaka, T., 2004. Venom of Euplectrus separatae causes hyperlipidemia by lysis of host fat body cells. J. Insect. Physiol. 50, 267–275.

Nurullahoglu, Z.U., Uckan, F., Sak, O., Ergin, E., 2004. Total lipid and fatty acid composition of Apanteles galleriae and its parasitized host. Ann. Entomol. Soc. Am. 97, 1000–1006.

Parkinson, N., Richards, E.H., Conyers, C., Smith, I., Edwards, J.P., 2002. Analysis of venom constituents from the parasitoid wasp Pimpla hypochondriaca and cloning of a cDNA encoding a venom protein. Insect. Biochem. Mol. Biol. 32, 729–735.

Parkinson, N., Smith, I., Audsley, N., Edwards, J.P., 2002. Purification of pimplin, a paralytic heterodimeric polypeptide from venom of the parasitoid wasp Pimpla hypochondriaca, and cloning of the cDNA encoding one of the subunits. Insect Biochem. Mol. Biol. 32, 1769–1773.

Parkinson, N., Smith, I., Weaver, R., Edwards, J.P., 2001. A new form of arthropod phenoloxidase is abundant in venom of the parasitoid wasp Pimpla hypochondriaca. Insect. Biochem. Mol. Biol. 31, 57–63.

Peiren, N., Vanrobaeys, F., de Graaf, D.C., Devreese, B., Van Beeumen, J., Jacobs, F.J., 2005. The protein composition of honeybee venom reconsidered by a proteomic approach. Biochim. Biophys. Acta Proteins and Proteomics 1752, 1–5.

Price, D.R.G., Bell, H.A., Hinchliffe, G., Fitches, E., Weaver, R., Gatehouse, J.A., 2009. A venom metalloproteinase from the parasitic wasp Eulophus pennicornis is toxic towards its host, tomato moth (Lacanobia oleracae). Insect Mol. Biol. 18, 195–202.

Quinton, L., Le Caer, J.P., Phan, G., Ligny-Lemaire, C., Bourdais-Jomaron, J., Ducancel, F., et al. 2005. Characterization of toxins within crude venoms by combined use of Fourier transform mass spectrometry and cloning. Anal. Chem. 77, 6630–6639.

Ratcliffe, N.A., King, P.E., 1967. Venom system of Nasonia vitripennis (Walker) (Hymenoptera – Pteromalidae). Proc. R. Entomol. Soc. London Ser. A, Gen. Entomol. 42, 49–61.

Ratcliffe, N.A., King, P.E., 1969. Morphological, ultrastructural, histochemical and electrophoretic studies on the venom system of Nasonia vitripennis walker (Hymenoptera — Pteromalidae). J. Morphol. 127, 177–203.

Reed, D.A., Loeb, M.J., Beckage, N.E., 1997. Inhibitory effects of parasitism by the gregarious endoparasitoid Cotesia congregata on host testicular development. Arch. Insect. Biochem. Physiol. 36, 95–114.

Ribeiro, S.P., Mendes, M.A., dos Santos, L.D., de Souza, B.M., Marques, M.R., de Azevedo, W.F., et al. 2004. Structural and functional characterization of N-terminally blocked peptides isolated from the venom of the social wasp Polybia paulista. Peptides 25, 2069–2078.

Richards, E.H., Edwards, J.P., 2000. Parasitization of Lacanobia oleracea (Lepidoptera) by the ectoparasitic wasp, Eulophus pennicornis, suppresses hemocyte-mediated recognition of non-self and phagocytosis. J. Invertebr. Pathol. 46, 1–11.

Richards, E.H., Edwards, J.P., 2001. Proteins synthesized and secreted by larvae of the ectoparasitic wasp, Eulophus pennicornis. Arch. Insect. Biochem. Physiol. 46, 140–151.

Richards, E.H., Edwards, J.P., 2002a. Larvae of the ectoparasitic wasp, Eulophus pennicornis, release factors which adversely affect hemocytes of their host, Lacanobia oleracea. J. Invertebr. Pathol. 48, 845–855.

Richards, E.H., Edwards, J.P., 2002b. Parasitism of *Lacanobia oleracea* (Lepidoptera) by the Ectoparasitic wasp, *Eulophus pennicornis*, disrupts the cytoskeleton of host hemocytes and suppresses encapsulation in vivo. Arch. Insect. Biochem. Physiol. 49, 108–124.

Rinehart, J.P., Denlinger, D.L., Rivers, D.B., 2002. Upregulation of transcripts encoding select heat shock proteins in the flesh fly *Sarcophaga crassipalpis* in response to venom from the ectoparasitoid wasp *Nasonia vitripennis*. J. Invertebr. Pathol. 79, 62–63.

Rivero, A., West, S.A., 2002. The physiological costs of being small in a parasitic wasp. Evol. Ecol. Res. 4, 407–420.

Rivers, D.B., Brogan, A., 2008. Venom glands from the ectoparasitoid Nasonia vitripennis (Walker) (Hymenoptera: Pteromalidae) produce a calreticulin-like protein that functions in developmental arrest and cell death in the flesh fly host, Sarcophaga bullata Parker (Diptera: Sarcophagidae). In: Maes, R.P. (Ed.), Insect Physiology: New Research (pp. 259–278). Nova Science Publishers, New York.

Rivers, D.B., Crawley, T., Bauser, H., 2005. Localization of intracellular calcium release in cells injured by venom from the ectoparasitoid *Nasonia vitripennis* (Walker) (Hymenoptera: Pteromalidae) and dependence of calcium mobilization on G-protein activation. J. Insect Physiol. 51, 149–160.

Rivers, D.B., Dani, M.P., Richards, E.H., 2009. The mode of action of venom from the endoparasitic wasp *Pimpla Hypochondriaca* (Hymenoptera: Ichneumonidae) involves Ca^{2+}-dependent cell death pathways. Arch. Insect Biochem. Physiol. 71, 173–190.

Rivers, D.B., Denlinger, D.L., 1994. Redirection of metabolism in the flesh fly, *Sarcophaga bullata*, following envenomation by the ectoparasitoid *Nasonia vitripennis* and correlation of metabolic effects with the diapause status of the host. J. Insect Physiol. 40, 207–215.

Rivers, D.B., Denlinger, D.L., 1995. Venom-induced alterations in fly lipid metabolism and its impact on larval development of the ectoparasitoid *Nasonia vitripennis* (Walker) (Hymenoptera, Pteromalidae). J. Invert. Pathol. 66, 104–110.

Rivers, D.B., Genco, M., Sanchez, R.A., 1999. In vitro analysis of venom from the wasp *Nasonia vitripennis*: Susceptibility of different cell lines and venom-induced changes in plasma membrane permeability. In Vitro Cell. Dev. Bio. Anim. 35, 102–110.

Rivers, D.B., Hink, W.F., Denlinger, D.L., 1993. Toxicity of the venom from *Nasonia vitripennis* (Hymenoptera, Pteromalidae) towards fly hosts, nontarget insects, different developmental stages, and cultured insect cells. Toxicon 31, 755–765.

Rivers, D.B., Ruggiero, L., Hayes, M., 2002. The ectoparasitic wasp *Nasonia vitripennis* (Walker) (Hymenoptera: Pteromalidae) differentially affects cells mediating the immune response of its flesh fly host, *Sarcophaga bullata* Parker (Diptera: Sarcophagidae). J. Insect Physiol. 48, 1053–1064.

Rivers, D.B., Uckan, F., Ergin, E., 2006. Characterization and biochemical analyses of venom from the ectoparasitic wasp *Nasonia vitripennis* (Walker) (Hymenoptera: Pteromalidae). Arch. Insect Biochem. Physiol. 61, 24–41.

Russell, D.H., Edmondson, R.D., 1997. High-resolution mass spectrometry and accurate mass measurements with emphasis on the characterization of peptides and proteins by matrix-assisted laser desorption/ionization time-of-flight mass spectrometry. J. Mass Spectrom. 32, 263–276.

Schrader, W., Klein, H.W., 2004. Liquid chromatography Fourier transform ion cyclotron resonance mass spectrometry (LC-FTICR MS): an early overview. Anal. Bioanal. Chem. 379, 1013–1024.

Seppälä, U., Selby, D., Monsalve, R., King, T.P., Ebner, C., Roepstorff, P., et al. 2009. Structural and immunological characterization of the N-glycans from the major yellow jacket allergen Ves v 2: The N-glycan structures are needed for the human antibody recognition. Mol. Immunol. 46, 2014–2021.

Shim, J.K., Ha, D.M., Nho, S.K., Song, K.S., Lee, K.Y., 2008. Upregulation of heat shock protein genes by envenomation of ectoparasitoid *Bracon hebetor* in larval host of Indian meal moth *Plodia interpunctella*. J. Invert. Pathol. 97, 306–309.

Souza, B.M., Mendes, M.A., Santos, L.D., Marques, M.R., Cesar, L.M.M., Almeida, R.N.A., et al. 2005. Structural and functional characterization of two novel peptide toxins isolated from the venom of the social wasp Polybia paulista. Peptides 26, 2157–2164.

Souza, G., Catharino, R.R., Ifa, D.R., Eberlin, M.N., Hyslop, S., 2008. Peptide fingerprinting of snake venoms by direct infusion nano-electrospray ionization mass spectrometry: potential use in venom identification and taxonomy. J. Mass Spectrom. 43, 594–599.

Steen, H., Mann, M., 2004. The ABC's (and XYZ's) of peptide sequencing. Nat. Rev. Mol. Cell Biol. 5, 699–711.

Stöcklin, R., Favreau, P., Thai, R., Pflugfelder, J., Bulet, P., Mebs, D., 2010. Structural identification by mass spectrometry of a novel antimicrobial peptide from the venom of the solitary bee *Osmia rufa* (Hymenoptera: Megachilidae). Toxicon 55, 20–27.

The Nasonia Genome Working Group 2010. Functional and Evolutionary Insights from the Genomes of Three Parasitoid Nasonia Species. Science 327, 343–348.

Turillazzi, S., Bruschini, C., Lambardi, D., Francese, S., Spadolini, I., Mastrobuoni, G., 2007. Comparison of the medium molecular weight venom fractions from five species of common social wasps by MALDI-TOF spectra profiling. J. Mass Spectrom. 42, 199–205.

Turillazzi, S., Mastrobuoni, G., Dani, F.R., Moneti, G., Pieraccini, G., la Marca, G., et al. 2006. Dominulin A and B: Two new antibacterial peptides identified on the cuticle and in the venom of the social paper wasp Polistes dominulus using MALDI-TOF, MALDI-TOF/TOF, and ESI-ion trap. J. Am. Soc. Mass. Spectrom. 17, 376–383.

Vincent, B., Kaeslin, M., Roth, T., Heller, M., Poulain, J., Cousserans, F., et al. 2010. The venom composition of the parasitic wasp Chelonus inanitus resolved by combined expressed sequence tags analysis and proteomic approach. BMC Genomics 11, 693.

Whiting, A.R., 1967. Biology of parasitic wasp *Mormoniella vitripennis = Nasonia brevicornis* (Walker). Q. Rev. Biol. 42, 333–406.

Wu, C.C., MacCoss, M.J., 2002. Shotgun proteomics: tools for the analysis of complex biological systems. Curr. Opin. Mol. Ther. 4, 242–250.

Ye, J.L., Zhao, H.W., Wang, H.J., Bian, J.M., Zheng, R.Q., 2010. A defensin antimicrobial peptide from the venoms of *Nasonia vitripennis*. Toxicon 56, 101–106.

Zhang, G.M., Lu, Z.Q., Jiang, H.B., Asgari, S., 2004. Negative regulation of prophenoloxidase (proPO) activation by a clip-domain serine proteinase homolog (SPH) from endoparasitoid venom. Insect Biochem. Mol. Biol. 34, 477–483.

Zhang, G.M., Schmidt, O., Asgari, S., 2004. A novel venom peptide from an endoparasitoid wasp is required for expression of polydnavirus genes in host hemocytes. J. Biol. Chem. 279, 41580–41585.

Zhang, G.M., Schmidt, O., Asgari, S., 2006. A calreticulin-like protein from endoparasitoid venom fluid is involved in host hemocyte inactivation. Dev. Comp. Immunol. 30, 756–764.

Zhu, J.Y., Ye, G.Y., Fang, Q., Hu, C., 2010. Alkaline phosphatase from venom of the endoparasitoid wasp, Pteromalus puparum. J. Insect Sci., 10, 14.

Chapter 20

Aphid Parasitoid Venom and its Role in Host Regulation

Francesco Pennacchio and Donato Mancini

Dipartimento di Entomologia e Zoologia Agraria 'F. Silvestri', Università di Napoli 'Federico II', Napoli, Italy

SUMMARY

The venom injected by aphidiine parasitoids (Hymenoptera, Braconidae) during oviposition plays a major role in the host regulation process. It is responsible for host castration, induced by a γ-glutamyl transpeptidase, which triggers apoptosis in the germaria of aphid ovarioles and prevents oocyte development. Moreover, venom also induces apterization of host aphids which, as for castration, is more pronounced when parasitism or envenomation take place on early instar nymphs, and causes the arrest of ovary development. The host castration process is completed by teratocytes, which perform an extra-oral digestion of aphid embryos already formed and facilitate the delivery of nutrients, in particular fatty acids, to the developing sister larvae. Moreover, a close interaction between teratocytes and bacteriocytes, harboring primary symbionts of aphids, is required to enhance and redirect the metabolic effort of *Buchnera* in favor of parasitoid larvae. The molecular analysis of this complex association is interesting from an evolutionary perspective and may provide new tools for aphid control.

ABBREVIATIONS

Ae-FABP	*Aphidius ervi* fatty acid binding protein
Ae-γ-GT	*Aphidius ervi* γ-glutamyl-transpeptidase
GSH	glutathione

INTRODUCTION

The evolutionary success of parasitic Hymenoptera has allowed the colonization of an astonishing number of ecological niches, and the definition of a wide array of developmental strategies to overcome the physiological and ecological constraints associated with the exploitation of living hosts (Pennacchio and Strand, 2006). Unlike many other insect parasitoids, the ovipositing females of parasitic wasps have the capacity to manipulate the host physiology, development and reproduction, in order to enhance their suitability for progeny development (Quicke, 1997). One of the most important traits of adult morphology conferring this unique capacity to parasitic wasps is the ovipositor: a very sophisticated injection tool delivering secretions produced by female reproductive tissues and associated glands (Quicke, 1997; Pennacchio and Strand, 2006). The structural diversity of the ovipositor is paralleled by an impressive molecular diversity of the factors injected at oviposition, which are encoded by the wasp genome or by associated polydnavirus symbionts (Pennacchio and Strand, 2006). In particular, the secretion products of venom glands represent one of the largest biodiversity reservoirs of molecules targeting different host species and developmental instars, which significantly accounts for the evolutionary success of these insects and may facilitate discovery of new insecticides and pharmaceutical leads (Danneels *et al.*, 2010; Pennacchio *et al.*, Chapter 22 of this volume)

The analysis of the functional role of the venom produced by parasitic wasps has stimulated considerable research efforts, especially in the last two decades, which have shed light on the mechanisms underlying the pathological alterations of parasitized or artificially envenomated hosts (see the recent review by Danneels *et al.*, 2010). The evident host paralysis and killing effects, induced by most idiobiont parasitoids at oviposition, are replaced by a wealth of more subtle and specialized physiological alterations in the case of koinobionts, which allow host survival until the completion of their preimaginal development (Gauld, 1988; Pennacchio and Strand, 2006). The neurotoxicity of the molecular components present in the venom of *Bracon* species stimulated the first research interests in this area, with the aim of isolating new agrochemical leads (reviewed in Weaver *et al.*, 2001). However, the increasing number of physiological and molecular studies on the venom produced by other parasitoid species have documented a remarkable variety of effects on host physiology, metabolism, development, and reproduction (Pennacchio and Strand, 2006; Danneels *et al.*, 2010). More recently, genomic, transcriptomic, and proteomic studies have allowed the in-depth analysis of the venom composition in selected parasitoid species (Crawford *et al.*, 2008; De Graaf *et al.*, 2010, and this volume; Vincent *et al.*, 2010). These studies have generated a number of insights on venom functions and have stimulated new research efforts aiming at the characterization and sustainable exploitation of a nearly untapped pharmacopoeia (Danneels *et al.*, 2010; Pennacchio *et al.*, Chapter 22 of this volume).

The venom is one of multiple factors controlling the host regulation process, as defined by Vinson and Iwantsch (1980). The integration of this maternal factor with those of embryo-derived and larval origin allows the definition of a physiological scenario and of a functional analysis of the host regulation process. To date, only a few host–parasitoid systems have been investigated from multiple perspectives. Among these, we have the aphid parasitoid *Aphidius ervi*, an endophagous braconid attacking different macrosiphine aphid species including the pea aphid, *Acyrthosiphon pisum*, which has been widely used as a model host in a number of ecological, behavioral, and physiological studies. Here we review the available information on the venom of this parasitic wasp, outlining its role, along with that of other host regulation factors, in the physiological redirection of parasitized host aphids.

HOST REGULATION BY APHID PARASITOIDS: THE ROLE OF VENOM

Host Castration

The most evident alteration in parasitized aphids is the remarkable reduction or suppression of their reproductive activity, observed in different host–parasitoid associations (Schlinger and Hall, 1960; Tremblay, 1964; Campbell and Mackauer, 1975; Rabasse and Shalaby, 1979; Soldán and Starý, 1981; Polaszek, 1986; Kring and Kring, 1988). The negative impact of parasitism on host reproduction was found to be associated with clear degenerative symptoms of ovaries, already observed 24 h after parasitoid oviposition, before parasitoid larval hatching, which suggested the possibility that female reproductive secretions injected at oviposition, in particular factors produced by venom glands (Fig. 1), were involved in these alterations (Polaszek, 1986).

This hypothesis was confirmed by studies on the host–parasitoid association *A. pisum–A. ervi*, which demonstrated that the venom of the parasitoid was responsible for host castration (Digilio *et al.*, 1998, 2000). Venom microinjections into non-parasitized host aphids reproduced the

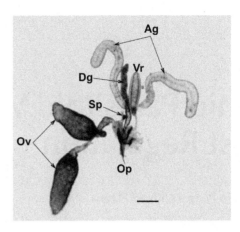

FIGURE 1   **Reproductive and venom apparatus of *Aphidius ervi* females.** Ag, acid gland; Dg, Dufour's (alkaline) gland; Op, ovipositor; Ov, ovary; Sp, spermatheca; Vr, venom reservoir. Scale bar: 0.2 mm.

alterations observed in the upper part of the ovarioles of naturally parasitized aphids, which showed ultrastructural details suggestive of apoptosis occurrence in the germarial cells and apical embryos (Digilio *et al.*, 2000). This alteration was induced by a dimeric protein, one of the most abundant components of *A. ervi* venom (Digilio *et al.*, 2000).

The apoptosis induction in the host germaria and ovariole sheath by this protein, comprised of 18- and 36-kDa subunits, was confirmed by TUNEL staining, and the purified protein subunits were internally sequenced, to get the information required for isolating their gene from a cDNA library of the venom gland (Falabella *et al.*, 2007). The isolated cDNA clone encoded a protein with a calculated molecular mass of 56.9 kDa, containing the sequences of both protein subunits, which are produced by post-translational processing of a single-chain precursor (Falabella *et al.*, 2007). This protein showed a significant level of sequence identity with γ-glutamyl transpeptidases, and therefore was named *Ae*-γ-GT. In a preliminary analysis of the transcriptome of the venom gland of the model species *A. ervi*, performed by extensive sequencing of a cDNA library, the gene encoding *Ae*-γ-GT was represented by a high proportion of clones, suggesting that it was highly expressed (Pennacchio *et al.*, unpublished data). The contribution of this protein to the venom blend is more significant if we consider the limited number of expressed genes coding for putatively secreted proteins (Pennacchio *et al.*, unpublished data), which is consistent with the relatively low number of protein bands in the electropherogram of *A. ervi* venom (Digilio *et al.*, 2000; Falabella *et al.*, 2007). Quite a few of the other putative components of the venom could have a potential role in the alteration of the innate immunity pathways of the host, such as those controlled by serine proteases (Pennacchio *et al.*, unpublished data).

These enzymes have been frequently reported to occur in parasitoid venoms, where they likely act as disruptors of the phenoloxidase cascade (Danneels *et al.*, 2010). Unfortunately, the functional analysis of their role in the context of aphid physiology remains quite difficult, as we know very little about the physiology of the immune system of these insects.

The molecular mechanism of action of *Ae-γ-GT* is still undefined. The role of γ-GT is essential in the glutathione (GSH) metabolism in all eukaryotic cells, where GSH, a tripeptide (L-γ-glutamyl-L-cysteinyl-glycine), acts as antioxidant and exerts protective functions (Meister, 1995; Karp *et al.*, 2001). Moreover, the catabolites deriving from the activity of this enzyme may favor ROS (reactive oxygen species) production, which can influence cell fate by controlling the proliferation/apoptosis balance (Carlisle *et al.*, 2003; Dominici *et al.*, 2005). The steep gradient of GSH drives its transport out of the cell, and the replenishment of the cellular pools is made possible only through the 'γ-glutamyl cycle'. The γ-GT catalyzes the first step of the cycle, by removing the γ-glutamyl group, which is transferred to an amino acid, and the resulting γ-glu-amino acid is internalized. The remaining cysteinylglycine is cleaved by a membrane dipeptidase and the released cysteine is transported inside the cell and used for GSH synthesis. Therefore, γ-GT plays an essential role in protecting the cells from oxidative stress, as it contributes to the reconstitution of GSH levels inside the cell. It is reasonable to speculate that the presence of γ-GT of parasitic origin may interfere with the regular enzymatic activity of endogenous γ-GT by competing for the substrate, and, therefore, disrupting the γ-glutamyl cycle. This may result in a depletion of GSH cellular levels, which exposes the cell to an oxidative stress triggering apoptosis.

It is intriguing to note that γ-GT has been found to act as virulence factor in the case of the human pathogen *Helicobacter pylori*, which causes apoptosis in gastric epithelial cells upon infection (Shibayama *et al.*, 2003; Kim *et al.*, 2010). This bacterial γ-GT (Shibayama *et al.*, 2003), like *Ae-γ-GT* (Falabella *et al.*, 2007), lacks the transmembrane domain, which in most cases anchors γ-GTs on the surface of the cell membrane (Zhang *et al.*, 2005). The adhesion on the host cell surface of *H. pylori* γ-GT is essential for apoptosis induction (Shibayama *et al.*, 2003). It is mediated by an ionic bond (Shibayama *et al.*, 2003) and requires the presence of specific molecular structures. The occurrence of these molecular structures on the surface of specific host tissues may account for the selective targeting by γ-GTs of parasitic origin. Therefore, it would be interesting to identify the molecular receptor that allows *Ae-γ-GT* to be anchored on the surface of the germaria of the pea aphid ovarioles, which are selectively disrupted in parasitized hosts.

Host Apterization and Developmental Arrest

Another marked phenotypic effect that aphid parasitoids cause to their hosts is the inhibition of wing development (apterization), observed in aphid nymphs that, when parasitized, develop into wingless or partially winged adults (Johnson, 1959; Christansen-Weniger and Hardie, 1998; Kati and Hardie, 2010). Moreover, these studies show that when parasitism takes place during first or second host stadium, aphids fail to reach the adult stage. The possible role of the venom in the induction of these alterations was suggested on the basis of indirect experimental evidence (Christansen-Weniger and Hardie, 2000), and more recently the key role of venom was demonstrated in a study on the host–parasitoid association *Aphis fabae-Aphidius colemani* (Kati and Hardie, 2010). The latter study provided evidence that host attack by the parasitoid or its artificial envenomation determines higher rates of apterization and developmental arrest when performed on very young host instars. It was proposed that the venom may contain a component directly active on wing-bud tissue, rather than interfering with the aphid endocrine system (Christansen-Weniger and Hardie, 2000; Kati and Hardie, 2010). However, it is difficult to explain how this toxic component does not exert any effect when parasitism takes place on later host instars, unless we invoke the occurrence of a very narrow sensitivity window.

A. ervi parasitism on pea aphid first-instar nymphs also induces host developmental arrest as fourth instar, which is associated with a significant reduction in ecdysteroid titres and with a total inhibition of reproduction, since the host does not reach the adulthood (Pennacchio *et al.*, 1995). Additionally, there is a complete suppression of ovary development, as previously reported also by Polaszek (1986). This can be easily explained by the effect of venom on the germaria, which blocks the differentiation of new oocytes (Digilio *et al.*, 2000; Falabella *et al.*, 2007). However, the inhibition of the molt to the adult stage is more difficult to interpret. Several aphid species show the degeneration of prothoracic glands during the fourth instar, which disappear at the final molt (Hardie, 1987). Therefore, the source of the ecdysteroid peak required to trigger the final molt to the adult stage and its neuroendocrine control remain undefined.

The hypothesis that the aphid ovary, and in particular the mature embryos, could be the source of the ecdysteroids required to trigger the final molt may allow the interpretation of the current data. According to this hypothesis, the complete castration observed only when first and second instars are parasitized would result in the block of the final molt, as a consequence of the venom-mediated suppression of ovary development and of its ecdysteroidogenic function. The resulting endocrine alterations might also partly account for the observed apterization process, as wing development in aphids seems to be both under environmental and endocrine control, even though details of this control are still poorly defined (Hartfelder and Emlen, 2005). The increasing rates of development to the adult stage and of regular wing formation when later instars are parasitized indicate that these traits are strictly related to the presence of formed aphid embryos. The possibility that the mature embryos in the aphid ovary are an important ecdysteroid source is suggested by the significant reduction of the ecydisteroid titre in fully castrated and developmentally arrested fourth instars, which is also reported in parasitized hosts attaining the adult stage, showing only partial castration and limited reproduction (Pennacchio *et al.*, 1995). Moreover, a large part of the ecdysteroid titer in reproducing pea aphid adults can be recovered from ovaries (Pennacchio *et al.*, 1995), which in fourth instar nymphs also show the accumulation of ecdysteroid conjugates (Pennacchio *et al.*, unpublished data). It would be very interesting to determine how the endocrine signal triggering the transition to the adult stage might be controlled by the ovaries, allowing the final molt to take place when reproductive maturity is achieved.

THE ROLE OF TERATOCYTES IN THE HOST REGULATION PROCESS

The host castration process mediated by aphid parasitoids is started by the injection of venom at oviposition which, as presented above, targets the upper part of the ovarioles, and prevents the development of new oocytes in the germaria. This early effect of parasitism on host reproductive tissues is complemented by a later disruption of embryos already formed, which, as described by Tremblay (1966), appear severely altered in the final part of parasitoid larval development when, upon dissection of parasitized aphids, only large cells originating from the wasp, called teratocytes, degenerating host tissues, and healthy bacteriocytes can be found. The degeneration process of aphid embryos is associated with evident signs of surface interactions with teratocytes (Fig. 2), which, on the basis of morphological and ultrastructural observations (Tremblay, 1966; Tremblay and Iaccarino, 1971), were interpreted by Tremblay (personal communication), as 'free intestinal cells' of the parasitoid, mediating external digestion of host tissues to favor the nutrition of their sister larvae.

Teratocytes are cells deriving from the dissociation of the embryonic membranes and have been described in several parasitoid species (Dahlman and Vinson, 1993; Beckage and Gelman, 2004; Pennacchio and Strand, 2006). A number of different roles have been attributed to these cells, found actively involved in the induction of physiological alterations of the host and in close interactions with its

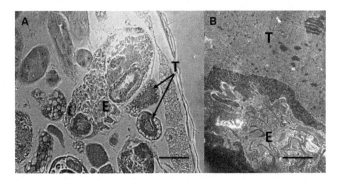

FIGURE 2 **A:** Light microscopy sections of pea aphid, *Acyrthosiphon pisum*, parasitized by *Aphidius ervi*, showing a close surface interaction between aphid embryos and teratocytes. **B:** A closer inspection by transmission electron microscopy evidences intense vesicular traffic at the interface and embryo degeneration. E, embryo; T, teratocyte. Scale bars: $50\,\mu m$ (A) and $10\,\mu m$ (B).

tissues, which are of key importance for the nutrition of the developing parasitoid larvae (Beckage and Gelman, 2004; Pennacchio and Strand, 2006).

The origin of embryonic membranes and teratocytes in braconid aphidiines has been studied in detail, and the involvement of polar bodies in their genesis described (Tremblay and Calvert, 1971; Tremblay and Caltagirone, 1973). The functional role of these cells has been investigated in the model species *A. ervi*, which has teratocytes producing and releasing abundant levels of two parasitism-specific proteins in the host hemocoel (p15 and p45), during the final part of parasitoid larval development (Falabella *et al.*, 2000). The p15 is a fatty acid binding protein (*Ae*-FABP) which shows high affinity for C_{14}–C_{18} saturated fatty acids (Falabella *et al.*, 2005). The unusual extracellular localization of this protein reflects its functional role, as it mediates the transport in the aqueous hemolymph environment of fatty acids deriving from the extra-oral digestion of host tissues, performed by the teratocytes of many parasitoid species (Tremblay and Iaccarino, 1971; Falabella *et al.*, 2000; Nakamatsu *et al.*, 2002; Nakamatsu and Tanaka 2004; Suzuki and Tanaka, 2007). This transport function is essential for effective delivery to parasitoid larvae of fatty acids, which can be readily absorbed through the epidermis and gut epithelia (Giordana *et al.*, unpublished data), as similarly observed for other nutrients (de Eguileor *et al.*, 2001). This peculiar aspect of parasitoid nutritional physiology represents an effective strategy to compete with host tissues, as these latter, like in most insects, can utilize acylglycerols transported by lipophorins, but not free fatty acids (Arrese *et al.*, 2001).

The p45 is an extracellular enolase (*Ae*-ENO), which is likely involved in the process of host tissue degradation, and may also play a role in the host immune suppression (Falabella *et al.*, 2009). Enolase is a key enzyme of the glycolysis, present virtually in all living organisms (Canback *et al.*, 2002). It has a multifunctional role in disease when present in the extracellular environment, on the surface of several prokaryotic and eukaryotic cells (Pancholi, 2001; Sun, 2006; Liu and Shih, 2007). There are several examples showing that enolases on the surface of pathogens or malignant tumor cells can recruit plasminogen, which, when bound, is converted to plasmin by plasminogen activators, and subsequently activates procollagenase to collagenase, leading to degradation of fibrin and of several extra-cellular matrix components (Liu and Shih, 2007). The vast accumulation of *Ae*-ENO on the surface of degenerating host tissues, in particular embryos, may account for selective degradation likely mediated by local activation of plasminogen-like precursors of lytic enzymes (Falabella *et al.*, 2009). Therefore, it seems that the two proteins produced by teratocytes of *A. ervi* have a complementary role as they mediate host tissue degradation and the transport of the resulting fatty acids from the site of digestion to the absorbing epithelia of the parasitoid larvae. The high abundance of lipids in the aphid body (Dillwith *et al.*, 1993) suggests that this mechanism is physiologically relevant for parasitoid nutrition and development.

HOST REGULATION BY APHID PARASITOIDS: A PHYSIOLOGICAL MODEL

The host regulation process by parasitic wasps is essential to suppress host immune response and to enhance nutritional suitability for the developing parasitoid larvae (Vinson and Iwantsch, 1980; Vinson *et al.*, 2001; Pennacchio and Strand, 2006). The impressive variety of parasitoid developmental strategies has been shaped to face these functional constraints and to maximize the parasitoid fitness in many different host environments. While the adaptive value of host immune suppression is obvious, the variety of alterations of host development and reproduction is more complex to characterize. However, a common theme arises when different biological systems are compared. In particular, the need to synchronize the most demanding part of parasitoid larval development with the highest availability of nutrients has become evident. This is achieved by redirecting the host metabolic effort, for its own physiological and reproductive needs, in favor of the developing parasitoid larvae (Vinson *et al.*, 2001; Pennacchio and Strand, 2006).

The analysis of the nutritional aspects of host–parasitoid interactions has received significant attention in the host–parasitoid association *A. pisum*–*A. ervi*. The peculiarity of this model system is the presence in the host of intracellular bacteria, belonging to the genus *Buchnera*, which are intimately associated in an obligate symbiosis of crucial nutritional importance (Douglas, 1998; Baumann, 2005; Brinza *et al.*, 2009). Moreover, aphids have secondary symbionts which, among the many phenotypic effects

they may control (thermal resistance, adaptation to the host plant, complementation of the nutritional role of *Buchnera*, resistance to pathogens), also have the capacity to influence the degree of resistance against parasitic wasps (Olivier *et al.*, 2010).

The presence of *Buchnera* has been found to be of considerable importance, not only for the nutrition of the aphids, but also for the parasitoid *A. ervi*, which develops poorly in aposymbiotic hosts (Pennacchio *et al.*, 1999). In particular, the significantly increased bacterial provision of essential amino acids, and especially the aromatic amino acids like tyrosine (Rahbé *et al.*, 2002), is crucial for larval development. This nutritional support of symbionts to the developing wasp larvae is also mediated by a close interaction with teratocytes, as indicated by the failure of these latter to produce abundant levels of *Ae*-ENO and *Ae*-FABP in aposymbiotic aphids (Pennacchio *et al.*, 1999). The same problem is observed in a resistant pea aphid clone, which by suppressing the teratocytes does not allow the development of *A. ervi* larvae (Li *et al.*, 2002). These results indicate that the amino acid supply by the primary symbionts is essential, and that a finely tuned interaction between teratocytes and aphid bacteriocytes is crucial for successful parasitism. A strong surface interaction between these cells is observed and, unlike other host tissue, bacteriocytes do not degenerate in the parasitized host but seem to be stimulated, as indicated by a significant increase in the bacterial biomass (Cloutier and Douglas, 2003).

The current information on the host regulation process by aphid parasitoids can be summarized in the model reported in Fig. 3, where the role of venom is placed in a more complete and meaningful physiological scenario. The major target of the parasitoid is the disruption of the host reproduction, which acts as a major metabolic sink in competition with parasitoid nutritional needs. This alteration is essential to redirect in favor of the parasitoid larva the host metabolic effort supporting reproduction.

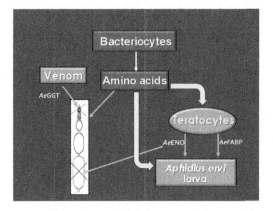

FIGURE 3 Schematic representation of the physiological model describing the functional integration of host regulation factors in the experimental model system *Acyrthosiphon pisum/Aphidius ervi*. See text for details.

Buchnera largely contributes to this effort, showing in parasitized host aphids a higher biomass and an increased biosynthetic activity.

The negative effect on host reproduction is mediated by the combined action of maternal and embryo-derived factors: venom and teratocytes. The venom acts on the upper part of the ovarioles (germaria) and, soon after oviposition, prevents the development of new oocytes, while teratocytes, in the final part of parasitoid larval development, start a process of extra-oral digestion of mature embryos already formed, which is in part mediated by *Ae*-ENO. This process allows the release of nutrients present in the host tissues in a soluble form that can be used by the parasitoid larva. Then, the role of the venom in host castration is preponderant when early host instars are parasitized, and becomes increasingly complemented by the teratocyte action when parasitism takes place on older instars or adults, which have ovaries already partially or fully developed. The negative impact of teratocyte activity on host reproductive tissues is associated with a mobilization of released fatty acids mediated by *Ae*-FABP, which are readily absorbed by the parasitoid larva. This represents an effective strategy that *A. ervi* larvae adopt to compete with host tissues for lipid nutrients.

CONCLUSIONS AND FUTURE DIRECTIONS

The venom of aphid parasitoids is one of the major host regulation factors involved in the host castration process. The bioactive component triggering the apoptosis of the germaria in the upper part of the ovarioles is a dimeric protein with γ-glutamyl transpeptidase activity (*Ae*-γGT), which prevents the development of new aphid embryos. How the selective targeting of ovaries is achieved remains to be elucidated. The analysis of host proteins interacting with *Ae*-γGT *in vivo* is needed to understand its mechanism of action, likely mediated by the selective anchoring on the surface of specific host tissues.

Current research efforts focusing on the transcriptome and proteome of the venom gland of the model species *A. ervi* will shed light on unknown roles of aphid parasitoid venom. The presence of putative immunosuppressive factors offers new tools for studying the functional details of the aphid immune system, which show a peculiar reduction of molecular barriers, compared to other insect species (The International Aphid Genomics Consortium, 2010).

The occurrence of intricate interactions among the aphid host and the associated symbionts, showing unique ecological and physiological characteristics, offers a valuable opportunity to test evolutionary hypotheses on aphid and parasitoid adaptive radiation.

Finally, an in-depth understanding of the molecular mechanisms of host castration and biochemical redirection will be also of considerable applied interest, as it can be profitably exploited for developing new insecticide molecules and to define/refine suitable artificial diets for the mass production of aphid parasitoids.

REFERENCES

Arrese, E.L., Canavoso, L.E., Jouni, Z.E., Pennington, J.E., Tsukida, K., Wells, M.A., 2001. Lipid storage and mobilization in insects: current status and future directions. Insect Biochem. Mol. Biol. 31, 7–17.

Baumann, P., 2005. Biology of bacteriocyte-associated endosymbionts of plant sap-sucking insects. Annu. Rev. Microbiol. 59, 155–189.

Beckage, N.E., Gelman, D.B., 2004. Wasp parasitoid disruption of host development implication for new biologically based strategies for insect control. Annu. Rev. Entomol. 49, 299–330.

Brinza, L., Viñuelas, J., Cotrett, L., Calevro, F., Rahbé, Y., Febvay, G., et al. 2009. Systematic analysis of the symbiotic function of *Buchnera aphidicola*, the primary endosymbiont of the pea aphid *Acyrtosiphon pisum*. C. R. Biol. 332, 1034–1049.

Campbell, A., Mackauer, M., 1975. The effect of parasitism by *Aphidius smithi* (Hymenoptera: Aphidiidae) on reproduction and population growth of the pea aphid, *Acyrtosiphon pisum* (Homoptera Aphididae). Can. J. Zool. 107, 919–1026.

Canback, B., Anderson, S.G.E., Kurland, C.G., 2002. The global phylogeny of glycolytic enzymes. Proc. Natl. Acad. Sci. U. S. A. 99, 6097–6102.

Carlisle, M.L., King, M.R., Karp, D.R., 2003. γ-Glutamyl transpeptidase activity alters the T cell response to oxidative stress and Fas-induced apoptosis. Int. Immunol. 109, 13–19.

Christansen-Weniger, P., Hardie, J., 1998. Wing development in parasitized male and female *Sitobion fragariae*. Physiol. Entomol. 23, 208–213.

Christansen-Weniger, P., Hardie, J., 2000. The influence of parasitism on wing development in male and female pea aphids. J. Insect Physiol. 46, 861–867.

Cloutier, C., Douglas, A.E., 2003. Impact of parasitoid on the bacterial symbiosis of its aphid host. Entomol. Exp. Appl. 109, 13–19.

Crawford, A.M., Brauning, R., Smolenski, G., Ferguson, C., Barton, D., Wheeler, T.T., et al. 2008. The constituents of *Microtonus sp.* parasitoid venoms. Insect. Mol. Biol. 17, 313–324.

Dahlman, D.L., Vinson, S.B., 1993. Teratocytes: developmental and biochemical characteristics In: Beckage, N.E., Thompson, S.N., Federici, B.A., (Ed.), Parasites and Pathogens of Insects. Vol. 1. Parasites (pp. 145–165). Academic Press, San Diego.

Danneels, E.L., Rivers, D.B., de Graaf, D.C., 2010. Venom proteins of the parasitoid wasp *Nasonia vitripennis*: recent discovery of an untapped pharmacopee. Toxins 2, 494–516.

de Eguileor, M., Gimaldi, A., Tettamanti, G., Valvassori, R., Leonardi, G., Giordana, B., et al. 2001. Larval anatomy and structure of absorbing epithelia in the aphid parasitoid *Aphidius ervi* Haliday (Hymenoptera, Braconidae). Arthropod Struct. Dev. 30, 27–37.

De Graaf, D.C., Brunain, M., Scharlaken, B., Peiren, N., Devreese, B., Ebo, D.G., et al. 2010. Two novel proteins expressed by the venom glands of Apis mellifera and Nasonia vitripennis share an ancient C1q-like domain. Insect. Mol. Biol. 19 (Suppl. 1), 1–10.

Digilio, M.C., Isidoro, N., Tremblay, E., Pennacchio, F., 2000. Host regulation by *Aphidius ervi* venom proteins. J. Insects Physiol. 46, 1041–1050.

Digilio, M.C., Pennacchio, F., Tremblay, E., 1998. Host regulation effects of ovary fluid and venom of *Aphidius ervi* (Hymenoptera: Braconidae). J. Insect Physiol. 44, 779–784.

Dillwith, J.W., Neese, P.A., Brigham, D.L., 1993. Lipid biochemistry in aphids pp. 389–434. In: Stanley-Samuelson, D.W., Nelson, D.R. (Eds.), Insect Lipids: Chemistry, Biochemistry and Biology. University of Nebraska Press, Lincoln, NE.

Dominici, S., Paolicchi, A., Corti, A., Maellaro, E., Pompella, A., 2005. Prooxidant reactions promoted by soluble cell-bound gamma-glutamyltransferase activity. Methods Enzymol. 401, 484–501.

Douglas, A.E., 1998. Nutritional interactions in insect–microbial symbioses: aphids and their symbiotic bacteria *Buchnera*. Annu. Rev. Entomol. 43, 17–37.

Falabella, P., Perugino, G., Caccialupi, P., Riviello, L., Varricchio, P., Tranfaglia, A., et al. 2005. A novel fatty acid binding produced by teratocytes of the aphid parasitoid *Aphidius ervi*. Insect Mol. Biol. 14, 195–205.

Falabella, P., Riviello, L., Caccialupi, P., Rossodivita, T., Valente, M.T., De Stradis, M.L., et al. 2007. A glutamyl transpeptidase of *Aphidius ervi* venom induces apoptosis in the ovaries of host aphids. Insect Biochem. Mol. Biol. 37, 453–465.

Falabella, P., Riviello, L., De Stradis, M.L., Stigliano, C., Varricchio, P., Grimaldi, A., et al. 2009. *Aphidius ervi* teratocytes release an extra-cellular enolase. Insect Biochem. Mol. Biol. 39, 801–813.

Falabella, P., Tremblay, E., Pennacchio, E., 2000. Host regulation by the aphid parasitoid *Aphidius ervi*: the role of teratocytes. Ent. Exp. Appl. 97, 1–9.

Gauld, I.D., 1988. Evolutionary patterns of host utilization by ichneumonoid parasitoids (Hymenoptera, Ichneumonidae and Braconidae). Biol. J. Linn. Soc. 35, 351–377.

Hardie, J., 1987. Neurosecretory and endocrine systems In: Minks, A.K., Harrewijn., P. (Eds.), Aphids. Their Biology, Natural Enemies and Control, vol. 2B. (pp. 139–152). Elsevier, Amsterdam.

Hartfelder, K., Emlen, D.J., 2005. Endocrine control of insect polyphenism. Compr. Mol. Insect Sci. 3, 651–703.

Johnson, B., 1959. Effect of parasitization by *Aphidius platensis* Bréthes on the developmental physiology of its host, *Aphis craccivora* Koch. Entomol. Exp. Appl. 2, 82–99.

Karp, D.R., Shimooku, K., Lipsky, P.E., 2001. Expression of gamma-glutamyl transpeptidase protects Ramos B cells from oxidation-induced cell death. J. Bio. Chem. 276, 3798–3804.

Kati, A., Hardie, J., 2010. Regulation of wing formation and adult development in an aphid host, *Aphis fabae*, by the parasitoid *Aphidius colemani*. J. Insect Physiol. 56, 14–20.

Kim, K.M., Lee, S.G., Kim, J.M., Kim, D.S., Song, J.Y., Kang, H.L., et al. 2010. *Helicobacter pylori* gamma-glutamyltranspeptidase induces cell cycle arrest at the G1-S phase transition. J. Microbiol. 48 (3), 372–377.

Kring, T.J., Kring, J.B., 1988. Aphid fecundity, reproductive longevity, and parasite development in the *Schizaphis graminum* (Rondani) (Homoptera: Aphididae) – *Lysiphlebus testaceipes* (Cresson) (Hymenoptera: Braconidae) system. Can. Entomol. 120, 1079–1083.

Li, S., Falabella, P., Giannantonio, S., Fanti, P., Battaglia, D., Digilio, M.C., et al. 2002. Pea aphid clonal resistance to the endophagous parasitoid *Aphidius ervi*. J. Insect Physiol. 48, 971–980.

Liu, K., Shih, N.J., 2007. The role of enolase in tissue invasion and metastasis of pathogens and tumor cells. J. Cancer Mol. 3, 45–48.

Meister, A., 1995. Glutathione metabolism. Methods Enzymol. 251, 3–7.

Nakamatsu, Y., Fuji, S., Tanaka, T., 2002. Larvae of an entomoparasitoid, *Cotesia kariyai* (Hymenoptera: Braconidae), feed on the host fat body directly in the second stadium with the help of teratocytes. J. Insect Physiol. 48, 1041–1052.

Nakamatsu, Y., Tanaka, T., 2004. Correlation between concentration of hemolymph nutrients and amount of fat body consumed in lightly and heavily parasitized hosts (*Pseudaletia separata*). J. Insect Physiol. 50, 135–141.

Olivier, K.M., Degnan, P.H., Burke, G.R., Moran, N.A., 2010. Facultative symbionts in aphids and the horizontal transfer of ecologically important traits. Annu. Rev. Entomol. 55, 247–266.

Pancholi, V., 2001. Multifunctional alpha-enolase: its role in diseases. Cell. Mol. Life Sci. 58, 902–920.

Pennacchio, F., Digilio, M.C., Tremblay, E., 1995. Biochemical and metabolic alterations in *Acyrthosiphon pisum* parasitized by *Aphidius ervi*. Arch. Insect Biochem. Physiol. 30, 351–367.

Pennacchio, F., Fanti, P., Falabella, P., Digilio, M.C., Bisaccia, F., Tremblay, E., 1999. Development and nutrition of the braconid wasp, *Aphidius ervi* in aposymbiotic host aphids. Arch. Insect Biochem. Physiol. 40, 53–63.

Pennacchio, F., Strand, M.R., 2006. Evolution of developmental strategies in parasitic Hymenoptera. Annu. Rev. Entomol. 51, 233–258.

Polaszek, A., 1986. The effect of two species of hymenopterous parasitoid on the reproductive system of the pea aphid, *Acyrtosiphon pisum*. Entomol. Exp. Appl. 40, 285–292.

Quicke, D.L.J., 1997. Parasitic Wasps. Chapman & Hall, London.

Rabasse, J.M., Shalaby, F.F., 1979. Incidence du parasite *Aphidius matricarie* Hal. (Hym.: Aphidiidae) sur la fécondité de son hôte *Myzus persicae* (Sulz) (Homoptera, Aphidiidae) á différentes temperatures. Ann. Zool. Ecol. Anim. 11, 359–369.

Rahbé, Y., Digilio, M.C., Febvay, G., Guillaud, J., Fanti, P., Pennacchio, F., 2002. Metabolic and symbiotic interactions in amino acid pools of the pea aphid, *Acyrthosiphon pisum*, parasitized by the braconid *Aphidius ervi*. J. Insect Physiol. Entomol. 22, 507–516.

Schlinger, E.I., Hall, J.C., 1960. The biology, behaviour and morphology of *Praon palitans* Muesebeck, an internal parasite of the spotted alfalfa aphid *Therioaphis maculata* (Buckton) (Hymenoptera: Braconidae, Aphidiinae). Ann. Entomol. Soc. Am. 53, 144–160.

Shibayama, K., Kamachi, K., Nagata, N., Yagi, T., Nada, T., Doi, Y., et al. 2003. A novel apoptosis-inducing protein from *Helicobacter pylori*. Mol. Microbiol. 47, 443–451.

Soldán, T., Starý, P., 1981. Parasitogenic effects of *Aphidius smithi* (Hymenoptera:, Aphidiidae) on the reproductive organs of the pea aphid, *Acyrthosiphon pisum* (Homoptera, Aphididae). Acta Ent. Bohemoslov. 78, 243–253.

Sun, H., 2006. The interaction between pathogens and host coagulation system. Physiology 21, 281–288.

Suzuki, M., Tanaka, T., 2007. Development of *Meteorus pulchricornis* and regulation of its noctuid host, *Pseudaletia separata*. J. Insect Physiol. 53, 1072–1078.

The International Aphid Genomics Consortium, Genome sequence of the pea aphid *Acyrtosiphon pisum*. PLoS Biol. 8 (2), e1000313.

Tremblay, E., 1964. Ricerche sugli imenotteri parassiti I. Studio morfo-biologico sul *Lysiphlebus fabarum* (Marshall) (Hymenoptera: Braconidae: Aphidiinae). Boll. Lab. Ent. Agr. 'F. Silvestri' Portici 22, 1–122.

Tremblay, E., 1966. Ricerche sugli Imenotteri parassiti II. Osservazioni sull'origine e sul destino dell'involucro embrionale degli Afidiini

(Hymenoptera: Braconidae Aphidiinae) e considerazioni sul significato generale delle membrane embrionali. Boll. Lab. Entomol. Agr. 'Filippo Silvestri' Portici 29, 119–166.

Tremblay, E., Calvert, D., 1971. Embryosistematics in the Aphidiines (Hymenoptera: Braconidae). Boll. Lab. Ent. Ent. Agr. Portici 29, 223–249.

Tremblay, E., Caltagirone, L.E., 1973. Fate of polar bodies in insects. Annu. Rev. Entomol. 18, 421–444.

Tremblay, E., Iaccarino, F.M., 1971. Notizie sull'ultrastruttura dei trofociti di *Aphidius matricariae* Hal. (Hymenoptera: Braconidae). Boll. Lab. Ent. Agr. 'Filippo Silvestri' Portici 29, 305–314.

Vincent, B., Kaeslin, M., Roth, T., Heller, M., Poulain, J., Cousserans, F., et al. 2010. The venom composition of the parasitic wasp *Chelonus inanitus* resolved by combined expressed sequence tags analysis and proteomic approach. BMC Genomics 11, 693.

Vinson, S.B., Iwantsch, G.F., 1980. Host regulation by insect parasitoids. Q. Rev. Biol. 55, 143–165.

Vinson, S.B., Pennacchio, F., Consoli, F.L., 2001. The parasitoid–host interaction from a nutrictional perspective. In: Edwards, J.P., Weaver, R.J. (Eds.), Endocrine Interaction of Insect Parasites and Pathogens (pp. 187–205). BIOS, Oxford.

Weaver, R.J., Marris, G.C., Bell, H.A., Edwards, J.P., 2001. Identity and mode of action of the host endocrine disruptors from the venom of parastoid wasps. In: Edwards, J.P., Weaver, R.J. (Eds.), Endocrine Interactions of Insect Parasites and Pathogens (pp. 33–58). BIOS, Oxford.

Zhang, H., Forman, H.J., Choi, J., 2005. Gamma-glutamyl transpeptidase in glutathione biosynthesis. Methods Enzymol. 401, 468–483.

When Parasitoids Lack Polydnaviruses, Can Venoms Subdue the Hosts? The Case Study of *Asobara* Species

Geneviève Prevost[*], Patrice Eslin[*], Anas Cherqui[*], Sébastien Moreau[†], and Géraldine Doury[*]

[*]*Laboratoire de Biologie des Entomophages, Université de Picardie–Jules Verne, 33 rue Saint Leu, 80039 Amiens cedex, France*

[†]*Institut de Recherche sur la Biologie de l'Insecte, 6035, CNRS, UMR, Université François-Rabelais, Faculté des Sciences et Techniques, Parc Grandmont, 37200 Tours, France*

SUMMARY

Hymenopteran Braconidae of the *Asobara* genus are endophagous parasitoids of *Drosophila* larvae. The few *Asobara* species that we have investigated so far are all lacking polydnaviruses and virus-like particles. However, they have developed other means of successfully overcoming their *Drosophila* hosts' defenses. Among them, factors present in the female wasps' reproductive tract, which are injected along with the parasitoid's egg during oviposition, play a major role. In *A. japonica* and *A. citri*, factors from the wasps' venom specifically attack the hemopoietic organ of their larval host while, unlike most parasitoid species carrying polydnaviruses or virus-like particles, the host hemocytes circulating in the hemocoel are not targeted. In addition to causing immunosuppresion, *A. japonica* venom provokes an overall disruption of the host's cells and tissues such that an 'antidote' of ovarian origin is necessary to counterbalance this venom's deadly effect. *A. japonica* venom and ovarian fluids act synergistically as a poison/antidote combination. In *A. tabida*, which is a nonimmunosuppressive species, some component in the egg chorion seems to be responsible for the evasion from encapsulation. Additionally, an aspartylglucosaminidase (AtAGA), the role of which is still unknown, accounts for 30% of the wasp's venom proteins. Venoms of the three species *A. japonica*, *A. citri*, and *A. tabida* have a paralyzing effect on the host larva, which could reflect some remnant of an ancestral ectoparasitic status of the genus *Asobara*.

ABBREVIATIONS

A1	the *Asobara tabida* strain which we use as the reference strain in our laboratory
AGA	aspartylglucosaminidase
AtAGA	*Asobara tabida* aspartylglucosaminidase
IO	interrupted oviposition
PDV	polydnavirus
PO	phenoloxidase
ProPO	prophenoloxidase
VLP	virus-like particle
WOPV	the parasitoid (*A. tabida*) strain collected in the vicinity of Woerdense Verlaat in the Netherlands

INTRODUCTION

Host regulation has been described as the many effects—mostly physiological changes—that parasitoids cause in their host which benefit their own development (Vinson and Iwantsch, 1980). It evokes developmental disruption usually via hormonal or neurohormonal pathways, like the endocrine signaling which coordinates development of the parasitoid with that of the host so that the two partners molt in synchrony (Beckage and Gelman, 2004). It also includes all the effects on the host immune system (Strand and Pech, 1995; Schmidt *et al.*, 2001; Pennacchio and Strand, 2006; Carton *et al.*, 2008; Eslin *et al.*, 2009), the first physiological barrier that endophagous parasitoids encounter after they enter the hemocoel of their host.

In order to regulate their host's immunity and physiology, parasitoids produce and release active factors in the host hemocoel. These factors may come from either the female wasp's reproductive apparatus and its associated glands, or the parasitic egg or larva itself. In many species of the ichneumonid and braconid families (Ichneumonoidea), symbiotic polydnaviruses (PDVs) or virus-like particles (VLPs) (Schmidt and Schumann-Feddersen, 1989; Strand and Pech, 1995; Beckage, 1998; Drezen *et al.*, 2003; Beckage and Gelman, 2004; Pennacchio and Strand, 2006; Bézier *et al.*, 2009) can act as infecting agents. PDVs multiply in the calyx cells of the female wasp's ovaries while

Parasitoid Viruses: Symbionts and Pathogens. DOI: 10.1016/B978-0-12-384858-1.00021-7

VLPs can be produced either in the ovaries or the venom apparatus (Barrat *et al.*, 1999; Suzuki and Tanaka, 2006, see Gatti *et al.*, Chapter 15 of this volume). Once injected along with the parasitoid egg into the host hemocoel, PDVs can specifically infect host tissues, while both PDVs and VLPs may enter the circulating hemocytes. In Ichneumonoidea, both were shown to be the main agents controlling the host's physiology and immunity (Webb and Luckhart, 1994; Schmidt *et al.*, 2001; Beck and Strand, 2005; Labrosse *et al.*, 2005).

Venoms from female parasitoids have also been reported to cause abnormalities of host development, with modes of action that vary according to the lifestyle of the parasitoid species (Pennacchio and Strand, 2006). Venoms of ectoparasitoids are often paralytic and may cause the developmental arrest of the host that can be beneficial to the externally developing parasite (Doury *et al.*, 1995). In most cases, the role of endoparasitoids' venoms in suppressing the host's immune defenses has not been clearly established. In some braconid species, however, the venom may contribute to the inhibition of the host's growth by acting synergistically with the calyx fluid or PDVs (Tanaka and Vinson, 1991a, b; Gupta and Ferkovich, 1998; Jones and Wache, 1998). A limited number of studies suggest that the venom of endoparasitoid species devoid of symbiotic viruses and VLPs (Richards and Parkinson, 2000; Asgari *et al.*, 2003; Cai *et al.*, 2004) may perturb the host's immune defenses. For example, venom from the pupal parasitoid *Pteromalus puparum* (Hymenoptera: Pteromalidae) suppresses the cellular immune responses of its host, *Pieris rapae* (Cai *et al.*, 2004), and venom from *Pimpla hypochondriaca*, which is cytotoxic to the host's hemocytes, can impair hemocyte-mediated immune responses (Richards and Parkinson, 2000).

Among larval parasitoids, species parasitizing *Drosophila* have been the object of many studies, first because of the extensive knowledge we have about their *Drosophila* hosts, in particular *Drosophila melanogaster*, and second because they are widely distributed, therefore allowing comparisons between geographical strains. The results from these studies showed that PDVs were never found in any of the parasitoids using *Drosophila* species as hosts. VLPs are present in the venom of the Figitidae of the *Leptopilina* genus (Rizki and Rizki, 1990; Dupas *et al.*, 1996; Labrosse *et al.*, 2003; Morales *et al.*, 2005) while neither PDVs nor VLPs were ever reported in any of the Braconidae Alysiinae *Asobara* species, differently to several other studied braconid endoparasitoids. This is not surprising since the Alysiinae subfamily does not belong to the microgastroid complex, the monophyletic group of bracovirus-associated wasps. However, it is worth noting that even in microgastroid species, venom components are often necessary to enhance the effects of polydnaviruses (Stoltz, 1986; Stoltz *et al.*, 1988).

The *Asobara* genus is represented by a complex of species, several of which have been well-studied. Both species, *A. tabida* and its sibling *A. rufescens*, occur in most of Europe (Kraaijeveld *et al.*, 1994; Kraaijeveld and van der Wel, 1994; Carton *et al.*, 1986), and *A. tabida* is also found in North America. Other *Asobara* species which have been studied include *A. citri*, a species originating from Africa; *A. japonica*, whose range is limited to Japan (Mitsui *et al.*, 2007; Ideo *et al.*, 2008); and *A. persimilis* which occurs in Australia. *Asobara* near *orientalis* from Indonesia was more recently collected and is still under investigation.

Wasps of the *Asobara* genus oviposit into first and second instars of *Drosophila* larvae. Two striking aspects of the development of *Asobara* parasitoids are: (1) the diversity of the strategies the different species have developed to deal with their respective host's immune defenses; and (2) the strong impact of the female wasp's secretions (from venom gland and ovaries) on the overall physiology of the parasitized host.

In the *Asobara* genus, the venom of the female wasps turned out to be a major factor exerting an active, regulative effect on *Drosophila* hosts' physiology and immunity (Mabiala-Moundoungou *et al.*, 2010). Other cases are known of wasp's venom playing a major role in disturbing host immunity (Richards and Parkinson, 2000; Cai *et al.*, 2004). However, the most striking observation in *Asobara* parasitoids is that while venoms of all tested species proved to be active on the host *D. melanogaster*, the different *Asobara* species induce very different physiological effects (Prévost *et al.*, 2009) together with showing species-specific venom components that facilitate different activities.

EFFECTS OF *ASOBARA* PARASITOIDS AND THEIR VENOM ON THE IMMUNE DEFENSES OF *D. MELANOGASTER* LARVAE

As in most endoparasitoid species, *Asobara* eggs are laid in the host hemocoel and are therefore exposed to attack by the host's hemocytes. Parasitism by *A. japonica* causes the overall suppression of the host's encapsulation ability. This immunosuppressive effect is very efficient in the host *D. melanogaster* since no encapsulation has been seen during any of our experiments. Immunosuppression by *A. japonica* is associated with a marked inhibitory effect on the host's hemopoietic organ (Fig. 1). The lymph glands, which are composed of two prominent anterior lobes and several posterior lobes in healthy nonparasitized *D. melanogaster* larvae (Fig. 1A), appear highly impaired in larvae parasitized by *A. japonica* (Fig. 1B). At 72h post parasitization, the two anterior lobes have totally collapsed and most posterior lobes have disintegrated (Fig. 1B). *Drosophila melanogaster* larvae parasitized by *A. japonica* also possess a lower concentration of circulating hemocytes in their hemolymph than unparasitized ones (Fig. 2A). Not only is the number of plasmatocytes reduced upon parasitization, but the burst of lamellocytes, the capsule-forming hemocytes which are normally released by the hemopoietic organ in the hemolymph

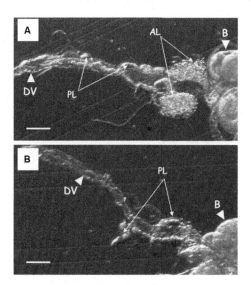

FIGURE 1 **Morphology of the hemopoietic organ from an unparasitized *D. melanogaster* larva (A) and from a larva parasitized by *A. japonica* (B) 72 h after parasitization, under a light microscope. A:** the hemopoietic organ from a control unparasitized larva is composed of two prominent anterior lobes (AL) and several posterior lobes (PL) ; B, brain; DV, dorsal vessel. **B:** the hemopoietic organ from a larva parasitized by *A. japonica* is characterized by anterior lobes that are totally altered (scale bars = 100 μm).

after parasitization (Eslin and Prévost, 1998; Moreau *et al.*, 2003; Labrosse *et al.*, 2005; Eslin *et al.*, 2009), never occurs (Fig. 2A). In order to test whether the inhibition of encapsulation was specifically limited to *A. japonica* eggs or reflected the overall capacity of *D. melanogaster* larvae to form hemocytic capsules, foreign bodies (here, paraffin oil drops) were injected into parasitized and unparasitized hosts (see protocol in Eslin and Doury, 2006). Results showed that *D. melanogaster* hemocytic reaction of encapsulation is totally inhibited upon parasitization by *A. japonica*, as demonstrated by the incapacity of the parasitized larvae to encapsulate injected foreign bodies (Fig. 3). However, neither circulating hemocytes presenting abnormal or transformed morphology nor cell fragments or remains could be observed in the hemolymph of the parasitized larvae. Therefore, it is believed that the depletion of the host's hemocytes population does not relate to an effect of the parasite on the circulating cells. Conversely, the host's hemopoietic organ appears to be the primary target of *A. japonica*, and this effect of the parasitoid must account for the suppression of the host's ability to mount hemocytic capsule formation (Prévost *et al.*, 2009; Mabiala-Moundoungou *et al.*, 2010) (Table 1).

Eggs and larvae of *A. japonica*, like injected foreign bodies, were never encapsulated by hemocytic cells, nor was there even a trace of melanin in *D. melanogaster* parasitized larvae. Experiments showed that parasitism of *D. melanogaster* by *A. japonica* first increases the host

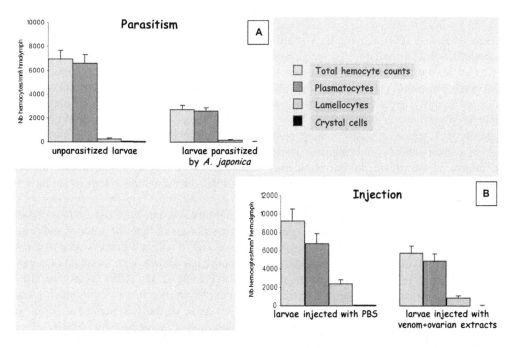

FIGURE 2 **Hemocyte count in *D. melanogaster* hosts parasitized by *A. japonica* (A) and hosts injected with a combination of *A. japonica* venom and ovarian extracts (B).** Controls consisted of unparasitized larvae (A) or larvae injected with PBS (B), respectively. Each parasitized, unparasitized or injected *D. melanogaster* larva was individually bled 72 h after treatment and the hemolymph collected on a Thoma hemocytometer slide. The number of hemocytes was counted individually for each larva immediatly after bleeding. Total hemocyte count (THC) and differential hemocyte count (i.e., number of plasmatocytes, lamellocytes, and crystal cells) were recorded. Values are expressed as mean numbers of cells per mm^3 of hemolymph (±standard error).

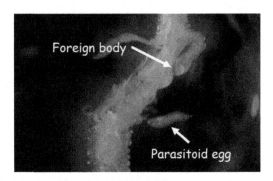

FIGURE 3 Paraffin oil drop and parasitoid egg not encapsulated in the hemolymph of a *D. melanogaster* host larva parasitized by *Asobara Japonica*. Please see color section at the back of the book.

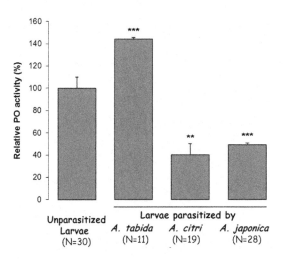

FIGURE 4 Hemolymph phenoloxidase (PO) activities in *D. melanogaster* larvae are affected 96 h after parasitization by *A. tabida*, *A. citri* or *A. japonica*. One hundred per cent of PO activities were allocated to total PO measured in the hemolymph of unparasitized *D. melanogaster* larvae. Phenoloxidase activities were measured in hemolymph extracted from a group of eight larva and rendered hemocyte-free following 10 min centrifugation at 10,000g. Absorbance values were read at 495 nm every minute for 15 min in a BioRad Multiplate reader by incubating active phenoloxidase with 5 mM DOPA (Di hydroxy-phenylalanine). Bars and vertical lines represent the mean per cent of total PO ± standard error.

TABLE 1 Rate of Encapsulation (Mean % ± Standard Error) of a Foreign Body (10 nl Paraffin Oil Drop) Injected in *D. melanogaster* Larvae 24 h Post Parasitization by *A. japonica*.

Encapsulation (%) of the foreign body by *D. melanogaster* larvae injected with paraffin oil drop	
Unparasitized larvae (N = 3 × 20)	Larvae parasitized by *A. japonica* (N = 3 × 20)
79 ± 4.8	0 ± 0

Controls consisted of unparasitized larvae injected in the same conditions. The encapsulation rate of the droplets was recorded after dissection 48 h post-injection. N = 3 × 20 injected larvae.

hemolymph prophenoloxidase (PO) activity, then significantly reduces both PO (Fig. 4) and ProPO, the precursor to active PO (not shown), within 72 h. These results demonstrate that not only hemocytic reactions but also humoral defenses are strongly impaired by the parasitoid *A. japonica* (Mabiala-Moundoungou *et al.*, 2010).

Manual injection of *A. japonica* wasps' venom combined with ovarian extract into *D. melanogaster* larvae reproduces the effects of parasitism on the circulating hemocytic cells, and the total number of circulating hemocytes decreases within 72 h post-injection (Fig. 2B). In contrast, ovarian extract injected alone into *D. melanogaster* larvae does not affect the population of hemocytes. Therefore, the wasp's venom must account for the immunosuppressive effect of the parasitoid. In addition, venom injected alone causes general histolysis such that the tissues of *D. melanogaster* larvae disintegrated within a few hours following injection. This shows that *A. japonica* venom must affect not only the host's immunity system but also the overall physiology of the parasitized larva and the integrity of its organs.

Asobara citri is rarely encapsulated in *D. melanogaster*, showing that this *Asobara* species too has developed

efficient means to disrupt the host's cellular defenses. Like *A. japonica*, *A. citri* causes an overall suppression of the host's cellular defenses, not only towards the developing parasitoid eggs but also towards foreign bodies injected in the parasitized host. It was shown from the hemocyte counts that *Drosophila* larvae parasitized by *A. citri* possessed fewer circulating hemocytes than unparasitized ones (Moreau *et al.*, 2003). The numbers of plasmatocytes and lamellocytes required for capsule formation are particularly reduced (Fig. 5), but no cell pathology or cell lysis can be observed among the circulating hemocytes. The study of the hemopoietic organs of *D. melanogaster* larvae parasitized by *A. citri* revealed that the size of the anterior lobes of the lymph glands is strongly reduced, while the posterior lobes are more developed than in unparasitized larvae (Moreau *et al.*, 2003; Prévost *et al.*, 2005). Increased sizes of both the anterior and posterior lobes are considered to reflect a primary step of the host's immune response, as shown with nonimmunosuppressive parasitoids (Prévost *et al.*, 2009). Therefore, the severe disruption of the anterior lobes of the lymph glands must be the major cause of the low hemocyte count recorded in the hemolymph of larvae parasitized by *A. citri*.

Electronic microscopy of hemocytes in the lymph gland of *D. melanogaster* larvae parasitized by *A. citri* clearly shows altered cells in the anterior lobes (Prévost *et al.*, 2005), supporting the hypothesis that, like with *A.*

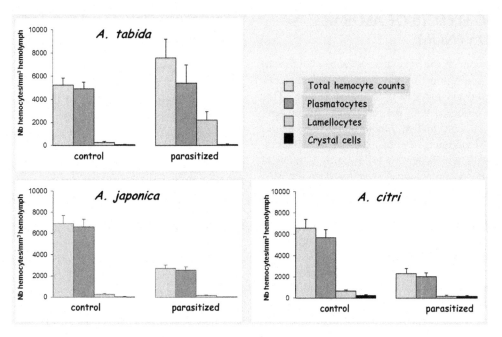

FIGURE 5   Hemocyte count in *D. melanogaster* hosts parasitized by *A. tabida*, *A. japonica*, or *A. citri*. Each parasitized or unparasitized control *D. melanogaster* larva was individually bled 72 h after parasitization and the hemolymph collected on a Thoma hemocytometer slide. The number of hemocytes was counted individually for each larva immediatly after bleeding. Total hemocyte count (THC) and differential hemocyte count (i.e., number of plasmatocytes, lamellocytes, and crystal cells) were recorded. Values are expressed as mean numbers of cells per mm³ of hemolymph (±standard error).

japonica, the host's hemopoietic organ is one main target of the immunosuppressive effect of the parasitoid. The destructive effect of *A. citri* on the anterior lobes of the host's hemopoietic organ is attributed to be necrosis.

Like the species *A. japonica*, *A. citri* significantly decreases the PO activity in the hemolymph of *D. melanogaster* larvae (Fig. 4), therefore inhibiting the humoral reaction of melanin deposition around the parasitic egg.

Some experiments were recently conducted on *A. persimilis*, the Australian *Asobara* species. Preliminary results suggested a regulator strategy, also targeting the host's hemopoietic organ.

In contrast to *A. japonica*, *A. citri* and *A. persimilis*, the species *A. tabida* has developed an original way to evade encapsulation. *A. tabida* does not impair either the hemopoietic organ nor reduce the circulating numbers of hemocytes of its hosts (Prévost *et al.*, 2005, 2009). Studies of the hemocyte population in the hemolymph of parasitized *Drosophila* larvae proved that *A. tabida* eggs are recognized by the host's immune system. Oviposition by *A. tabida* is followed by a burst of lamellocytes (Fig. 5) and a significant increase in PO activity (Fig. 4) in the host hemolymph, both hemocytic and humoral reactions, which typically indicate that the host immune system is responding to the presence of the parasite (Eslin and Prévost, 1996, 1998; Eslin *et al.*, 2009). In the A1 strain of *A. tabida*, the exochorion of the parasitic egg possesses adhesive properties such that it can attach to almost any host tissue floating in the hemocoel (fat body, digestive system, tracheal cells) (Kraaijeveld and van Alphen, 1994; Monconduit and Prévost, 1994; Eslin *et al.*, 1996). Movements of the host larva probably contribute to create many contact areas between the parasitic egg (at first floating free in the host hemocoel) and the host tissues, therefore resulting in the embedment of the parasitoid within host tissues (Prévost *et al.*, 2005, 2009). Embedment can be so well completed that the presence of a parasitoid egg may not be detectable in the dissected host. It is therefore understandable that attachment of the parasitoid egg's chorion to the host tissues protects the parasite from any contact with the host's cellular and humoral defenses, i.e., spreading hemocytes associated with the concomitant cytotoxic products like melanin, reactive oxygen and nitrogen intermediates, and hydroxyl radicals. *Asobara tabida* eggs benefit from an efficient protection by adhering to host tissues. Proof is given by the existence of a nonvirulent strain of *A. tabida*, the WOPV strain from the Netherlands. WOPV wasps lay nonsticky eggs floating free in the host hemocoel which are always readily encapsulated by *D. melanogaster* larvae (Kraaijeveld, 1994). WOPV wasps can thus develop only in *D. subobscura* hosts, one of the *Drosophila* species of the *obscura* group presenting an innate immunity deficiency (Eslin and Doury, 2006; Eslin *et al.*, 2009; Havard *et al.*, 2009).

PARALYZING EFFECTS OF *ASOBARA* PARASITOID VENOM

Another expression of the regulative effects of *Asobara* species on the physiology of the host *D. melanogaster* is the host's paralysis induced upon parasitization. This effect has been observed with the three well-studied species *A. tabida*, *A. citri* and *A. japonica* (Moreau *et al.*, 2002; Mabiala-Moundoungou *et al.*, 2010). The induction of a total paralysis, which may be followed by a transient immobility, is unusual compared to what has been reported for venom properties of other known endoparasitoid species which are koinobiont, i.e., whose egg is layed into a host which continues to develop normally. By contrast, this type of effect is frequently observed in hosts parasitized by idiobiont ectoparasitoids (Doury *et al.*, 1997). In this lifestyle, the egg-laying female paralyzes the host before laying an egg and the host does not recover from paralysis during the development of the ectoparasitoid. *Asobara* species inducing paralysis could reflect their ancestral ectoparasitic status (Moreau *et al.*, 2002). Indeed, most of the Alysiinae are solitary koinobionts of Diptera larvae, but the subfamily is included in a complex of subfamilies (the braconoids) that are predominantly idiobiont ectoparasitoids (Dowton *et al.*, 1998).

In order to have *Asobara* females stinging *Drosophila* larvae without laying an egg, we designed controlled conditions for what we named 'interrupted oviposition' (IO). Parasitoid females are offered *Drosophila* larvae for parasitization and observed under a stereomicroscope. As soon as an oviposition behavior is initiated, it is quickly interrupted by disturbing the female wasp. The stung larva is then immediately removed before egg laying can take place and prevention of egg laying is confirmed later by dissection.

Fig. 6 describes the effects of the interrupted oviposition (IO) by *A. tabida*, *A. citri*, or *A. japonica* wasps, on *D. melanogaster* larvae. Larvae are paralyzed within a few seconds after IO by an *A. japonica* female. The IO by *A. japonica* turned out to be lethal within 2 h. After IO either by *A. tabida* or *A. citri* females, *D. melanogaster* larvae are paralyzed too, but all recover within 2–3 min and 12–15 min, respectively.

Asobara japonica venom extract injected at the lower dose of 0.2 ng of total proteins per larva have no effect on *D. melanogaster* larvae. For all the other doses tested (i.e., 1–40 ng of total protein per larva), *D. melanogaster* larvae become progressively and irreversibly paralyzed, or even die, in a dose-dependent manner, within 2 h following venom injection (Fig. 7). These results, compared with the ones obtained with the IO, suggest that *A. japonica* female wasps must release their venom in the host hemocoel just before they lay an egg. After injection of 2 ng venom extract per larva, mortality occurs within the same space of time as the one observed after an IO, suggesting that an equivalent venom dose is injected by the female wasp during the IO.

Injection of *A. tabida* or *A. citri* venom extract into *D. melanogaster* larvae induces a transient paralysis which takes place within a few minutes. However, the paralysis is

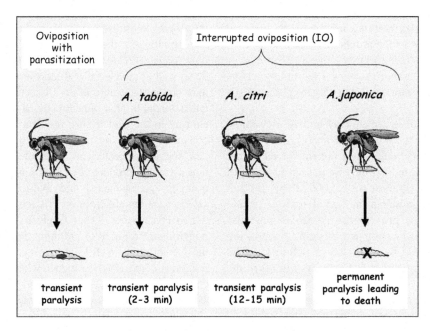

FIGURE 6 Effect of parasitism (i.e. oviposition with parasitization) or interrupted oviposition by *A. tabida*, *A. citri* or *A. japonica* on *D. melanogaster* larvae.

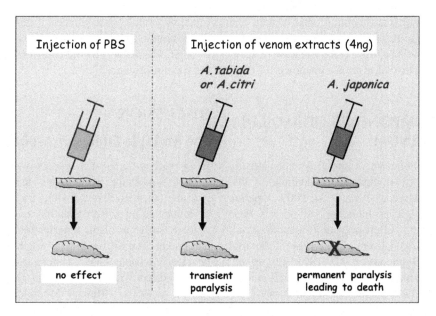

FIGURE 7 Effect of the injection of 20 nl (4 ng proteins) of venom extract from *A. tabida*, *A. citri* or *A. japonica* on *D. melanogaster* larvae. Controls consisted of larvae injected with PBS.

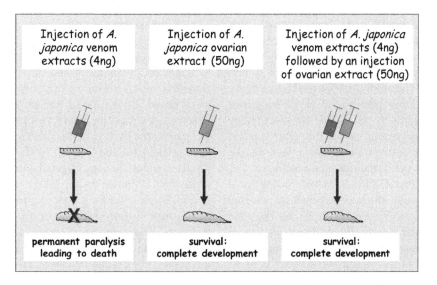

FIGURE 8 Effect of the injection of venom extract, ovarian extract, or a combination of venom and ovarian extract from *A. japonica* on *D. melanogaster* larvae.

never lethal and the injected larvae recover normal activity within 2 or 15 min, respectively, even at the highest delivered doses. In control larvae injected with PBS, neither the mobility nor the survival of the larvae is affected (Fig. 7).

The injection of *A. japonica* ovarian extract has no particular effect on *D. melanogaster* larvae, even at the dose of 50 ng of total protein per larva. However, when the injection of ovarian extract is performed in the minute following an IO, it totally reverses the effects induced by the sting of the female parasitoid, i.e., paralysis and mortality.

The same result is obtained when the ovarian extract is injected immediately after a previous injection of venom extract (Fig. 8).

Strikingly, when *Drosophila* larvae are submitted first to an IO by *A. japonica* wasp or an injection of *A. japonica* venom extract and then receive an ovarian extract injection, they all develop normally and emerge as adult flies. This surprising effect demonstrates that an extract from *A. japonica* ovaries totally reverses the deadly effect of the wasp venom delivered either by the female during an interrupted oviposition or manually using microinjections. Venom and ovarian fluids act thus, in this species, similarly to a poison/antidote combination. To date, the molecular factors responsible for these antagonistic effects remain unknown.

MOLECULAR COMPONENTS OF *ASOBARA* PARASITOIDS' VENOM

In *A. tabida*, the most abundant venom protein has already been well studied: it is an aspartylglucosaminidase that was named AtAGA (Moreau *et al.*, 2004, 2009; Vinchon *et al.*, 2010). Aspartylglucosaminidase (AGA) has been described in a wide range of prokaryotic (Tarentino *et al.*, 1994) and eukaryotic (McGovern *et al.*, 1983; Tollersrud and Aronson, 1992; Tenhunen *et al.*, 1995; Liu *et al.*, 1996) organisms. The human AGA has been well studied since deleterious mutations in the gene cause a rare lysosomal storage disorder known as aspartylglucosaminuria (Ikonen *et al.*, 1991; Saarela *et al.*, 2001).

AGA is commonly known as a lysosomal enzyme that hydrolyses linkage between an asparagine residue and the N-acetylglucosamine moiety. Based on analogies to other enzymes to which it is related, it has been suggested that this protein, once injected into the host, could be responsible for the production of aspartate, which is a known excitatory neurotransmitter of the *Drosophila* nervous system (Besson *et al.*, 2000). Therefore, AtAGA could potentially be involved in the transient paralysis of *D. melanogaster* larvae upon parasitization by *A. tabida*. However, AGA was not found in any venom of all the other studied *Asobara* species, despite their paralysing effect on the host, suggesting it may not be the true active factor.

It was shown that the cDNA of AtAGA codes for a proαβ precursor molecule which is preceded by a signal peptide of 19 amino acids (Vinchon *et al.*, 2010). The gene products can be detected specifically in the wasp's venom gland in which it is found under two forms: an active heterotetramer composed of two α and two β subunits of 18 and 30 kDa, respectively, corresponding to the active form, and an homodimer of 44 kDa that might correspond to a secreted precursor related to AtAGA. AtAGA does not exhibit any glycopeptide N-glycosidase activity towards complete glycoproteins. Therefore, its activity seems to be restricted to the deglycosylation of free glycosylasparagines like human AGA. The role played by the AGA in the host–parasite interaction is still unknown but is likely to be important since AGA production represents a physiological investment for the wasp because this protein is by far the most abundant in the venom gland, representing up to 30% of the protein content. The abundant secretion of this typically lysosomal hydrolase by venom gland cells raises several interesting questions concerning (1) the mechanisms by which AtAGA has acquired a venomous

function, (2) the consequences of this original evolution on the sequence and structural features of AtAGA compared to other AGAs, and (3) the contribution of this enzyme to the wasp's virulence.

DISCUSSION

The Multiple Effects of *Asobara* Venoms

In *D. melanogaster* larvae parasitized by *Asobara* parasitoids, two major 'pathologies' associated with parasitism are considered responsible for the host's incapacity to mount cellular encapsulation reactions. One of them is the destruction of the anterior lobes of the larval host's hemopoietic organ, an effect which is strong enough to be clearly visible under a stereomicroscope. This effect was reported with both *A. citri* (Moreau *et al.*, 2003) and *A. japonica* (Mabiala-Moundoungou *et al.*, 2010), although disruption of the host's lymph gland was more pronounced with *A. japonica*. In *D. melanogaster* parasitized by parasitoids of the *Leptopilina* genus, hemocytes in the hemopoietic organ are also the targets of the parasites (Chiu and Govind, 2002). However, the reported effects of *Leptopilina* species differ in several aspects from what is observed with *A. japonica* and *A. citri*. Firstly, cell lysis provoked by *Leptopilina* species has been described as apoptosis, while electronic microscopy suggested that the effect of *A. citri* on the host's lymph gland was necrosis (Prévost *et al.*, 2005). Secondly, cell lysis has been observed in circulating hemocytes of *D. melanogaster* larvae parasitized by *Leptopilina* and *Ganaspis* species (Russo *et al.*, 2001; Chiu and Govind, 2002), while the effects of *A. japonica* and *A. citri* are targeted on the host's hemopoietic organ only (Moreau *et al.*, 2003, Mabiala-Moundoungou *et al.*, 2010). Attacking the anterior lobes of the host's hemopoietic organ (with no concomitant destruction of the circulating hemocytes), thus, seems to be specific to the regulation effects of *Asobara* species. However, these effects on host cellular defenses could not be proven to be directly induced by the venom of *Asobara* female wasps.

The second important change is the drop in phenoloxidase activity in the hemolymph of parasitized larvae, another effect of *Asobara* parasitoids which can account, at least partially, for the inhibition of the encapsulation reaction. The effect of an endoparasite on the host's phenoloxidase system is rather common to many host–parasitoid interactions. Recently, in the venom of *Leptopilina boulardi*, another *Drosophila* parasitoid, protease inhibitors were shown to target the phenoloxidase system (Colinet *et al.*, 2009). Therefore, it is not so surprising that *A. japonica* and *A. citri*, showing a strong immuno-suppressive effect, also alter their host's phenoloxidase activity, although the mechanism and agent responsible for this effect are not yet known.

Parasitization of *D. melanogaster* larvae gives rise to more or less pronounced paralysis in the three tested *Asobara* species. Like many other parasitic wasps, *Asobara* parasitoids tend to locate their larval host using vibrotaxis and ovipositor searching (Sokolowski and Turlings, 1987). Paralyzing the hosts thus could be a method developed by these solitary wasp species to prevent superparasitism, at least for some period of time following parasitization. It is also possible that this trait could have been inherited from endoparasitic ancestors of the Alysiinae, which belong to the braconoids, a complex of subfamilies predominantly paralyzing idiobiont ectoparasitoids.

In contrast to immune defenses, paralysis of host larvae can be undoubtedly attributed to the effect of the female wasps' venoms in the three tested *Asobara* species. Manual injection of wasp venom into unparasitized *D. melanogaster* larvae gives rise to different degrees of paralysis depending on the *Asobara* species. Both *A. tabida* and *A. citri* possess the least harmful venoms, while *A. japonica* venom can be considered a 'lethal weapon', since all larvae die within a few hours following injection (Mabiala-Moundoungou *et al.*, 2010). The capacity of *A. japonica* venom to cause almost immediate, irreversible paralysis of *D. melanogaster* larvae suggests the presence of strong neurotoxic factors. Beard (1978) suggested that venoms of Braconidae affect neuromuscular synapses and block the nerve impulses controlling muscle tone and movement. Also, spider and wasp venoms are known to possess various acyl-polyamine toxins which paralyse insects by blocking the nerve–muscle signal transduction of glutamatergic synapses (Itagaki *et al.*, 1997).

The venom of *A. japonica* is also potentially involved in the disruption of host cells and tissues and it is possible that other functions than cellular immune defenses are targeted. This could result from the effects of hydrolase enzymes, which often present cytotoxic and histolytic effects once injected into host or prey. By their function of cell and tissue disruption, these enzymes could constitute self-supporting factors or facilitate the action of other components (like neurotoxic, paralytic, immunesuppressive, anticlotting, antimicrobial or cytotoxic factors) (Kuhn-Nentwig *et al.*, 2004; Piek, 1986). Virulence factors provoking paralysis and/or destruction of the host's hemopoietic organ are still under investigation.

Concerning *A. tabida* AGA, which has been recently sequenced, it was revealed to be slightly more divergent to *Apis mellifera* sequence than the AGA insect homologs (Vinchon *et al.*, 2010). In particular, AtAGA displays relatively low amino acid conservation with lysosomal AGAs from parasitoid (*Nasonia vitripennis*) and free-living (*Apis mellifera*) hymenopteran species. In the braconid wasp, *Cotesia congregata,* where the divergence levels of cellular and polydnavirus genes were compared, it was shown that a greater divergence level was a specific hallmark of the genes involved in the parasitoid's virulence (Bézier *et al.*, 2008). The slightly greater sequence variation observed in AtAGA is thus possibly due to specific evolutionary constraints that applied to this venomous enzyme in the context of host–parasitic relationships. AtAGA is a functional enzyme which could have evolved from a lysosomal origin and may be under specific selective pressures to acquire an original function in the venom of *A. tabida* (Vinchon *et al.*, 2010). The study of this enzyme may allow a better understanding of the functional evolution of venom enzymes in hymenopteran parasitoids.

The most remarkable feature in *Asobara* parasitoids is this unique interaction between venom and ovarian substances in the species *A. japonica*. *Asobara japonica* female wasps possess an extremely powerful, deadly venom which kills the host at low concentrations, such that the counter action of an 'antidote' present in the wasp ovaries is absolutely necessary to regulate the effects of the venom and permit the development of the parasite. This is the first report of (1) an endophagous parasitoid injecting into the host such 'deadly' venom, and (2) wasp's venom and ovary substances working in a poison/antidote manner to ensure regulation of the host's physiology (Mabiala-Moundoungou *et al.*, 2010). *Asobara japonica* venom acts synergistically with ovarian secretions to alter the host's cellular immune response (by destroying a major part of the hemopoietic organ) without killing the host. Therefore, *A. japonica* ovaries must contain factors antagonizing the lethal effect of the crude venom. Venom synergizes the effect of ovary secretions (Digilio *et al.*, 1998), calyx fluid (Wago and Tanaka, 1989), or polydnavirus present in calyx fluid (Soller and Lanzrein, 1996) in several host–parasitoid systems. However, ovarian secretions which counterbalance the deadly effect of venom, as it occurs in *A. japonica*, remain an exceptional case among parasitoids.

SUMMARY OF *ASOBARA'S* STRATEGIES AND TOOLKITS

Several questions arise from the study of *Asobara* parasitoids. One question concerns the diversity of the mechanisms developed by the various species to overcome their host defenses. Active inhibition of host defenses seems to be the main rule among *Asobara* species, while passive evasion is clearly established only in *A. tabida*. We may wonder if *A. japonica* and *A. citri* are sharing the same molecular tools to overcome host defenses. As a matter of fact, both species are targeting the host hemopoietic organ instead of attacking the circulating hemocytes, a much more common strategy among endophagous parasitoids. The more pronounced effects of *A. japonica* then could result from a dose dependent effect. However, the 'antidote' of ovarian origin that is necessary to regulate the deadly effects of *A. japonica* venom rather suggests that this species has developed specific factors of virulence.

This diversity among *Asobara* species may reflect their different evolutionary histories, each of them inhabiting different continents. We also now know that there is an amazing array of molecular factors developed by endoparasitoids to overcome host defenses (Beckage and Gelman, 2004). Whether the nature and diversity of these molecules are more likely to reflect the lifestyle of the parasitoids, or their habitat and ecological niche, or the genus or family of their host, or rather their ancestral origin, is not well understood. It is possible that among the large toolkit of gene products inherited from Apocrita Hymenoptera, the virulence factors which are retained by one given family, genus, or even species of parasitoids, is partially fortuitous and is affected by factors other than the host ranges. This could explain the large diversity of the virulence strategies met in parasitoids which are either phylogenetically related or exploiting the same hosts.

The various strategies developed by parasitoids to overcome their hosts' immunity and overall physiology raise the question of the relative physiological cost for the parasitoid. *Asobara tabida* seems to 'passively' avoid encapsulation but sticky eggs may require specific chorionic proteins and AtAGA (its role still unknown) accounts for 30% of the wasp's venom proteins. *Asobara citri* and *A. japonica,* which attack the host's lymph gland, must produce in their venom, ovaries, or embryos an active factor responsible for the impairment of the host's cellular defense system. Also, *A. japonica* produces an overvirulent venom which needs to be regulated by a factor of ovarian origin. In *A. japonica*, a possible hypothesis would be that active factors in the wasp's venom are produced at low cost because they are active at low concentrations and attack any host tissue. It is also possible that venom and ovarian active factors simultaneously play several roles in the regulation of the host physiology, besides the identified effect of immunosuppression. The nature and functions of the active factors in both venom and ovaries of *Asobara* parasitoids are presently under investigation. Identification of these factors should reveal the molecular machinery that *Asobara* parasitoids have developed to overcome and regulate the immunity and physiology their host.

Parasitoids may inherit their virulence factors from regulatory molecules present in ancestral species. How this basal toolkit of gene products has changed with the developmental strategies of the parasitoid species needs to be investigated. For this approach, the model *Asobara* is of particular interest because it shows some variability between *Asobara* species—and even within species—in the strategies the parasitoids have developed to circumvent host defenses. In addition, the *Asobara* genus is atypical because it shares some properties with ectoparasitoid

braconids, like the paralyzing effects of the wasps' venoms (Moreau *et al.*, 2009; Mabiala-Moundoungou *et al.*, 2010). The hypothesis that the Alysiinae belong to a predominantly ectoparasitic clade and may have recently reverted to endoparasitism (Dowton *et al.*, 1998) would explain why *Asobara* species present several peculiar traits compared to what is known in other endophagous braconids studied, mostly associated with bracoviruses. Therefore, these wasps may be of great interest in studying the evolutionary relationship between ecto- and endoparasitic species.

ACKNOWLEDGMENTS

We thank Yvelise Fourdrain for rearing the insects and contributing to experiments involving host infestations under controlled conditions. We also thank Dr. Roland Allemand for graciously providing the A1 strain of *Asobara tabida*, and Prof. J.J.M. van Alphen for graciously providing the parasitoid (*A. tabida*) strain collected in the vicinity of Woerdense Verlaat in the Netherlands. This work has been supported by the 'Agence Nationale de la Recherche' (ANR: projects Evparasitoid ans Paratoxose) and by the GDR-CNRS 2153.

REFERENCES

Asgari, S., Zareie, R., Zhang, G., Schmidt, O., 2003. Isolation and characterization of a novel venom protein from an endoparasitoid, *Cotesia rubecula* (Hym: Braconidae). Arch. Insect Biochem. Physiol. 53, 92–100.

Barratt, B.I.P., Evans, A.A., Stoltz, D.B., Vinson, S.B., Easingwood, R., 1999. Virus-like particles in the ovaries of *Microctonus aethiopoides* Loan (Hymenoptera: Braconidae), a parasitoid of adult weevils (Coleoptera: Curculionidae). J. Invertebr. Pathol. 73, 182–188.

Beard, R.L., 1978. Venoms of Braconidae. Handbuch der Experimentellen pharmakologie, Arthropods Venoms 48, 773–800.

Beck, M., Strand, M.R., 2005. Glc1.8 from *Microplitis demolitor* bracovirus induces a loss of adhesion and phagocytosis in insect high five and S2 cells. J. Virol. 79, 1861–1870.

Beckage, N.E., 1998. Modulation of immune responses to parasitoids by polydnaviruses. Parasitology 116, 57–64.

Beckage, N.E., Gelman, D.B., 2004. Wasp parasitoid disruption of host development: implications for new biologically based strategies for insect control. Annu. Rev. Entomol. 49, 299–330.

Besson, M.T., Soustelle, L., Birman, S., 2000. Selective high-affinity transport of aspartate by a *Drosophila* homologue of the excitatory amino acid transporters. Curr. Biol. 10, 207–210.

Bézier, A., Herbinière, J., Serbielle, C., Lesobre, J., Wincker, P., Huguet, E., et al. 2008. Bracovirus gene products are highly divergent from insect proteins. Arch. Insect Biochem. Physiol. 67, 172–187.

Bézier, A., Annaheim, I., Herbinière, J., Wetterwald, C., Gyapay, G., Bernard-Samain, S., et al. 2009. Polydnaviruses of braconid wasps derive from an ancestral nudivirus. Science 323, 926–930.

Cai, J., Ye, G-y., Hu, C., 2004. Parasitism of *Pieris rapae* (Lepidoptera: Pieridae) by a pupal endoparasitoid, *Pteromalus puparum*

(Hymenoptera: Pteromalidae): effects of parasitization and venom on host hemocytes. J. Insect Physiol. 50, 315–322.

Carton, Y., Boulétreau, M., van Alphen, J.J.M., van Lenteren, J.C., 1986. The *Drosophila* parasitic wasps. In: Ashburner, M., Thomson, J. (Eds.), The Genetics and Biology of *Drosophila*, vol. 3C. Academic Press, London.

Carton, Y., Poirié, M., Nappi, A.J., 2008. Insect immune resistance to parasitoids. Insect Sci. 15, 67–87.

Chiu, H., Govind, S., 2002. Natural infection by virulent parasitic wasps induces apoptotic depletion of hematopoietic precursors. Cell Death Differ. 9, 1379–1381.

Colinet, D., Dubuffet, A., Cazes, D., Moreau, S., Drezen, J.-M., Poirié, M.A., 2009. Serpin from the parasitoid wasp *Leptopilina boulardi* targets the *Drosophila* phenoloxidase cascade. Dev. Comp. Immunol. 33, 681–689.

Digilio, M.C., Pennacchio, F., Tremblay, E., 1998. Host regulation effects of ovary fluid and venom of *Aphidius ervi* (Hymenoptera: Braconidae). J. Insect Physiol. 44, 779–784.

Doury, G., Bigot, Y., Periquet, G., 1997. Physiological and biochemical analysis of factors in the female venom gland and larval salivary secretions of the ectoparasitoid wasp *Eupelmus orientalis*. J. Insect Physiol. 43, 69–81.

Doury, G., Rojas-Rousse, D., Periquet, G., 1995. Ability of *Eupelmus orientalis* ectoparasitoid larvae to develop on an unparalysed host in the absence of female stinging behaviour. J. Insect Physiol. 41, 287–296.

Dowton, M., Austin, A.D., Antolin, M.F., 1998. Evolutionary relationships among the Braconidae (Hymenoptera: Ichneumonoidea) inferred from partial 16S rDNA gene sequences. Insect Mol. Biol. 7, 129–150.

Drezen, J.-M., Provost, E., Espagne, E., Cattolico, L., Dupuy, C., Poirié, M., et al. 2003. Polydnavirus genome: integrated vs. free virus. J. Insect Physiol. 49, 407–417.

Dupas, S., Brehélin, M., Frey, F., Carton, Y., 1996. Immune suppressive virus-like particles in a *Drosophila* parasitoid: significance of their intraspecific morphological variations. Parasitology 113, 207–212.

Eslin, P., Doury, G., 2006. The fly *Drosophila subobscura*: a natural case of innate immunity deficiency. Dev. Comp. Immunol. 30, 977–983.

Eslin, P., Giordanengo, P., Fourdrain, Y., Prévost, G., 1996. Avoidance of encapsulation in the absence of VLP by a braconid parasitoid of *Drosophila* larvae: an ultrastructural study. Can. J. Zool. 74, 2193–2198.

Eslin, P., Prévost, G., 1996. Variation in *Drosophila* concentration of hemocytes associated with different ability to encapsulate *Asobara tabida* larval parasitoid. J. Insect Physiol. 42, 549–555.

Eslin, P., Prévost, G., 1998. Hemocyte load and immune resistance to *Asobara tabida* are correlated in species of *Drosophila melanogaster* subgroup. J. Insect Physiol. 44, 807–816.

Eslin P., Prévost G., Havard S., Doury G., 2009. Immune resistance of *Drosophila* hosts against *Asobara* parasitoids: cellular aspects. In: G. Prévost (Ed.), Parasitoids of *Drosophila*, Advances in Parasitology, vol. 70 (pp. 189–215). Academic Press, Amsterdam.

Gupta, P., Ferkovich, S.M., 1998. Interaction of calyx fluid and venom from *Microplitis croceipes* (Braconidae) on developmental disruption of the natural host, *Heliocoverpa zea*, and two atypical hosts, *Galleria mellonella* and *Spodoptera exigua*. J. Insect Physiol. 44, 713–719.

Havard, S., Eslin, P., Prévost, G., Doury, G., 2009. Encapsulation ability: are all *Drosophila* species equally armed? An investigation in the obscura group. Can. J. Zool. 87, 635–641.

Ideo, S., Watada, M., Mitsui, H., Kimura, M.T., 2008. Host range of *Asobara japonica* (Hymenoptera: Braconidae), a larval parasitoid of drosophilid flies. Entomol. Sci. 11, 1–6.

Ikonen, E., Baumann, M., Grön, K., Syvänen, A.C., Enomaa, N., Halila, R., et al. 1991. Aspartylglucosaminuria: cDNA encoding aspartylglucosaminidase and the missense mutation causing the disease. EMBO J. 10, 51–58.

Itagaki, Y., Naoki, H., Fujita, T., Hisada, M., Nakajima, M., 1997. Characterization of spider venom by mass spectrometry construction of analytical system. Yakugaku Zasshi 117, 715–728.

Jones, D., Wache, S., 1998. Preultimate 4th/5th instar *Trichoplusia ni* naturally injected with venom/calyx fluid from *Chelonus curvimaculatus* precociously metamorphose, rather than obey the metamorphic size threshold that would normally compel molting to a 5th/6th instar. J. Insect Physiol. 44, 755–765.

Kraaijeveld A.R., 1994. Local adaptations in a parasitoid–host system – a coevolutionary arms race? Ph.D. thesis, University of Leiden.

Kraaijeveld, A.R., van der Wel, N.N., 1994. Geographical vartiation in reproductive success of the parasitoid *Asobara tabida* in larvae of several *Drosophila* species. Ecol. Entomol. 19, 221–229.

Kraaijeveld, A.R., Voet, S., van Alphen, J.J.M., 1994. Geographical variation in habitat choice and host suitability in the parasitoid *Asobara rufescens*. Entomologia Experimentalis et Applicata 72, 109–114.

Kraaijeveld, A.R., van Alphen, J.M., 1994. Geographic variation in resistance against encapsulation by *Drosophila melanogaster* larvae: the mechanism explored. Physiol. Entomol. 19, 9–14.

Kuhn-Nentwig, L., Schaller, J., Nentwig, W., 2004. Biochemistry, toxicology and ecology of the venom of the spider *Cupiennius salei* (Ctenidae). Toxicon 43, 543–553.

Labrosse, C., Carton, Y., Dubuffet, A., Drezen, J.-M., Poirié, M., 2003. Active suppression of *D. melanogaster* immune response by long gland products of the parasitic wasp *Leptopilina boulardi*. J. Insect Physiol. 49, 513–522.

Labrosse, C., Stasiak, K., Lesobre, J., Grangeia, A., Huguet, E., Drezen, J.-M., et al. 2005. A RhoGAP protein as a main immune suppressive factor in the *Leptopilina boulardi* (Hymenoptera, Figitidae)–*Drosophila melanogaster* interaction. Insect Biochem. Mol. Biol. 35, 93–103.

Liu, Y., Dunn, G.S., Aronson Jr., N.N., 1996. Purification, biochemistry and molecular cloning of an insect glycosylasparaginase from *Spodoptera frugiperda*. Glycobiology 6, 527–536.

Mabiala-Moundoungou, A.D.N., Doury, G., Eslin, P., Cherqui, A., Prévost, G., 2010. Deadly venom of *Asobara japonica* needs ovarian antidote to regulate host physiology. J. Insect Physiol. 56, 35–41.

McGovern, M.M., Aula, P., Desnick, R.J., 1983. Purification and properties of human hepatic aspartylglucosaminidase. J. Biol. Chem. 258, 10743–10747.

Mitsui, H., Van Achterberg, K., Nordlander, G., Kimura, M.T., 2007. Geographical distributions and host associations of larval parasitoids of frugivorous Drosophilidae in Japan. J. Nat. His. 41, 1731–1738.

Monconduit, H., Prévost, G., 1994. Avoidance of encapsulation by *Asobara tabida*, a larval parasitoid of *Drosophila* species. Norwegian J. Agric. Sci. 16, 301–309.

Morales, J., Chiu, H., Oo, T., Plaza, R., Hoskins, S., Govind, S., 2005. Biogenesis, structure, and immunesuppressive effects of virus-like particles of a *Drosophila* parasitoid, *Leptopilina victoriae*. J. Insect Physiol. 51, 181–195.

Moreau, S.J.M., Cherqui, A., Doury, G., Dubois, F., Fourdrain, Y., Sabatier, L., et al. 2004. Identification of an aspartylglucosaminidase-like protein in the venom of the parasitic wasp *Asobara tabida* (Hymenoptera: Braconidae). Insect Biochem. Mol. Biol. 34, 485–492.

Moreau, S.J.M., Dingremont, A., Doury, G., Giordanengo, P., 2002. Effects of parasitism by *Asobara tabida* (Hymenoptera: Braconidae) on the development, survival and activity of *Drosophila melanogaster* larvae. J. Insect Physiol. 48, 337–347.

Moreau, S.J.M., Eslin, P., Giordanengo, P., Doury, G., 2003. Comparative study of the strategies evolved by two parasitoids of the *Asobara* genus to avoid immune response of the host, *Drosophila melanogaster*. Dev. Comp. Immunol. 27, 273–282.

Moreau S.J.M., Vinchon S., Cherqui A., Prévost G., 2009. Components of *Asobara* venoms and their effects on hosts. In: G. Prévost (Ed.), Parasitoids of *Drosophila*, Advances in Parasitology, vol. 27 (pp. 217–232). Academic Press, Amsterdam.

Pennacchio, F., Strand, M.R., 2006. Evolution of developmental strategies in parasitic hymenoptera. Annu. Rev. Entomol. 51, 233–258.

Piek, T., 1986. Venoms of Bumble-bees and Carpenter-bees. Venoms of Hymenoptera, 417–424.

Prévost, G., Doury, G., Mabiala-Moundoungou, A., Cherqui, A., Eslin, P., 2009. Strategies of avoidance of host immune defenses in *Asobara* species. In: G. Prévost (Ed.), Parasitoids of *Drosophila*, Advances in Parasitology, vol. 70 (pp.235–255). Academic Press, Amsterdam.

Prévost, G., Eslin, P., Doury, G., Moreau, S.J.M., Guillot, S., 2005. *Asobara*, braconid parasitoids of *Drosophila* larvae: unusual strategies to avoid encapsulation without VLPs. J. Insect Physiol. 51, 171–179.

Richards, E.H., Parkinson, N.M., 2000. Venom from the endoparasitic wasp *Pimpla hypochondriaca* adversely affects the morphology, viability, and immune function of hemocytes from larvae of the tomato moth, *Lacanobia oleracea*. J. Invertebr. Pathol. 76, 33–42.

Rizki, T.M., Rizki, R.M., 1980. Properties of the larval hemocytes of *Drosophila melanogaster*. Experientia 36, 1223–1226.

Russo, J., Brehélin, M., Carton, Y., 2001. Hemocyte changes in resistant and susceptible strains of *D. melanogaster* caused by virulent and avirulent strains of the parasitic wasp *Leptopilina boulardi*. J. Insect Physiol. 47, 167–172.

Saarela, J., Laine, M., Oinonen, C., Jalanko, A., Rouvinen, J., von Schantz, C., et al. 2001. Molecular pathogenesis of a disease: structural consequences of aspartylglucosaminuria mutations. Hum. Mol. Genet. 10, 983–995.

Schmidt, O., Schumann-Feddersen, I, 1989. Role of virus-like particles in parasitoid–host interactions in insects In: Harris, J.R. (Ed.). Subcellular Biochemistry, vol. 15 (pp. 91–118). Plenum Publishing Corporation.

Schmidt, O., Theopold, U., Strand, M.R., 2001. Innate immunity and its evasion and suppression by hymenopteran endoparasitoids. BioEssays 23, 344–351.

Sokolowski, M.B., Turlings, T.C.J., 1987. *Drosophila* parasitoid–host interactions: vibrotaxis and ovipositor searching from the host's perspective. Can. J. Zool. 65, 461–464.

Soller, M., Lanzrein, B., 1996. Polydnavirus and venom of the egg–larval parasitoid *Chelonus inanitus* (Braconidae) induce developmental arrest in the prepupa of its host *Spodoptera littoralis* (Noctuidae). J. Insect Physiol. 42, 471–481.

Stoltz, D.B., 1986. Interaction between parasitoid derived products and host insects: an overview. J. Insect Physiol. 32, 347–350.

Stoltz, D.B., Guzo, D., Belland, E.R., Lucarotti, C.J., MacKinnon, E.A., 1988. Venom promotes uncoating *in vitro* and persistence *in vivo* of a braconid polydnavirus. J. Gen. Virol. 69, 903–907.

Strand, M.R., Pech, L.L., 1995. Immunological basis for compatibility in parasitoid–host relationships. Annu. Rev. Entomol. 40, 31–56.

Suzuki, M., Tanaka, T., 2006. Virus-like particles in venom of *Meteorus pulchricornis* induce host hemocyte apoptosis. J. Insect Physiol. 52, 602–613.

Tanaka, T., Vinson, S.B., 1991a. Depression of prothoracic gland activity of *Heliothis virescens* by venom and calyx fluids from the parasitoid, *Cardiochiles nigriceps*. J. Insect Physiol. 37, 139–144.

Tanaka, T., Vinson, S.B., 1991b. Interaction of venoms with the calyx fluids of three parasitoids, *Cardiochiles nigriceps*, *Campoletis croceipes* (Hymenoptera: Braconidae), and *Campoletis sonorensis* (Hymenoptera: Ichneumonidae) in effecting a delay in the pupation of *Heliothis virescens*. Ann. Entomol. Soc. Am. 84, 87–92.

Tarentino, A.L., Quinones, G., Hauer, C.R., Changchien, L.M., Plummer Jr., T.H., 1994. Molecular cloning and sequence analysis of *Flavobacterium meningosepticum* glycosylasparaginase: a single gene encodes the α and β subunits. Arch. Biochem. Biophys. 316, 399–406.

Tenhunen, K., Laan, M., Manninen, T., Palotie, A., Peltonen, L., Jalanko, A., 1995. Molecular cloning, chromosomal assignment, and expression of the mouse aspartylglucosaminidase gene. Genomics 30, 244–250.

Tollersrud, O.K., Aronson Jr., N.N., 1992. Comparison of liver glycosylasparginases from six vertebrates. Biochem. J. 282, 291–297.

Vinchon, S., Moreau, S.J.M., Drezen, J.-M., Prévost, G., Cherqui, A., 2010. Molecular and biochemical analysis of an aspartylglucosaminidase from the venom of the parasitoid wasp *Asobara tabida* (Hymenoptera: Braconidae). Insect Biochem. Mol. Biol. 40, 35–48.

Vinson, S.B., Iwantsch, G.F., 1980. Host regulation by insect parasitoids. Q. Rev. Biol. 55, 143–165.

Wago, H., Tanaka, T., 1989. Synergic effects of calyx fluid and venom of *Apanteles kariyai* Watanabe (Hymenoptera: Braconidae) on the granular cells of *Pseudaletia separata* Walker (Leptoptera: Noctuidae). Zool. Sci. 6, 691–696.

Webb, B.A., Luckhart, S., 1994. Evidence for an early immunosuppressive role for related *Campoletis sonorensis* venom and ovarian proteins in *Heliothis virescens*. Arch. Insect Biochem. Physiol. 26, 147–163.

Futuristic Visions

Chapter 22

Applications of Parasitoid Virus and Venom Research in Agriculture

Francesco Pennacchio[*], Barbara Giordana[†], and Rosa Rao[‡]

[*]*Dipartimento di Entomologia e Zoologia Agraria 'F. Silvestri', Università di Napoli 'Federico II', Napoli, Italy*

[†]*Dipartimento di Protezione dei Sistemi Agroalimentare e Urbano e Valorizzazione della Biodiversità, Università di Milano, Italy*

[‡]*Dipartimento di Scienza del Suolo, della Pianta dell'Ambiente e della Produzione Animale, Università di Napoli 'Federico II', Italy*

| TnBV | *Toxoneuron nigriceps* bracovirus |
| TSP | teratocyte secreted proteins |

SUMMARY

Insect parasitoids are able to suppress the immune response of the host and to disrupt its development and reproduction. The virulence and host regulation factors triggering these alterations are injected by the ovipositing females and are encoded either by the genome of the wasp or of its associated polydnavirus. The astonishing richness of species in the parasitic Hymenoptera provides a unique reservoir of molecular biodiversity for new bioinsecticide molecules, targeting a number of insect species and developmental stages. The growing information on parasitoid and polydnavirus functional genomics opens new avenues of research for developing innovative pest control technologies based on the use of new molecules and of effective delivery strategies. These latter have to be designed in order to allow the ingested bioinsecticide molecule to reach its receptor, often located in the hemocoel, at the intracellular level. Delivery strategies meeting these requirements are of crucial importance for successful use of parasitoid-derived bioinsecticides.

ABBREVIATIONS

Aea-TMOF	*Aedes aegypti*-Trypsin Modulating Oostatic Factor
Ae-FABP	*Aphidius ervi* fatty acid binding protein
Bt	*Bacillus thuringiensis*
CPPs	cell penetrating peptides
CsIV	*Campoletis sonorensis* ichnovirus
dsRNA	double strand RNA
eGFP	enhanced Green Fluorescent Protein
GM	genetically manipulated
GNA	*Galanthus nivalis* agglutinin
IPM	integrated pest management
PDV	polydnavirus
PEG	polyethylene glycol
PM	peritrophic membrane
RNAi	RNA interference
SJ	septate junction
TJ	tight junction

INTRODUCTION

Crop destruction and disease transmission by insects have a remarkable impact on human economy and health. Nearly 20% of the annual crop production is destroyed by insects (Oerke and Dehne, 2004), and about the same percentage of loss is registered for stored food grains (Bergvinson and Garcia-Lara, 2004). Disease transmission to animals and to the human population are growing problems in different areas of the globe, and malaria still remains one of the major causes of death in the world, largely concentrated in Africa (Breman, 2001; Gubler 2002; Snow *et al.*, 2005).

The need for insect control has ancient roots, dating back to the origins of agriculture, and has been always pursued with little use of chemicals, until the introduction of the synthetic insecticides, which started in 1940 with DDT, and continued with the development of a growing number of neurotoxic compounds (Casida and Quistad, 1998). The illusion of a 'global' and effective chemical control of insects was gradually challenged over the years by a number of problems, such as insecticide resistance, environmental pollution and toxicity for nontarget organisms, which promoted the need for sustainable insect control methodologies, based on the use of more benign insecticides of natural origin (Horowitz and Ishaaya, 2004), in the framework of integrated pest management (IPM) strategies (Dent, 1995).

The enormous diversity of living organisms and of the molecular 'tools' they use to face the environmental challenges posed by competitors and antagonists have allowed the discovery of a number of novel chemical structures, pharmaceutical and agronomical leads, with diverse

Parasitoid Viruses: Symbionts and Pathogens. DOI: 10.1016/B978-0-12-384858-1.00022-9

biological activities (Demain and Sanchez, 2009; Dayan et al., 2009). Insecticides of natural origins have been largely obtained from plants and microorganisms, by exploiting at best the molecular bases of their natural antagonism with insects (Roh et al., 2007; Bravo et al., 2007; Dayan et al., 2009; Kirst, 2010). The growing interest in this research area has permitted significant expansion of our knowledge of natural insecticidal toxins (Nicholson, 2007a) which now includes, besides the well-established *Bacillus thuringiensis (Bt)* toxin, a number of new good candidate molecules for insect control. These molecules have different biological origins, such as bacteria (*Photorhabdus* and *Xenorhabdus*) associated with entomopathogenic nematodes (ffrench-Constant et al., 2007), and predatory arthropods like scorpions and spiders, which release venom secretions rich in potent neurotoxins, used to paralyze their prey (Gordon et al., 2007; Gurevitz et al., 2007; Nicholson, 2007b; Rohou et al., 2007).

The diversity of entomophagous organisms reaches its highest level within insects, with the large majority of parasitoids belonging to the order Hymenoptera, and representing nearly 10–20% of all insects (Godfray, 1994; Quicke, 1997; Whitfield, 2003). This is one of the largest biodiversity reservoirs existing in nature, which provides a rich array of different molecules, used by insect natural enemies to kill and/or manipulate their victims (Pennacchio and Strand, 2006). We can reasonably assume that this group of arthropods offers the opportunity to identify a number of novel compounds targeting different host insects. Moreover, because the level of host specificity of parasitic wasps is rather high in a significant number of species, especially in koinobionts (i.e., those that do not kill or permanently paralyse their victims during oviposition) (Pennacchio and

Strand, 2006), we may reasonably speculate that these molecules could be characterized by a considerable level of selectivity, thus ensuring environmental safety.

In the present chapter, we focus our attention on what we know about molecules produced by parasitoids and associated polydnaviruses, which are potential candidates for the development of innovative biotechnologies for insect control. The future exploitation of these molecules in sustainable strategies of pest control is largely dependent on an in-depth knowledge of their molecular and functional characteristics, and a thorough understanding of the physiological processes which regulate the absorption of macromolecules by the insect gut. Ability to cross the gut tissues to gain entry to the hemocoel is of crucial importance for the development of effective and safe delivery strategies for most parasitoid-derived molecules; this will be discussed in the second part of this chapter.

THE USE OF INSECT NATURAL ANTAGONISTS: BEYOND CLASSICAL BIOLOGICAL CONTROL

Classical biological control is based on the use of parasites, predators, and pathogens for the regulation of host pest densities (DeBach, 1964). This control strategy has been successfully implemented for more than a century, and still represents one of the key tools in sustainable IPM (Bale et al., 2008) (Fig. 1). For many beneficial organisms, there is a growing body of information on the intimate molecular mechanisms they adopt to circumvent the insect defenses developed against their virulence strategies, dynamically shaped by a continuous coevolutionary arms race. The identification of the genes which regulate the molecular interactions between insect hosts and their natural antagonists offers the opportunity to isolate new virulence factors and their cognate target receptors in the host, that can be profitably used to propose innovative biotechnologies for insect control, through the development of new natural insecticides and resistant plants (Fig. 1). These biotechnologies, based on molecules and genes that are produced and delivered by natural enemies, could fall into an expanded definition of biological control, which includes not only the use of the organisms, but also of genes and molecules derived from them. A well-known example of this concept is offered by the entomopathogen *Bacillus thuringiensis*, which was first discovered more than a century ago. Since then, the most active toxins responsible for host death have been identified. The genes coding these toxins were used for developing the first generation of genetically manipulated (GM) plants, which reached the market about fifteen years ago (James, 2005; Roh et al., 2007).

The beneficial arthropods are active insect antagonists, and their strategy of host attack is very often based

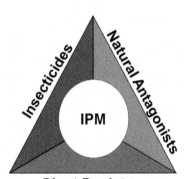

FIGURE 1 The IPM (integrated pest management) triangle, showing the main categories of insect control technologies. Their use is strongly biased, as the synthetic insecticides still represent by far the most consistent part of plant protection costs. The study of insect antagonism, at different trophic levels (plant–insect–natural antagonists), can offer new insights and tools for sustainable management of natural populations, and may allow the identification of new natural molecules and genes that can be used as natural insecticides or to enhance both direct and indirect defense mechanisms of plants.

on the use of secretions involved in the capture/colonization and subsequent physiological/biochemical redirection to enhance the nutritional suitability of the victim and maximize the developmental success of their own progeny (Vinson *et al.*, 2001). The most advanced information on molecules causing prey paralysis has been obtained for scorpions and spiders, because of their medical importance. These studies have allowed the identification and molecular characterization of potent neurotoxins, which are able to disrupt in a very selective way different kinds of voltage-gated ion channels in arthropods (Zlotkin, 2005; Maggio *et al.*, 2005; Gordon *et al.*, 2007; Gurevitz *et al.*, 2007; Nicholson, 2007b; Rohou *et al.*, 2007). The research work on these neurotoxins has attracted considerable attention in the agrochemical industry, due to the possibility that structure–activity studies may allow the development of new families of synthetic insecticides, starting from *natural* molecules, generated by long coevolutionary processes.

The high biodiversity of insect parasitoids and their need to solve specific problems posed by the colonization of different ecological niches has favored the development of a number of virulence strategies in different phylogenetic lineages (Pennacchio and Strand, 2006). The resulting host physiological alterations, often triggered by factors of parasitic origin injected by the ovipositing wasp females, are of key importance for successful parasitism. These physiological changes are commonly referred to as host regulation (Vinson and Iwantsch, 1980). Thus, the ovipositor and the injection of a complex mixture of virus, venom, and ovarian factors has conferred on parasitic Hymenoptera the unique capacity of finely modulating the recipient environment of their progeny. The ovipositor is undoubtedly one of the most important morphological adaptations that has significantly promoted the intense adaptive radiation process and evolutionary success of these wasps (Quicke, 1997).

In the more basal lineages, we have the idiobiont parasitoids, which deliver during oviposition venom blends rich in neurotoxins, causing host paralysis and death (Pennacchio and Strand, 2006) (Fig. 2). However, in these basal evolutionary lineages, we do also observe the presence of virulence factors disrupting the immune response, which, if unaltered, would negatively affect the larval

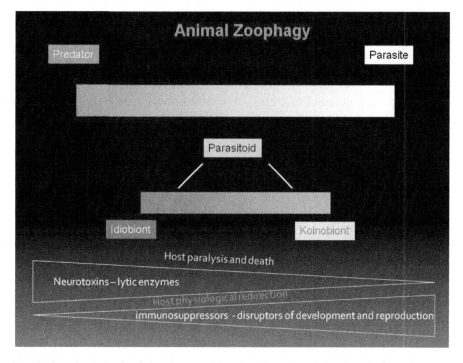

FIGURE 2 **The zoophagy in the animal kingdom is a widespread life habit.** There are a number of adaptive solutions in a continuum range delimited by two extremes: predators and parasites. The very sophisticated strategies adopted by the latter, to finely redirect the host physiology to their own benefit, in order to complete several generations without killing the host, are absent in predators, which exploit multiple victims to complete their development and sustain their reproductive activity as adults. The parasitoids, largely belonging to the class of Insecta, have intermediate characteristics, as they eventually kill (like predators) their host, when completing the preimaginal development, even though juvenile stages may show a wide range of specialized life habits. This degree of specialization is very sophisticated in koinobionts, which finely interact with victims surviving for a relatively long time interval, while it is considerably lower in idiobionts, which rapidly paralyze and/or kill their hosts. The different strategies of host colonization and exploitation adopted are modulated by different categories of virulence and host regulation factors. The idiobionts are a very good source of fast acting neurotoxins and lytic enzymes impacting different host tissues, while the koinobionts offer an astonishing variety of immunosuppressors and molecules disrupting host development and reproduction. Please see color plate section at the back of the book.

feeding activity with clotting and melanization reactions (Strand and Pech, 1995). These factors are key components of the ancestral 'toolkit' that have made possible the evolution of very sophisticated immunosuppressive strategies exhibited by the endophagous wasp species.

The transition from ectoparasitism to endoparasitism has occurred multiple times during the evolution of parasitic Hymenoptera, leading in the majority of cases to koinobiosis (development of the parasitoid in a developing host) (Whitfield, 1998, 2003). For koinobionts it is possible to observe the most striking diversification of host regulation strategies and factors (Fig. 2) which allow a prolonged modulation of the host's physiology. This is an important requirement to effectively meet the changing needs of the developing parasitoid progeny, which finely interact with surviving hosts (Gauld, 1988). However, in most cases, the injection of venom alone during oviposition is not sufficient for prolonged host regulation, which requires the release of bioactive factors, in a timely and tissue-specific manner, as it characteristically occurs in true metazoan host–parasite interactions (Pennacchio and Strand, 2006). This need is one of the functional constraints that has promoted the evolution of new delivery strategies of host regulation factors, which include unusual cell types of embryonic origin, namely teratocytes (Dahlman and Vinson, 1993; Beckage and Gelman, 2004), and parasitoid-associated symbionts, such as the members of the viral family Polydnaviridae (Kroemer and Webb, 2004; Webb and Strand, 2005). These latter are unique insect viruses, associated with ichneumonoid and braconid wasps, which parasitize larvae of lepidopteran species. They are stably integrated as proviruses in the genome of the wasp, where they exclusively replicate in the ovary, and are delivered with the egg at the parasitization. When injected in the host caterpillar, they enter various tissues, where they do not replicate but they do express virulence factors, which exert a potent immunosuppressive activity and disrupt the endocrine balance. These molecules are viewed as parasitoid-derived virulence factors, delivered by finely tuned systems of genetic secretion of viral origin (Webb and Strand, 2005), which derive from ancestral viral pathogens of the caterpillar host (Bézier et al., 2009).

The major host pathologies observed in naturally parasitized larvae are in part or completely induced by viral infection. Polydnaviruses (PDVs) are, therefore, a natural source of potential bioinsecticide molecules, which are of particular economic interest, as they are expected to be especially active against Lepidoptera, one of the most economically important insect groups. Indeed, the larval stages of moths are widespread and highly destructive. Thus, considerable chemical control measures are commonly required, with about 40% of all chemical insecticides being used against heliothine species (Brooks and Hines, 1999).

The brief overview above sheds light on the significant, and nearly unexploited, potential offered by insect parasitoids and associated symbionts. The use of these organisms as biological control agents can be profitably expanded if we understand the molecular mechanisms they use to disrupt the host's physiology. This offers the opportunity to develop new pest control technologies that exploit tools and mimic strategies that parasitoids use to kill their hosts.

CANDIDATE MOLECULES AND GENES OF PARASITIC ORIGIN FOR INSECT CONTROL APPLICATIONS

Case Studies on Parasitoid- and PDV-Derived Molecules

The use of parasitoid-derived molecules for insect control has already received attention. There are cases of unexpected oral activity exerted by molecules targeting hemocoelic receptors, which appear to be very promising, and certainly deserve further research efforts.

The first attempt in this direction was pursued by Dahlman and co-workers, focusing on a braconid parasitoid of larval stages of Heliothis/Helicoverpa species, Microplitis croceipes. This wasp harbors a bracovirus (Microplitis corceipes, McBV), and its embryonic membrane, when the larva hatches, dissociates into isolated peculiar cells, called teratocytes, which grow enormously in size without undergoing division (Dahlman and Vinson, 1993). These cells synthesize a number of peptides and proteins which are released into the host hemocoel where they play an important role in the regulation of its physiology. Some alterations observed in naturally parasitized hosts, such as reduced growth and lowered hemolymphatic titre of juvenile hormone esterase, ecdysone, and of several proteins, were duplicated by teratocyte injections, as well as by injection of some teratocyte secreted proteins (TSP) (Jarlfors et al., 1997). A reduction in the protein synthesis at post-transcriptional level was triggered by a specific TSP, named TSP14, which showed a fat body specific activity. TSP14 was not observed on other host tissues, nor on insect and mammalian cell lines, likely due to the lack of the appropriate receptor (Dahlman et al., 2003). It is interesting to note that TSP14 has a cysteine-rich motif and shares significant sequence similarity with proteins of the Cys-motif gene family, which are encoded by the ichnovirus associated with Campoletis sonorensis (CsIV) (Dib-Haij et al., 1993).

The strong impact on host physiology of TSP14, and its putative lack of activity in nontarget organisms, stimulated the idea of assessing its oral activity. This was performed by developing transgenic tobacco plants which were used for feeding bioassays with tobacco budworm larvae,

Heliothis virescens (Maiti *et al.*, 2003). The experimental moth larvae maintained on transgenic plants showed, compared to controls, reduced feeding and slower growth and development, which eventually resulted in higher mortality rates and reduced plant damage (Maiti *et al.*, 2003). The quantitative differences, even though not impressive, were statistically significant and feeding on transgenic plants certainly had a negative effect on moth fitness. Moreover, the observed sensitivity of *Manduca sexta* larvae to these transgenic plants that were expressing TSP14, demonstrated the potential of this protein to effectively target other lepidopteran pest species (Maiti *et al.*, 2003).

An interesting level of insecticide activity was also observed for different members of the *CsIV* Cys-motif gene family (Fath-Goodin *et al.*, 2006). In this study, the effects of five Cys-motif proteins on *H. virescens* growth, development, and immunity were evaluated. The recombinant proteins were either fed or injected, and the protein named rVHv1.1 was found to be the most active of the five tested. Oral intake of this protein resulted in a marked reduction of larval growth (50–70%) and gave rise to a fairly large number of nonviable pupae (36%). Interestingly, marked inhibitory effects of this protein were also observed in *Spodoptera exigua*, a non-permissive host of *C. sonorensis*, suggesting a broader range of biological activity not confined to the habitual hosts of the parasitoid. Moreover, when the protein rVHv1.1 was injected or fed at higher concentrations on artificial diet to *H. virescens* larvae, death was observed in many experimental insects, associated with clear symptoms of baculovirus infection (Fath-Goodin *et al.*, 2006). The intake of this immunosuppressive protein likely rendered the tobacco budworm larvae more susceptible to a cryptic viral pathogen that was asymptomatic in the insect colony used. The observed higher susceptibility to disease agents may well result in the enhancement of insecticide activity of natural or artificially sprayed pathogens, and indicates that these and other parasitoid-derived molecules targeting the insect immune system have the potential to increase the impact of biological control agents. The observed pleiotropic effects on host physiology are likely due to the negative impact that these host regulation factors have on protein synthesis at the post-transcriptional level, which has been documented in several species of parasitized host larvae (Shelby and Webb, 1997; Shelby *et al.*, 1998; Kim, 2005).

These results stimulated efforts to produce transgenic tobacco plants expressing some of the Cys-motif proteins used for *in vitro* bioassays, which were fed to lepidopteran larvae (Gill *et al.*, 2006). Similarly to that which was observed for transgenic tobacco plants expressing TSP14 (Maiti *et al.*, 2003), several independent transgenic lines showed a significant reduction in leaf damage caused by caterpillar feeding (Gill *et al.*, 2006).

Collectively, these results indicate that Cys-motif proteins have a very good potential for insect control

applications, as they have both direct and indirect effects on different noctuid target species of considerable economic importance.

These pioneering studies were paralleled by similar approaches in our laboratories. We developed transgenic tobacco plants expressing *Tn*BVank1, a member of the viral ankyrin gene family of the bracovirus associated with *Toxoneuron nigriceps* (*Tn*BV) (Rao *et al.*, unpublished data; Falabella *et al.*, 2007b). The viral ankyrin gene products correspond to a truncated version of the Cactus/iκB gene having the ankyrin repeats but lacking the N- and C-terminal regulatory domains controlling Cactus/iκB signal-induced and basal degradation, respectively (Kroemer and Webb, 2005; Thoetkiattikul *et al.*, 2005; Falabella *et al.*, 2007b). These virulence factors irreversibly bind to NF-κB transcription factors, thus preventing their entrance into the nucleus and the activation of immune genes under control of κB promoters (Thoetkiattikul *et al.*, 2005; Falabella *et al.*, 2007b). However, the ankyrin domains may mediate other molecular interactions, likely triggering additional functional alterations. In a recent study, using transgenic strains of *Drosophila melanogaster*, we have shown that *Tn*BVANK1 interferes with cytoskeleton organization, by disrupting the microtubule network and microtubule motor protein activity (Duchi *et al.*, 2010). This alteration most likely impairs the cytoskeleton and actin polymerization, thus disrupting functionality of hemocytes and preventing encapsulation, as the microtubules play a crucial role in cortical activation of Rac1 and lamellipodium formation during polarized cell migration (Siegrist and Doe, 2007). This process has been well described in leukocyte chemotaxis and is highly conserved in eukaryotic cells (Affolter and Weijer, 2005; Etienne-Manneville, 2006). These multiple effects, potentially targeting a number of cellular functions, suggests that *Tn*BVANK1 may have pleiotropic effects on host vital functions and, therefore, may prove to be a promising bioinsecticide molecule.

Feeding on leaf-disks of transgenic tobacco plants expressing *Tn*BVank1 negatively influenced the growth of tobacco budworm larvae, which attained a significantly lower weight before pupation and pupated about three days later than controls (Rao *et al.*, unpublished data). Moreover, these growth alterations were associated with a slight but significant increase (30%) of cumulative larval/pupal mortality. More interestingly from an applied perspective, the expression of *Tn*BVank1 altered the plant transcriptome, as a few genes involved in defense against pathogens were up-regulated, and this enhanced *in vitro* the degree of plant resistance to some fungal pathogens (Rao *et al.*, unpublished data). The functional bases of these multiple tolerance or resistance traits remain to be elucidated, but indicate that the use of these PDV-encoded molecules may have unexpected impacts on plant pathogens.

New Candidate Bioinsecticide Molecules

It is difficult to predict on a theoretical basis which PDV or parasitoid genes will prove to be valuable candidate molecules to be profitably considered for implementation in insect control. The choice of a good candidate is determined not only by its intrinsic molecular and functional characteristics, but also by the oral activity exerted when those compounds are ingested with food. This is largely influenced by the presence of a molecular target in the gut and/or by the capacity of the ingested molecule to cross the gut barrier and enter the hemocoel to be transported to target tissues in an active form.

Based on the targeted function in the host, it is possible to define groups of molecules acting as immunosuppressors, disruptors of physiology, development and/or reproduction, and neurotoxins. These latter molecules, largely deriving from idiobiont parasitoids, were the main focus of the pioneering studies on *Bracon* species, which eventually led to the partial characterization of three proteins that block glutaminergic transmission and are responsible for the paralytic activity of the venom (reviewed in Weaver et al., 2001). Unfortunately, this promising research area has not been further significantly developed yet, in spite of the large number of species that could be used as source of new potent neurotoxins. It would be valuable to focus on species characterized by a fast induction of host paralysis and death.We are currently studying some species in the genus *Pnigalio* (Bernardo et al., 2006) that appear quite promising in this respect.

To date, the immunosuppressive molecules are characterized in more detail. The stringent need for disrupting the immune response by ectoparasitic wasps, in order to facilitate their feeding activity, and the even more stringent requirement for endoparasitoids to circumvent the immune barriers of the host in order to survive, have generated a large diversity of bioactive molecules (Pennacchio and Strand, 2006). The venom glands and PDVs are the source of a number of molecules that interfere with both the humoral and cellular components of the immune response. The phenoloxidase cascade is one of the main physiological pathways that are targets of different wasp virulence factors. Disruption of hemocyte function, for which alterations of the cystoskeleton structure/dynamics, surface molecular characteristics and apoptosis induction have been widely reported, is another critical process targeted by PDVs and/or venom (Webb and Strand, 2005; Pennacchio and Strand, 2006; Danneels et al., 2010). The degree of functional and molecular knowledge we currently have on these immunosuppressive factors is by far more detailed for those encoded by PDVs, for which several gene families have been characterized and their mechanisms of action elucidated (Webb and Strand, 2005). Moreover, since the genomes of many PDVs have been fully sequenced, we expect a growing number of studies addressing important issues of functional genomics. In contrast, for venom molecules, the degree of information is comparatively much less detailed, even though the recent molecular work made possible by the completion of the genome sequence of *Nasonia vitripennis*, and extensive molecular analyses in other parasitic wasps, certainly provide new functional insights and tools to characterize bioactive toxins with immunosuppressive activity (de Graaf et al., 2010; Danneels et al., 2010; Vincent et al., 2010; Formesyn et al., Chapter 19 of this volume).

The observed direct effects of these molecules are complemented by the indirect mortality they induce by enhancing the impact and biocontrol efficiency of natural pathogens including baculoviruses, as a consequence of the immune disruption syndrome of the host (Washburn et al., 2000; Fath-Goodin et al., 2006). The immunosuppressed host is made more susceptible to a wide range of other pathogens including fungi, viruses, and bacteria. This offers the possibility to develop new integrated control strategies not exclusively relying on toxic molecules, but aimed at achieving sustainable manipulation of biocontrol agents.

Quite a few of the immunosuppressive molecules identified so far exert their action by interacting with conserved molecular components, which are often shared by different cellular pathways. This accounts for the pleiotropic activity they show when artificially delivered in host insects (Gill et al., 2006). Moreover, the expression of the same PDV genes, or of members of the same gene family, in different host tissues, including the nervous system, suggests that they are part of complex gene networks, which may modulate different parasitoid-induced disruptions of the endocrine balance, development, feeding activity, metabolism, and reproduction (Pennacchio and Strand, 2006). These are multifaceted and diverse alterations with cases of developmental delay or block, as well as of premature metamorphosis, associated with a large variety of endocrine changes, triggered by host regulation factors, both of maternal (venom, PDV, ovarian secretions) and embryonic/larval origin (serosa, teratocytes and larval secretions) (Pennacchio and Strand, 2006). The impressive heterogeneity of these alterations provides evidence for the existence of a wide array of bioactive molecules. However, the information currently available on their mechanisms of action and cognate receptors is still limited compared to what we know about virulence factors acting as immunosuppressors. This will certainly limit their use in the near future as candidate bioinsecticides.

The impact of parasitism on host reproduction has been reported in different biological systems (Beckage and Gelman, 2004; Pennacchio and Strand, 2006), but the most interesting direct effect is known for the aphid parasitoid species *Aphidius ervi*. The venom of this braconid induces host castration by triggering apoptosis in the germarial cells of the ovarioles (Digilio *et al.*, 2000). This alteration is induced by a γ-glutamyl transpeptidase, which likely interferes with the glutathione cycle, causing an oxidative stress (Falabella *et al.*, 2007a). Even though the functional basis accounting for the observed tissue-specificity of apoptosis induction remains to be elucidated, the strong negative impact of this protein on the reproduction of the pea aphid, *Acyrthosiphon pisum*, opens new perspectives for aphid control. This is a particularly interesting opportunity as there are very few molecules and genes that can be used for biotechnological applications directed against sucking insect pests (Gatehouse, 2008). Effects of expression of these molecules in transgenic plants on aphid feeding have not yet been evaluated.

The study of the molecular details of how parasitoids, and in particular their venom and associated PDV, interfere with host physiology, and the discovery of which molecules/genes are specifically targeted by virulence factors offers the possibility of developing new control strategies based on the use of RNA interference (RNAi) technology (Price and Gatehouse, 2008). The basic idea we are currently pursuing is to select target genes, to be silenced by RNAi, on the basis of the phenotypic effect associated with their downregulation mediated by natural parasitism or selective treatment with a specific virulence factor. The recent discovery of PDV noncoding transcripts, which down-regulate the expression of host genes of key importance (Pennacchio *et al.*, unpublished data), provides new insights in this direction.

The mild but evident effects that candidate bioinsecticide molecules from parasitoids and PDVs have on insect mortality, even though not always sufficient for an ultimate control, could cause significant levels of fitness reduction in phytophagous insects. This makes some parasitoid-derived molecules particularly amenable for inclusion in IPM plans, because the weakening of the pest natural populations they induce, even though not sufficient *per se* as a single control method, may be effectively complemented by other additional control measures, including the use of natural antagonists. The reduction rather than eradication of pest populations allows better preservation of the ecological balance and exploitation of the services offered by beneficial organisms included in the complex food-webs we try to manipulate.

MOLECULES RELEVANT FOR PARASITOID REARING ON ARTIFICIAL DIETS

The different strategies of host physiology redirection in favor of the developing parasitoid progeny share a common objective: the synchronization between the exponential growth phase of wasp juveniles and the availability of nutritional resources in the host microenvironment

(Pennacchio and Strand, 2006). This is achieved at the expense of host growth (Vinson *et al.*, 2001) or reproduction (Falabella *et al.*, 2007a, 2009). As a result, the nutritional needs of parasitoid larvae are finely tuned with host metabolic and biochemical changes over time, through a number of physiological adaptations that permit the optimal exploitation of their victims (Vinson *et al.*, 2001). An in-depth understanding of these aspects is crucial in the development of artificial diets for mass production of endophagous parasitoids, which, to date, still remains problematic, compared to the success that has been achieved in rearing several egg and pupal parasitoids *in vitro* (Grenier, 2009).

The frustrating results in this research area are also due to the limited knowledge we have of the nutritional physiology of the parasitoids, which show unusual nutrient absorption pathways (de Eguileor *et al.*, 2001; Giordana *et al.*, 2003). Moreover, this complex scenario can be further complicated by subtle interplays between host tissues and teratocytes, which are of pivotal nutritional importance to the developing wasp larvae, as studied in detail for the aphid parasitoid *A. ervi* (Falabella *et al.*, 2000, 2005, 2007a, 2009). The isolation of a gene highly expressed in the teratocytes of this species that encodes an extracellular fatty acid binding protein (*Ae*-FABP) (Falabella *et al.*, 2005), has recently allowed us to study a few aspects of the lipid nutrition in the mature larvae of this braconid. *Ae*-FABP shows a high affinity for C_{14}–C_{18} saturated fatty acids and acts as a vector protein in the hemolymph, insuring their delivery at the absorbing epithelia of the midgut and integument of *A. ervi* larvae (Giordana *et al.*, unpublished data). Then, the lipid nutrition *in vitro* of the larvae of this species very likely requires the presence in the medium of specific fatty acids, along with *Ae*-FABP, which is necessary to ensure their transport/delivery and absorption.

All these pieces of molecular information, which are unfortunately lacking for most parasitoid species used in biocontrol, are essential for attempting the formulation of suitable artificial diets, meeting the changing nutritional needs and the physiological requirements of parasitoid larvae. Significant advancements in our current knowledge in this area are very much needed to generate the technology required for industrial production of these beneficial organisms on artificial diet.

BIOINSECTICIDE DELIVERY STRATEGIES

Delivery Vectors

The delivery of bioinsecticide macromolecules requires the use of biological vectors expressing their coding genes, or of suitable formulations allowing the environmental distribution of recombinant proteins. There is a substantial dichotomy between bioactive molecules targeting receptors exposed in the gut lumen and those that have to reach receptors located within or behind the gut epithelium. Those in the first category, exerting their activity in the gut lumen, can allow the use of oral delivery strategies, while the others are much more efficiently delivered via insect-specific symbionts and pathogens (Inceoglu et al., 2006; Whetstone and Hammock, 2007). Insect pathogenic viruses belonging to the family of Baculoviridae have received considerable attention, as the relatively easy manipulation of their genome allows enhancement of the virulence of transgenic strains by expressing toxic molecules which increase their speed of killing, or that may broaden the host range (Kamita et al., 2005; Inceoglu et al., 2006). However, in spite of the fact that this is a mature technology, that could exploit a wide range of natural bioinsecticide molecules (in particular those targeting endocellular and/or hemocoelic receptors), the use of engineered pathogens is strongly limited by the concerns of the public about the potential risks to the environment and nontarget organisms posed by recombinant baculoviruses (Whetstone and Hammock, 2007).

The oral intake of macromolecules with insecticide activity remains the main route of entry in most cases. Therefore, the direct application of recombinant molecules or the use of recombinant organisms expressing them requires the presence of cognate receptors in the gut lumen. That is why the first GM plants active against herbivore pests, which were developed more than twenty years ago and made commercially available in U.S. about a decade later, expressed an orally active toxin produced by the entomopathogen *B. thuringiensis* (Vacek et al., 1987). The increasing use of *Bt*-plants has allowed a significant reduction of pesticide applications, with positive effects on environment and production costs (Toenniessen et al., 2003; Brookes and Barfoot, 2005), and no adverse effects on beneficial insects (Bale et al., 2008). However, the potential of GM plants is only partially exploited due to the limited availability of alternative bioinsecticide molecules. This has promoted the development of novel *Bt* toxins and of additional new strategies (Christou et al., 2006; Gatehouse, 2008). Therefore, the possibility of using alternative sources of bioactive molecules, such as natural antagonists of insects, is very attractive, as this represents a nearly unexploited and very valuable reservoir of molecular biodiversity. A general analysis of these molecules and of their targets clearly indicates that the major problem to face after ingestion is their stability in the gut environment and the absorption of suitable amounts in an active form. The use of epiphytic bacteria as delivery vectors is an interesting alternative to transgenic plants, which has been successfully used for controlling the pine processionary caterpillar (*Thaumetopoea pityocampa*), using a *Bt* toxin (Alberghini et al., 2005, 2006). Because these bacteria

are ingested along with the leaf substrate they colonize, any molecule expressed by them, like for GM plants, has to target a gut receptor or should be able to overcome the gut barrier. Therefore, the objective of enhancing gut permeation rates of macromolecules is currently a key goal to achieve, in order to facilitate the development of new biotechnologies for insect control.

The gut absorption issue seems to be relatively less relevant in the case of GM plants producing double strand RNAs (dsRNAs) interfering with the expression of insect genes, which represent a promising new way of plant manipulation for crop protection (Price and Gatehouse, 2008). It has been recently demonstrated that transgenic plants expressing dsRNA targeting a vacuolar ATPase expressed in insect midgut are more resistant to insect attack (Baum et al., 2007). Similarly, transgenic cotton plants expressing dsRNA targeting a P450 gene expressed in the midgut of the cotton bollworm, related with the detoxification of gossypol, are resistant to *Helicoverpa armigera* (Mao et al., 2007). As briefly outlined above, parasitoid research will likely offer new opportunities to develop plant protection strategies based on this technology. If and how the manipulation of the gut barrier's permeability would increase the efficiency of this method remains to be studied, but will largely depend on the uptake mechanisms of dsRNA in different insect species still largely unknown (Huvenne and Smagghe, 2010).

ENHANCING ORAL TOXICITY

The ingested macromolecules, in order to reach internal targets, have to cross the peritrophic membrane (PM) and the gut epithelium. The PM is an extracellular thin sheet lining the midgut lumen, largely made of chitin and proteins, which protects the midgut columnar epithelium and plays an important role in digestive physiology (Terra, 2001). Structural alterations of the PM induced by chitinases, that disrupt chitin microfibrils, or by some metalloproteases that degrade PM proteins, promote an increased permeability of the PM which results in higher mortality rates of insect pests (Ding et al., 1998; Cao et al., 2002; Rao et al., 2004; Corrado et al., 2008). GM tobacco and rice plants expressing a gene of *Trichoplusia ni* granulovirus (*Tn*GV) coding for enhancin (Cao et al., 2002), a metalloprotease originally isolated from a granulosis virus (Tanada et al., 1973), enhanced the infection of lepidopteran larvae by *Spodoptera exigua* nucleopolyhedrovirus and resulted in growth inhibition and higher mortality rates (Hayakawa et al., 2000; Cao et al., 2002). These data indicate that the alteration of PM permeability could be a powerful tool to facilitate the contact of intact macromolecules with gut epithelium and most likely, their uptake. We have recently produced experimental evidence supporting this hypothesis. Transgenic tobacco plants coexpressing the Chitinase A of the *Autographa californica* nuclear polyhedrosis virus (*Ac*MNPV) and the *Aedes aegypti*-Trypsin Modulating Oostatic Factor (*Aea*-TMOF), a peptide that impairs insect digestive function by inhibiting gut trypsin synthesis, have a significant stronger negative effect on growth rate, developmental time, and mortality of *Heliothis virescens* (Lepidoptera, Noctuidae) larvae, compared to parental lines, separately expressing one of the two transgenes (Fiandra *et al.*, 2010). This higher insecticide activity of GM plants co-expressing the two molecules was associated with an increased PM permeability and more intense binding of *Aea*-TMOF to a putative receptor located on the basolateral membrane of the midgut epithelial cells (Fiandra *et al.*, 2010).

Passing through the PM is just the first step. A large part of the proteins and peptides ingested by the insect are expected to be denatured when exposed to the quite extreme physicochemical conditions of the gut luminal fluids (pH, ionic strength, redox conditions), and degraded by the activity of endo- and exopeptidases secreted by the epithelium. In addition, their final permeation across the gut should be largely limited by their hydrophilic properties. However, it is now well acknowledged that orally administered peptides and proteins that reach the absorbing part of the insect gut undegraded can cross the lining epithelium and reach the hemocoel in their native form (recently reviewed by Jeffers and Roe, 2008). Even though the mechanisms involved in this transepithelial transfer are largely unknown, a number of efforts have been made to increase the amount of the insecticidal peptides/proteins that effectively reach their targets within the gut epithelium or insect hemocoel.

An increased delivery of polypeptides can be obtained by designing analogs able to resist peptidase attack. Nachman *et al.* (2002) reported an increased bioavailability of an amphiphilic analog of the pyrokinin/PBAN neuropeptide orally administered to adult *H. virescens* moths. The chemical modification conferred peptidase resistance to the native neuropeptide molecules, increasing their survival time in the digestive tract and, thus, favoring their enhanced penetration across the gut wall.

The administration of insect neuropeptides or insect-specific toxins as fusion proteins with *Galanthus nivalis* agglutinin (GNA), a mannose-specific lectin, has proved to be a very effective approach to increase the amount of a poorly permeable peptide transferred to the insect hemocoel. The fusion protein GNA-allatostatin, but not allatostatin alone, reached the hemolymph of the tomato moth, *Lacanobia oleracea*, undegraded (Fitches *et al.*, 2002). Likewise, the fusion protein formed by GNA and an insecticidal spider venom toxin (*Segestria florentina* toxin 1, SFI1) fed to *L. oleracea* larvae caused acute toxicity, while mortality was drastically reduced when larvae were separately fed with GNA or SFI1 (Fitches *et al.*, 2004). GNA

increased the virulence of the fused spider venom toxin also in the aphid *Myzus persicae* and in the planthopper *Nilaparvata lugens* (Down *et al.*, 2006), proving that mannose lectins can increase gut protein permeability in different insect orders. Lectins have been largely used for drug delivery in the mammalian intestine, where the ability to improve transepithelial drug release is largely due to their mucoadhesion properties and/or to their ability to enter the epithelial cell by receptor mediated endocytosis (Gabor *et al.*, 2004).

PEGylation is a pharmaceutical drug modification and delivery technology that has, in the last decade, greatly improved the clinical efficacy of otherwise unstable molecules (Kang *et al.*, 2009; Ryan *et al.*, 2008). PEGylation is based on the covalent binding to peptides, proteins, oligonucleotides or small organic compounds of polyethylene glycol (PEG) molecules, opportunely tailored to improve their physicochemical properties, in order to confer to the active biomolecule an enhanced resistance to proteolysis and an increased circulating life in the internal environment. Roe and Brandt (2003) applied this technology for protein delivery in insects and proved that insulin, a polypeptide unable to cross the insect gut, can be found in the hemolymph in detectable amounts when PEGylated. Jeffers and Roe (2008) report in their review that the same approach increased the delivery of the decapeptide TMOF, used as a bioinsecticide. The PEGylated hormone increased 10-fold its toxicity on mosquito larvae, and the molecule accumulated in the hemolymph when fed to *H. virescens* larvae. Experiments *in vitro* proved that PEGylation effectively reduced the rate of degradation of TMOF by leucine aminopeptidase, and that an inverse correlation was seen between the molecular weight of the PEG polymers and the rate of degradation (Shen *et al.*, 2009).

Cell Penetrating Peptides (CPPs) are short peptides of less than 30 amino acids that do not share common sequence motifs but are all characterized by the presence of numerous arginine and lysine residues. These peptides have the intriguing ability to cross the plasma membrane of mammalian cells, and can act as delivery systems because they carry across the membrane different types of cargo molecules, like small therapeutic organic drugs, proteins, antibodies, nucleic acids, and other compounds (for recent reviews, see Patel *et al.*, 2007; Prochiantz, 2008; Fonseca *et al.*, 2009). The first CPPs identified were the third helix of the *Drosophila* Antennapedia homeodomain (Derossi *et al.*, 1994) and the HIV-type 1 transactivator protein (Frankel and Pabo, 1988; Green and Loewenstein, 1988). In the last two decades the number of peptides in the CPPs family that have been characterized has impressively increased (Fonseca *et al.*, 2009). The family now includes the so-called protein-derived CPPs, i.e., the minimal effective amino acid sequence of the parent translocation protein, like the two peptides Tat (48–60), derived from the HIV-1 protein, and Penetratin (43–58), derived from the Antennapedia homeodomain, as well as new specific peptides designed *de novo* from the structure–activity relationship of known CPPs, like the oligoarginine peptides. The mechanism of CPPs' cell penetration, alone or fused with their cargo, has been the subject of a large number of studies. It is now established that a CPP that penetrates the cell membrane follows either the energy-dependent endocytic pathway(s) or the energy-independent transmembrane translocation across the lipid bilayer, according to its length and charge localization, the physicochemical properties of the cargo, the cell type, and the experimental conditions (Patel *et al.*, 2007; Fonseca *et al.*, 2009).

We have recently proved in our laboratories that the CPP Tat could be a good vector for the delivery of proteins to the lepidopteran midgut (Casartelli *et al.*, unpublished observations). We found that the recombinant fusion protein Tat-enhanced Green Fluorescent Protein (eGFP) was internalized by columnar cells in culture of *Bombyx mori* larval midgut (Cermenati *et al.*, 2007) more readily than the recombinant eGFP alone. No inhibition of the fusion protein uptake was observed after treatment with metabolic inhibitors or drugs that interfere with the endocytic uptake, indicating that Tat-eGFP protein translocation is an energy-independent nonendocytic process. Most interestingly, the CPP Tat enhanced eGFP internalization also when added to the luminal side of the intact midgut *in vitro*, mounted in a suitable perfusion apparatus. After 3 h of incubation, a strong fluorescent signal was observed in columnar cells by confocal microscopy, while the signal was weak in midgut incubated with eGFP alone. These results open new possibilities for the delivery of selective bioinsecticide molecules with a direct effect on midgut cells, while further studies are necessary to elucidate whether CPPs could also be used to increase the transepithelial transport of a cargo.

Physiology of Gut Absorption as a Basis for Developing New Delivery Strategies

As briefly summarized above, it is now well established that peptides and proteins may move across the insect gut and their crossing rate may be enhanced by using technologies that successfully increase rates of protein delivery in mammalian intestine. However, the insect midgut, and the lepidopteran larval midgut in particular (Giordana *et al.*, 1998), has functional peculiarities that, if thoroughly understood, may provide new insights into how to enhance peptides/proteins translocation. We have therefore focused our current research on the study of the functional properties of the two routes that can be followed by these molecules when crossing the lepidopteran midgut epithelium: the

cellular path, or transcytosis, and the paracellular pathway, through the aqueous space delimited by the intercellular junctions between adjacent epithelial cells. Albumin, which we chose as a model protein, crossed *B. mori* larval midgut epithelium *in vitro* by transcytosis, a process strongly dependent on the integrity of the cell cytoskeletal architecture (Casartelli *et al.*, 2005). Transcytosis is an extremely complex route (reviewed by Tuma and Hubbard, 2003). We used confocal microscopy to dissect the entire albumin transcellular pathway (schematically represented in Fig. 3) in isolated midgut columnar cells maintained in culture (Cermenati *et al.*, 2007). We found that this protein was internalized by clathrin-dependent endocytosis, mediated by a megalin-like multiligand receptor (Casartelli *et al.*, 2008), homologous to that expressed in many mammalian absorptive epithelia (Moestrup and Verroust, 2001). The insect megalin-like receptor, which considerably increases the transepithelial transport of the bound protein, recognizes also insulin, transferrin, and the polybasic drug gentamicin (Casartelli *et al.*, unpublished data). Therefore, it might act as a scavenger receptor, like its mammalian homolog (Christensen and Birn, 2002), offering a new opportunity for the delivery of proteins across the insect gut. After internalization, albumin was found, as expected, in early endosomes, but it was then directed only in small amounts to the basolateral membrane, to be transferred to the hemolymph, while its major part was directed to lysosomes for intracellular degradation (Casartelli *et al.*, unpublished data). We are now examining whether the amount of protein released across the midgut may be enhanced by inhibiting the pathway to lysosomes with specific drugs.

The paracellular route across the intestinal tight junction (TJ) is a crucial path for drug delivery in mammals (Salama *et al.*, 2006; Deli, 2009) because it lacks proteolytic enzymes and its permeability can be modulated by various enhancers, allowing the passage of large peptides (Cano-Cebrian *et al.*, 2005; Deli, 2009). Despite the well-established differences in morphology, localization, and overall molecular composition between the vertebrate TJ and the insect septate junction (SJ) (Furuse and Tsukita, 2006), it is now clear that claudins, the integral proteins that confer to TJs their specific permeability properties (reviewed by Van Itallie and Anderson, 2006), have homolog counterparts in SJs (Behr *et al.*, 2003; Wu *et al.*, 2004; Furuse and Tsukita, 2006), in accordance with the vital barrier function imparted by both intercellular junctions to the epithelial layers. Since little is known about insect SJ functional properties, we started a characterization *in vitro* of the paracellular pathway in lepidopteran larval midgut. We found that it is cation-selective and that small organic molecules and peptides, like proctolin and fluorescein, cross the epithelium exclusively by this route (Fiandra *et al.*, 2009). Interestingly, its permeability is modulated by cAMP and/

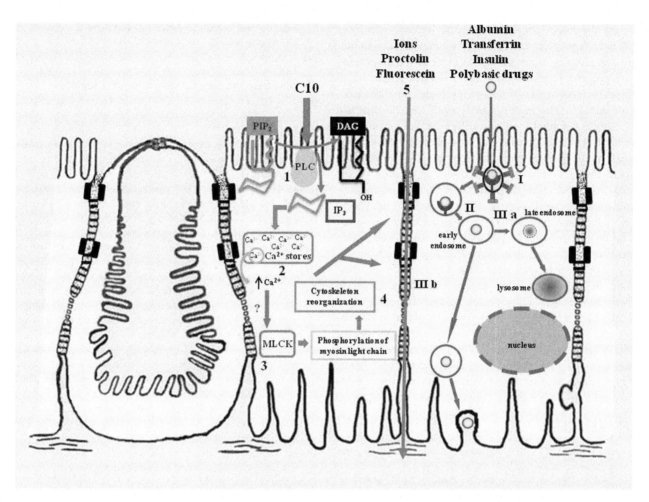

FIGURE 3 **Schematic representation of the lepidopteran larval midgut in which one goblet cell (left) is linked to two columnar cells (right).** The mechanism involved in albumin transcytosis (right columnar cell; Casartelli *et al.*, 2005) and the signaling pathway that leads to an increase in the paracellular permeability to ions and organic molecules (left columnar cell; Casartelli *et al.*, unpublished data) are illustrated. Right cell: after binding to a megalin-like receptor (Casartelli *et al.*, 2008), albumin (and other macromolecules) is internalized by clathrin-mediated endocytosis (I); then, from early endosomes (II) the major part of the internalized protein is directed to the degradative pathway (III a), while a minor but significant amount is delivered to the basolateral membrane and exocytosed (III b) (Casartelli *et al.*, unpublished data). Left cell: a medium-chain fatty acid (C_{10}) activates phospholipase C (PLC, 1), triggering the IP3-dependent signaling cascade (PIP: phosphaditilinositol (4,5)-biphosphate; DAG: diacylglycerol); Ca^{2+} is released from the intracellular stores (2) and myosin light chain is phosphorylated by myosin light chain kinase (MLCK, 3); phosphorylation apparently causes a reorientation of the cytoskeletal elements connected to the septate junctions (4), eliciting the increase in the paracellular permeability to ions and small organic molecules (5). Please see color plate section at the back of the book.

or by a fine regulation of cytosolic Ca^{2+} concentration (Fiandra *et al.*, 2006). We have now studied one of the signaling pathway(s) that may lead to the intracellular release of Ca^{2+}, responsible for the increase of the junction permeability, by using a medium-chain fatty acid (C_{10}), known to modulate the mammalian TJ, through the activation of a Ca^{2+} mediated intracellular signaling pathway (Cano-Cebrian *et al.*, 2005; Deli, 2009). Lepidopteran larval midguts were isolated in conventional Ussing chambers and C_{10} was added to the luminal side. The fatty acid caused an increase of the junction conductance to ions and of the paracellular fluxes of proctolin and fluorescein. C_{10} triggered an IP3-dependent signaling cascade in midgut cells by activating phospholipase C in the apical membrane, with

a consequent release of Ca^{2+} from its intracellular stores. The Ca^{2+} concentration rise in the cytosol was followed by the phosphorylation of myosin light chains by myosin light chain kinases (see Fig. 3; Casartelli, Giordana, Pennacchio *et al.*, unpublished data). It is not known yet how myosin light-chain phosphorylation causes the rearrangement of the cytoskeletal elements connected to the numerous multidomain scaffolding proteins which interact with SJ integral proteins (Wu and Beitel, 2004).

CONCLUSIONS

Parasitoids and associated polydnaviruses are an underexploited source of valuable biomolecules of relevant interest for insect control. The many different virulence and host regulation factors used to impair host immunity and redirect its physiology in favor of the developing wasp progeny represents one of the largest reservoirs of molecular biodiversity of natural molecules with bioinsecticide activity. The in-depth analysis of their mechanism of action is essential to evaluate the potential benefits and problems associated with their environmental delivery, which has to be effective and targeted. Most of the biological delivery vectors, or the direct application of the recombinant bioinsecticides on crops, require that the molecules used are orally active. Because most of the receptors of the parasitoid-derived bioinsecticides are expected to be located behind the gut wall, in the hemocoel, often at intracellular level, it is crucial to develop new strategies of delivery to overcome these barriers. The growing efforts in this area of study, largely inspired by the recent advances in pharmaceutical research, offer the opportunity to define new effective delivery strategies aimed at increasing selectivity and enhancing the effectiveness of the bioinsecticide molecule used. The exploitation of insect natural enemies and their associated symbionts as a source of novel molecules for insect control significantly expands and reinforces the beneficial services of biocontrol agents.

REFERENCES

Affolter, M., Weijer, C.J., 2005. Signaling to cytoskeletal dynamics during chemotaxis. Dev. Cell 9, 19–34.

Alberghini, S., Filippini, R., Marchetti, E., Dindo, M.L., Shevelev, A.B., Battisti, A., et al. 2005. Construction of a *Pseudomonas* sp. derivative carrying the cry9Aa gene from *Bacillus thuringiensis* and a proposal for new standard criteria to assess entomocidal properties of bacteria. Res. Microbiol. 156, 690–699.

Alberghini, S., Filippini, R., Shevelev, A.B., Squartini, A., Battisti, A., 2006. Extended plant protection by an epiphytic *Pseudomonas* sp. derivative carrying the cry9Aa gene from *Bacillus thuringiensis galleriae* against the pine processionary moth *Thaumetopoea pityocampa*. Biocontrol Sci. Technol. 16, 709–715.

Bale, J.S., van Lenteren, J.C., Bigler, F., 2008. Biological control and sustainable food production. Philos. Trans. R. Soc. Lond. B. 363, 761–776.

Baum, J.A., Bogaert, T., Clinton, W., Heck, G.R., Feldmann, P., Ilagan, O., et al. 2007. Control of coleopteran insect pests through RNA interference. Nat. Biotech. 25, 1322–1326.

Beckage, N.E., Gelman, D.B., 2004. Wasp parasitoid disruption of host development: implications for new biologically based strategies for insect control. Annu. Rev. Entomol. 49, 299–330.

Behr, M., Riedel, D., Schuh, R., 2003. The claudin-like megatrachea is essential in septate junctions for the epithelial barrier function in *Drosophila*. Dev. Cell 5, 611–620.

Bergvinson, D., Garcia-Lara, S., 2004. Genetic approaches to reducing losses of stored grain to insects and diseases. Curr. Opin. Plant Biol. 7, 480–485.

Bernardo, U., Pedata, P.A., Viggiani, G., 2006. Life history of *Pnigalio soemius* (Walker) (Hymenoptera: Eulophidae) and its impact on a leafminer host through parasitization, destructive host-feeding and host-stinging behavior. Biol. Control 37, 98–107.

Bézier, A., Annaheim, M., Herbinière, J., Wetterwald, C., Gyapay, G., Bernard-Samain, S., et al. 2009. Polydnaviruses of braconid wasps derive from an ancestral nudivirus. Science 323, 926–930.

Bravo, A., Gill, S.S., Soberón, M., 2007. Mode of action of *Bacillus thuringiensis* Cry and Cyt toxins and their potential for insect control. Toxicon 49, 423–435.

Breman, J.G., 2001. The ears of the hippopotamus: manifestations, determinants, and estimates of the malaria burden. Am. J. Trop. Med. Hyg. 64, 1–11.

Brooks, E.M., Hines, E.R., 1999. Viral biopesticides for heliothine control—fact or fiction. Today's Life Sci. January/February, 38–44.

Brookes, G., Barfoot, P., 2005. GM crops: the global economic and environmental impact: the first nine years 1996–2004. AgBioForum 8, 15.

Cano-Cebrian, M.J., Zornoza, T., Granero, L., Polache, A., 2005. Intestinal absorption enhancement via the paracellular route by fatty acids, chitosans and others: a target for drug delivery. Curr. Drug Deliv. 2, 9–22.

Cao, J., Ibrahim, H., Garcia, J.J., Mason, H., Granados, R.R., Earle, E.D., 2002. Transgenic tobacco plants carrying a baculovirus enhancin gene slow the development and increase the mortality of *Trichoplusia ni* larvae. Plant Cell Rep. 21, 244–250.

Casartelli, M., Cermenati, G., Rodighiero, S., Pennacchio, F., Giordana, B., 2008. A megalin-like receptor is involved in protein endocytosis in the midgut of an insect (*Bombyx mori*, Lepidoptera). Am. J. Physiol. 295, R1290–R1300.

Casartelli, M., Corti, P., Leonardi, M.G., Fiandra, L., Burlini, N., Pennacchio, F., et al. 2005. Absorption of albumin by the midgut of a lepidopteran larva. J. Insect Physiol. 51, 933–940.

Casida, J.E., Quistad, G.B., 1998. Golden age of insecticide research: past, present, or future? Annu. Rev. Entomol. 43, 1–16.

Cermenati, G., Corti, P., Caccia, S., Giordana, B., Casartelli, M., 2007. A morphological and functional characterization of *Bombyx mori* larval midgut cells in culture. Invert. Surv. J. 4, 119–126.

Christensen, E.I., Birn, H., 2002. Megalin and cubilin: multifunctional endocytic receptors. Nat. Rev. Mol. Cell. Biol. 3, 258–268.

Christou, P., Capell, T., Kohli, A., Gatehouse, J.A., Gatehouse, A.M.R., 2006. Recent developments and future prospects in insect pest control in transgenic crops. Trends Plant Sci. 11, 302–308.

Corrado, G., Arciello, S., Fanti, P., Fiandra, L., Garonna, A., Digilio, M.G., et al. 2008. The Chitinase A from the baculovirus AcMNPV enhances resistance to both fungi and herbivorous pests in tobacco. Transgenic Res. 17, 557–571.

Dahlman, D.L., Rana, R.L., Schepers, E.J., Schepers, T., DiLuna, F.A., Webb, B.A., 2003. A teratocyte gene from a parasitic wasp that is associated with inhibition of insect growth and development inhibits host protein synthesis. Insect Mol. Biol. 12, 527–534.

Dahlman, D.L., Vinson, S.B., 1993. Teratocytes: developmental and biochemical characteristics. In: Beckage, N.E., Thompson, N.S., Federici, B.A. (Eds.), vol 1, Parasites and Pathogens of Insects (pp. 145–165). Academic Press, New York.

Danneels, E.L., Rivers, D.B., de Graaf, D.C., 2010. Venom proteins of the parasitoid wasp *Nasonia vitripennis*: recent discovery of an untapped pharmacopee. Toxins 2, 494–516.

Dayan, F.E., Cantrell, C.L., Duke, S.O., 2009. Natural products in crop protection. Bioorg. Med. Chem. 17, 4022–4034.

de Eguileor, M., Grimaldi, A., Tettamanti, G., Valvassori, R., Leonardi, M.G., Giordana, B., et al. 2001. Larval anatomy and structure of absorbing epithelia in the aphid parasitoid *Aphidius ervi*, Haliday (Hymenoptera, Braconidae). Arthropod Struct. Dev. 30, 27–37.

De Graaf, D.C., Aerts, M., Brunain, M., Desjardins, C.A., Jacobs, F.J., Werren, J.H., et al. 2010. Insights into the venom composition of the ectoparasitoid wasp *Nasonia vitripennis* from bioinformatics and proteomics studies. Insect Mol. Biol. 19, 11–26.

DeBach, P., 1964. Biological Control of Insect Pests and Weeds. Chapman and Hall, London.

Deli, M.A., 2009. Potential use of tight junction modulators to reversibly open membranous barriers and improve drug delivery. Biochim. Biophys. Acta 1788, 892–910.

Demain, A.L., Sanchez, S., 2009. Microbial drug discovery: 80 years of progress. J. Antibiot. 62, 5–16.

Dent, D.R. (Ed.), 1995. Integrated Pest Management. Chapman & Hall, London.

Derossi, D., Joliot, A.H., Chassaing, G., Prochiantz, A., 1994. The 3rd helix of the Antennapedia homeodomain translocates through biological membranes. J. Biol. Chem. 269, 10444–10450.

Dib-Hajj, S.D., Webb, B.A., Summers, M.D., 1993. Structure and evolutionary implications of a 'cysteine-rich' *Campoletis sonorensis* polydnavirus gene family. Proc. Natl. Acad. Sci. U.S.A. 90, 3765–3769.

Digilio, M.C, Isidoro, N., Tremblay, E., Pennacchio, F., 2000. Host castration by *Aphidius ervi* venom proteins. J. Insect Physiol. 46, 1041–1050.

Ding, X., Gopalakrishnan, B., Johnson, L.B., White, F.F., Wang, X., Morgan, D.T., et al. 1998. Insect resistance of transgenic tobacco expressing an insect chitinase gene. Transgenic Res. 7, 77–84.

Down, R.E., Fitches, E., Wiles, D.P., Corti, P., Bell, H.A., Gatehouse, J.A., et al. 2006. Insecticidal spider venom toxin fused to snowdrop lectin is toxic to the peach-potato aphid, *Myzus persicae* (Hemiptera: Aphididae) and the rice brown planthopper, *Nilaparvata lugens* (Hemiptera: Delphacidae). Pest Manage. Sci. 62, 77–85.

Duchi, S., Cavaliere, V., Fagnocchi, L., Grimaldi, M.R., Falabella, P., Graziani, F., et al. 2010. The impact on microtubule network of a bracovirus IκB-like Protein. Cell. Mol. Life Sci. 67, 1699–1712.

Etienne-Manneville, S., 2006. In vitro assay of primary astrocyte migration as a tool to study Rho GTPase function in cell polarization. Methods Enzymol. 406, 565–578.

Falabella, P., Perugino, G., Caccialupi, P., Riviello, L., Varricchio, P., Tranfaglia, A., et al. 2005. A novel fatty acid binding protein produced by teratocytes of the aphid parasitoid *Aphidius ervi*. Insect Mol. Biol. 14, 195–205.

Falabella, P., Riviello, L., De Stradis, M.L., Stigliano, C., Varricchio, P., Grimaldi, A., et al. 2009. *Aphidius ervi* teratocytes release an extracellular enolase. Insect Biochem. Mol. Biol. 39, 801–813.

Falabella, P., Riviello, L., Caccialupi, P., Rossodivita, T., Valente, M.T., De Stradis, M.L., et al. 2007. A γ-glutamyl transpeptidase of *Aphidius ervi* venom induces apoptosis in the ovaries of host aphids. Insect Biochem. Mol. Biol. 37, 453–465.

Falabella, P., Tremblay, E., Pennacchio, F., 2000. Host regulation by the aphid parasitoid *Aphidius ervi*: the role of teratocytes. Entomol. Exp. Appl. 97, 1–9.

Falabella, P., Varricchio, P., Provost, B., Espagne, E., Ferrarese, R., Grimaldi, A., et al. 2007. Characterization of the IkB-like gene family in polydnaviruses associated with wasps belonging to different Braconid subfamilies. J. Gen. Virol. 88, 92–104.

Fath-Goodin, A., Gill, T.A., Martin, S.B., Webb, B.A., 2006. Effect of *Campoletis sonorensis* cys-motif proteins on *Heliothis virescens* larval development. J. Insect Physiol. 52, 576–585.

ffrench-Constant, R.H., Dowling, A., Waterfield, N.R, 2007. Insecticidal toxins from *Photorhabdus* bacteria and their potential use in agriculture. Toxicon 49, 436–451.

Fiandra, L., Casartelli, M., Cermenati, G., Burlini, N., Giordana, B., 2009. The intestinal barrier in lepidopteran larvae: Permeability of the peritrophic membrane and of the midgut epithelium to two biologically active peptides. J. Insect Physiol. 55, 10–18.

Fiandra, L., Casartelli, M., Giordana, B., 2006. The paracellular pathway in the lepidopteran larval midgut: modulation by intracellular mediators. Comp. Biochem. Physiol. 144A, 464–473.

Fiandra, L., Terracciano, I., Fanti, P., Garonna, A., Ferracane, L., Fogliano, V., et al. 2010. A viral chitinase enhances oral activity of TMOF. Insect Biochem. Mol. Biol. 40, 533–540.

Fitches, E., Audsley, N., Gatehouse, J.A., Edwards, J.P., 2002. Fusion proteins containing neuropeptides as novel insect control agents: snowdrop lectin delivers fused allatostatin to insect hemolymph following oral ingestion. Insect Biochem. Mol. Biol. 32, 1653–1661.

Fitches, E., Edwards, M.G., Mee, C., Grishin, E., Gatehouse, A.M.R., Edwards, J.P., et al. 2004. Fusion protein containing insect-specific toxins as pest control agents: snowdrop lectin delivers fused insecticidal spider venom toxin to insect hemolymph following oral ingestion. J. Insect Physiol. 50, 61–71.

Fonseca, S.B., Pereira, M.P., Kelley, S.O., 2009. Recent advances in the use of cell-penetrating peptides for medical and biological applications. Adv. Drug Deliv. Rev. 61, 953–964.

Frankel, A.D., Pabo, C.O., 1988. Cellular uptake of the tat protein from human immunodeficiency virus. Cell 55, 1189–1193.

Furuse, M., Tsukita, S., 2006. Claudins in occluding junctions of humans and flies. Trends Cell Biol. 16, 181–188.

Gabor, F., Bogner, E., Weissenboeck, A., Wirth, M., 2004. The lectin–cell interaction and its implications to intestinal lectin-mediated drug delivery. Adv. Drug Deliv. Rev. 56, 459–480.

Gatehouse, J.A., 2008. Biotechnological prospects for engineering insect-resistant plants. Plant Physiol. 146, 881–887.

Gauld, I.D., 1988. Evolutionary patterns of host utilization by ichneumonid parasitoids. Biol. J. Linn. Soc. 35, 351–377.

Gill, T.A., Fath-Goodin, A., Maiti, I.B., Webb, B.A., 2006. Potential uses of cys-motif and other polydnavirus genes in biotechnology. Adv. Virus Res. 68, 393–426.

Giordana, B., Leonardi, M.G., Casartelli, M., Parenti, P., 1998. K^+-neutral amino acid symport of *Bombyx mori* larval midgut: a system operative in extreme conditions. Am. J. Physiol. 74, R1361–R1371.

Giordana, B., Milani, A., Grimaldi, A., Farneti, R., Casartelli, M., Ambrosecchio, M.R., et al. 2003. Absorption of sugars and amino acids by the epidermis of *Aphidius ervi* larvae. J. Insect Physiol. 49, 1115–1124.

Godfray, H.J.C., 1994. Parasitoids: Behavioural and Evolutionary Ecology. Princeton University Press, Princeton.

Gordon, D., Karbat, I., Ilan, N., Cohen, L., Kahn, R., Gilles, N., et al. 2007. The differential preference of scorpion α-toxins for insect or mammalian sodium channels: Implications for improved insect control. Toxicon 49, 452–472.

Green, M., Loewenstein, P.M., 1988. Autonomous functional domains of chemically synthesized human immunodeficiency virus tat trans-activator protein. Cell 55, 1179–1188.

Grenier, S., 2009. In vitro rearing of entomophagous insects – past and future trends: a minireview. Bull. Insectol. 62, 1–6.

Gubler, D.J., 2002. The global emergence/resurgence of arboviral diseases as public health problems. Arch. Med. Res. 33, 330–342.

Gurevitz, M., Karbat, I., Cohen, L., Ilan, N., Kahn, R., Turkov, M., et al. 2007. The insecticidal potential of scorpion β-toxins. Toxicon 49, 473–489.

Hayakawa, T., Shimojo, E., Mori, M., Kaido, M., Furusawa, I., Miyata, S., et al. 2000. Enhancement of baculovirus infection in *Spodoptera exigua* (Lepidoptera: Noctuidae) larvae with *Autographa californica* nucleopolyhedrovirus or *Nicotiana tabacum* engineered with a granulovirus enhancin gene. Appl. Entomol. Zool. 35, 163–170.

Horowitz, A.R., Ishaaya, I. (Eds.), 2004. Insect Pest Management. Springer-Verlag, Berlin.

Huvenne, H., Smagghe, G., 2010. Mechanisms of dsRNA uptake in insects and potential of RNAi for pest control: a review. J. Insect Physiol. 56, 227–235.

Inceoglu, A.B., Kamita, S.G., Hammock, B.D., 2006. Genetically modified baculoviruses: a historical overview and future outlook. Adv. Virus Res. 68, 323–360.

James, C., 2005. Preview: Global Status of Commercialized Biotech/GM Crops: 2005. ISAAA Briefs No 34. Ithaca, NY.

Jarlfors, U.E., Dahlman, D.L., Zhang, D., 1997. Effects of *Microplitis croceipes* teratocytes on host hemolymph protein content and fat body proliferation. J. Insect Physiol. 43, 577–585.

Jeffers, L.A., Roe, R.M., 2008. The movement of proteins across the insect and tick digestive system. J. Insect Physiol. 54, 319–332.

Kamita, S.G., Kang, K.-D., Hammock, B.D., Inceoglu, A.B., 2005. Genetically modified baculoviruses for pest insect control. In: Gilbert, L.I., Iatrou, K., Gill, S.S. (Eds.), Comprehensive Molecular Insect Science, vol. 6. (pp. 271–322). Elsevier, San Diego.

Kang, J.S., DeLuca, P.P., Lee, K.C., 2009. Emerging PEGylated drugs. Expert Opin. Emerg. Drugs 14, 363–380.

Kim, Y., 2005. Identification of host translation inhibitory factor of *Campoletis sonorensis* ichnovirus on the tobacco budworm, *Heliothis virescens*. Arch. Insect Biochem. Physiol. 59, 230–244.

Kirst, A.H., 2010. The spinosyn family of insecticides: realizing the potential of natural products research. J. Antibiot. 63, 101–111.

Kroemer, J.A., Webb, B.A., 2005. Iκβ-Related *vankyrin* Genes in the *Campoletis sonorensis* Ichnovirus: Temporal and Tissue-Specific Patterns of Expression in Parasitized *Heliothis virescens* Lepidopteran Hosts. J. Virol. 79, 7617–7628.

Maggio, F., Sollod, B.L., Tedford, H.W., King, G.F., 2005. Spider toxins and their potential for insect control. In: Gilbert, L.I., Iatrou, K., Gill, S.S. (Eds.), Comprehensive Molecular Insect Science, vol. 5. (pp. 221–238). Elsevier, San Diego.

Maiti, I.B., Dey, N., Dahlman, D.L., Webb, B.A., 2003. Antibiosis-type resistance in transgenic plants expressing teratocyte secretory peptide (TSP) gene from hymenopteran endoparasite (*Microplitis croceipes*). Plant Biotechnol. 1, 209–219.

Mao, Y.-B., Cai, W.-J., Wang, J.-W., Hong, G.-J., Tao, X.-Y., Wang, L.-J., et al. 2007. Silencing a cotton bollworm P450 monooxygenase gene by plant-mediated RNAi impairs larval tolerance of gossypol. Nat. Biotechnol. 25, 1307–1313.

Moestrup, S.K., Verroust, P.J., 2001. Megalin- and cubilin-mediated endocytosis of protein-bound vitamins, lipids, and hormones in polarized epithelia. Annu. Rev. Nutr. 21, 407–428.

Nachman, R.J., Teal, P.E.A., Strey, A., 2002. Enhanced oral availability/pheromonotropic activity of peptidase-resistant topical amphiphilic analogs of pyrokinin/PBAN insect neuropeptides. Peptides 23, 2035–2043.

Nicholson, G.M., 2007a. Fighting the global pest problem: preface to the special *Toxicon* issue on insecticidal toxins and their potential for insect pest control. Toxicon 49, 413–422.

Nicholson, G.M., 2007b. Insect-selective spider toxins targeting voltage-gated sodium channels. Toxicon 49, 490–512.

Oerke, E.-C., Dehne, H.-W., 2004. Safeguarding production-losses in major crops and the role of crop protection. Crop Prot. 23, 275–285.

Patel, L.N., Zaro, J.L., Shen, W., 2007. Cell penetrating peptides: intracellular pathways and pharmaceutical perspectives. Pharmacol. Res. 24, 1977–1992.

Pennacchio, F., Strand, M.R., 2006. Evolution of developmental strategies in parasitic Hymenoptera. Annu. Rev. Entomol. 51, 233–258.

Price, D.R.G., Gatehouse, J.A., 2008. RNAi-mediated crop protection against insects. Trends Biotech. 26, 393–400.

Prochiantz, A., 2008. Protein and peptide transduction, twenty years later a happy birthday. Adv. Drug Deliv. Rev. 60, 448–451.

Quicke, D.L.J., 1997. Parasitic Wasps. Chapman & Hall, London.

Rao, R., Fiandra, L., Giordana, B., de Eguileor, M., Congiu, T., Burlini, N., et al. 2004. AcMNPV ChiA protein disrupts the peritrophic membrane and alters midgut physiology of *Bombyx mori* larvae. Insect Biochem. Mol. Biol. 34, 1205–1213.

Roe, R.M., Brandt, A.E., 2003. Polymer conjugates of insecticidal peptides or nucleic acids or insecticides and methods of use thereof. United States Patent Application #20030108585. June 12, 2003.

Roh, J.Y., Choi, J.Y., Li, M.S., Jin, B.R., Je, Y.H., 2007. *Bacillus thuringiensis* as a specific, safe, and effective tool for insect pest control. J. Microbiol. Biotechnol. 17, 547–559.

Rohou, A., Nield, J., Ushkaryov, Y.A., 2007. Insecticidal toxins from black widow spider venom. Toxicon 49, 531–549.

Ryan, S.M., Mantovani, G., Wang, X., Haddleton, D.M., Brayden, D.J., 2008. Advances in PEGylation of important biotech molecules: delivery aspects. Expert Opin. Drug Deliv. 5, 371–383.

Salama, N.N., Eddington, N.D., Fasano, A., 2006. Tight junction modulation and its relationship to drug delivery. Adv. Drug Deliv. Rev. 58, 15–28.

Shelby, K.S., Cui, L., Webb, B.A., 1998. Polydnavirus-mediated inhibition of lysozyme gene expression and the antibacterial response. Insect Mol. Biol. 7, 265–272.

Shelby, K.S., Webb, B.A., 1997. Polydnavirus infection inhibits translation of specific growth-associated host proteins. Insect Biochem. Mol. Biol. 27, 263–270.

Shen, H., Brandt, A., Witting-Bissinger, B.E., Gunnoe, T.B., Roe, R.M., 2009. Novel insecticide polymer chemistry to reduce the enzymatic digestion of a protein pesticide, trypsin modulating oostatic factor (TMOF). Pestic. Biochem. Physiol. 93, 144–152.

Siegrist, S.E., Doe, C.Q., 2007. Microtubule-induced cortical cell polarity. Genes Dev. 21, 483–496.

Snow, R.W., Guerra, C.A., Noor, A.M., Myint, H.Y., Hay, S.I., 2005. The global distribution of clinical episodes of *Plasmodium falciparum* malaria. Nature 434, 214–217.

Strand, M.R., Pech, L.L., 1995. Immunological compatibility in parasitoid–host relationships. Annu. Rev. Entomol. 40, 31–56.

Tanada, Y., Himeno, M., Omi, E.M., 1973. Isolation of a factor, from the capsule of a granulosis virus, synergistic for a nuclear-polyhedrosis virus of the armyworm. J. Invertebr. Pathol. 21, 81–90.

Terra, W.R., 2001. The origin and function of the insect peritrophic membrane and peritrophic gel. Arch. Insect Biochem. Physiol. 47, 47–61.

Thoetkiattikul, H., Beck, M.H., Strand, M.R., 2005. Inhibitor kB-like proteins from a polydnavirus inhibit NF-kB activation and suppress the insect immune response. Proc. Natl. Acad. Sci. U.S.A. 102, 11426–11431.

Toenniessen, G.H., O'Toole, J .C., DeVries, J., 2003. Advances in plant biotechnology and its adoption in developing countries. Curr. Opin. Plant Biol. 6, 191–198.

Tuma, P.L., Hubbard, A.L., 2003. Transcytosis: crossing cellular barriers. Physiol. Rev. 83, 871–932.

Vacek, M., Reynaets, A., Hofte, H., Jansens, S., Beukeleer, M.D., Dean, C., et al. 1987. Transgenic plants protected from insect attack. Nature 328, 33–37.

Van Itallie, C.M., Anderson, J.M., 2006. Claudins and epithelial paracellular transport. Annu. Rev. Physiol. 68, 403–429.

Vincent, B., Kaeslin, M., Roth, T., Heller, M., Poulain, J., Cousserans, F., et al. 2010. The venom composition of the parasitic wasp *Chelonus inanitus* resolved by combined expressed sequence tags analysis and proteomic approach. BMC Genomics 11, 693.

Vinson, S.B., Iwantsch, G.F., 1980. Host regulation by insect parasitoids. Q. Rev. Biol. 55, 143–165.

Vinson, S.B., Pennacchio, F., Consoli, F., 2001. The parasitoid–host endocrine interaction from a nutritional perspective. In: Edwards, J. (Ed.), Manipulating Hormonal Control (Endocrine Interactions) (pp. 263–283). BIOS, Oxford.

Washburn, J.O., Haas-Stapleton, E.J., Tan, F.F., Beckage, N.E., Volkman, L.E., 2000. Coinfection of *Manduca sexta* larvae with polydnavirus from *Cotesia congregata* increases susceptibility to fatal infection by *Autographa californica* M Nucleopolyhedrovirus. J. Insect Physiol. 46, 179–180.

Weaver, R.J., Marris, G.C., Bell, H.A., Edwards, J.P., 2001. Identity and mode of action of the host endocrine disrupters from the venom of parasitoid wasps. In: Edwards, J.P., Weaver, R.J. (Eds.), Endocrine Interactions of Insect Parasites and Pathogens (pp. 33–85). BIOS, Oxford.

Webb, B.A., Strand, M.R., 2005. The biology and genomics of polydnaviruses. In: Gilbert, L.I., Iatrou, K., Gill, S.S. (Eds.), Comprehensive Molecular Insect Science, vol. 6, (pp. 323–360). Elsevier, San Diego.

Whetstone, P.A., Hammock, B.D., 2007. Delivery methods for peptide and protein toxins in insect control. Toxicon 49, 576–596.

Whitfield, J.B., 1998. Phylogeny and evolution of host–parasitoid interactions in Hymenoptera. Annu. Rev. Entomol. 43, 129–151.

Whitfield, J.B., 2003. Phylogenetic insights into the evolution of parasitism in Hymenoptera. Adv. Parasitol. 54, 69–100.

Wu, V.M., Beitel, G.J., 2004. A junctional problem of apical proportions: epithelial tube-size control by septate junctions in the *Drosophila* tracheal system. Curr. Opin. Cell Biol. 16, 493–499.

Wu, V.M., Schulte, J., Hirschi, A., Tepass, U., Beitel, G.J., 2004. Sinuous is a *Drosophila* claudin required for septate junction organization and epithelial tube size control. J. Cell Biol. 164, 313–323.

Zlotkin, E., 2005. Scorpion venoms. In: Gilbert, L.I., Iatrou, K., Gill, S.S. (Eds.), Comprehensive Molecular Insect Science, vol. 5. (pp. 173–220). Elsevier, San Diego.

The Legacy of George Salt, Pioneer in Parasitoid Virology, and Prospects for the Future of Parasitoid Polydnavirus and Venom Research

Jean-Michel Drezen and Nancy Beckage

THE LEGACY OF GEORGE SALT IN PARASITOID VIRUS RESEARCH

Dr George Salt of Cambridge University studied, ironically, a system in which virus-like particles were found in the reproductive tract of *Nemeritis* (now *Venturia*) *canescens* but the particles, upon later examination, were found to have no DNA or RNA, very much unlike most of the viruses described in this book. The mechanism whereby the virus particles of *Venturia* protect the parasitoid eggs from encapsulation by host hemocytes is also unique. His enormous body of work inspired investigators around the world, including Drs. Bradleigh Vinson (see Foreword) and Donald Stoltz (Chapter 1), to screen the reproductive tracts and calyx fluid of many species of parasitic wasps for virus particles, which at the time could only be found in electron micrographs of the wasps' ovaries. Later, the first functional studies were carried out using parasitoid eggs that were washed free of virus and then injected into the unparasitized host larva. These 'clean' eggs without virus particles were always encapsulated, whereas eggs injected with virions escaped encapsulation. The immunological role of polydnaviruses was thus discovered. These very simple experiments inspired many laboratories to examine particles in other species. It would then be discovered later that each species carries a unique virus found only in that species.

As documented in the many chapters of this book, the field of parasitoid virus and venom research has expanded enormously in the decades since Dr Salt made his initial observations. This field is now a mature discipline, hence the inspiration for this book.

PROSPECTS FOR THE FUTURE OF PARASITOID POLYDNAVIRUS AND VENOM RESEARCH

Origins of Virus Associations with Parasitoids

While we now have amassed much information about the origin and evolution of bracoviruses, we know significantly less about ichnoviruses and virus-like particles (VLPs). The data already obtained have been limited by the size of the molecules amenable to analysis using former cloning techniques. A broader view of the organization of the PDV genomes within wasp genomes will most probably be available in the near future, thanks to whole genome sequencing of the wasps' genomes, which might be undertaken more easily with the decrease of sequencing costs. In particular for bracoviruses, it is already known that a major locus comprises most of the virus segments in the wasp genome and it would be interesting to determine whether the nudiviral cluster comprising genes involved in virus particle production is located in the same region of the wasp genome corresponding to the integration site of the nudivirus initially captured in the ancestor wasp genome. Ichnovirus sequences appear scattered and surrounded by wasp genes but a comprehensive view of the wasp genome might similarly reveal that they are restricted to particular chromosomal regions.

Moreover, high-quality sequencing of parasitoid genomes will allow understanding of the evolution of PDV sequences in wasp genomes, by comparing the organization of these sequences in related species. Are the different loci located at homologous chromosomal positions?

What is the dynamic of duplications creating new versions of virulence genes? Is there some evidence that circles can sometimes reintegrate into the wasp genome?

An important challenge is the identification of a pathogenic virus related to the ichnovirus ancestor. The genome of such a virus would be characterized by the presence of homologs of ichnovirus genes coding for particle components (IVSPER genes). Expanding our knowledge of the biology of this virus would give insight into the early events that have favored the association(s) leading to ichnoviruses as we know them today. Deep sequencing of whole body transcriptomes of a variety of arthropods, such as recently undertaken for phylogenetic analyses, might be sufficient to identify ichnovirus-related pathogenic viruses, since IVSPER gene expression was detected by using such an approach on a PDV-associated wasp. More generally, deep sequencing approaches are expected to allow the detection of the diversity of viral symbionts or pathogens associated with parasitoid wasps or other organisms. This would give clues as to whether symbiotic viruses are a particular feature of parasitoid wasps or are fairly common but identified to date only in parasitoids because they have been well studied, both as models for evolutionary biology, and for their importance as biological control agents of lepidopteran populations.

It is still unclear whether PDVs in ichneumonid wasps originated from a single or multiple events of virus capture since the PDVs described in the Banchinae subfamily have unique features that are different from those of ichnoviruses described in the Campopleginae subfamily. Moreover, the subfamilies associated with PDVs are separated by groups not bearing PDVs, which suggests that PDVs have been lost in these groups or that two independent associations occurred, one with the Campopleginae, the other with the Banchinae ancestor. A more comprehensive analysis of these groups looking for the presence of PDVs or of their remnants in wasp genomes will be required to answer this question.

Another unsolved question concerns the DNA replication mechanism used for the production of the DNA packaged in the particles. No gene involved in DNA replication has been identified yet among those expressed in the ovaries of PDV-associated wasps; this could be due to their lower level of expression compared with those coding for particle components and a deeper sequencing using new approaches would facilitate their identification. Another possibility is that cellular genes are involved.

Evolution of Packaged Genome Content

Another interesting question concerns the gene content of the packaged genome that appears to be determined mostly by the interaction with the host. Indeed, no common genes have been found across all bracovirus genomes while the machinery producing the particles is well conserved. With the new sequencing techniques it is easier to get viral segments sequenced from a limited number of wasps. It will thus be possible to have access to the high diversity of species while most of the analyses have been limited to date to the few species reared in laboratories. When data will be available on wasps distributed through known phylogenies with detailed life history information, it will be possible to design analyses which could identify features of the wasp–PDV–caterpillar interaction that are functionally significant in determining the wasps' host ranges. Moreover, identifying the genes involved in defining host specificity will allow a more appropriate choice of populations of parasitoids to be used for biological control.

Since some parasitoid genera are comprised of more than a thousand species, comparison of homologous genes in these species is possible and would permit us to pinpoint the genes currently involved in the parasite–host arms race. In particular, by measuring selection pressure operating on PDV-packaged genes, it will be possible to identify those evolving under diversifying selection.

Function of PDV-Encoded Products

The study of the function of PDV-encoded products is less amenable to high throughput approaches. Among the packaged genes already identified, few have a predictable function and even fewer have been studied by functional approaches such that their biological function in the host–parasite relationship has been convincingly demonstrated. While parasitoid viral genomics has significantly advanced in recent years, viral proteomics studies require much more attention.

The difficulty is to identify the target of PDV products not only by showing protein–protein interaction *in vitro* or in heterologous systems but we need *in vivo* experimental approaches with the actual host molecules. Much effort should be targeted to designing new tools such as RNAi approaches giving reproducible results in lepidopteran hosts, otherwise the progress in genomic data will not translate into progress in understanding the biology and functional roles of virally encoded molecules in host–parasite interactions.

While we have some information about the role of parasitoid viruses in causing host developmental arrest, we currently have no information about what viral products or venom components cause feeding inhibition and anorexia in parasitized hosts. Usually the host stops feeding one to three days before its parasitoids emerge from the host, and never resumes feeding following emergence. Identifying the genes whose products induce host feeding cessation and arrest will have major impacts on agricultural biotechnology and design of transgenic plants expressing those genes. Proteins secreted by the parasitoids may also figure prominently in inducing host arrest.

Progress in Parasitoid Venom Research

Concerning venoms, we can expect that deep sequencing approaches will allow comprehensive studies of the contents of the venom gland, which represents a source of new molecules for future applications in agronomy and agricultural pest control. Comparative studies can also be carried out. Preliminary results suggest the molecules secreted in the venom gland of parasitoid species are different even within different strains of the same species. Future studies will aim at determining why and how this versatility of strategies and differences in venom gland proteins is achieved. Studies of the biological functions of venom factors will require the same approaches as used in studies of PDV gene products. Some of these venom proteins may induce host developmental arrest similar to PDV gene products. Venom components sometimes facilitate or synergize the action of PDVs due to unknown mechanisms, which would be interesting to explore.

Nature of VLPs

The VLPs of *Venturia canescens* were the first entities identified as resembling viruses in parasitoids although they do not contain nucleic acids. As PDVs they are involved in host immunosuppresion and many VLPs have been described since George Salt published his pioneering studies. However, their morphology varies from particles resembling viruses to vesicles and their relation with viruses has not been firmly established. Strikingly, none of the proteins purified from VLPs to date are related to known viruses. Future studies are required to determine, depending on the parasitoid species, which VLPs are produced by viral machineries and which represent secretory vesicles.

Medical Applications

In addition to obvious applications of isolation and characterization of new viral molecules that could be used as biopesticides, PDVs have the potential to inspire the design of vectors for human gene therapy. Indeed PDVs have naturally evolved to have 'jumping genes', with the capacity to deliver more than a hundred and fifty genes, thus exceeding by far the capacity of retroviruses that have been already used experimentally as vectors in medicine. Nature has provided us with PDV mobile elements, which have multiple potential uses including genetic engineering of insects. Knowledge of the PDV replication mechanism and the development of cell lines that could be induced to produce the particles *ex vivo* might allow us someday to manipulate the content of the DNA packaged and the content of the particles, opening the way to study the feasibility of using modified PDVs as vectors for gene transfer. Viral promoters may also be exploited in agriculture and medicine.

In conclusion, in the decades since parasitoid viruses were first discovered, exciting progress has been made and research on these elements now spans several continents. We are now entering the phase where applications to agricultural pest control, biotechnology, and medicine can be tested. Our goal in producing this book is to capture the attention of investigators in other fields and attract them to parasitoid virology in both basic and applied science, and we will let the readers judge whether we will be successful in our endeavor.

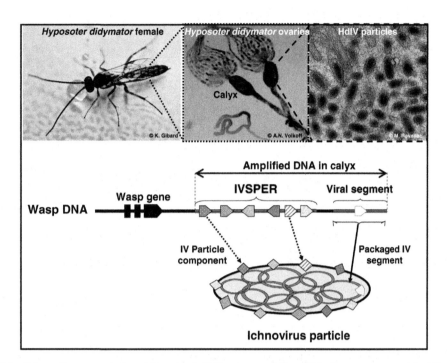

CHAPTER 3: FIGURE 5   Schematic representation of the extended IV genome residing within the wasp genome. The IV genome is composed of (1) the IVSPERs, specialized regions containing genes encoding viral structural proteins and (2) the viral segments. Both regions are amplified in calyx cells but only the viral segments are incorporated into virus particles following DNA excision and circularization.

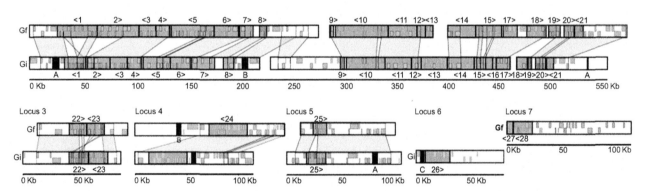

CHAPTER 4: FIGURE 2   Structure and organization of bracovirus proviral genome segments within the parasitoid genome: macrolocus and additional loci. Bracoviruses *Glyptapanteles flavicoxis* (Gf) and *G. indiensis* (Gi) proviral genome segments (gray) assembled from BAC clone data (Desjardins *et al.*, 2008) and their synteny are shown. For each, the corresponding segment number of the encapsidated viral genome segment form is given above the proviral genome segment, with the symbols > or < depicting the directionality of the segment excision regulatory signals. Protein coding genes within proviral genome segments are indicated by light blue boxes; genes encoded in parasitoid flanking DNA at each locus are indicated by light purple boxes. Regions of synteny between proviral segment and flanking DNA are shaded in light gray. Three different long tandem repeat classes (A, B, and C) are denoted by black boxes. Locus sizes are given in kilobase pairs. Figure modified from Desjardins *et al.*, 2008.

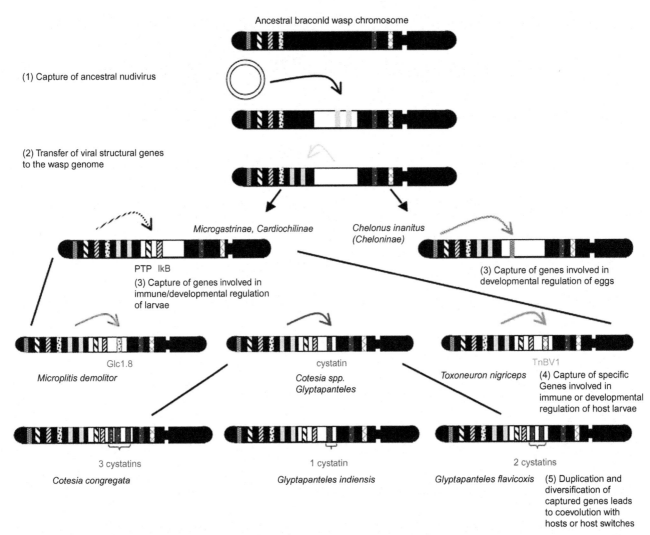

CHAPTER 5: FIGURE 1   Proposed evolutionary scenario for the origin and evolution of bracoviruses. Glc1.8, cystatins and TnBV1 are genes specific to each bracovirus associated with *M. demolitor, Cotesia* and *Glyptapanteles*, and *Toxoneuron nigriceps*, respectively.

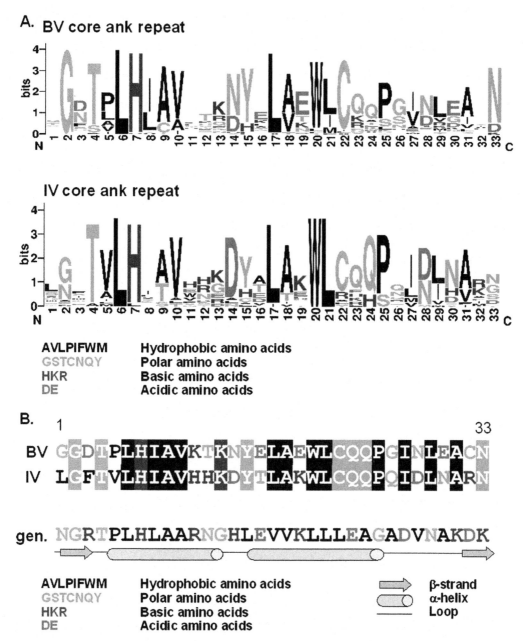

A. BV core ank repeat

IV core ank repeat

AVLPIFWM	Hydrophobic amino acids
GSTCNQY	Polar amino acids
HKR	Basic amino acids
DE	Acidic amino acids

B.

1   33

BV   GGDTPLHIAVKTKNYELAEWLCQQPGINIEACN

IV   LGFTVLHIAVHHKDYTLAKWLCQQPOIDLNARN

gen.   NGRTPLHLAARNGHLEVVKLLLEAGADVNAKDK

AVLPIFWM	Hydrophobic amino acids		β-strand
GSTCNQY	Polar amino acids		α-helix
HKR	Basic amino acids		Loop
DE	Acidic amino acids		

CHAPTER 5: FIGURE 3 Consensus sequences of the core ank repeat of BV and IV ankyrin-like proteins. A: Consensus sequences of the core ank repeats of BV and IV ankyrin-like proteins were drawn using the WebLogo application (http://weblogo.berkeley.edu) after multiple sequence alignment of 45 BV and 23 IV ankyrin-like proteins. **B:** Comparison between the consensus sequences of BV and IV core ank repeats. Identical amino acids are highlighted. The general consensus sequence of ank repeats, adapted from Mosavi *et al.* (2002) is shown for comparison.

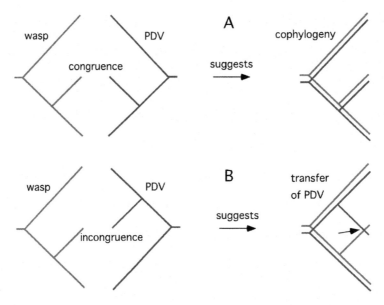

CHAPTER 7: FIGURE 2 Simple schematic showing logic of co-phylogeny inferences from examining congruence among wasp and virus gene trees.

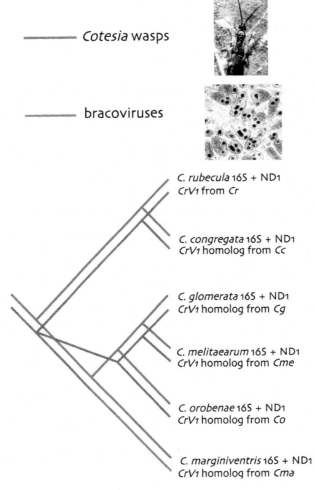

Cotesia wasps

bracoviruses

C. rubecula 16S + ND1
CrV₁ from _Cr_

C. congregata 16S + ND1
CrV₁ homolog from _Cc_

C. glomerata 16S + ND1
CrV₁ homolog from _Cg_

C. melitaearum 16S + ND1
CrV₁ homolog from _Cme_

C. orobenae 16S + ND1
CrV₁ homolog from _Co_

C. marginiventris 16S + ND1
CrV₁ homolog from _Cma_

CHAPTER 7: FIGURE 3 Inferred co-phylogeny of BV genes with genes of wasps in the genus _Cotesia_ (**from Whitfield and Asgari, 2003, re-used with permission from Elsevier**). The wasp phylogeny was estimated based on partial sequences from the 16S and ND1 mtDNA genes; BV relationships were estimated based on partial sequences from the CrV1 gene and its orthologs in other _Cotesia_ species.

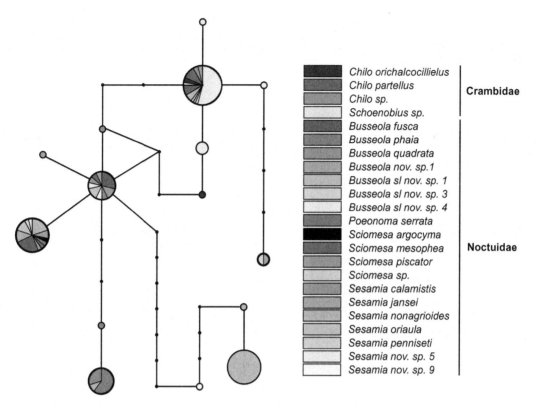

CHAPTER 10: FIGURE 2 Parsimonious network of association between CrV1 variants and host species *(from Branca et al., in press).*

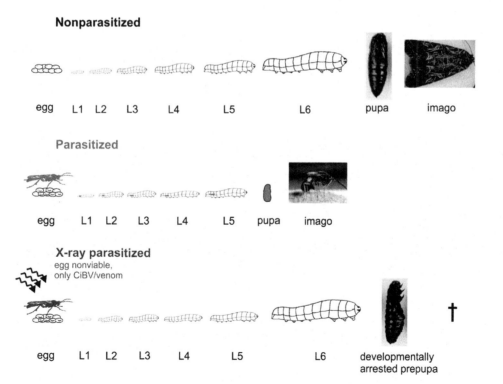

CHAPTER 14: FIGURE 1 Life-cycles of nonparasitized *S. littoralis* (top) and when parasitized by *C. inanitus* (middle). On the bottom the method to study CiBV/venom functions in the absence of a developing parasitoid (X-ray treatment of wasps) is shown along with the effect on *S. littoralis* development.

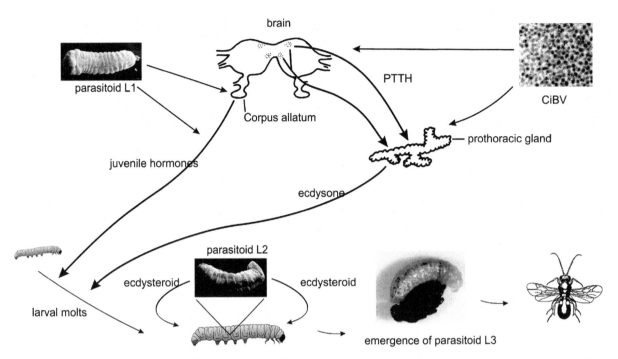

CHAPTER 14: FIGURE 3 The effects of CiBV and the parasitoid L1 larva on the endocrine system of the host, of release of ecdysteroid by L2 parasitoid, and of emergence of L3 parasitoid from host.

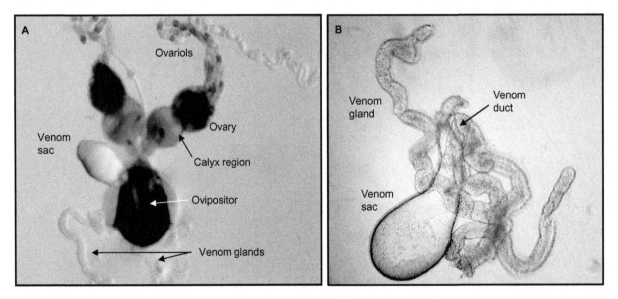

CHAPTER 18: FIGURE 1 A representative image of the type 2 venom apparatus from the endoparasitoid, *Cotesia rubecula* (Hymenoptera: Braconidae). **A**: Female reproductive organ including venom glands and venom sac. **B**: Venom apparatus.

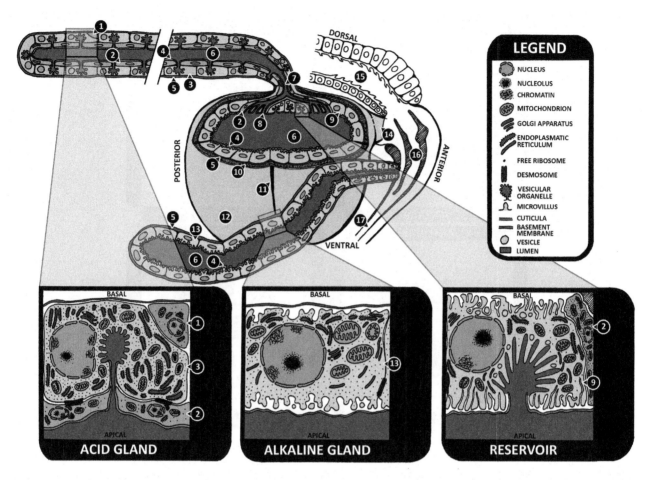

CHAPTER 19: FIGURE 1   Overview of the *Nasonia vitripennis* venom apparatus. (1) Interstitial cell (acid gland), (2) chitogenous cell (acid gland), (3) columnar cell (acid gland), (4) cuticula, (5) basement membrane, (6) lumen, (7) duct, (8) ductile, (9) secretory cell (reservoir), (10) striated muscle (reservoir), (11) medial constriction (reservoir), (12) squamous region (reservoir), (13) secretory cell (alkaline gland), (14) efferent duct, (15) vagina, (16) funnel, (17) ovipositor shaft.

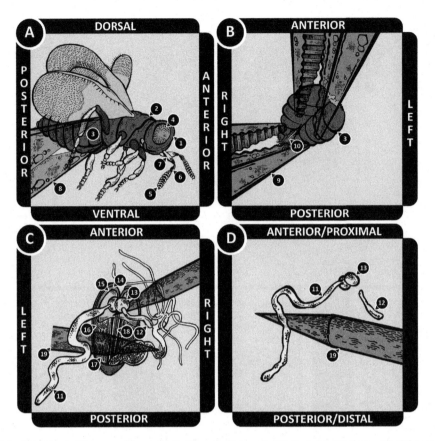

CHAPTER 19: FIGURE 2 **Dissection of the venom apparatus in *Nasonia vitripennis*.** (1) Cephalon, (2) thorax, (3) abdomen, (4) compound eye, (5) flagellum (antenna), (6) pedicel (antenna), (7) scapus (antenna), (8) scissors, (9) forceps, (10) ovipositor, (11) acid gland, (12) alkaline gland, (13) reservoir, (14) stylet, (15) ovipositor ramus, (16) inner ovipositor plate, (17) outer ovipositor plate, (18) sheat and stylets, (19) needle.

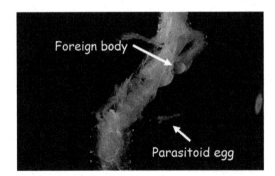

CHAPTER 21: FIGURE 3 **Paraffin oil drop and parasitoid egg not encapsulated in the hemolymph of a *D. melanogaster* host larva parasitized by *Asobara Japonica*.**

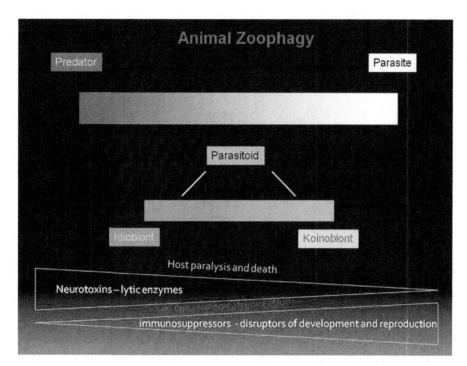

CHAPTER 22: FIGURE 2 **The zoophagy in the animal kingdom is a widespread life habit.** There are a number of adaptive solutions in a continuum range delimited by two extremes: predators and parasites. The very sophisticated strategies adopted by the latter, to finely redirect the host physiology to their own benefit, in order to complete several generations without killing the host, are absent in predators, which exploit multiple victims to complete their development and sustain their reproductive activity as adults. The parasitoids, largely belonging to the class of Insecta, have intermediate characteristics, as they eventually kill (like predators) their host, when completing the preimaginal development, even though juvenile stages may show a wide range of specialized life habits. This degree of specialization is very sophisticated in koinobionts, which finely interact with victims surviving for a relatively long time interval, while it is considerably lower in idiobionts, which rapidly paralyze and/or kill their hosts. The different strategies of host colonization and exploitation adopted are modulated by different categories of virulence and host regulation factors. The idiobionts are a very good source of fast acting neurotoxins and lytic enzymes impacting different host tissues, while the koinobionts offer an astonishing variety of immunosuppressors and molecules disrupting host development and reproduction.

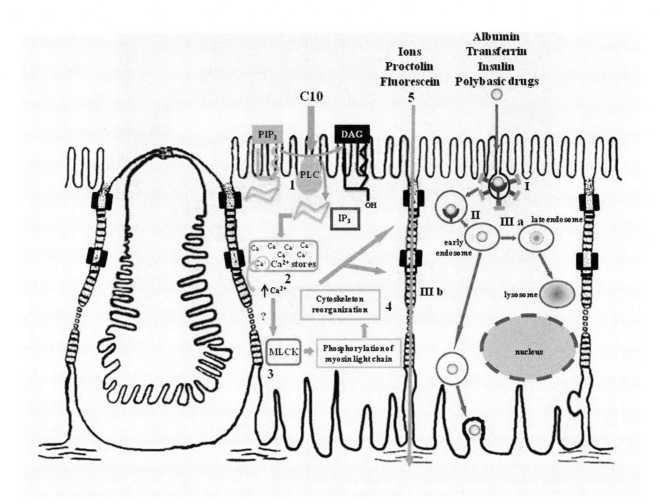

CHAPTER 22: FIGURE 3 **Schematic representation of the lepidopteran larval midgut in which one goblet cell (left) is linked to two columnar cells (right).** The mechanism involved in albumin transcytosis (right columnar cell; Casartelli *et al.*, 2005) and the signaling pathway that leads to an increase in the paracellular permeability to ions and organic molecules (left columnar cell; Casartelli *et al.*, unpublished data) are illustrated. Right cell: after binding to a megalin-like receptor (Casartelli *et al.*, 2008), albumin (and other macromolecules) is internalized by clathrin-mediated endocytosis (I); then, from early endosomes (II) the major part of the internalized protein is directed to the degradative pathway (III a), while a minor but significant amount is delivered to the basolateral membrane and exocytosed (III b) (Casartelli *et al.,* unpublished data). Left cell: a medium-chain fatty acid (C_{10}) activates phospholipase C (PLC, 1), triggering the IP3-dependent signaling cascade (PIP: phosphaditilinositol (4,5)-biphosphate; DAG: diacylglycerol); Ca^{2+} is released from the intracellular stores (2) and myosin light chain is phosphorylated by myosin light chain kinase (MLCK, 3); phosphorylation apparently causes a re-orientation of the cytoskeletal elements connected to the septate junctions (4), eliciting the increase in the paracellular permeability to ions and small organic molecules (5).

CHAPTER 4: TABLE 1 Number of Genes Present in the Related Gene Families in BVs (A), Campoplegine IVs (B) and Banchine IV (C) Sequenced to Date[1]

A. BVs

Present in:		5 BVs							4 BVs						3 BVs		2 BVs		1 BV	
												Number of genes/family/genome								
Virus	# of ORFs	PTP[2]	Vank	BEN	BV4	Cys	CrV1-like	P94	Cystatin	RNase T2	EP1-like	C-type lectin	BV2	BV3	Duffy	BV1	ST	H4	Glc	Egf-like
CcBV[3]	165	27	8	12	2	4	1	2	3	2	6	1	6	2	1	2	0	1	0	0
CvBV[4]	125	36	8	11	3	0	1	4	3	3	7	1	3	1	1	5	0	1	0	0
GiBV[5]	197	42	9	9	2	1	3	1	1	2	2	2	6	2	1	0	3	1	0	0
GfBV[5]	193	32	8	6	2	2	2	1	2	2	2	5	5	4	0	0	5	1	0	0
MdBV[6]	61	13	12	1	1	2	0	0	0	0	0	0	0	0	0	0	0	0	2	6

B. Campoplegine IVs

Present in:		3 IVs						1 IV	
					Number of genes/family/genome				
Virus	# of ORFs	Rep	Vank	Cys	Inx	PRRP	N-gene	TrV	OSSP
CsIV[6]	101	30	7	10	4	5	2	0	0
HfIV[7]	150	38	9	5	11	11	3	0	0
TrIV[7,8]	86	17	2	1	3	1	4	7	4

C. Banchine IVs

Present in:						Recombinase-like (MULE)
				Number of genes/family		
Virus	# of ORFs	PTP	Vank	NTPase-like	BV-like	
GfIV[9,10]	64	23	4	9	4	1

[1] Data presented in this table are restricted to those reported for fully or near-fully sequenced genomes. Additional data may be gleaned from various other studies and the NCBI database where individual or groups of genes were cloned and characterized for other polydnaviruses, including Toxoneuron nigriceps BV (TnBV), Chelonus inanitus BV (CiBV), Cotesia karyai BV (CkBV), Cotesia glomerata BV (CgBV), Cotesia rubicula BV (CrBV), Campoletis chlorideae IV (CcIV); see text for details.

[2] Gene family abbreviations: **PTP**, protein tyrosine phosphatase; **Vank**, viral ankyrin; **BEN**, proteins with BEN domain (CcBV hp2 in Espagne et al., 2004); **BV4**, BV family 4 (new family discovered since publication of Espagne et al., 2004); **Cys**, Cys-motif (or cystein-rich) proteins; **CrV1-like**, homologs of a gene first identified in CrBV; **P94**, related to P94 baculovirus protein; **EP1-like**, homologs of "early-expressed protein 1" of CcBV; **BV2**, BV family 2 (CcBVf2 in Espagne et al., 2004); **BV3**, BV family 3 (CcBV hp1 in Espagne et al., 2004); **Duffy**, proteins with Duffy-binding-like domain; **BV1**, BV family 1 (CcBVf1 in Espagne et al., 2004); **ST**, sugar transporter; **H4**, histone-4-like; **Glc**, glycosylated central domain proteins; **Egf-like**, epidermal growth factor-like proteins; **Rep**, repeat-element; **Inx**, innexins; **PRRP**, polar-residue-rich proteins; **TrV**, homologs of TrV1 from TrIV; **OSSP**, ovary-specific secreted proteins; **BV-like**, homologs of a CcBV-specific hypothetical protein; **Recombinase-like**, homolog of a hypothetical protein from CiBV; PTP, Vank and Cys proteins are color-coded to highlight their presence in two or three of the virus groups shown here.

[3] Cotesia congregata BV: Espagne et al., 2004; A. Bézier, personal communication.
[4] Cotesia plutellae BV (= C. vestalis BV): Choi et al., 2009; NCBI.
[5] Desjardins et al., 2008; NCBI.
[6] Webb et al., 2006.
[7] Tanaka et al., 2007.
[8] Rasoolizadeh et al., 2009.
[9] Lapointe et al., 2007.
[10] Cusson et al., Chapter 6 of this volume.

Printed in the United States
By Bookmasters